Photothermal Spectroscopy Methods for Chemical Analysis

CHEMICAL ANALYSIS

A SERIES OF MONOGRAPHS ON ANALYTICAL CHEMISTRY AND ITS APPLICATIONS

Editor
J. D. WINEFORDNER

VOLUME 134

A WILEY-INTERSCIENCE PUBLICATION

JOHN WILEY & SONS, INC.

New York / Chichester / Brisbane / Toronto / Singapore

Photothermal Spectroscopy Methods for Chemical Analysis

STEPHEN E. BIALKOWSKI

Utah State University
Department of Chemistry and Biochemistry
Logan, Utah

A WILEY-INTERSCIENCE PUBLICATION

JOHN WILEY & SONS, INC.

New York / Chichester / Brisbane / Toronto / Singapore

This text is printed on acid-free paper.

Library of Congress Cataloging in Publication Data:

Bialkowski, E. Stephen.

Photothermal spectroscopy methods for chemical analysis/by Stephen E. Bialkowski.

p. cm. -- (Chemical analysis; v. 134)

"A Wiley-Interscience publication."

Includes bibliographical references.

ISBN 0-471-57467-8 (cloth: alk. paper)

1. Photothermal spectroscopy. I. Title. II. Series.

QD96.P54B53 1996

543'. 0858--dc20 95-34049

Printed in the United States of America

10 9 8 7 6 5 4 3 2 1

To my beautiful sons,
Stephen Edward and Daniel Charles

CONTENTS

PREFACE

Photothermal spectroscopy is an ultrasensitive means to measure optical absorbance. As such, knowledge of its operating principles should be included in the arsenal of any analytical chemist. Photothermal spectroscopy includes a variety of ingenious instrumental methods, including photothermal lens, photothermal deflection, photothermal diffraction, and photothermal interferometry spectroscopies, and a host of complex experimental techniques that do not have names. Some photothermal spectrometers are sensitive enough to measure optical absorbance due to few analyte molecules, molecules that do not fluoresce. Unlike mass spectrometry, which when skillfully operated has comparable detection limits, there are no commercial photothermal spectrometers. All wishing to use photothermal spectroscopy must construct their own apparatus. On the other hand, photothermal spectroscopy is a nondestructive, in situ technique. It is more easily adapted to environments and applications that elude mass spectroscopy due both to its destructive nature and its finicky sample matrix requirements.

Most of the apparatus constructed for photothermal spectroscopy analysis incorporates laser light sources. Many use lasers both for sample excitation and for monitoring changes in refractive index that occur due to the nonradiative energy loss of the target analyte to the supporting matrix. Signal generation is a second-order process, depending not only on the analyte but also on the sample matrix and laser optics. Because of this, some argue that photothermal spectroscopy is too complicated for practical application to chemical analysis. However, this viewpoint cannot be taken after critical examination of the operating principles governing optical methods of chemical analysis in general. All the processes that occur in photothermal spectroscopy occur in conventional spectroscopies. Lasers allow only for magnification of specific effects through more accurate control of experimental parameters. It is this magnification that allows photothermal spectroscopy to be used for ultrasensitive chemical analysis. Nonetheless, successful application of this and any other analytical procedure utilizing laser-based instrumentation, requires a little more caution and knowledge on the part of the analyst. Experimental observations can be controlled by processes that in a conventional spectrometer are simply regarded as artifacts. To be sure, a working knowledge of laser optics, principles governing absorption, excited-state energy transfer, fluid dynamics, and measurement electronics are all necessary ingredients required to solve problems that might arise in laser-based analysis. This knowledge constitutes a practical and consistent theoretical description of the photothermal effect.

This book on photothermal spectroscopy concentrates on the theoretical basis and practical considerations required for successful application of this technique to the analysis of samples in homogeneous media. The relatively short final chapter addresses some considerations for heterogeneous sample analysis. Photothermal spectroscopy is treated from a fundamental standpoint. Perhaps the greatest strength of this book is that it provides a nearly complete description of photothermal spectroscopy using a common mathematical language. It is hoped that this systematic approach to a description of the physical basis for the photothermal signal generation processes will result in a more complete understanding of why certain problems may be encountered in analytical applications and how the effects can either be avoided or compensated for. The information gathered to produce this description comes from many forms of analytical spectroscopy, measurement physics, physical optics, and chemical dynamics. It is information that should be known to persons doing any type of spectroscopy. As such, this book should be useful to all spectroscopists.

The book is arranged in the order of the theoretical basis and practical aspects of photothermal spectroscopy. The first chapter deals with a general description of photothermal spectroscopy and related techniques. The techniques are defined and described, and subtle distinctions between photothermal and related techniques, such as photoacoustic and photopyroelectric spectroscopies, are made. The history and important features of the various techniques are given. Applications-oriented literature is not reviewed extensively, although pertinent citations and examples are given to illustrate the points. Each part includes a discussion of the operating principles of signal generation and applications of the various techniques for materials analysis. Much of this chapter is based on previous review material. In particular, the many reviews by A. C. Tam were invaluable in constructing this chapter.

Regardless of the means by which the thermal perturbation is detected, or the phase of the sample, the treatment of the operating principles of photothermal spectroscopy must include discussions of the optical excitation and exited-state relaxation and of the thermal or hydrodynamic relaxation processes. These processes are not unrelated, although for purposes of description, they are treated in separate chapters. Optical excitation and excited-state relaxation are intimately coupled to energy transfer throughout the sample matrix. Chapter 2 addresses some basic principles of optical absorption, excited-state relaxation, scattering loss, and nonlinear optical absorption effects that can occur. A great deal of attention is paid to nonlinear processes — in particular, those arising due to excited-state dynamics. It is shown that these processes cannot only affect the energy absorbed by the sample but also distort the spatial distribution of energy density. This spatial distortion can result in experimental artifacts in chemical analysis and so should be recognized by the analyst. These effects must be addressed for successful interpretation of nonlinear optical spectroscopies in general.

Chapter 3 addresses the hydrodynamic equations that govern energy transfer in the sample matrix. It is a necessary exercise in nonequilibrium fluid dynamics. After reviewing the literature in this area, it was decided that a new approach was needed to bring the concepts of thermal diffusion and acoustic relaxation together. The approach used is based on the solution to the energy, mass, and momentum conservation equations used by Berne and Pecora in their textbook describing dynamic light scattering. The approach is no more complex than others used to describe photothermal signals, but the results are more agreeable in the sense that a separation of acoustic and thermal relaxation impulse responses results. The impulse-response approach lends a more general appeal to this chapter since the results can be applied to almost any situation encountered in photothermal spectroscopy. This chapter holds some surprises. The connection between the hydrodynamic relaxation resulting from a photothermal perturbation and that resulting from statistical thermal fluctuations was not neglected. This connection allowed the calculation of detection limits based only on fundamental properties of matter. The treatment is not complete, owing to a lack of time, but it should serve as a basis for a better understanding of ultimate detection limits of detection available using optical spectroscopy and to guide the analytical chemist in optimization of experimental parameters.

A description of the optical principles required for an understanding of photothermal spectroscopy is the topic of Chapter 4. Again, a fundamental approach is taken. This was done to make connections between (1) refractive and diffractive descriptions of experimental apparatuses, and (2) density, excited-state populations, and refractive index changes. Students studying lasers and lasers optics have always been curious about the apparent separation of refraction and diffraction-based physical optics and of the strange procedures used to calculate the beam propagation characteristics of laser beams. There is little in the way of help in this regard since the information is scattered throughout the literature and certain aspects appear in various books on optical and electrical engineering techniques. This chapter is an attempt to bring these seemingly unrelated concepts together. It covers the basic principles required to understand the propagation of light, linear and nonlinear changes in refractive index, and propagation of laser radiation. Fourier optics principles, a good approximation to diffraction theory, are used to describe the effect of optical elements on laser beams. Dynamic diffraction gratings are described briefly at the end of the chapter.

Chapter 5 addresses temperature change and photothermal optical element generation. It is somewhat embarrassing that many of the high-brow theoretical aspects developed in the preceding three chapters had to be discarded, or at least approximated, to provide a comprehensible description of photothermal spectroscopy. On the other hand, the relatively mundane approach used does accomplish certain goals. Not only can an understanding of optical element generation be obtained, but since the approximations are introduced and quantified, the limitations of this approach may also be recognized. Again, these may help in a description of certain artifacts that have been observed in photothermal spectroscopy. Several improvements to a quantitative description of photothermal spectroscopy are obtained in this chapter despite the approximations used. In particular, the role that excited states play in the strength of the optical elements as recognized by Terazima, and the sensitivity of the technique to temperature changes, as measured and reported by Harris and Tran, may help future researchers understand and identify sources of inaccuracy when using laser-based analysis.

In Chapter 6 a simple mathematical description of signals produced in photothermal spectroscopy used for homogeneous samples is shown. There has been steady development in the theory since the first descriptions of the photothermal lens, and an understanding of the physical principles underlying the photothermal spectroscopy signal generation process in homogeneous samples has been addressed by several authors. These works are sufficiently separated in time, and the approaches used are different, owing to subtle differences in the areas in which these researchers worked. In this chapter we attempt to consolidate this work by developing a consistent mathematical basis for signal description. These consolidated results are then used to make predictions to extend the understanding of photothermal spectroscopy. It is hoped that the consolidated theory can serve as a platform from which further studies into improving photothermal spectroscopy can be launched.

Successful applications of photothermal spectroscopy to analytical measurements are considered in Chapters 7 and 8. Chapter 7 is a generalized treatment of considerations for chemical analysis. Many of the considerations are routine. In addition to the usual considerations for sampling, sample preparation, separations, reagent and solvent purity, and the like, one must also consider the thermal and optical properties of the solvents and optical materials used in the apparatus. Because of the extreme sensitivity to absorbance and the spatial dependent nature of the measurement, one must also be aware of how the environmental affects the measurement. In this chapter we summarize some of these effects and address means by which accurate measurements can be obtained. Various means of discriminating analyte of solvent or matrix are described. The instrumental techniques for measuring the photothermal signals are then described. The instrumental apparatuses, data collection, signal processing, and data reduction steps are addressed and the virtues and limitations of the various techniques are highlighted. The literature of analytical applications of photothermal spectroscopy is examined in Chapter 8.

The last chapter is included to introduce the reader to differences between homogeneous and heterogeneous sample photothermal spectroscopy. There are now several books devoted to photothermal spectroscopy of heterogeneous materials and this chapter we only outline the processes. This is done by examining the thermal diffusion equations that must be used to obtain a working description of the photothermal signals. A small literature review of the application of this technique to chemical analysis is included at the end.

STEPHEN BIALKOWSKI

Logan, Utah

ACKNOWLEDGMENTS

I would like to thank some people who although they did not contribute directly to the book, helped me to complete the work. Andrew Tam has brought important works in photothermal and photoacoustic spectroscopies to my attention over the past several years through unsolicited reprints and tips. Norman Dovichi has shared insight and information regarding analytical applications photothermal spectroscopy. Joel Harris has been a stimulating source of personal and scientific support for my endeavors involving pulsed-laser-excited photothermal and nonlinear spectroscopies. The positive attitude of these scientists inspired me to continue writing in times when this book seemed hopelessly wrong. In addition to these three, Totaro Imasaka, Andreas Mandelis, Richard Russo, Ted Eyring, Chieu Tran, Agnes Chartier, Donald Bobbitt, Yen Yang, Robert Swofford, John R. Barker, R. Gupta, Joan Power, M. Terazima, W. Tong, and others have participated in, and were stimulating contributors, to the several symposia addressing analytical aspects of photothermal spectroscopy that I had the pleasure to attend; have all openly shared timely information regarding the strengths and weaknesses of their particular techniques and applications; have never had unkind words to say; and have all participated in forwarding the art of analytical photothermal spectroscopy. Discussions with my students; George Long, Scott Nickolaisen, Chris Erickson, and Shashi Kalaskar, have helped me refine my views of photothermal spectroscopy. There is a need to acknowledge my computer, to which I have become well acquainted. The software I used to produce this text was WordPerfect. The Macsyma symbolic language processor was indispensible, as were the Borland products, Turbo C and Quattro Pro. Finally, I thank my friends and colleagues Pete Poston, Xiaoyun Gu, and Agnès Chartier for forgiving me my distraction with the book, helping me through the "real world" of condensed-phase analysis, and not laughing too loud as I stumbled into the aqueous phase. This book should show that I *did* know what you were saying and that photothermal spectroscopy *does* work after all.

CHEMICAL ANALYSIS

A SERIES OF MONOGRAPHS ON ANALYTICAL CHEMISTRY AND ITS APPLICATIONS

J. D. Winefordner, *Series Editor*

Vol. 1. **The Analytical Chemistry of Industrial Poisons, Hazards, and Solvents.** *Second Edition.* By the late Morris B. Jacobs

Vol. 2. **Chromatographic Adsorption Analysis.** By Harold H. Strain (*out of print*)

Vol. 3. **Photometric Determination of Traces of Metals.** *Fourth Edition*
Part I: General Aspects. By E. B. Sandell and Hiroshi Onishi
Part IIA: Individual Metals, Aluminum to Lithium. By Hiroshi Onishi
Part IIB: Individual Metals, Magnesium to Zirconium. By Hiroshi Onishi

Vol. 4. **Organic Reagents Used in Gravimetric and Volumetric Analysis.** By John F. Flagg (*out of print*)

Vol. 5. **Aquametry: A Teatise on Methods for the Determination of Water.** *Second Edition (in three parts).* By John Mitchell, Jr. and Donald Milton Smith

Vol. 6. **Analysis of Insecticides and Acaricides.** By Francis A. Gunther and Roger C. Blinn (*out of print*)

Vol. 7. **Chemical Analysis of Industrial Solvents.** By the late Morris B. Jacobs and Leopold Schetlan

Vol. 8. **Colorimetric Determination of Nonmetals.** *Second Edition.* Edited by the late David F. Boltz and James A. Howell

Vol. 9. **Analytical Chemistry of Titanium Metals and Compounds.** By Maurice Codell

Vol. 10. **The Chemical Analysis of Air Pollutants.** By the late Morris B. Jacobs

Vol. 11. **X-Ray Spectrochemical Analysis.** *Second Edition.* By L. S. Birks

Vol. 12. **Systematic Analysis of Surface-Active Agents.** *Second Edition.* By Milton J. Rosen and Henry A. Goldsmith

Vol. 13. **Alternating Current Polarography and Tensammetry.** By B. Breyer and H. H. Bauer

Vol. 14. **Flame Photometry.** By R. Herrmann and J. Alkemade

Vol. 15. **The Titration of Organic Compounds** (*in two parts*). By M. R. F. Ashworth

Vol. 16. **Complexation in Analytical Chemistry: A Guide for the Critical Selection of Analytical Methods Based on Complexation Reactions.** By the late Anders Ringbom

Vol. 17. **Electron Probe Microanalysis.** *Second Edition.* By L. S. Birks

Vol. 51. **Trace Element Analysis of Geological Materials.** By Roger D. Reeves and Robert R. Brooks

Vol. 52. **Chemical Analysis by Microwave Rotational Spectroscopy.** By Ravi Varma and Lawrence W. Hrubesh

Vol. 53. **Information Theory As Applied to Chemical Analysis.** By Karel Eckschlager and Vladimir Štěpánek

Vol. 54. **Applied Infrared Spectroscopy: Fundamentals, Techniques, and Analytical Problem-solving.** By A. Lee Smith

Vol. 55. **Archaeological Chemistry.** By Zvi Goffer

Vol. 56. **Immobilized Enzymes in Analytical and Clinical Chemistry.** By P. W. Carr and L. D. Bowers

Vol. 57. **Photoacoustics and Photoacoustic Spectroscopy.** By Allan Rosencwaig

Vol. 58. **Analysis of Pesticide Residues.** Edited by H. Anson Moye

Vol. 59. **Affinity Chromatography.** By William H. Scouten

Vol. 60. **Quality Control in Analytical Chemistry.** *Second Edition.* By G. Kateman and L. Buydens

Vol. 61. **Direct Characterization of Fineparticles.** By Brian H. Kaye

Vol. 62. **Flow Injection Analysis.** By J. Ruzicka and E. H. Hansen

Vol. 63. **Applied Electron Spectroscopy for Chemical Analysis.** Edited by Hassan Windawi and Floyd Ho

Vol. 64. **Analytical Aspects of Environmental Chemistry.** Edited by David F. S. Natusch and Philip K. Hopke

Vol. 65. **The Interpretation of Analytical Chemical Data by the Use of Cluster Analysis.** By D. Luc Massart and Leonard Kaufman

Vol. 66. **Solid Phase Biochemistry: Analytical and Synthetic Aspects.** Edited by William H. Scouten

Vol. 67. **An Introduction to Photoelectron Spectroscopy.** By Pradip K. Ghosh

Vol. 68. **Room Temperature Phosphorimetry for Chemical Analysis.** By Tuan Vo-Dinh

Vol. 69. **Potentiometry and Potentiometric Titrations.** By E. P. Serjeant

Vol. 70. **Design and Application of Process Analyzer Systems.** By Paul E. Mix

Vol. 71. **Analysis of Organic and Biological Surfaces.** Edited by Patrick Echlin

Vol. 72. **Small Bore Liquid Chromatography Columns: Their Properties and Uses.** Edited by Raymond P. W. Scott

Vol. 73. **Modern Methods of Particle Size Analysis.** Edited by Howard G. Barth

Vol. 74. **Auger Electron Spectroscopy.** By Michael Thompson, M. D. Baker, Alec Christie, and J. F. Tyson

Vol. 75. **Spot Test Analysis: Clinical, Environmental, Forensic and Geochemical Applications.** By Ervin Jungreis

Vol. 76. **Receptor Modeling in Environmental Chemistry.** By Philip K. Hopke

Vol. 77. **Molecular Luminescence Spectroscopy: Methods and Applications** (*in three parts*). Edited by Stephen G. Schulman

Vol. 78. **Inorganic Chromatographic Analysis.** Edited by John C. McDonald

Vol. 79. **Analytical Solution Calorimetry.** Edited by J. K. Grime

Vol. 80. **Selected Methods of Trace Metal Analysis: Biological and Environmental Samples.** By Jon C. VanLoon

Vol. 81. **The Analysis of Extraterrestrial Materials.** By Isidore Adler

Vol. 82. **Chemometrics.** By Muhammad A. Sharaf, Deborah L. Illman, and Bruce R. Kowalski

Vol. 83. **Fourier Transform Infrared Spectrometry.** By Peter R. Griffiths and James A. de Haseth

Vol. 84. **Trace Analysis: Spectroscopic Methods for Molecules.** Edited by Gary Christian and James B. Callis

Vol. 85. **Ultratrace Analysis of Pharmaceuticals and Other Compounds of Interest.** Edited by S. Ahuja

Vol. 86. **Secondary Ion Mass Spectrometry: Basic Concepts, Instrumental Aspects, Applications and Trends.** By A. Benninghoven, F. G. Rüdenauer, and H. W. Werner

Vol. 87. **Analytical Applications of Lasers.** Edited by Edward H. Piepmeier

Vol. 88. **Applied Geochemical Analysis.** By C. O. Ingamells and F. F. Pitard

Vol. 89. **Detectors for Liquid Chromatography.** Edited by Edward S. Yeung

Vol. 90. **Inductively Coupled Plasma Emission Spectroscopy: Part I: Methodology, Instrumentation, and Performance; Part II: Applications and Fundamentals.** Edited by J. M. Boumans

Vol. 91. **Applications of New Mass Spectrometry Techniques in Pesticide Chemistry.** Edited by Joseph Rosen

Vol. 92. **X-Ray Absorption: Principles, Applications, Techniques of EXAFS, SEXAFS, and XANES.** Edited by D. C. Konnigsberger

Vol. 93. **Quantitative Structure-Chromatographic Retention Relationships.** By Roman Kaliszan

Vol. 94. **Laser Remote Chemical Analysis.** Edited by Raymond M. Measures

Vol. 95. **Inorganic Mass Spectrometry.** Edited by F. Adams, R. Gijbels, and R. Van Grieken

Vol. 96. **Kinetic Aspects of Analytical Chemistry.** By Horacio A. Mottola

Vol. 97. **Two-Dimensional NMR Spectroscopy.** By Jan Schraml and Jon M. Bellama

Vol. 98. **High Performance Liquid Chromatography.** Edited by Phyllis R. Brown and Richard A. Hartwick

Vol. 99. **X-Ray Fluorescence Spectrometry.** By Ron Jenkins

Vol. 100. **Analytical Aspects of Drug Testing.** Edited by Dale G. Deutsch

Vol. 101. **Chemical Analysis of Polycyclic Aromatic Compounds.** Edited by Tuan Vo-Dinh

Vol. 102. **Quadrupole Storage Mass Spectrometry.** By Raymond E. March and Richard J. Hughes

Vol. 103. **Determination of Molecular Weight.** Edited by Anthony R. Cooper

Vol. 104. **Selectivity and Detectability Optimizations in HPLC.** By Satinder Ahuja

Vol. 105. **Laser Microanalysis.** By Lieselotte Moenke- Blankenburg

Vol. 106. **Clinical Chemistry.** Edited by E. Howard Taylor

Vol. 107. **Multielement Detection Systems for Spectrochemical Analysis.** By Kenneth W. Busch and Marianna A. Busch

Vol. 108. **Planar Chromatography in the Life Sciences.** Edited by Joseph C. Touchstone

Vol. 109. **Fluorometric Analysis in Biomedical Chemistry: Trends and Techniques Including HPLC Applications.** By Norio Ichinose, George Schwedt, Frank Michael Schnepel, and Kyoko Adochi

Vol. 110. **An Introduction to Laboratory Automation.** By Victor Cerdá and Guillermo Ramis

Vol. 111. **Gas Chromatography: Biochemical, Biomedical, and Clinical Applications.** Edited by Ray E. Clement

Vol. 112. **The Analytical Chemistry of Silicones.** Edited by A. Lee Smith

Vol. 113. **Modern Methods of Polymer Characterization.** Edited by Howard G. Barth and Jimmy W. Mays

Vol. 114. **Analytical Raman Spectroscopy.** Edited by Jeannette Graselli and Bernard J. Bulkin

Vol. 115. **Trace and Ultratrace Analysis by HPLC.** By Satinder Ahuja

Vol. 116. **Radiochemistry and Nuclear Methods of Analysis.** By William D. Ehmann and Diane E. Vance

Vol. 117. **Applications of Fluorescence in Immunoassays.** By Ilkka Hemmila

Vol. 118. **Principles and Practice of Spectroscopic Calibration.** By Howard Mark

Vol. 119. **Activation Spectrometry in Chemical Analysis.** By S. J. Parry

Vol. 120. **Remote Sensing by Fourier Transform Spectrometry.** By Reinhard Beer

Vol. 121. **Detectors for Capillary Chromatography.** Edited by Herbert H. Hill and Dennis McMinn

Vol. 122. **Photochemical Vapor Deposition.** By J. G. Eden

Vol. 123. **Statistical Methods in Analytical Chemistry.** By Peter C. Meier and Richard Zund

Vol. 124. **Laser Ionization Mass Analysis.** Edited by Akos Vertes, Renaat Gijbels, and Fred Adams

Vol. 125. **Physics and Chemistry of Solid State Sensor Devices.** By Andreas Mandelis and Constantinos Christofides

Vol. 126. **Electroanalytical Stripping Methods.** By Khjena Z. Brainina and E. Neyman

Vol. 127. **Air Monitoring by Spectroscopic Techniques.** Edited by Markus W. Sigrist

Vol. 128. **Information Theory in Analytical Chemistry.** By Karel Eckschlager and Klaus Danzer

Vol. 129. **Flame Chemiluminescence Analysis by Molecular Emission Cavity Detection.** Edited by David Stiles, Anthony Calokerinos, and Alan Townshend

Vol. 130. **Hydride Generation Atomic Absorption Spectrometry.** By Jiri Dedina and Dimiter L. Tsalev

Vol. 131. **Selective Detectors: Environmental, Industrial, and Biomedical Applications.** Edited by Robert E. Sievers

Vol. 132. **High Speed Countercurrent Chromatography.** By Yoichiro Ito and Water D. Conway

Vol. 133. **Particle-Induced X-Ray Emission Spectrometry.** By Sven A. E. Johansson, John L. Campbell, and Klas G. Malmqvist

Vol. 134. **Photothermal Spectroscopy Methods for Chemical Analysis.** By Stephen Bialkowski

Photothermal Spectroscopy Methods for Chemical Analysis

CHAPTER

1

INTRODUCTION

1.1. PHOTOTHERMAL SPECTROSCOPY

Photothermal spectroscopy comprises a group of high-sensitivity methods used to measure optical absorption and thermal characteristics of a sample. The basis of photothermal spectroscopy is a *photo*-induced change in the *thermal* state of the sample. Light energy absorbed and not lost by subsequent emission results in sample heating. This heating results in a temperature change as well as changes in thermodynamic parameters of the sample which are related to temperature. Measurement of the temperature, pressure, or density changes that occur due to optical absorption are ultimately the basis for the photothermal spectroscopic methods.

Ingle and Crouch (1988) classify photothermal spectroscopy as one of several indirect methods for optical absorption analysis. Indirect methods do not measure the transmission of light used to excite the sample directly, but rather, measure an effect that optical absorption has on the sample. The term *indirect* applies to the light measurement, not to the optical absorbance. Photothermal spectroscopy is, in a sense, a more direct measure of optical absorption than are optical transmission–based spectroscopies. Sample heating is a direct consequence of optical absorption, so photothermal spectroscopy signals are directly dependent on light absorption. Scattering and reflection losses do not produce photothermal signals. Subsequently, photothermal spectroscopy more accurately measures optical absorption in scattering solutions, in solids, and at interfaces. This aspect makes it particularly attractive for application to surface and solid absorption studies and studies in scattering media.

The indirect nature of the measurement also results in photothermal spectroscopy being more sensitive than optical absorption measured by transmission methods. There are two reasons for this. First, photothermal effects can amplify the optical signal measured. This amplification, referred to as the *enhancement factor* (Dovichi and Harris 1979, Mori et al. 1982), is the ratio of the signal obtained using photothermal spectroscopy to that of conventional transmission spectroscopy. Enhancement factors depend on the thermal and optical properties of the sample, the power or energy of the light source used to excite the sample, and on the optical geometry used to excite the sample. Since the optical excitation power or energy and the geometry are variable, the enhancement can be made very large, even for samples with relatively poor thermal and optical properties. In fact, the problem with photothermal spectroscopy is not the absorption detection limit; the problem is the detection of analyte absorbance in the presence of a relatively large (10^{-5} cm^{-1}) absorbance of the solvent. The second reason that photothermal spectroscopy is more sensitive than transmission is that the precision of the measurement is inherently better than that of the direct transmission method. The fundamental limitation of conventional absorption spectroscopy, shot noise, may be partially circumvented (Bialkowski et al. 1992). Because of the increased fundamental signal-to-noise ratios, the problem of being able to detect the analyte in the presence of a relatively large background absorption should be resolvable with perseverance.

The high sensitivity of photothermal spectroscopy methods has led to applications for analysis of low-absorbance samples. Dovichi (1987) reviewed the literature regarding the use of photothermal spectroscopy for chemical analysis. The magnitude of the photothermal spectroscopy signal depends on the specific method used to detect the photothermal effect and on the type of sample being analyzed. Many different detection limits have been reported, and it is difficult to specify an absolute lower limit of detection since the method may be used to measure the background absorption of the solvent itself. But it is safe to say that optical absorbances of less than 10^{-6} can be detected with optimized experimental designs. Subsequently, photothermal spectroscopy is often characterized as a trace analysis method. The concentration limits of detection measurements can be impressive. Electronic transitions of strongly absorbing chromophores have molar absorptivities exceeding $10^4\ M^{-1}\ cm^{-1}$. Using photothermal methods, concentrations lower than $10^{-10}\ M$ of these strongly absorbing chromophores may be measured in standard cuvettes. These limits of detection are slightly higher than those obtained using laser-excited fluorescence spectroscopy and are two or three orders of magnitude better than that obtained using conventional transmission spectroscopy. The low molar absorption detection limits, coupled with the fact that the volume being probed can be very small results in extremely small numbers of molecules being detected. The high absorbance sensitivity of these methods has opened up new areas of trace chemical analysis based on optical absorption spectroscopy.

Photothermal signals depend on the thermodynamic and energy transfer properties of the sample. Temperature changes resulting from optical absorption are directly related to heat capacity and thermal conductivity. This makes absolute sample absorption measurements difficult. The thermal and optical properties of the sample must be known to high accuracy, or the instrument response must be calibrated with samples of known composition and absorbance. However, this dependence on thermodynamic and energy transfer properties allows for analysis of the thermal structure of materials. With calibrated apparatus, the static and dynamic thermal properties of the sample can be measured. Photothermal spectroscopy has been used to measure acoustic velocities, thermal diffusion coefficients, sample temperatures, bulk sample flow rates, specific heats, volume expansion coefficients, and heterogeneous thermal conductivities in solids. In particular, a technique called thermal wave imaging allows nondestructive material inspection by measuring the rate of heat transfer in heterogeneous materials.

Photothermal spectroscopy is usually performed using laser light sources. There are two main reasons for this. The first is the high spectral purity and power. For an excitation of a sample with a given absorption coefficient, the temperature change will be proportional to the optical power, in the case of continuous excitation, or energy, in the case of pulsed excitation. The photothermal spectroscopy signal is generally proportional to the temperature change. Thus the greater the power or energy, the greater the resulting signal. Lasers can deliver high powers or pulse energies over very narrow optical bandwidths, thereby enhancing the photothermal signals. The second reason is spatial coherence. The temperature change is not only proportional to the optical power or energy but is also inversely proportional to the volume over which the light is absorbed since heat capacity changes with the amount of substance. The spatial coherence properties of laser sources allow the light to be focused to small diffraction-limited volumes. The small volumes used in photothermal spectroscopy enhance signal magnitudes, allow photothermal spectroscopy to be used in small-volume sample analysis, and allow for microscopic analysis of heterogeneous materials.

1.2. BASIC PROCESSES IN PHOTOTHERMAL SPECTROSCOPY

The basic processes responsible for photothermal spectroscopy signal generation are shown in Figure 1.1. Optical radiation, usually from a laser, is used to excite a sample. The sample absorbs some of this radiation, resulting in an increase in the internal energy. The internal energy is dispersed in two different modes of hydrodynamic relaxation. The increased internal energy results in a temperature change in the sample or the coupling fluid placed next to the sample. This temperature change results in a change in sample or coupling fluid density. If the photothermal-induced temperature change

Optical Excitation

⇓

Absorption

↕

Excited State Relaxation

⇓

Temperature Change (Thermal Diffusion) ⇔ ***Density Change*** ⇔ ***Pressure Change (Acoustic Wave)***

⇓

Optical Probe ⇒ ***Refractive Index Change*** ⇒ ***Photothermal Signal***

Figure 1.1. Processes involved in photothermal spectroscopy. Absorption of radiation from the excitation source followed by nonradiative excited-state relaxation results in changes in the sample temperature, pressure, and density. The density change is primarily responsible for the refractive index change, which can be probed by a variety of methods.

occurs faster than the time required for the fluid to expand or in a few cases contract, the rapid temperature change will result in a pressure change. The pressure perturbation will disperse in an acoustic wave. Once the pressure has relaxed to the equilibrium pressure, a density change proportional to the temperature will remain. In either case there will be a change in temperature induced by the absorption of optical energy. This temperature change will in turn result in a density change in the sample. In combination, temperature and density changes affect other properties of the sample. Photothermal spectroscopy is based on a measurement of these properties. In particular, the sensitive photothermal methods are based on measurement of the refractive index change that occurs with changes in temperature and density of the sample.

There are three main areas that must be considered when attempting to obtain a quantitative description of the photothermal spectroscopy signal. The first is a description of the optical absorption and excited-state relaxation processes. Optical excitation followed by excited-state relaxation results in sample heating. The rates and amounts of excited-state excitation and relaxation will control the rate and magnitude of heat production. The energy transfer steps that need be accounted for are shown in Figure 1.2. Energy can be transferred to the sample by optical absorption and inelastic scattering process such as Raman. Scattering is inefficient and the amount of energy lost to sampling is usually small enough to be neglected. After absorption, the molecules are in an excited state. Excited-state relaxation transfers energy to the solvent or sample matrix. Radiative relaxation does not result in complete loss of the absorbed energy to the sample. Some of the energy is lost in the form of the radiated light. Thermal relaxation transfers the energy to the sample matrix and results in sample heating. Excited species may also form long-lived metastable states that trap energy

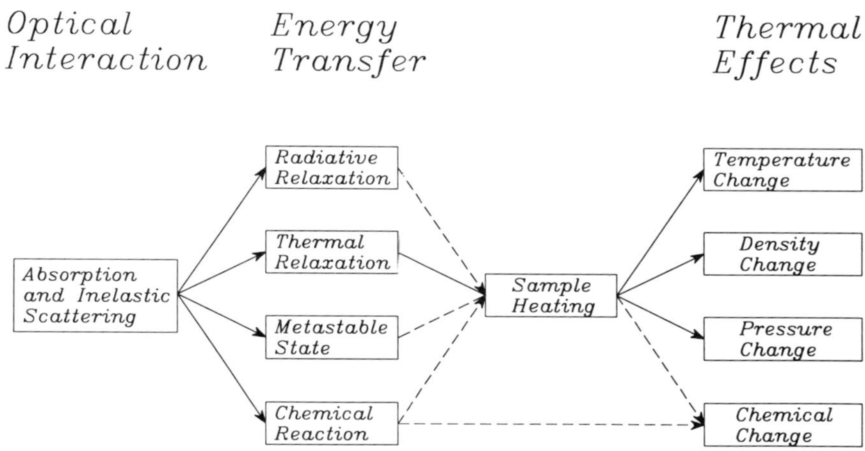

Figure 1.2. Several of the mechanisms for excited-state relaxation are illustrated. The primary steps are optical interaction, energy transfer, sample heating, and thermal effects. Radiative relaxation, metastable-state production, and photochemical reaction may result in some sample heating. Energy transfer may result in fast or slow kinetic energy production.

and prevent further optical absorption. This will result in delayed heating of the sample. The excited-state species may also participate in photochemical reactions. Photochemical reaction can produce heat but also produce new chemical species that alter the thermal and optical characteristics of the sample.

These relaxation processes may all produce excess energy in the form of heat. The heat increases the internal energy of the sample. The sample will respond to this increased energy. The second area is that of the hydrodynamic relaxation. After optical heating, the sample is not at thermal equilibrium with itself or with the surrounding environment during a measurement. Heat generated by the optical excitation and relaxation processes will result in thermal gradients between the excited sample and the surroundings. The thermal gradients result in heat transport. Heat is transferred within the sample in a fashion such as to move toward thermal equilibrium. Hydrodynamic relaxation produces changes in the temperature, pressure, and density of the sample.

The third area is that of the signal generation process. Photothermal spectroscopy signals are based on changes in sample temperature or related thermodynamic properties of the sample. These are usually monitored through the refractive index of the sample or a thermal coupling fluid placed in contact with the sample. Several properties may affect the refractive index of the medium. The most common is the density. However, the refractive

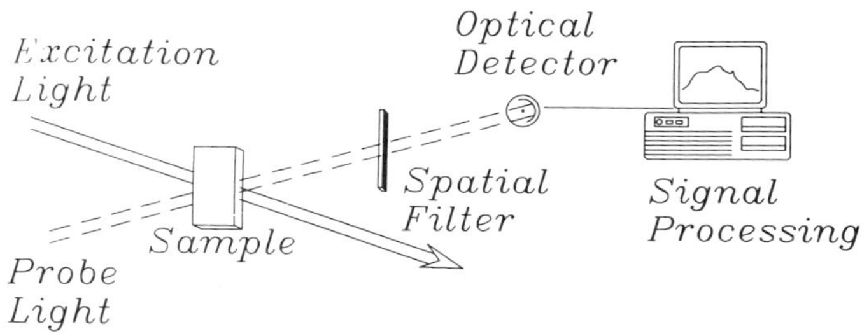

Figure 1.3. Generic photothermal spectrometer showing essential features.

index may also change with temperature, with population in optically excited states, and with chemical composition if photochemical reaction occurs. A variety of instrumental methods are used to probe the changes in the sample's refractive index. Other instrumental methods used for photothermal spectroscopy probe the temperature or related thermodynamic properties directly, but the most sensitive methods probe the spatial or temporal gradients of these properties.

A schematic diagram illustrating the main components of apparatus used for photothermal spectroscopy is shown in Figure 1.3. Most apparatus consists of six main components: (1) light used for sample excitation; (2) sample; (3) light used to monitor refractive index perturbations; (4) a mask, aperture, or other form of spatial filter for the probe light; (5) an optical detector used to detect the optically filtered probe light; and (6) electronic signal processing equipment. The excitation light heats the sample. The probe light monitors changes in the refractive index of the sample resulting from heating. The spatial and propagation characteristics of the probe light will be altered by the refractive index. The spatial filter selects those components of the altered probe light that change with the sample's refractive index. The optical detector monitors changes in the probe light power past the spatial filter. In some apparatus, a spatial filter and a single-channel detector are combined using an image detector. Signals generated by the photodetector are processed to enhance the signal-to-noise ratio.

Apparatus may also be equipped with detectors to monitor the excitation and probe light power, a thermostatic sample holder, and optical spatial filters to control the spatial profiles of the excitation and probe light. This additional equipment is used to control the experiment environment and to measure the optical power required to quantify accurately changes that occur in the sample. These components are necessary when the data must be used to determine absolute absorption of the sample. In theory, the photothermal spectroscopy signal can be calculated accurately based on knowledge of the experimental apparatus, the parameters that characterize light propagation, and the optical parameters of the sample. The following items must all be accounted for in the calculations; (1) the optical absorption resulting in sample heating; (2) the rate of heat production; (3) the temporal and spatial temperature and density change; (4) the refractive index change as related to the temperature or density change using the thermal–optical parameters of the sample; (5) the strength of the optical element formed from the spatially-dependent refractive index change; and (6) the optical and electronic signal resulting from passage of light through apertures or using specialized detectors.

1.3. PHOTOTHERMAL SPECTROSCOPY METHODS

A variety of methods are used to monitor the thermal state of the analytical sample (Harris 1986, Tam 1986, Dovichi 1987, Tam 1989). Direct calorimetric or thermometric methods use temperature transducers to measure analytical sample temperature. Pressure transducers are used to monitor the pressure wave associated with rapid sample heating. Photothermal interferometry, photothermal deflection spectroscopy, photothermal lensing spectroscopy (also known as thermal lens spectroscopy), photothermal diffraction spectroscopy, and methods based on sample reflection changes are all based on monitoring refractive index changes associated with sample heating. Infrared detectors can be used to monitor changes in the samples infrared emission associated with heating. Each of these methods is based on a measurement of temperature change associated with increasing the energy of the analytical sample.

Photothermal methods have been reported by people working in several areas of science and technology, and the particular methods are known by various names. The temperature changes resulting from the photothermal effect can be detected using a variety of methods, summarized in Table 1.1. Temperature can be measured directly using thermocouples, thermistors, or pyroelectric devices in the method of *photothermal calorimetry*. Temperature changes can also be measured indirectly using methods that monitor infrared emission since the thermal infrared emission is related to sample temperature. The method of *thermal emission* or *photothermal radiometry* of infrared radiation can be used to monitor relatively large temperature changes that occur as a consequence of optical absorption. Although not very sensitive, this method has great potential for application in nondestructive materials analysis and testing. Using infrared-sensitive cameras, it can be used for imaging the thermal properties of large samples. Two other temperature-dependent thermodynamic parameters that are commonly exploited in photothermal spectroscopy are pressure and density. The pressure

Table 1.1. Common Detection Used in Photothermal Spectroscopy

Thermodynamic Parameter	Measured Property	Detection Technique
Temperature	Temperature	Calorimetry
	Infrared emission	Infrared emission
		Photothermal radiometry
Pressure	Acoustic wave	Photoacoustic spectroscopy
Density	Refractive index	Photothermal lens
		Photothermal interferometry
		Photothermal deflection
		Photothermal refraction
		Photothermal diffraction
	Surface deformation	Surface deflection

changes that occur upon periodic or pulsed sample heating can be detected by using a microphone or other pressure transducer to monitor the acoustic wave. The method of optoacoustic or *photoacoustic spectroscopy* is based on the measurement of this pressure wave.

Although produced by the same photothermal effects, photoacoustic, infrared radiometry, and photothermal spectroscopies are typically treated as separate methods. Photothermal spectroscopy refers to methods that monitor the temperature-dependent refractive index changes, usually with a probe laser. Nonetheless, it is apparent from hydrodynamic relaxation that the photoacoustics cannot be avoided in a treatment of photothermal spectroscopy. The photoacoustic pressure wave generated by the photothermal effect is observed in photothermal spectroscopy, and the rate of sample relaxation is controlled by the rate at which the sample can approach isobaric conditions. Moreover, infrared emission is another method of thermal heat transfer that should at least be quantified in terms of the effect that it may have on the photothermal signal magnitude. All these effects should be considered in a comprehensive treatment of the photothermal effect.

Under steady-state isobaric conditions, the density is related to the temperature through the volume expansion coefficient. Temperature-dependent density changes are difficult to measure directly. But density changes can affect samples in several different ways. In solid samples the density changes alter physical dimensions at sample surfaces. Sample dimension changes give rise to two optical methods for monitoring the temperature change based on surface deformation. A homogeneous deformation (expansion or contraction) displaces the surface of the sample. Interferometry can be used on reflective samples. Since small displacements, on the order of a few parts per million of the wavelength of probe beam light, can be measured using interferometry, this method may be used for sensitive measurement solid sample absorption. Spatially heterogeneous expansion (or contraction) can also cause the surface angle to change. A probe beam reflected from the surface will change angle when heterogeneous expansion occurs. Measurement of the probe beam angle gives rise to the method of *photothermal surface deflection spectroscopy.*

The majority of studies addressing the use of photothermal spectroscopy for chemical analysis have been based on refractive index measurements. In transparent samples, the temperature-dependent refractive index of the sample itself is probed. For opaque or scattering surfaces, temperature-dependent changes in the refractive index of fluid that couples heat out of a solid sample are measured. Several methods are used to detect the resulting refractive index change. Several of these are shown in Figure 1.4. Publications in photothermal spectroscopy come from researchers working in the fields of analytical and physical chemistry, physics, and optical engineering. Subsequently, there is a wide range of nomenclature used to describe methods for refractive index change detection in the photothermal spectroscopy literature. But all these methods rely on a few basic principles of light propagation: optical pathlength changes, diffraction, and refraction. Light refraction can result in a direction change and/or focusing. The optical pathlength changes that occur due to the photothermal-induced refractive index change can be measured with interferometry. Using interferometry, the phase of monochromatic light passing through the heated sample relative to the phase passing through the reference arm results in a change in power at a photoelectric detector. There are several different interferometric schemes that can be used to detect changes in the optical pathlength induced by the photothermal effect. These methods may all be classified as being *photothermal interferometry.*

Spatial gradients in refractive index result in a direction change in the propagation of a ray of light. Thus light will exit a medium with a refractive index gradient at an angle relative to the incident ray. This bending of light path is commonly called *photothermal deflection spectroscopy.* Spatially dependent refractive index profiles can also result in focusing or defocusing of light. This occurs when the refractive index profiles are curved. Thus the thermally perturbed sample can act as a lens. Light transmitted through an aperture placed beyond the photothermal lens will vary with the strength of the lens. Photothermal methods based on measurement of the strength of this lens are called *photothermal lensing spectroscopy.* Some experimental

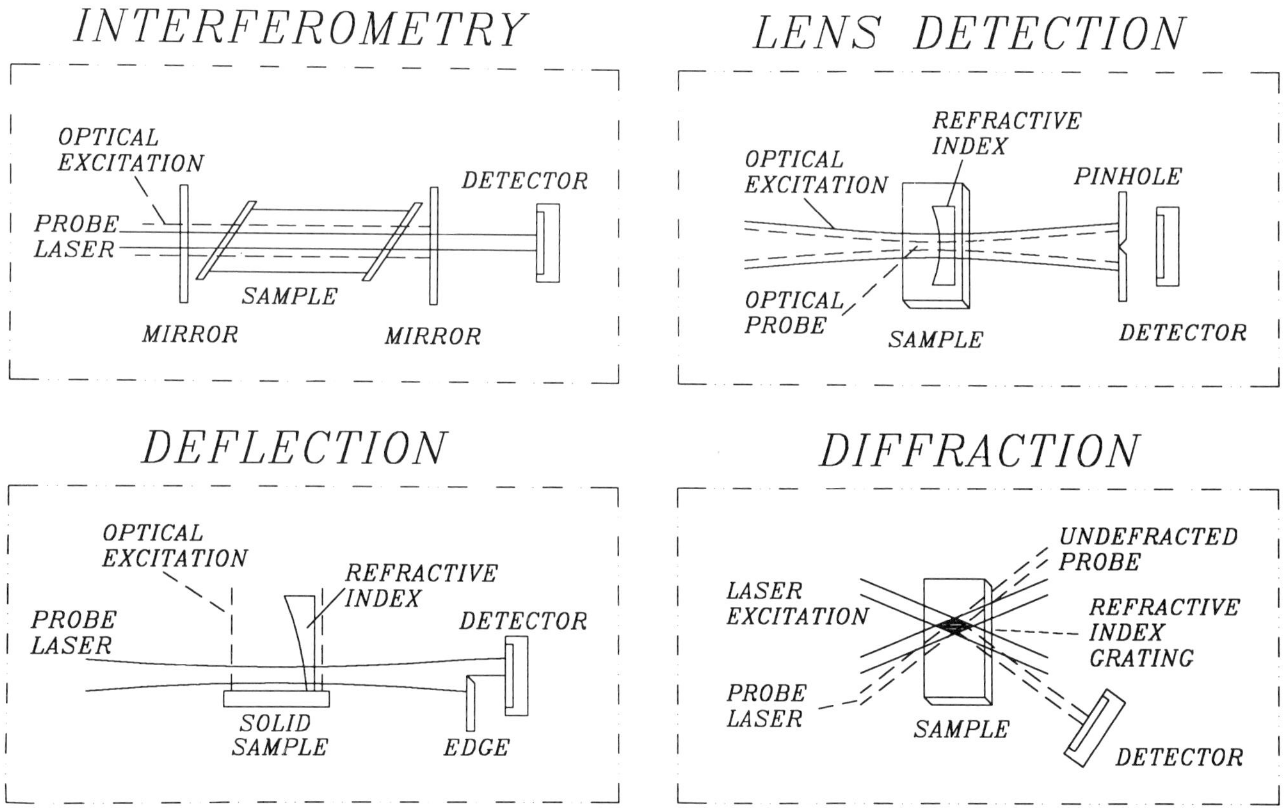

Figure 1.4. Four methods used for photothermal spectroscopy. Interferometry measures the refractive index directly. Deflection measures the gradient. Photothermal lens spectroscopy is based on beam focusing or defocusing. Diffraction methods measure the power of a beam diffracted by the periodic index.

apparatuses measure a signal that is due to the combined effects of deflection and lensing. These may be generally classified as *photothermal refraction spectroscopy* methods. Finally, a periodic spatial refractive index modulation results in a volume-phase diffraction grating. The grating will diffract light at an angle that meets requirements from Bragg's law. The amount of light diffracted is proportional to the refractive index change. The diffracted light is measured with a photoelectric detector. Methods used to measure spectroscopic signals based on volume-phase diffraction gratings formed by the photothermal effects are called *photothermal diffraction spectroscopy.*

The key to the success of sensitive photothermal apparatus lies in measurement of a thermal *change*, not of the thermal state itself. Although apparatus could measure directly or indirectly thermodynamic parameters such as temperature, pressure, density, and energy state, the limiting absorption that could be measured would be imposed by thermodynamic fluctuations. Sensitive photothermal spectroscopy methods circumvent direct measurements by measuring refractive index changes due to a non-equilibrium change in the energy of the sample. The change occurs in both space and time. Photothermal spectroscopy methods measure some effect that the spatially or temporally dependent refractive index change has on the propagation characteristics of light used to monitor the refractive index. Each piece of apparatus detects the change in refractive index that accompanies optical absorption. Photodetectors are used to monitor probe power changes. These power signals are time dependent. The analytical signal is usually related to the change in detected power relative to the incident power of the probe. Analytical signals can have three principal types of time dependence. These, in turn, depend on the temporal character of the excitation source. The main excitation and detection schemes are listed in Table 1.2

Pulsed excitation sources produce transient signals. These signals are at a maximum immediately following sample excitation and decay as the sample approaches equilibrium through thermal diffusion. The transient signals last from a few microseconds in the gas phase to several milliseconds in condensed phases. The time duration is inversely proportional to the thermal conductivity of the media since thermal diffusion or conduction removes energy from the sample and, more important, distributes the energy throughout the sample. Photothermal lens, deflection, and diffraction apparatus respond to spatial variations in the refractive index. Thus homogeneous distribution of energy throughout the sample does not result in a signal. Interferometric measurements may be able to detect the refractive index change after thermal diffusion has distributed the energy. However, environmental thermal stability is usually not good enough to allow this. Sensitive

Table 1.2. Principal Sample Excitation Schemes Used in Photothermal Spectroscopy

Excitation	Signal	Detection
Pulsed	Short-lived transient; magnitude decreases with time	Peak magnitude estimation and transient waveform analysis
Continuous	Long-lived transient; magnitude increases with time	Steady-state magnitude estimation and transient waveform analysis
Modulated	Periodic modulation; magnitude and phase are functions of frequency	Periodic-wave magnitude and phase analysis using frequency-selective filters or lock-in amplifiers

interferometric apparatus relies on the detection of a temporal change in refractive index.

Continuous excitation produces signals that are initially small but increase in magnitude as the irradiation time progresses. Initially, thermal diffusion removes heat slower than the heat produced by optical excitation. The Fourier law of heat diffusion states that the heat flux, $\mathbf{j}_H$, is proportional to the temperature gradient:

$$\mathbf{j}_H = -k \nabla T \tag{1}$$

The proportionality constant is the thermal conductivity. As the sample absorbs radiation and converts the energy to heat, the temperature gradient increases. When the radiative heating flux equals the energy flux due to thermal conduction, a steady-state spatially dependent temperature change is attained. Thus the photothermal signals eventually reach a steady-state value. The signals develop over the course of from milliseconds to seconds, the time required to attain the steady-state value being proportional to the thermal conductivity.

For analytical (e.g., concentration) measurements, both pulsed and continuous excitation require estimation of the signal magnitude. Signal magnitudes are directly proportional to the sample absorbance in a first-order approximation. Signal magnitudes can be measured directly (e.g., using an oscilloscope or ammeter), or the signal transient can be recorded and subsequently processed to enhance measurement precision.

Excitation sources may also be modulated. Chopped or oscillatory excitation produces oscillating signals. The resulting signals can be processed using bandpass filters or lock-in amplifiers. The magnitudes of the oscillating signals depend on sample absorbance, frequency of excitation, and the thermal conductivity of the medium. With modulated excitation, signal magnitudes are proportional to sample absorbance but decrease with increasing frequency. In addition to the signal amplitude information, phase-sensitive lock-in analyzers also produce signal-to-excitation phase-shift information. The frequency-dependent phase-shift information is essentially equivalent to that contained in the time-dependent signal transients obtained using pulsed excitation.

1.4. APPLICATION OF PHOTOTHERMAL SPECTROSCOPY

There have been many applications of photothermal methods for chemical and material analysis. Tam (1983, 1986, 1989) is perhaps primarily responsible for sorting through the vast amount of literature and characterizing the applications of these methods. Many of these applications are covered in the book edited by Sell (1989). These applications fall under four main categories.

1. *Photothermal Spectroscopy.* The signal magnitude is measured as a function of wavelength in this application. The photothermal signal is proportional to the absorbed light, so the spectrum is technically an excitation spectrum. The resulting excitation spectrum can be an accurate measure of the absorption spectrum if the thermal quantum yield and fraction of light transmitted to the absorber do not change with wavelength. This technique has found widespread use for solid sample analysis where incoherent excitation light sources can be used. Applications to liquid and gas sample analysis has been limited because of the difficulties encountered when attempting to scan the wavelengths of lasers while keeping them focused at a particular position.
2. *Photothermal Detection.* This method is similar to photothermal spectroscopy except that a single-wavelength source is used to excite the sample. The signal magnitude can be related to sample absorbance or analyte concentration. Samples must be prepared and separated so that there is no interference absorption and so that the sample matrix is the same for all samples measured. The principal application is for trace analysis. Although not restricted to coherent sources, this application is normally performed using laser excitation sources to enhance the limits of detection. The application is also suited for effluent detection in chromatography. The spatial coherence of lasers allows the use of small-volume detection cells or on-column detection.

3. *Photothermal Monitoring of Excitation and Relaxation Process.* In this application the signal magnitude is measured as a function of time or excitation irradiance. The time-dependent data are used to deduce photophysical and photochemical parameters such as excited-state lifetimes, enthalpies of formation, lifetimes of metastable states, and thermalization times. The excitation irradiance-dependent data can be used to calculate multiphoton absorption cross sections and parameters relating to optical saturation and bleaching.

4. *Photothermal Probing of the Physical Properties.* Many of the physical properties of a sample can be determined using photothermal methods. These methods have been used to measure temperature, thermal diffusivities, sound velocity, bulk flow velocities, surface thickness, and specific heats. In homogeneous samples, the full photothermal transient is typically analyzed to obtain this information. However, some of these parameters can be determined by measuring signal magnitudes, signal decay times, and signal onset times for carefully designed experiments. Thermal properties of heterogeneous samples can be obtained by raster scanning the optical excitation source over the sample surface. In this case the signal magnitude and phase are measured as a function of spatial coordinate.

1.5. ILLUSTRATIVE HISTORY OF PHOTOTHERMAL SPECTROSCOPY

1.5.1. Nature of the Photothermal Effect

Most of us observe the photothermal effect in our lives. On the beach, sand is too hot to walk on with bare feet in midday summer. This is because the sand absorbs sun's radiation and converts this energy to heat. The added heat results in a temperature increase because of the finite heat capacity of the sand. When the heat is generated faster than it can be dissipated by radiative or diffusive mechanisms, the temperature of the sand increases. However, the rate of heat dissipation increases with the temperature difference between the surface sand, and soil below or air above it. Under constant-illumination conditions, the sand reaches an equilibrium temperature wherein the rate of heat generated by the photothermal effect is balanced by the rate at which the heat is dissipated. Another way we utilize the photothermal effect is to warm ourselves by the radiation of a campfire. Here our skin is the absorber and the campfire is the source of the infrared radiation.

A concrete example of the photothermal effect, which is also the basis for a photothermal spectroscopy method, is the shimmering surface or optical mirage effect. This effect is illustrated in Figure 1.5. A hot highway

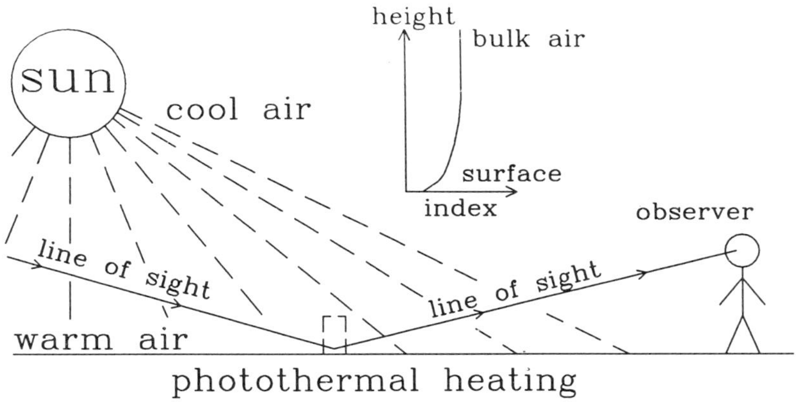

Figure 1.5. Early photothermal deflection apparatus for measuring absorbance of the earth's surface. Since the signal depends on meteorological and solar conditions, it is difficult to obtain accurate numbers using a human detector.

sometimes looks like a reflective surface. It appears as if it were a puddle of water. We come to understand that the apparently shiny surface is not due to reflection. It is just a mirage. In fact, the mirage effect is one of the photothermal effects that have been exploited for chemical and materials analysis. Radiation from the sun is absorbed by the concrete or asphalt, resulting in surface heating. The hot surface transfers energy to the air above the surface. A temperature gradient develops between the air near the surface and the bulk air above. Air expands when it is heated. The density of the air at the surface is less than that in the bulk. The decreased density results in a decreased refractive index. Since the speed of light is faster in the low-refractive-index media, light incident at an acute tangent angle is refracted upward. An observer looking at the surface at an acute tangent angle does not see the surface but rather, sees the rays coming from the sky above the surface.

It is likely that our predecessors had a working knowledge of the photothermal effect long before they could apply more abstract concepts, such as optical transmission, color, and other factors leading to modern theories of spectroscopy. But although photothermal effects may have been recognized in the prehistoric past, it took an understanding of the photothermal process to apply the photothermal effect for spectrochemical measurements. Much of what is now known about photothermal spectroscopy has been developed over the past century. Many of the advances came about as a result of the developments in laser technology about 25 years ago. Other advances were made simple by the recognition and understanding of what is now called the *photothermal effect*.

1.5.2. Photoacoustic Spectroscopy

The oldest technical application of the photothermal effect is believed to be the communication device, the photophone, invented by Bell (1880, 1881). Bell found that audible sound could be heard coming from a tube filled with various materials when the light shining on the transparent tube was modulated. The sound was loud when the tube was filled with radiation-absorbing gases or solids, and weak when filled with a liquid. The operational principles are now well understood. Modulation of the light impinging on an absorbing substance will produce a similar modulation in temperature through the photothermal effect. In a gas of restricted volume, temperature modulation produces a pressure modulation. The periodic pressure modulation is an acoustic signal.

Some time later Viengerov (1938) used the photoacoustic effect to study light absorption in gases and obtained quantitative estimates of concentration in gas mixtures based on signal magnitudes. This may have been the first use of photoacoustic spectroscopy. Sensitive chemical measurement applications followed the work of Kerr and Atwood (1968), who used a laser to excite the samples. More interest in the method was generated when Kreuzer (1971) demonstrated part-per-billion (ppb) detection sensitivities of methane in nitrogen using a 3.39-μm helium–neon laser excitation source, and later (Kreuzer et al. 1972) sub-ppb detection of ammonia and other gases using infrared CO and CO_2 lasers. These high-sensitivity measurements were possible because of the laser source used for excitation. Large photoacoustic spectroscopy signals resulted from the high spectral brightness and the spatial coherence of the lasers used for sample excitation. The photoacoustic measurement methods came at about the same time as the recognition that trace species could have a major impact on the environment.

In the time since the first chemical measurements by Viengerov (1938), the theory and practice has been developed to a high degree. The theories for sound generation, propagation, and interaction with matter were developed though the mid-twentieth century (Landau and Lifshitz 1959, Herzfeld and Litovitz 1959) and acoustics were applied to physical chemical analysis. The theories are complex, and exact solutions for sample excitation and signal generation are often difficult to interpret and verify. Nonetheless, the principles of photoacoustic spectroscopy are now commonly understood and photoacoustic spectroscopy is being applied to a wide range of analysis problems.

The essential components for apparatus used for photoacoustic spectroscopy is shown in Figure 1.6. The light source, either pulsed or modulated, periodically heats the sample by the photothermal effect. Periodic sample

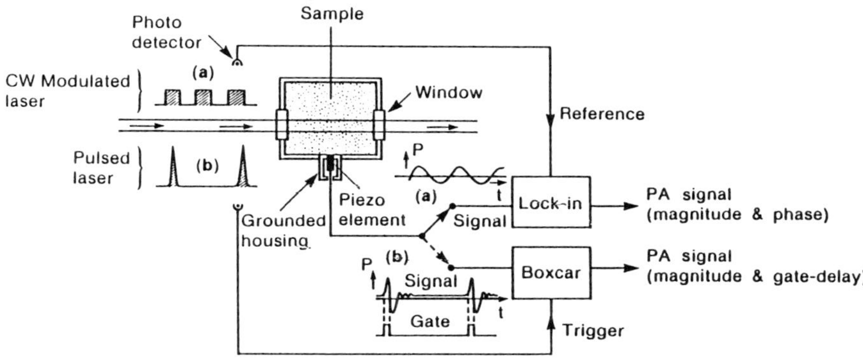

Figure 1.6. Schematic of a photoacoustic spectrometer based on direct acoustic-wave detection. Chopped (*a*) or pulsed (*b*) sample excitation results in acoustic-pressure-wave generation. The signal is detected with a piezoelectric pressure transducer and processed with either a lock-in or a sampling (boxcar) amplifier. (Reprinted with permission from Tam 1989. Copyright 1989, Academic Press.)

heating followed by expansion causes a periodic pressure wave which is detected with the pressure transducer. The pressure transducer signal is proportional to the amplitude of the pressure wave. Consider a sample that has a low enough absorption coefficient that the transmission can be approximated by,

$$T(l) = e^{-\alpha l} \approx 1 - \alpha l \tag{2}$$

where $T(l)$ is the optical pathlength, l (m), which depends on transmission, and α (m^{-1}) is the absorption coefficient. The amount of energy absorbed from a laser source with an optical energy of Q (J) is $Q[1 - T(l)] \approx Q\alpha l$. If the quantum yield for heat production is unity, all the optical energy absorbed is converted into heat. The peak pressure change, $\delta P_{\text{acoustic}}$ (Pa), is proportional to (Lai and Young 1982, Tam 1986)

$$\delta P_{\text{acoustic}} \propto \tau^{-3/2} \frac{\alpha \beta Q}{C_P} \left(\frac{c}{r}\right)^{1/2} \tag{3}$$

where c ($m\,s^{-1}$) is the sound velocity, β (K^{-1}) the volume expansion coefficient, r (m) the radial distance between the transducer and the source, C_P ($J\,kg^{-1}\,K^{-1}$) the specific heat, Q (J) the pulse energy, and the pressure perturbation time, τ (s), is the root-mean-square average of the relaxation times and the pulse or modulation width. Relaxation times may include contributions from the excited-state relaxation time and the acoustic relaxation time

$$\tau_a^2 = \frac{w^2}{2c^2} \tag{4}$$

where τ_a (s) is the acoustic relaxation time and w (m) is the radius of the beam used for sample excitation. The acoustic relaxation time is that required for the heated sample to expand.

The important points to be deduced from the acoustic pressure equation are that (1) the signal scales as the αQ product; (2) the signal falls off as the pressure transducer is moved away from the excited region as $r^{-1/2}$; (3) the signal is inversely proportional to the pressure perturbation time, favoring short pulse excitation and small beam waists; and (4) the signal magnitude is proportional to the thermodynamic properties of the sample through the $\beta c^{1/2}/C_P$ term. In general, β is much smaller for liquids and solids than it is for gases. Not only does this explain the early observations of Bell (1881) but also why direct photoacoustic spectroscopy is most sensitive for gas sample analysis.

The spectra of solid or liquid samples can be measured by directly coupling the acoustic wave to a transducer or by coupling the heat generated at the surface to a gas coupling fluid. This principle was used in Bell's original photophone but was not rediscovered until Parker (1973) noticed that optical energy absorbed by the gas sample cell windows would transfer heat to a gas, thereby causing a significant photoacoustic signal. This effect was developed by Rosencwaig (1977, 1980) and is now commonly used for obtaining spectra of strongly absorbing solids and liquids. A modern version of a device for photoacoustic spectroscopy of condensed samples is shown in Figure 1.7. A solid or liquid sample is placed in the sealed photoacoustic cell. The excitation source is absorbed at or near the surface. Absorbed radiation is randomized increasing the surface temperature. The heated surface heats the gas, causing it to expand. Periodic heating of the surface creates an acoustic wave that is monitored with the sound transducer.

There have been scores of publications on the uses of photoacoustic spectroscopy for chemical and material analysis. Absorption detection limits (α) are about $10^{-10}\,\mathrm{cm}^{-1}$ for gases (Patel et al. 1977) and $10^{-6}\,\mathrm{cm}^{-1}$ for liquids (Beitz et al. 1990). These are very close to the theoretical detection limits (Zharov and Letokhov 1986). Many review articles and books have been written on this method. Some of the more recent reviews of general applications are those of Tam (1983, 1986) and Hutchins and Tam (1986). Patel and Tam (1981) reviewed applications of photoacoustic spectroscopy for condensed matter. Betteridge and Meylor (1984) have reviewed the applications of photoacoustic spectroscopy in chemical analysis. Zharov

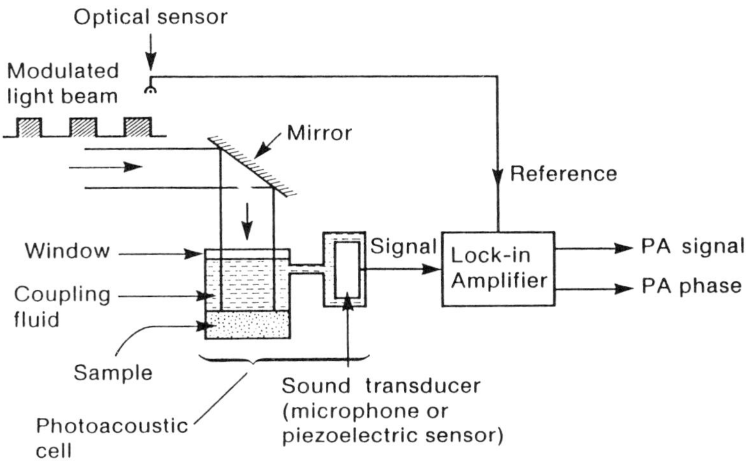

Figure 1.7. Schematic of an indirect photoacoustic spectrometer based on chopped excitation. The thermal perturbation generated in the sample is coupled to the fluid, usually a gas, and sensed with a microphone pressure transducer. The microphone signal is then processed with a lock-in amplifier to enhance the signal. (Reprinted with permission from Tam 1989. Copyright 1989, Academic Press.)

(1986) reviewed photoacoustic applications to chromatography. Meyer and Sigrist (1990) have reviewed applications to gas analysis. General books on photoacoustic spectroscopy include those of Pao (1977), Rosencwaig (1980), Zharov (1986), and Zharov and Letokhov (1986). Mandelis (1987) has edited a book on application of photoacoustic and photothermal spectroscopy methods for semiconductor analysis. Hess (1989a,b) has edited a book regarding the application of photoacoustic and photothermal spectroscopy methods for gas and surface analysis. Nyquist et al. (1990) and Putzig et al. (1992) have reviewed photoacoustic and photothermal spectroscopies in their *Analytical Chemistry Fundamental Reviews* of infrared analysis. Kitamori and Sawada (1991) have discussed unconventional applications in their review.

1.5.3. Photothermal Lens Spectroscopy

The first photothermal spectroscopic method to be applied for sensitive chemical analysis was photothermal lens spectroscopy. The photothermal lens effect was discovered when Gordon et al. (1964, 1965) observed transient power and beam divergence changes in the output of a helium–neon laser after placing transparent samples in the laser cavity. Their apparatus, shown in Figure 1.8, was originally intended to be used as a

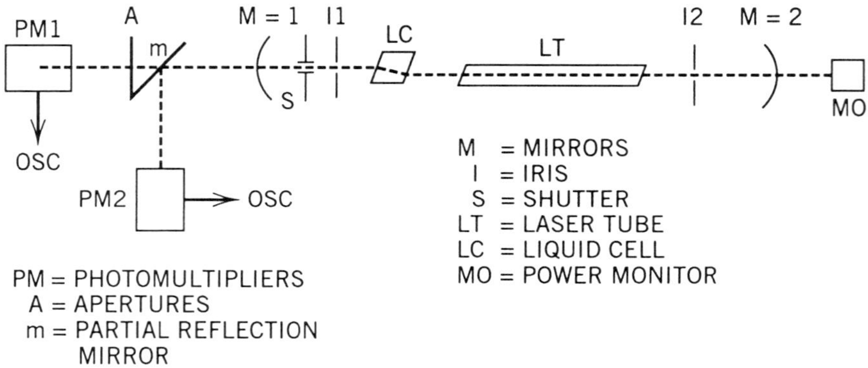

Figure 1.8. First photothermal lens apparatus. The sample was placed in the cavity of the laser. Irises were used to restrict the laser to single (TEM_{00})-mode operation. Detectors were used to measure the laser power and the laser output both with and without an external pinhole. (Reprinted with permission from Gordon et al. 1965. Copyright 1965, American Physical Society.)

high-irradiance source for Raman spectroscopy. They observed the photothermal lens effect when pure organic liquids and solids, glass and Lucite, were placed in the laser cavity. A theory describing the effect was developed to account for their observations. This theory was an accurate description of the physics of thermal lens formation and signal generation and is essentially the same as that used to this day (Whinnery 1974). The photothermal lens results from optical absorption and heating of the sample in regions localized to the extent of the excitation source. The lens is created through the temperature dependence of the sample refractive index. The lens usually has a negative focal length since most materials expand upon heating and the refractive index is proportional to the density. This negative lens causes beam divergence and the signal is detected as a time-dependent decrease in power at the center of the beam.

Laser output power transients for the first apparatus are shown in Figure 1.9. Although the theory was accurate, these transients were difficult to interpret. The transients arose due to the interaction between the intracavity beam propagation altering character of the photothermal lens element and the intracavity apertures. Nonetheless, Solimini (1966) refined the apparatus and measured the absorption coefficients of 27 organic liquids using this method. The first extracavity sample photothermal lens apparatus was used by Grabiner et al. (1972) to measure vibrational relaxation rate constants. Hu and Whinnery (1973) recognized that the extracavity sample configuration would be more flexible and could also result in sensitive absorbance measurements. The apparatus and beam analysis, shown in Figure 1.10, is essentially the same as that used for single-laser photothermal lens spectro-

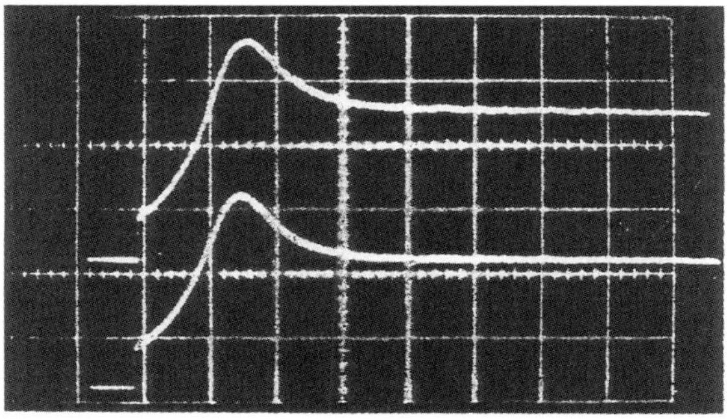

Figure 1.9. Transient signals observed using the intracavity photothermal lens apparatus. The top trace was obtained with an extracavity pinhole, and the bottom trace was obtained without the pinhole. These data were used to confirm the premise that the laser was operating in single mode and that the signal was generated by the internal apertures. These signals were difficult to analyze because of the interrelationships between laser power and cavity losses. (Reprinted with permission from Gordon et al. 1965. Copyright 1965, American Physical Society.)

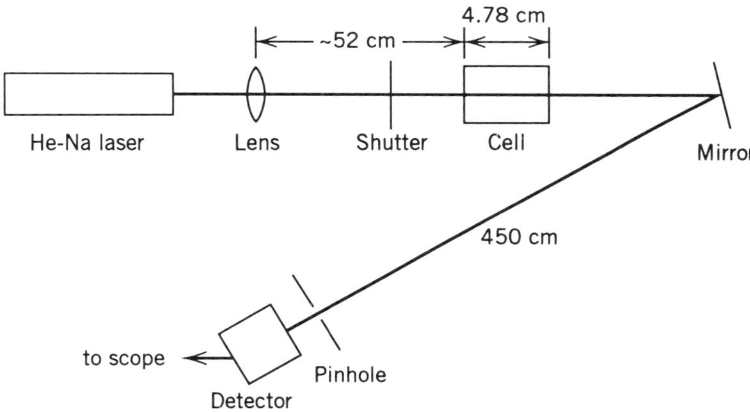

Figure 1.10. Schematic of the extracavity photothermal lens spectrometer used by Hu and Whinnery (1973) to measure the optical absorbances in transparent fluids. The lens focuses the laser beam one confocal distance in front of the sample cell. The pinhole and detector are placed in the far field of the focus. (Reprinted with permission from Whinnery 1974. Copyright 1974, American Chemical Society.)

scopy today. The transient signals produced extracavity are less complicated than those of the intracavity configuration and the theory describing the transients is more tractable. The essential components of the apparatus are (1) the coherent, laser excitation source, which can deliver high optical powers over a small cross-sectional area of the sample, (2) a low-absorbance sample, (3) a spatial filter or pinhole placed in the far field, and (4) a photodetector to measure the power past the pinhole.

The extracavity photothermal lens spectroscopy signal can be described in terms of the focal length of the thermal lens formed within the sample. The simplest form of the focal length is found by assuming that $\alpha l \ll 1$ and unit quantum efficiency for heat production. A sample excited by a laser beam with an irradiance of

$$E(r) = \frac{2\Phi_0}{\pi w^2} e^{-2r^2/w^2} \tag{5}$$

where $E(r)$ ($W\,m^{-2}$) is the radially dependent irradiance and Φ_0 (W) is the incident radiant power, will produce a time-dependent photothermal lens with a focal length, $f(t)$:

$$f(t) = f(\infty)\left(1 + \frac{t_c}{2t}\right) \tag{6}$$

where $f(\infty)$ (m) is the steady-state focal length formed at infinite time:

$$f(\infty) = \frac{n_0 \pi k w^2}{\Phi_0 \alpha l (dn/dT)} \tag{7}$$

and t_c (s) is the characteristic thermal time constant

$$t_c = \frac{w^2 \rho C_P}{4\kappa} \tag{8}$$

where κ ($J\,cm^{-1}\,s\,K^{-1}$) is the thermal conductivity, n_0 the refractive index of the medium where detection takes place (normally air), n the refractive index of the sample, T (K) the temperature, ρ ($kg\,m^{-3}$) the density, and C_P ($J\,kg^{-1}\,K^{-1}$) the specific heat. The lens is formed because the optically heated sample has a different refractive index from that of the bulk of the sample. The differential term $(dn/dT)_P$ is the temperature-dependent refractive index change at constant pressure. The shape of the temperature change produced by a Gaussian excitation source is parabolic near the center. The parabolic refractive index perturbation is equivalent in form to a simple lens.

The photothermal lens signal is obtained by monitoring the laser power that passes through a pinhole placed far from the sample (Figure 1.11). The photothermal lens will either focus or defocus the laser. When this happens, the power at the center of the beam will either increase or decrease. This change in power is maximized when the sample is placed one confocal

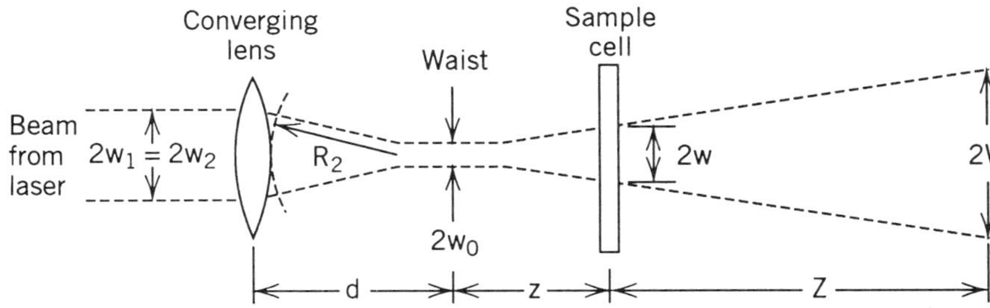

Figure 1.11. Beam geometry and definitions used for extracavity photothermal lens spectroscopy. (Reprinted with permission from Whinnery 1974. Copyright 1974, American Chemical Society.)

distance to either side of the laser's focus. In this case the relative change in power monitored past the pinhole aperture is

$$\frac{\Phi_d(0) - \Phi_d(t)}{\Phi_d(t)} \approx \frac{\pm 2z_0}{f(t)} \tag{9}$$

where $\Phi_d(t)$ is the time-dependent power and the confocal distance is $z_0 = n_0 \pi w_0^2/\lambda$, w_0 being the beam waist radius at the focus, and λ (m) is the wavelength of the laser. The $+$ sign applies to samples placed before the focus, the $-$ sign to samples behind the focus. The time-dependent signal observed past the pinhole is

$$\frac{\Phi_d(0) - \Phi_d(t)}{\Phi_d(t)} \approx \pm\left(\frac{dn}{dT}\right)\frac{\alpha l \Phi_0}{\lambda \kappa}\frac{1}{1 + t_c/2t} \tag{10}$$

The essential components to interpreting the signal are that (1) the time-dependent signal increases or decreases the power past the pinhole; (2) the time constant for signal evolution, t_c, is proportional to the square of the beam waist radius in the sample; (3) the signal magnitude is proportional to the absorption coefficient, pathlength, and excitation power; (4) the signal magnitude also depends on the thermal, κ, and optical, dn/dT, properties of the sample; and (5) for times much greater than t_c, the steady-state power change is related to absorption coefficient by

$$\pm\alpha = \frac{\Phi_d(0) - \Phi_d(\infty)}{\Phi_d(\infty)} \frac{\lambda \kappa}{\Phi_0 l(dn/dT)} \tag{11}$$

The absorption coefficient can be obtained by measuring the power change with knowledge of the temperature-dependent refractive index.

It is difficult from these equations to see how photothermal lens spectroscopy method can enhance absorbance measurements. Dovichi and Harris (1979) introduced the concept of the enhancement factor, the ratio of the photothermal lens signal magnitude to that which would be obtained using conventional transmission spectroscopy. For weakly absorbing samples, the transmission spectroscopy signal can be cast in a form similar to that for the photothermal lens spectroscopy signal:

$$\frac{\Phi_0 - \Phi_l}{\Phi_0} \approx \alpha l \tag{12}$$

where Φ_l is the power after passing through the sample. The ratio of the photothermal lens signal to this signal yields the enhancement factor

$$E = \left(\frac{dn}{dT}\right)_P \frac{\Phi_0}{\lambda\kappa} \tag{13}$$

The enhancement is a function of the thermodynamic and optical properties of the solvent and on the power used to excite the sample. Nonpolar solvents are particularly useful for trace analysis because of their relatively high $(dn/dT)_P$ and low κ. For example, CCl_4 has a temperature-dependent refractive index of $-6.12 \times 10^{-4}\,K^{-1}$ and a thermal conductivity of $0.103\ W\,m^{-1}\,K^{-1}$ (Dovichi 1987). The theoretical enhancement factor is $11{,}560\ W^{-1}$ for the 514-nm line of an argon ion laser. Of course, the higher the power, the greater the enhancement. Even a modest 10-mW laser will yield signals that are over 100-fold better than those of the conventional transmission spectrophotometer. Absorption coefficient detection limits in 1-cm cuvettes are about $10^{-7}\ cm^{-1}$. This detection limit was reported by Dovichi and Harris (1981) for 514.5-nm excitation of samples in CCl_4 solvent using 160 mW of laser power. The enhancement factor under these conditions is about 1850. Based on this, the absorbance detection limits calculated for the equivalent conventional transmission spectrophotometer would be 2×10^{-4} absorbance unit. Although it is a matter for discussion, this is about what one might expect from a double dispersing transmission spectrophotometer.

The characteristic time constant, t_c, should also be considered in the experimental design. With a shorter time constant, more measurements can be made in a given time. Since replicate measurements can be used to increase the precision of the estimate, the shorter time constants resulting from smaller focus spot sizes, are favored. For example, CCl_4 has a thermal diffusivity of $7.5 \times 10^{-8}\ m^2\,s^{-1}$. A laser with a beam waist radius of 1 mm in the sample cell will produce a signal with a characteristic thermal time constant of 3.3 s, whereas using a 10-μm beam waist radius, $t_c = 0.33$ ms.

The 10-μm beam would allow 10^4 replicate measurements in the same time required to obtain one measurement with a 1-mm beam waist. The measurement precision would increase by 100 using the smaller beam waist and equivalent measurement times.

The first analytical application of photothermal spectroscopy was the trace-level determination of Cu(II) with an EDTA complex reported by Dovichi and Harris (1979). They used a single-laser extracavity photothermal lens apparatus. This method is perhaps the best known and most used of all photothermal spectroscopy methods. The relative simplicity of the apparatus coupled with the low-solution absorption detection limits, $10^{-7}\,\text{cm}^{-1}$ (Dovichi and Harris 1981), make it highly attractive for trace analysis applications.

1.5.4. Photothermal Interferometry

Shortly after the discovery of the photothermal lens effect, researchers found that the photothermal-induced refractive index change could be measured by more direct means. McLean et al. (1968) and Longaker and Litvak (1969) recognized that optical absorption resulting in sample heating and subsequent changes in refractive index would cause a phase shift in light passing through the heated region. The optical phase shift can be detected with an interferometer. The method of using optical interferometry to measure refractive index changes was not in itself new, but using an excitation laser to heat the sample while monitoring the refractive index change was. Most photothermal interferometry apparatus is based on laser excitation sources. Stone (1972, 1973) showed that both coherent and wideband incoherent sources could be used. Stone used the modified Jamin interferometer apparatus shown in Figure 1.12 to obtain the absorption spectrum of chlorobenzene shown in Figure 1.13. Using this apparatus, 2 to 3 mW of excitation source power could be used to measure absorption coefficient of about $2 \times 10^{-5}\,\text{cm}^{-1}$.

The conventional approach to measuring small absorption coefficients is to increase the optical pathlength. The data in Figure 1.13 compare results obtained using long-pathlength transmission spectrophotometry to those of the photothermal interferometer. Transmission losses may be due to reflection, scattering, and absorption. The finite transmission losses seen in the bromobenzene spectrum are not necessarily due to optical absorption. On the other hand, the photothermal interferometer responds only to absorption. The resulting spectrum is technically an excitation spectrum since the heat is generated by optical absorption of the excitation light. An almost astonishing feature of the interferometric method is its sensitivity. Davis and Petuchowski (1981) have measured absorption coefficient detection limits as low as $10^{-10}\,\text{cm}^{-1}$ for gaseous samples in windowless absorption cells using

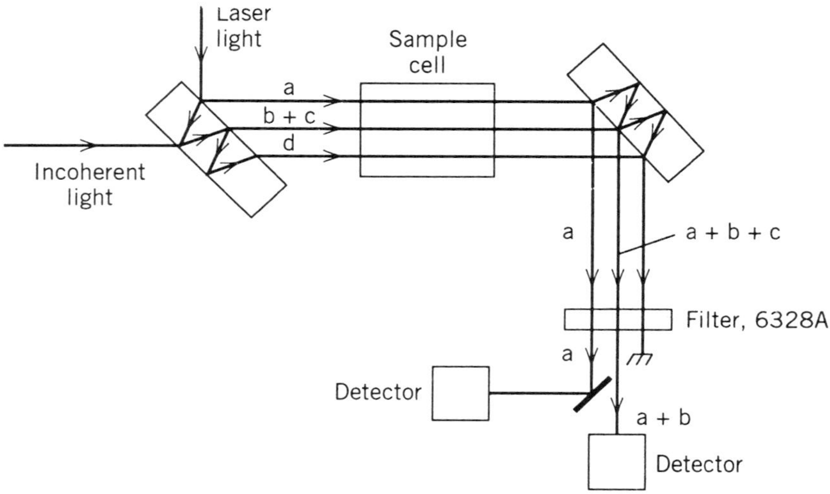

Figure 1.12. Modified Jamin interferometer apparatus used by Stone (1973). Incoherent light from a xenon arc is collimated and filtered by a series of bandpass filters before passage through the center of the sample cell along (*c*). Helium–neon laser light detects the optical phase shift. The laser light is split by the optical flat and passes through a reference path (*a*) and a probe path (*b*). The two laser beams are combined at the second optical flat. One detector monitors the power of the reference beam and the other the power in the interfering beams ($a + b$). A phase shift in the two interfering beams results in a power change at the signal detector. Phase shifts are found from the ratio of the signal to reference powers. (Reprinted with permission from Whinnery 1974. Copyright 1974, American Chemical Society.)

chopped infrared excitation lasers at irradiances of 2.5 MWm^{-2}. Other sensitive interferometric methods for measuring the photothermal effect are discussed by Friedrich (1983), and Dovichi (1987) has reviewed the applications to chemical analysis.

The interferometric studies of Longaker and Litvak (1969) used cameras to obtain images of phase-shift patterns resulting from the refractive index perturbation produced by pulsed Nd glass laser sample excitation. This classic and innovative work revealed a wealth of information regarding photothermal effects. The apparatus used for these studies is shown in Figure 1.14. The photographic camera was used to obtain pictures of the fringe patterns for visual analysis, and the vidicon camera was used to obtain quantitative information for critical evaluation of the data. Photographic images shown in Figure 1.15 reveals some of the effects they observed. For absorbing samples, the refractive index perturbation had two components with different space and time behaviors. A long-lived transient was observed near the region excited by the pulsed laser. This component was the thermal perturbation produced by the photothermal effect.

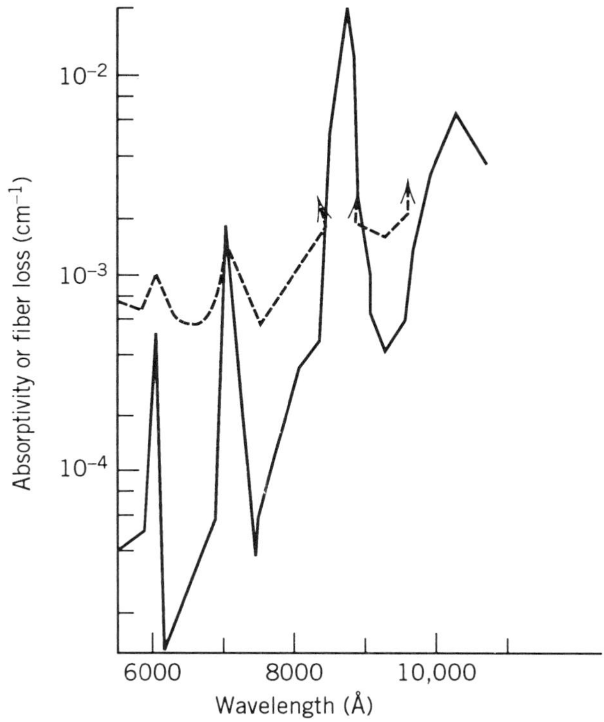

Figure 1.13. Data obtained for chlorobenzene using the photothermal interferometer of Figure 1.12 (solid) and that of bromobenzene in a glass capillary (dashed line) obtained with a transmission spectrophotometer. The structured absorption features are C—H stretch vibrational overtones. (Reprinted with permission from Whinnery 1974. Copyright 1974, American Chemical Society.)

The phase shift, $\delta\phi$ (rad), produced from the thermal component is related to the density change through

$$\delta\phi = \frac{2\pi l}{\lambda}\left(\frac{dn}{d\rho}\right)\delta\rho \tag{14}$$

where λ is the wavelength of the laser used to measure the refractive index change. The theory developed by Longaker and Litvak predicts that for weakly absorbing samples with rapid excitation and excited-state relaxation times, the on-axis time-dependent density change for pulsed radiation is

$$\delta\rho(t) \propto \frac{2\beta\alpha Q}{\pi w^2 C_P}\left(e^{-t^2/\tau_a^2} - 1\right) \tag{15}$$

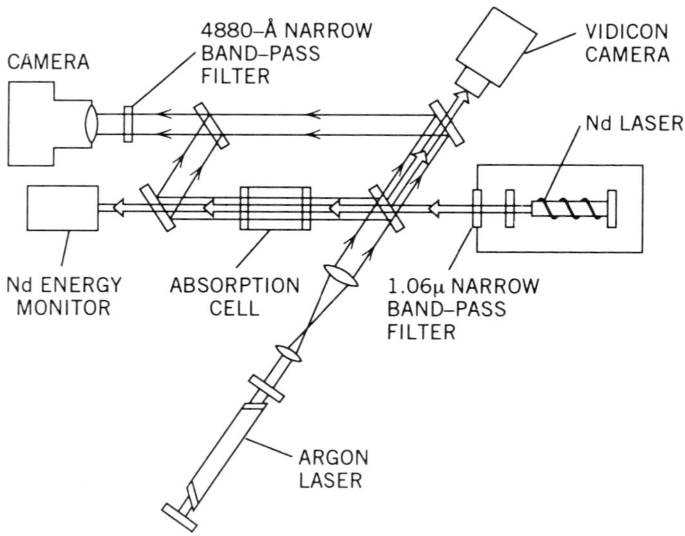

Figure 1.14. Interferometer used by Longaker and Litvak (1969) to obtain images of the density perturbation in gas and liquid samples. The pulsed Nd laser is used to excite the sample and the continuous Ar^+ laser is used to probe the refractive index changes. The camera records the fringe shift of the Ar^+ laser beam. (Reprinted with permission from Longaker and Litvak 1969. Copyright 1969, American Physical Society.)

for times much shorter than t_c. Thus the signal rise time is limited by the same acoustic relaxation time that limits the signal magnitude in photoacoustic spectroscopy. The spatial density change could be determined quantitatively by counting interference fringes. Vidicon camera data were analyzed in terms of the thermal-induced phase shifts and the focal length of the photothermal lens resulting from the thermal perturbation. These data were found to agree with the theory developed by Gordon et al. to describe the photothermal lens.

In addition to the thermal component, a short-lived transient component was found. This component propagated away from the heated region as a wave. This was identified as an acoustic pressure wave. Referring to the ammonia gas data in Figure 1.15, the thermal perturbation can be seen at the center and the dark ring around the central perturbation is due to the propagating pressure or acoustic wave. The acoustic wave is produced by the rapidly expanding sample heated by the pulsed laser. The ammonia gas absorbs energy from the pulsed excitation source. Excited-state ammonia rapidly relaxes, thereby increasing the temperature of the sample. The heated sample then expands to produce an acoustic compression wave. The compression wave propagates out away from the excited region. The

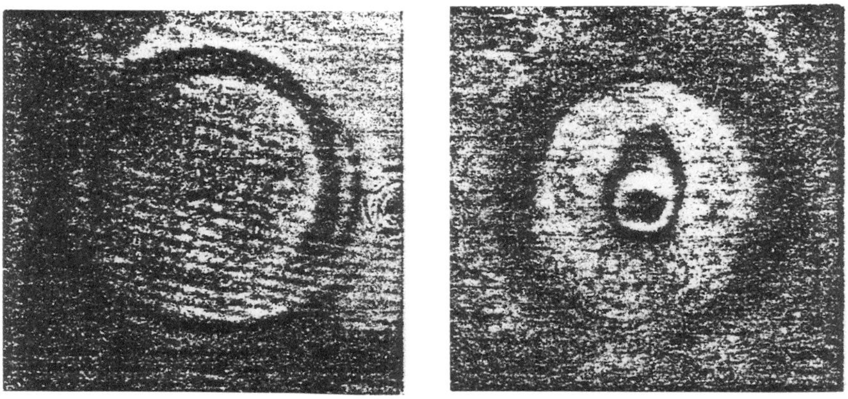

Figure 1.15. Image data obtained with the apparatus illustrated in Figure 1.14. The picture on the left is that of a 5-cm sample of nonabsorbing liquid CS_2. No thermal perturbation is observed and the fringe shift is due only to the acoustic wave generated by electrostriction. The image on the right is for a weakly absorbing sample and has both a thermal perturbation (in the center) and an acoustic-wave component (the dark ring). (Reprinted with permission from Longaker and Litvak 1969. Copyright 1969, American Physical Society.)

compression increases the density of the gas, thereby causing an increase in the refractive index. Thus the acoustic wave also results in a photothermal signal. Although the acoustic wave carries away some of the energy, most of the thermal energy remains in the region local to the excitation laser irradiation (Bialkowski 1988). Although Longaker and Litvak were not the first to observe this effect, their pictorial observations clearly demonstrate the principles of photothermal and photoacoustic spectroscopies and showed the connection between the two.

Photoacoustic wave generation by the photothermal effect is only one of several mechanisms for acoustic wave generation. Figure 1.15 also shows data obtained for CS_2, a nonabsorbing, highly polarizable liquid. The CS_2 data illustrate acoustic waves created without a photothermal perturbation. The acoustic waves are generated by an effect called electrostriction, wherein polarizable media are compressed by the electric field of the optical radiation. Electrostriction has not, to date, been observed using photothermal spectroscopy methods.

1.5.5. Two-Laser Photothermal Lens Spectroscopy

The two-laser photothermal lens apparatus was used before the extracavity single-laser method was found. Grabiner et al. (1972) used a helium–neon laser to probe the photothermal lens produced by a pulsed, infrared laser.

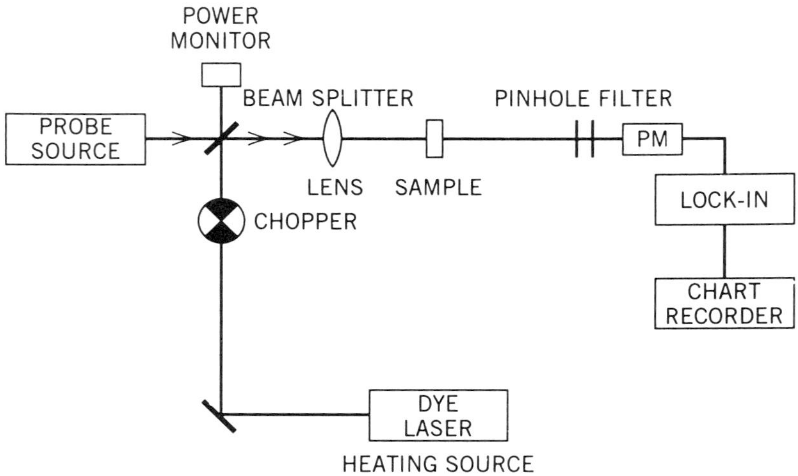

Figure 1.16. Dual-beam photothermal lens spectrometer. The dye laser excites the sample and the probe source monitors the resulting refractive index change through the photothermal lens effect. The diverging probe beam passes through a pinhole spatial filter to develop the signal. The wavelength filter rejects the excitation wavelength. The chopped signal is processed with a lock-in amplifier to improve the signal-to-noise ratio. The dye laser can be scanned to obtain photothermal excitation spectra. (Reprinted with permission from Swofford and Morrell 1978. Copyright 1978, American Physical Society.)

They used this two-laser photothermal lens apparatus to determine the vibrational relaxation rate constants for methyl chloride and methyl fluoride gases. Later, Siebert et al. (1974) used the technique to study relaxation of vibrationally excited CD_4, SO_2, and OCS. Of interest was the rise time of the photothermal lens signal. The rise times were measured as a function of added gas pressure, and the vibrational relaxation rate constants were deduced from these measurements. The technique was found to be quite satisfactory for relaxation times that were greater than the acoustic limited rise times. The vibrational relaxation rate constants compared well to those obtained using other methods. Although not exploited in this work, Grabiner et al. and Siebert et al. showed that by using this photothermal lens method, infrared absorption could be measured using visible detectors. This would be used to advantage later in short-pathlength infrared absorption studies.

Long et al. (1976) used the two-laser photothermal lens apparatus shown in Figure 1.16 to measure absorption spectra due to vibrational overtones in pure solvents. A repetitively chopped continuous dye laser was used to form the photothermal lens in the sample and a continuous helium–neon laser probed the resulting lens element. The equations that describe the temperature change and focal length of the photothermal lens are the same as those given above. However, several advantages of using separate excitation and probe light sources in photothermal lens spectroscopy can be realized in this configuration: (1) The dye laser can be scanned to produce excitation spectra of the sample without having to account for photodetector wavelength response; (2) the excitation source can be focused directly into the sample, which increases the irradiance and the resulting photothermal lens signal by decreasing the beam waist radius in the sample; and (3) a lock-in amplifier can be used to decrease the bandwidth of the measurement, thereby enhancing the signal-to-noise ratio.

Twarowski and Kliger (1977a) developed a quantitative theory to describe the pulsed-laser-excited photothermal lens spectroscopy signals and applied this theory to study the two-photon absorption of benzene (1977b). This was the first derivation of the time-dependent thermal lens given for pulsed laser excitation. Basically, a pulsed laser with an integrated irradiance $H(r, t)$ ($\mathrm{J\,m^{-2}}$) of

$$H(r, t) = \frac{2Q}{\pi w^2} e^{-2r^2/w^2} \tag{16}$$

will produce a temperature change of

$$\delta T(r, t) = \frac{2\alpha Q}{\pi w^2 \rho C_P} \frac{e^{-2r^2/w^2(1+2t/t_c)}}{1 + 2t/t_c} \tag{17}$$

for a single-photon absorption process and for times greater than those required for acoustic relaxation. The inverse focal length was found to be

$$\frac{1}{f(t)} = \frac{1}{n_0}\left(\frac{dn}{dT}\right)\frac{8\alpha l Q}{\pi w^4 \rho C_P}\frac{1}{(1 + 2t/t_c)^2} \tag{18}$$

The main characteristics of the pulsed-laser photothermal lens spectroscopy signal are that (1) the signal magnitude is greatest at zero time, just after acoustic relaxation of the sample, which allows the pulsed laser technique to be used to study excited-state relaxation kinetics; (2) the signal is inversely proportional to w^4, favoring tighter focused beams; (3) the signal decays in a time that is inversely proportional to t^2; (4) as with the chopped continuous excitation laser method, the pulsed laser method can use dye lasers to obtain excitation spectra and the excitation laser can be focussed into the sample cell resulting in greater signal magnitudes; (5) the high irradiance at the focus can be high enough to induce nonlinear absorption effects. The multiphoton absorption signal is essentially the same but with the caveat that the absorbed energy is proportional to the integrated irradiance raised to the power of the number of photons absorbed. Thus the effective squared beam waist radius is decreased by a factor inversely proportional to the number of photons absorbed per transition, $w^2/p \to w^2$. This further enhances the signal magnitude and has led to the belief that photothermal lens spectroscopy is very useful for multiphoton spectroscopy.

Barker and Rothem (1982) pointed out that the simple theoretical description of the photothermal lens shown above does not yield quantitative results in the early times of the signal. They point out an apparent dilemma wherein Grabiner et al. (1972) use an acoustic-wave equation to model results whereas Twarowski and Kliger (1977a) use a thermal diffusion equation. Barker and Rothem developed a quantitative theory for predicting the photothermal lens signal that takes into account several hydrodynamic relaxation effects. This theory predicts that all but the first of the five points given above hold, but that the signal rise time is limited by the rate at which the density can change. The latter is related to the sound velocity and the radius of the excitation source (Barker and Toselli 1989).

Fang and Swofford (1983) have written an excellent overview of the theory and developments in photothermal lensing spectroscopy. Dovichi (1987) has reviewed the literature and has commented on analytical applications of the technique, including absorbance detection limits of about 10^{-7} to 10^{-8} cm^{-1} for liquids and gases using 10- to 200-mW continuous sources. Sell (1989) has collected together a number of chapters addressing many important applications of photothermal spectroscopy. Morris and Fotiou (1989) have reviewed applications to chromatography detection. Dovichi (1990) has included this technique in his review of laser-based microanalysis.

1.5.6. Photothermal Deflection, Refraction, and Diffraction

The mirage is a common and well-understood example of the photothermal effect. However, the analytical method based on this principle, photothermal deflection spectroscopy, was somehow overlooked until Boccara et al. demonstrated probe laser beam deflection in 1979. The method was applied to surface analysis. A typical experimental setup for photothermal deflection analysis of surfaces is shown in Figure 1.17*c*. Like the indirect photoacoustic spectroscopy method, this method may be used to examine optical absorptions at or near the surface of solid samples. The sample absorbs optical radiation and heats the gas or liquid above the surface. The heated gas acts like a prism and deflects the probe laser incident tangent to the surface.

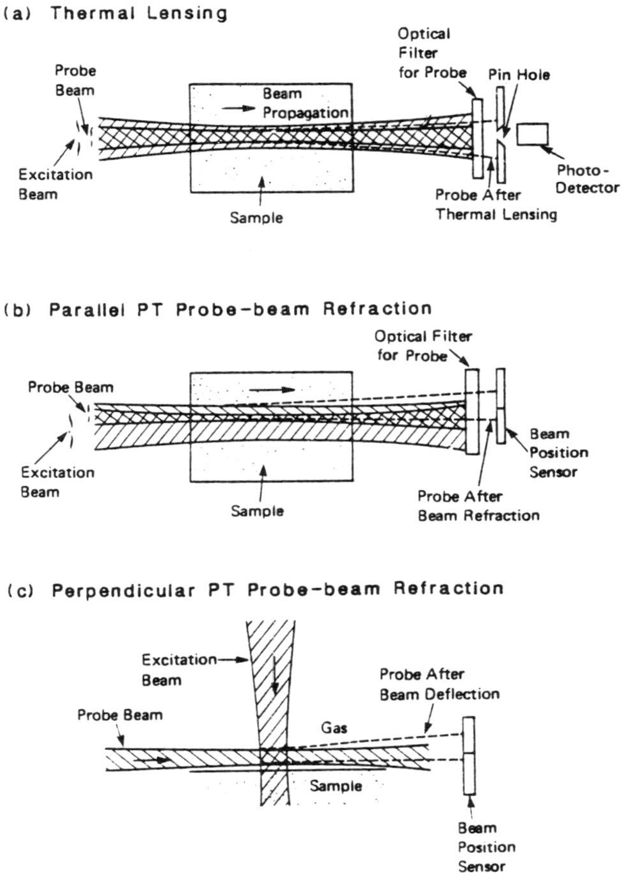

Figure 1.17. Various excitation and probe beam geometries used in photothermal spectroscopy. (Reprinted with permission from Tam 1986. Copyright 1986, American Physical Society.)

Probe laser beam deflection is monitored with a position-sensing detector. The apparatus is very easy to set up and can produce very sensitive measurements of surface absorption.

The theory for describing the photothermal deflection signal has been worked out for both chopped and pulsed excitation sources. This theory is more complicated than that describing homogeneous fluids because the thermal conduction in the solid and the fluid must both be accounted for. The temperature change that occurs upon pulsed irradiation of a surface with an adsorbed absorbing species is

$$\delta T_f(x, t) = \frac{\kappa_S H}{\rho_s C_{P,s} D_{T,s}^{1/2} + \rho_f C_{P,f} D_{T,f}^{1/2}} \frac{1}{(4\pi t)^{1/2}} e^{-x^2/4D_{T,f}t} \tag{19}$$

where the x direction is normal to the surface, κ_S is the unitless surface absorption coefficient, and D_T and ρC_P are the thermal diffusivities and heat capacities of the solid (s) and fluid coupling medium (f), respectively. The deflection angle of a probe being refracted by the temperature gradient produced by the heated surface is

$$\phi(x, t) = -\left(\frac{dn}{dT}\right) \frac{\kappa_S H}{\rho_s C_{P,s} D_{T,s}^{1/2} + \rho_f C_{P,f} D_{T,f}^{1/2}} \frac{x}{4D_{T,f}\sqrt{\pi}\, t^{3/2}} e^{-x^2/4D_{T,f}t} \tag{20}$$

The deflection angle is monitored using a position-sensing detector placed a short distance from the surface. A change in angle at the sample results in a displacement of the probe laser spot on the detector. For small angles, the linear displacement of the probe laser beam spot is directly proportional to the deflection angle. Equation 20 shows that the magnitude of the signal will be a function of the offset, x, of the probe laser beam from the surface. There is an optimum offset for maximum signal. This optimum offset is a function of time. The time is that required for the temperature change to diffuse to the region probed by the laser. The temperature diffusion process is often called the thermal wave. This equation also shows that at a particular offset, the time-dependent deflection signal will rise and then fall with time. The time to the maximum is $t_{\max} = x^2/6D_{T,f}$. So the time to the maximum signal and the magnitude of the maximum signal are both functions of the displacement of the probe laser beam relative to the surface. This distance is difficult to measure, so photothermal deflection cannot be used to measure absolute absorption coefficients.

One application of this technique caught on rapidly. It was apparent that photothermal deflection could be used for topographic and thermal characterization of samples. The signal magnitude depends on the surface topography, surface absorption coefficient, the thermal properties of the fluid, and the thermal properties of the solid. All other parameters being equal, the signal dependence on the surface to probe laser beam offset allows the surface topography to be measured. For relatively flat surfaces, signal dependence on the solid's thermodynamic parameters allows a *thermal image* of the solid to be obtained (Murphy and Aamodt 1980, 1981). A solid with a constant surface absorption or an optically dense solid will result in

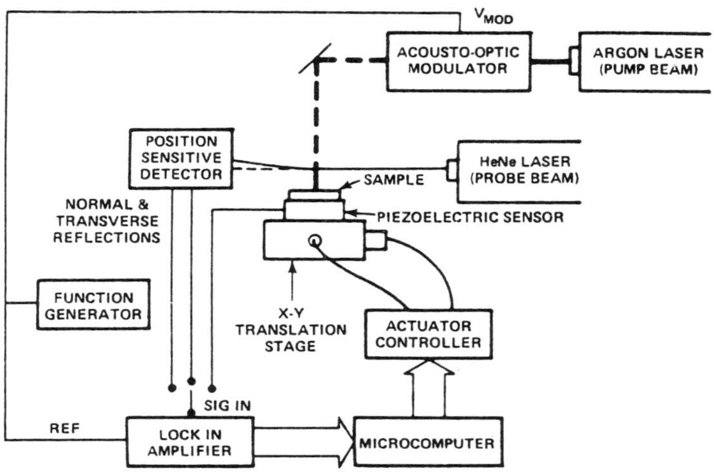

Figure 1.18. Photothermal deflection apparatus used to measure the photothermal image of a surface. The magnitude and phase of the photothermal deflection signal are measured at each position of sample excitation. The sample is raster scanned using an x–y translational stage. The microcomputer records the data and performs image analysis. (Reprinted with permission, from Murphy et al. 1986. Copyright 1986, IEEE.)

a signal that is inversely proportional to the solid's thermal conductivity. An example of a thermal imaging apparatus is shown in Figure 1.18. When the solid sample is raster scanned under a focused excitation laser source, the photothermal deflection signal magnitude will be inversely proportional to the solid's thermal parameters. This thermal imaging technique has been used to determine sample thickness, inclusions in metals (McDonald 1986), coating quality (Busse 1989), and imaging boundaries at crystal domains (Murphy et al. 1986). The thermal image shown in Figure 1.19 is of aluminum metal. The lighter regions are thought to be due to subsurface inclusions in the metal. Several applications of photothermal deflection spectroscopy have been discussed in the recent chapter by Fournier and Boccara (1988).

A useful extension of this technique is to irradiate the entire surface with a series of patterns instead of scanning the excitation and probe lasers across the surface. Fotiou and Morris (1986) use a moving Hadamard encoded mask to analyze the spatial distribution of absorption on stationary thin-layer-chromatography plates. This method is described more fully in Morris and Fotiou (1989). A typical apparatus is shown in Figure 1.20. This apparatus was used to measure band positions and absorptions on dyed plates. Imaging thus far has been one-dimensional, but there is no apparent reason why two-dimensional images could not be obtained.

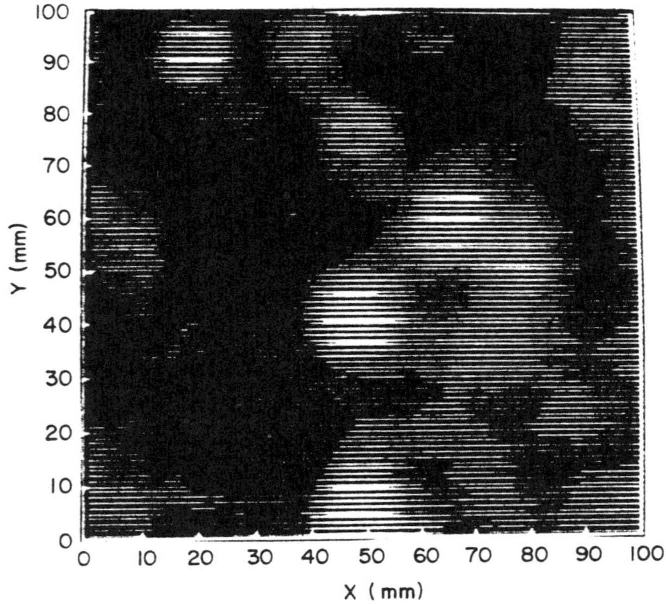

Figure 1.19. Photothermal deflection image of an aluminum surface. (Reprinted with permission from McDonald 1986. Copyright 1986, NRC Canada.)

Jackson et al. (1980, 1981) extended the photothermal deflection method to include optically transmitting gas and liquid analysis. In these experiments the excitation and probe lasers propagate collinearly through the sample cell. In this case the pulsed laser–induced temperature change results in the deflection of a collinear probe beam

$$\phi(r, t) = -\left(\frac{dn}{dT}\right) \frac{8r\alpha lQ}{\pi w^4 \rho c_P} e^{-2r^2/w^2(1+2t/t_c)} \tag{21}$$

There is a subtle distinction between the photothermal methods used for surface and transparent sample analysis. For surface analysis, the probe laser is used to detect a refractive index gradient formed in the media above the surface. In transparent samples, the refractive index is changed within the sample itself. Thus the deflection-angle signal is essentially the same as the pulsed laser photothermal lens inverse focal length. In fact, the signal strengths observed are about the same (Jackson et al., 1981). This method is very similar to photothermal lens spectroscopy. The similarity between the photothermal lens method and the beam deflection technique has been noticed by many authors; see, for example, Tam (1983, 1986, 1989) and

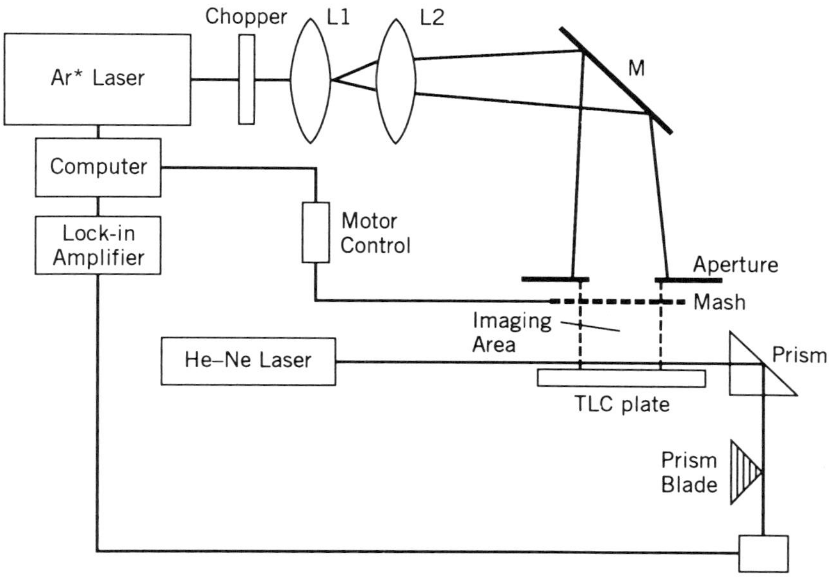

Figure 1.20. One-dimensional imaging apparatus developed for plate chromatography based on a time-multiplexed Hadamard encoded mask. Each position of the mask results in a unique surface excitation pattern. A photothermal deflection signal is recorded for each position of the mask and the resulting data set is transformed. The transformed data yield the absorption image of the thin-layer-chromatography plate. (Reprinted with permission from Morris and Fotiou 1989. Copyright 1989, American Chemical Society.)

Dovichi (1987). Photothermal lens and photothermal deflection methods both rely on the generation of a refractive index gradient in the sample itself. Collectively, they have become known as refractive index gradient detection or photothermal refraction spectroscopy methods (Zharov and Letokhov 1986, Tam 1986). The different geometries for sample excitation and monitoring of the photothermal response are shown in Figure 1.17.

The main advantage of photothermal deflection spectroscopy is the versatility. The same method can be used for solid, surface, and liquid- and gas-phase analysis. Excitation sources can be either pulsed or chopped continuous. The absorption coefficient detection limits for these methods are about the same as those of the two-laser photothermal lens method. Fournier et al. (1980) demonstrated absorption coefficient detection limits of 10^{-7} cm^{-1} for gas-phase samples in a windowless flow cell using a 1-W modulated infrared carbon dioxide excitation laser. Long and Bialkowski (1985) used a 10-mJ pulsed infrared laser to obtain gas-phase absorption coefficient detection limits equivalent to 10^{-8} cm^{-1}. Bialkowski and He

(1988) later used an etalon to amplify the deflection angle signal and found a 100-fold signal-to-noise ratio improvement, or about a 10^{-10} cm^{-1} detection limit for a 10-mJ pulse. Jackson et al. (1981) demonstrated 10^{-6} cm^{-1} absorption coefficient detection of benzene in CCl_4 using a 1-mJ pulsed dye laser. The solvent itself had an absorption coefficient of 10^{-6} cm^{-1}, and absorption due to the benzene analyte was found by scanning the wavelength of the pulsed pump laser. The estimated limit of absorption coefficient detection was 10^{-7} cm^{-1}. Dovichi (1987) pointed out that the photothermal refraction methods are advantageous when there is a significant signal due to sample cell window absorbance. Since the excitation and probe lasers do not have to pass into the sample at the same spot, photothermal perturbations due to the window can be ignored. There has been several reviews on probe beam deflection techniques. These reviews are often compiled along with those for photoacoustic spectroscopy. The reviews by Tam (1983, 1986, 1989), Murphy et al. (1986), Dovichi (1987), and Fournier and Boccara (1988) all cover aspects of this method. The books edited by Mandelis (1987) and Hess (1989a,b) have chapters devoted to this method.

Another method based on the generation of refractive index changes within the sample is photothermal diffraction spectroscopy. Laser-induced gratings have been known for quite some time (Eichler et al. 1986) and are the basis of optical holography (Collier et al. 1971). However, the first analytical application was by Pelletier et al. (1982), who demonstrated that a refractive index grating could be formed in a weakly absorbing sample by interfering two beams from a single excitation laser within the sample. The apparatus is shown in Figure 1.21. The grating diffracts a probe laser beam at a specific angle that satisfies the Bragg condition.

For pulsed laser excitation, the diffracted probe beam power is (Pelletier and Harris 1983)

$$\frac{\Phi_+}{\Phi_0} = \frac{2\pi}{\sqrt{3}} \left[\left(\frac{dn}{dT} \right)_P \frac{Q\alpha}{\rho C_P w \lambda \sin 2\theta} \right]^2 \tag{22}$$

Here Φ_+ the diffracted and Φ_0 the incident probe laser power, Q the total (combined) pulse energy, λ the wavelength of the probe laser, and 2θ the angle between the two pulsed pump laser beams. The diffraction signal is proportional to $(\alpha Q)^2$, thus apparently limiting the sensitivity at low concentrations. However, unlike infrared emission, the background is very small. The background noise limitation is essentially the same as those of laser-excited fluorescence spectroscopy. Current absorbance detection limits are about $\sim 10^{-6}$ cm^{-1}. Although not exploited to any great extent, this

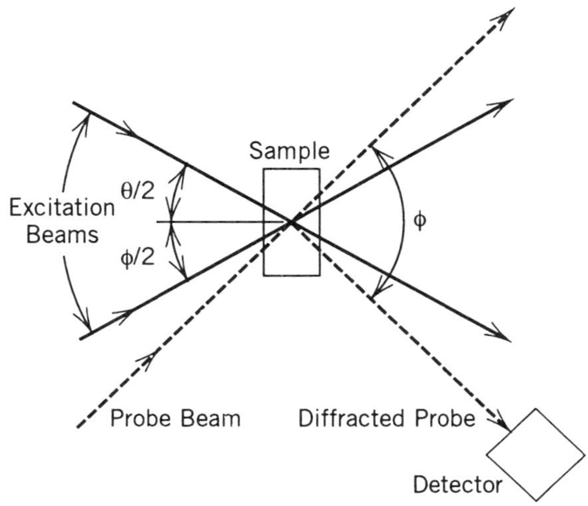

Figure 1.21. Schematic of an apparatus used for photothermal diffraction. The two excitation beams are mutually coherent, arising from the same laser excitation source. Interference of the two beams produced a periodic irradiance and subsequently a periodic refractive index perturbation, or phase grating, in a weakly absorbing sample. The phase grating is probed with a second probe laser. (Reprinted with permission from Pelletier et al. 1982. Copyright 1982, American Chemical Society.)

method has potential for trace analysis. The main advantage of this technique is apparently in the relatively simple data that result when the sample undergoes nonlinear absorption. In this case the distorted grating formed by nonlinear absorption can be decomposed by Fourier analysis into a series of orthogonal gratings, each with a different spatial period. Each grating then produces a different diffraction angle. The type of nonlinear absorption can be determined by analysis of the magnitude and irradiance dependence of the probe laser at each diffraction angle. The connection between photothermal lens, photothermal refraction, and photothermal diffraction spectroscopies has recently been given by Harris (1986), and the principles and applications of the photothermal diffraction method have recently been reviewed by Zhu et al. (1992).

1.5.7. Photothermal Radiometry

Another photothermal method is photothermal radiometry. In photothermal radiometry, the sample is excited with an optical source and the infrared emission is monitored. The infrared emission is related to the sample temperature by the blackbody radiation law (Ingle and Crouch 1988),

$$B(\lambda, T) = \varepsilon(\lambda)\, \frac{2hc^2}{\lambda^5} \frac{1}{e^{hc/\lambda kT} - 1} \tag{23}$$

where $B(\lambda, T)$ ($\mathrm{W\,sr^{-1}\,m^{-2}}$) is the radiant emissivity, h (J s) is Planck's constant, k ($\mathrm{J\ K^{-1}}$) is Boltzmann's constant, and $\varepsilon(\lambda)$ is the sample emissivity. For samples at thermal equilibrium, $\alpha(\lambda) = \varepsilon(\lambda)$. The relative change in spectral radiance with respect to temperature is

$$\frac{\delta B(\lambda, T)}{B(\lambda, T)} = \frac{ch}{\lambda k T^2} \frac{\alpha(\lambda)}{1 - e^{-ch/\lambda kT}} \tag{24}$$

In the mid-infrared region, $hc/\lambda kT < 1$, and for pulsed laser excitation,

$$\frac{\delta B(\lambda_{\mathrm{em}}, T)}{B(\lambda_{\mathrm{em}}, T)} \approx \frac{ch\alpha(\lambda_{\mathrm{em}})}{\lambda_{\mathrm{em}} k T^2} \frac{2\alpha(\lambda_{\mathrm{ex}})Q}{\pi w^2 \rho C_P} \tag{25}$$

where λ_{em} is the emission wavelength and λ_{ex} is the excitation wavelength. The important features of this equation are that the relative radiance (1) increases with excitation laser energy and absorption coefficient, (2) decreases with equilibrium temperature, and (3) decreases with increasing wavelength, and (4) since $\alpha(\lambda_{\mathrm{em}})$ and $\alpha(\lambda_{\mathrm{ex}})$ are both proportional to number density or concentration, the relative emission signal decreases as the square of the number density. The latter suggests that this method will not be as sensitive as other photothermal techniques since the sensitivity decreases with decreasing α.

This method of analysis has a long and somewhat obscured history. The advent of lasers and cryogenic infrared detectors resulted in several studies of infrared emission from excited gas samples. The method used to study gas samples is called laser-induced fluorescence (Hocker et al. 1966, Yardley and Moore 1966, and Stephenson 1968, to name a few). These pioneering studies used pulsed infrared lasers to excite specific vibrational modes while monitoring infrared fluorescence at wavelengths different from that used for excitation. Time-resolved fluorescence emission studies were performed to reveal the excited-state vibrational lifetimes of the low-pressure gas species. These lifetimes are a direct measure of the thermalization time of the sample. With the notable exception of the work by Belz et al. (1987), infrared emission methods have been neglected for gas analysis, although the utility is apparent.

Photothermal radiometry has been recognized as an important tool for surface studies and material analysis (Nordal and Kanstad 1979). The solid being analyzed is treated as a blackbody emitter. The total surface emission will follow the Stefan–Boltzmann law. The relative temperature-dependent change in surface emittance (e.g., integrating over wavelength) is simply

$$\frac{\delta M(T)}{M(T)} = 4\,\frac{\delta T}{T} \tag{26}$$

where $M(T)$ ($\mathrm{W\,m^{-2}}$) is the temperature-dependent emittance of the surface and ΔT is induced by the photothermal effect. The temperature change can be produced by either pulsed or chopped excitation sources. This method is not very sensitive, and rather large temperature changes have to be induced in the samples to obtain good thermal images. The thermal images are not a function of surface topography (unless three-dimensional imaging optics are used). The images will depend only on the optical absorption coefficient and the heat capacity of the sample. Time- or phase-dependent analysis of the emittance yields information regarding the thermal conductivity. If the integrated emission data are not processed as indicated in equation (26), the thermal image will also be a function of the topographical emissivity of the material. Tam (1985) has examined the use of photothermal radiometry for solid sample analysis and has reviewed the literature regarding applications of the method (Tam 1983, 1986, 1989). Busse (1989) has reviewed this in reviews of nondestructive materials evaluation using photothermal methods.

1.5.8. Historic Summary

The important discoveries in the history of photothermal spectroscopy are given in Table 1.3. Many of these discoveries were made possible by using lasers. Although lasers do not have to be used to excite samples, the signals thereby obtained are much greater than those obtained using incoherent light sources. Historically, the birth of high-sensitivity photoacoustic and photothermal spectroscopies can be traced back to the laser. The first application of the photoacoustic effect was that of Bell (1880), while chemical analysis applications of photoacoustic spectroscopy can be traced back to Viengerov (1938). The first photothermal method was discovered by Leite et al. (1964) and Gordon et al. (1964) when they found that an intracavity sample, laser-based apparatus gave rise to photothermal blooming, the photothermal lens. Some time later Kreuzer (1971) showed that photoacoustic spectroscopy could be used for sensitive analysis when laser light sources were utilized.

The reasons lasers have made such an impact on high-sensitivity spectroscopy are because they possess a high spectral brightness and have outputs that are coherent. The high spectral brightness allows high powers

Table 1.3. Major Developments in the Early History of Photothermal Spectroscopy

1880	Photoacoustic effect used for photophone by Bell.
1938	Viengerov uses photoacoustic effect for gas-phase chemical analysis.
1964	Photothermal lens used to measure optical absorptions by Leite et al.
1968	McLean et al. use interferometry to monitor photothermal effect.
1969	Longaker and Litvak find thermal and acoustic components in photothermal effect. Use of pulsed excitation laser.
1971	Kreuzer demonstrates sensitive gas-phase detection with photoacoustic spectroscopy.
1972	Grabiner et al. use two-laser pulsed excitation photothermal lens method for excited-state relaxation kinetics measurements.
1973	Hu and Whinnery develop extracavity single-laser photothermal lens spectroscopy.
1976	Long et al. measure excitation spectra with two-laser photothermal lens spectroscopy.
1977	Twarowski and Kliger develop theory for pulsed laser photothermal lens and measure two-photon absorption spectra of benzene. Patel et al. report 10^{-10}-cm^{-1} detection limits for gas-phase photoacoustic measurements.
1979	Dovichi and Harris use single-laser photothermal lens for chemical analysis and introduce the enhancement factor. Boccara et al. report photothermal deflection.
1981	Murphy and Aamodt report thermal images using photothermal deflection. Davis and Petuchowski report 10^{-10}-cm^{-1} detection limits for gas-phase photothermal interferometry measurements. Dovichi and Harris report 10^{-7}-cm^{-1} detection limits with single-laser solution-phase photothermal lens.

or energies to be imparted to the sample. The coherence, in particular the spatial coherence, allows this power or energy to be delivered to small volumes. The signals in both photoacoustic and photothermal spectroscopies are enhanced with smaller excitation volumes.

1.6. SOME IMPORTANT FEATURES OF PHOTOTHERMAL SPECTROSCOPY

We have now come to regard photothermal spectroscopy as a group of high-sensitivity techniques that can be used for chemical and materials analysis. Photothermal signals arise from optical absorption in a sample. However, photothermal spectroscopy techniques have sensitivities far exceeding those of conventional absorption spectrophotometry. The reasons for the high sensitivity of photothermal spectroscopy is that it is an indirect technique for measuring optic absorption. For an analyte with less than unit fluorescence quantum yield, electromagnetic energy absorbed and not lost by subsequent emission results in an increase in the energy of the sample. Energy absorbed and not subsequently lost by emission is usually randomized, resulting in sample heating. The photothermal spectroscopy signal is derived from this heating.

A variety of methods are used to monitor sample heating. Calorimetric or thermometric methods use temperature transducers to measure sample temperature. The method of photoacoustic spectroscopy uses a pressure transducer to monitor the pressure wave associated with rapid sample heating. Photothermal emission radiometry uses photometric transducers to monitor changes in the samples' infrared emission associated with heating. Photothermal interferometry, photothermal deflection, photothermal lensing, and photothermal diffraction spectrometries are all photothermal techniques based on monitoring refractive index changes associated with sample heating.

The distinctions between all but the calorimetric and photothermal radiometry techniques are lessening. The connection between photothermal and photoacoustic spectroscopies is apparent. Signals attributable to the photoacoustic effect are seen in photothermal spectroscopy experiments, although the converse is not true. A pressure transducer placed far from the excitation source will not respond to the thermal perturbation. Photoacoustic spectroscopy apparatus may use probe lasers to detect the acoustic wave. This eliminates the use of pressure transducers that may have low response times. However, photoacoustic deflection spectroscopy is orders of magnitude less sensitive than photothermal lensing or deflection.

Accurate theories for describing photothermal spectroscopy signals have been developed. In most cases these theories take into account the thermodynamics, hydrodynamics, and optics of the experimental apparatus. In some cases, for example in the single-laser photothermal lens apparatus, the absorption coefficient may be obtained directly from the signal if the thermodynamic and optical parameters are well enough known. In most other cases the signal magnitude has an instrumental factor that must be determined using samples of known absorbance in sample matrices that are identical to those of the unknown sample.

A summary of the best absorption coefficient detection limits for photothermal and photoacoustic methods are given in Table 1.4. The detection limits are given in inverse energy or power units since all photothermal methods scale proportional to the excitation. For continuous excitation, lock-in amplifier signal processing is generally used to recover the oscillatory signal. The noise power is proportional to the bandwidth of the measurement; these detection limits are inversely proportional to the

Table 1.4. Absorption Coefficient Detection Limits Using Photothermal and Photoacoustic Methods for Pulsed α ($J^{-1}\,cm^{-1}$) and Chopped α ($W^{-1}\,Hz^{-1/2}$) Sources

	Photothermal Spectroscopy[a]		Photoacoustic Spectroscopy[b]		Theoretical Photoacoustic[c]		
	Pulsed	CW	Pulsed	CW	Pulsed	CW	Typical Background[d]
Gas	10^{-12}	10^{-10}	10^{-10}	10^{-10}	10^{-12}	10^{-11}	10^{-5}(IR)
Liquid	10^{-10}	10^{-8}	10^{-9}	10^{-6}	10^{-10}	10^{-9}	10^{-6}(vis)
Solid	—	—	10^{-7}	10^{-5}	—	10^{-6}	10^{-5}(vis)

[a]Section 1.5 of this book and Dovichi (1987) α detection limits scaled to W^{-1} or J^{-1}.
[b]Tam (1983), and Zharov and Letokhov (1986).
[c]Theoretical absorption detections limits summarized by Zharov and Letokhov (1986).
[d]Background absorptions typical of water vapor at 10 μm, solvent overtones in liquids, and impurities in fused silica in the visible.

square-root measurement bandwidth. The theoretical detection limits are based on thermodynamic fluctuations in sample pressure (Slatkine 1981). The noise equivalent power, NEP ($W\,Hz^{-1/2}$), is

$$\mathrm{NEP} \approx \left(4k_B T^2\,\delta f\,\frac{2\pi r l k}{\sqrt{4D_T/f}}\right)^{1/2} \tag{27}$$

where k_B is the Boltzmann constant, f the frequency, δf the measurement bandwidth, r and l the radius and length of the sample cell, k the thermal conductivity, and D_T the thermal diffusivity. The fluctuations that ultimately limit photoacoustic spectroscopy should also place a lower bound on photothermal spectroscopy since pressure and density are related:

$$\delta\rho = \left(\frac{\partial P}{\partial\rho}\right)_T \delta P = \rho_0 K_T\,\delta T \tag{28}$$

where K_T (Pa^{-1}) is the isothermal compressibility.

For gas- and solution-phase analysis, photothermal and photoacoustic spectroscopy apparatuses have been developed which yield signals that are close (one to two orders of magnitude) to the theoretical limits of absorbance detection. Quantitative work in solid analysis is nearly impossible because of the difficulty in preparing standard samples. These detection limits are lower than the background absorbance of water and other trace gases in the atmosphere and of solvents used to host the analytes. The lower detection limits are obtained for gas samples because of reduced matrix absorptions and favorable thermodynamic parameters. The problem with the sensitive absorbance measurement methods is not so much the measurement of low absorption coefficients, but rather, that of discriminating the low analyte from that of the solvent or other species in the sample.

From these and other examples it is clear that photothermal spectroscopy is a valuable tool that can be used to solve a variety of chemical and materials analysis problems. Some salient features are:

1. The sensitivity of photothermal spectroscopy is theoretically enhanced over that of conventional absorption spectrophotometry. This theoretical enhancement has been realized and absorbance detection limits of about 10^{-7} in liquids and about 10^{-10} in gases are possible. High sensitivity has allowed the measurement of weak optical absorbances in pure samples and trace analysis, and in volume-restricted samples.
2. Photothermal spectroscopy signal contains both magnitude and dynamic (time-dependent) components. The magnitude component is most important for quantitative absorption studies, spectroscopy, trace analysis, chromatography, and so on. The dynamic component is important in qualitative analysis, material composition, chemical kinetics, and so on.
3. Photothermal signals are inversely proportional to the excitation volume. This arises not only because of the higher temperature changes that can be induced with a given power or energy, but also because photothermal signals are usually derived from a spatial gradient in the resulting refractive index change. The small volume character of photothermal spectroscopy has lead to its use for microanalysis and effluent detection in chromatography.

REFERENCES

Barker, J. R. and Rothem, T. *Chem. Phys.* **68** 331 (1982).

Barker, J. R. and Toselli, B. M. In *Photothermal Investigations in Solids and Fluids*, Sell, J. A. ed., Academic Press, New York (1989).

Beitz, J. V.; Doxtader, M. M.; Maroni, V. A.; Okajima, S.; and Reed, D. T. *Rev. Sci. Instrum.* **61** 1395 (1990).

Bell, A. G. *Am. J. Sci.* **20** 305 (1880).

Bell, A. G. *Philos. Mag.* **11** 510 (1881).

Belz, H. H.; Gutberlet, H.; Schallert, B.; and Schrader, B. *Appl. Spectrosc.* **41** 1009 (1987).

Betteridge, C. M. and Meylor P. J. *CRC Crit. Rev. Anal. Chem.* **14** 267 (1984).

Bialkowski, S. E. *Chem. Phys. Lett.* **151** 88 (1988).

Bialkowski, S. E. and He, Z.-F. *Anal. Chem.* **60** 2674 (1988).

Bialkowski, S. E.; Gu, X.; Poston, P. E.; and Powers, L. S. *Appl. Spectrosc.* **46** 1335 (1992).

Boccara, A. C.; Fournier, D.; and Badoz, *J. Appl. Phys. Lett.* **36** 130 (1979).

Busse, G. In *Photoacoustic, Photothermal and Photochemical Processes at Surfaces and Thin Films*, Hess, P., ed., Springer-Verlag, New York (1989).

Collier, R. J.; Burckhardt, C. B.; and Lin, L. H. *Optical Holography*, Academic Press, New York (1971).

Davis, C. C. and Petuchowski, S. J. *Appl. Opt.* **20** 2539 (1981).

Dovichi, N. J. *CRC Crit. Rev. Anal. Chem.* **17** 357 (1987).

Dovichi, N. J. *Rev. Sci. Instrum.* **61** 3653 (1990).

Dovichi, N. J. and Harris, J. M. *Anal. Chem.* **51** 728 (1979).

Dovichi, N. J. and Harris, J. M. *Anal. Chem.* **53** 106 (1981).

Eichler, H. J.; Günter, P.; and Pohl, D. W. *Laser-Induced Dynamic Gratings*, Springer-Verlag, New York (1986).

Fang, H. L. and Swofford, R. L. In *Ultrasensitive Laser Spectroscopy*, Kliger, D. S., ed., Academic Press, New York (1983).

Fotiou, F. K. and Morris, M. D. *Appl. Spectrosc.* **40** 704 (1986).

Fournier, D. and Boccara, A. C. In *Photothermal Investigations in Solids and Fluids*, Sell, J. A. ed., Academic Press, New York (1988).

Fournier, D.; Boccara, A. C.; Amer, N. M.; and Gerlach, R. *Appl. Phys. Lett.* **37** 519 (1980).

Friedrich, D. M. In *Ultrasensitive Laser Spectroscopy*, Kliger, D. S., ed., Academic Press, New York (1983).

Gagne, M. C.; Galarneau, P.; and Chin, S. L. *Can. J. Phys.* **64** 1117 (1986).

Gordon, J. P.; Leite, R. C. C.; Moore, R. S.; Porto, S. P. S.; and Whinnery, J. R. *Bull. Am. Phys. Soc.* **9** 501(1964).

Gordon, J. P.; Leite, R. C. C.; Moore, R. S.; Porto, S. P. S.; and Whinnery, J. R. *J. Appl. Phys.* **36** 3 (1965).

Grabiner, F. R.; Siebert, D. R.; and Flynn, G. W. *Chem. Phys. Lett.* **17** 189 (1972).

Harris, J. M. *Optics News*, October, p. 8 (1986).

Herzfeld, K. F. and Litovitz, T. A. *Absorption and Dispersion of Ultrasonic Waves*, Academic Press, New York (1959).

Hess, P., ed. *Photoacoustic, Photothermal and Photochemical Processes in Gases*, Springer-Verlag, New York (1989a).

Hess, P., ed. *Photoacoustic, Photothermal and Photochemical Processes at Surfaces and Thin Films*, Springer-Verlag, New York (1989b).

Hocker, L. O.; Kovacs, M. A.; Rhodes, C. K.; Flynn, G. W.; and Javan, A. *Phys. Rev. Lett.* **17** 233 (1966).

Hu, C. and Whinnery, J. R. *Appl. Opt.* **12** 72 (1973).

Hutchins, D. A. and Tam, A. C. *IEEE Trans. Ultrason. Ferroelectr. Frequency Control* **UFFC-33** 429 (1986).

Ingle, J. D., Jr. and Crouch, S. R. *Spectrochemical Analysis*, Prentice Hall, Englewood Cliffs, NJ (1988).

Jackson, W. B.; Amer, N. M.; Boccara, A. C.; and Fournier, D. *Opt. Lett.* **5** 337 (1980).

Jackson, W. B.; Amer, N. M.; Boccara, A. C.; and Fournier, D. *Appl. Opt.* **20** 1333 (1981).

Kerr, E. L. and Atwood, J. G. *Appl. Opt.* **7** 915 (1968).

Kitamori, T. and Sawada, T. *Spectrochem. Acta Rev.* **14** 275 (1991).

Kreuzer, L. B. *J. Appl. Phys.* **42** 2934 (1971).

Kreuzer, L. B.; Kenyon, N. D.; and Patel, C. K. N. *Science* **177** 347 (1972).

Lai, H. M. and Young, K. *J. Acoust. Soc. Am.* **76** 2000 (1982).

Landau, L. D. and Lifshitz, E. M. *Fluid Mechanics*, Addison-Wesley, Reading, MA (1959).

Leite, R. C.; Moore, R. S.; and Whinnery, J. R. *Appl. Phys. Lett.* **5** 141 (1964).

Long, G. R. and Bialkowski, S. E. *Anal. Chem.* **57** 1079 (1985).

Long, M.E.; Swofford, R. L.; and Albrecht, A. C. *Science* **191** 183 (1976).

Longaker, P. R. and Litvak, M. M. *J. Appl. Phys.* **40** 4033 (1969).

Mandelis, A., ed. *Photoacoustic and Thermal Wave Phenomena in Semiconductors* North-Holland, New York (1987).

Mandelis, A. In *Photoacoustic, Photothermal and Photochemical Processes at Surfaces and Thin Films*, Hess, P., ed., Springer-Verlag, New York (1989).

McDonald, F. A. *Can. J. Phys.* **64** 1023 (1986).

McLean, E. A.; Sica, L.; and Glass, A. J. *Appl. Phys. Lett.* **13** 369 (1968).

Meyer, P. L. and Sigrist, M. W. *Rev. Sci. Instrum.* **61** 1779 (1990).

Mori, K.; Imashaka, T.; and Ishibashi, N. *Anal. Chem.* **54** 2034 (1982).

Morris, M. D and Fotiou, F. K. In *Photothermal Investigations in Solids and Fluids*, Sell, J. A. ed., Academic Press, New York (1989).

Murphy, J. C. and Aamodt, L. C. *J. Appl. Phys.* **51** 4580 (1980).

Murphy, J. C. and Aamodt, L. C. *Appl. Phys. Lett.* **38** 196 (1981).

Murphy, J. C.; Maclachlan, J. W.; and Aamodt, L. C. *IEEE Trans. Ultrason. Ferroelectr. Frequency Control* **UFFC-33** 529 (1986).

Nordal, P.-E. and Kanstad, S. O. *Phys. Scr.* **20** 659 (1979).

Nyquist, R. A.; Leugers, M. A.; McKelvy, M. L.; Papenfuss, R. R.; Putzig, C. L.; and Yurga, L. *Anal. Chem.* **62** 223R (1990).

Pao, Y.-H., ed. *Opto-acoustic Spectroscopy and Detection*, Academic Press, New York (1977).

Parker, J. G. *Appl. Opt.* **12** 2974 (1973).

Patel, C. K. N. and Tam, A. C. *Rev. Modern Phys.* **53** 517 (1981).

Patel, C. K. N.; Kerl, R. J.; and Burkhardt, E. G. *Phys. Rev. Lett.* **38** 1204 (1977).

Pelletier, M. J. and Harris, J. M. *Anal. Chem.* **54** 1537 (1983).

Pelletier, M. J.; Thornsheim, H. R.; and Harris, J. M. *Anal. Chem.* **54** 239 (1982).

Putzig, C. L.; Leugers, M. A.; McKelvy, M. L.; Mitchell, G. E.; Nyquist, R. A.; Papenfuss, R. R.; and Yurga, L. *Anal. Chem.* **64** 270R (1992).

Rosencwaig, A. In *Opto-acoustic Spectroscopy and Detection*, Pao, Y.-H., ed., Academic Press, New York (1977).

Rosencwaig, A. *Photoacoustics and Photoacoustic Spectroscopy*, Wiley, New York (1980).

Sell, J. A., ed., *Photothermal Investigations in Solids and Fluids*, Academic Press, New York (1989).

Siebert, D. R.; Grabiner, F. R.; and Flynn, G. W. *J. Chem. Phys.* **60** 1564 (1974).

Slatkine, M. *Appl. Opt.* **20** 2880 (1981).

Solimini, D. J. *Appl. Phys.* **12** 3314 (1966).

Stephenson, J. C.; Wood, R. E.; and Moore, C. B. *J. Chem. Phys.* **48** 4790 (1968).

Stone, J. *J. Opt. Soc. Am.* **62** 327 (1972).

Stone, J. *Appl. Opt.* **12** 1828 (1973).

Swofford, R. L. and Morrell, J. A. *J. Appl. Phys.* **49** 3667 (1978).

Tam, A. C. In *Ultrasensitive Laser Spectroscopy*, Kliger, D. S., ed., Academic Press, New York (1983).

Tam, A. C. *Infrared Phys.* **25** 305 (1985).

Tam, A. C. *Rev. Modern Phys.* **58** 381 (1986).

Tam, A. C. In *Photothermal Investigations in Solids and Fluids*, Sell, J. A. ed., Academic Press, New York (1989).

Twarowski, A. J. and Kliger, D. S. *Chem. Phys.* **20** 253 (1977a).

Twarowski, A. J. and Kliger, D. S. *Chem. Phys.* **20** 259 (1977b).

Viengerov, M. L. *Dokl. Akad. Nauk SSSR* **19** 687 (1938).

Whinnery, J. R. *Acc. Chem. Res.* **7** 225 (1974).

Yardley, J. T. and Moore, C. B. *J. Chem. Phys.* **45** 1066 (1966).

Zharov, V. P. In *Laser Analytical Spectrochemistry*, Letokhov, V. S., ed., Adam Hilger, Boston (1986).

Zharov, V. P. and Letokhov, V. S. *Laser Optoacoustic Spectroscopy*, Springer-Verlag, New York (1986).

Zhu, X. R.; McGraw, D. J.; and Harris, J. M. *Anal. Chem.* **64** 710A (1992).

CHAPTER

2

ABSORPTION, ENERGY TRANSFER, AND EXCITED-STATE RELAXATION

2.1. FACTORS AFFECTING OPTICAL ABSORPTION

Absorption, emission, and inelastic scattering are the three main basic ways that light can interact with matter and result in energy transfer to or from the sample. In addition, reflection, refraction, diffraction, and elastic scattering can alter the light path and thereby decrease the amount of power transmitted through the sample. These processes are shown in Figure 2.1. In quantitative analysis, the purpose of the spectroscopic measurement is to determine the absorbance or absorption coefficient of the analyte. The absorption coefficient can rarely be obtained by a simple transmission measurement. All processes resulting in optical attenuation must be measured to determine sample absorbance accurately.

Figure 2.2 illustrates three optical experiments that can be used to determine optical absorption. Transmission spectroscopy is a direct method to measure optical energy loss in the form of an attenuation in light used to excite the sample. Optical emission spectroscopy is an indirect method by which the optical energy transferred from the sample to the surroundings is measured. This emission can result from photothermal heating, photochemical transformation giving rise to chemiluminescence, or optical excitation resulting in photoluminescence, both fluorescence and phosphorescence. Photothermal spectroscopy measures the energy than goes into sample heating as a consequence of optical energy transfer to the sample.

In transmission spectroscopy measurements, an excitation source with a radiant power of $\Phi_0(\mathrm{W})$ at an optical frequency of $\nu(\mathrm{s}^{-1})$ passes through a sample with a transmission, T. Light exiting the sample has been attenuated by several factors affecting transmission and the power is measured with a detector. Two rules apply to the power loss in the sample. First, total power must be conserved. The sum of powers either lost or transmitted through the sample must equal the power used to excite the sample. Second, the relative attenuation due to each factor is constant in homogeneous media. The total power loss due to a particular factor is the integral over the attenuation pathlength.

Attenuation processes include optical absorption, reflection, refraction, diffraction, elastic scattering, and inelastic or Raman scattering. Refraction,

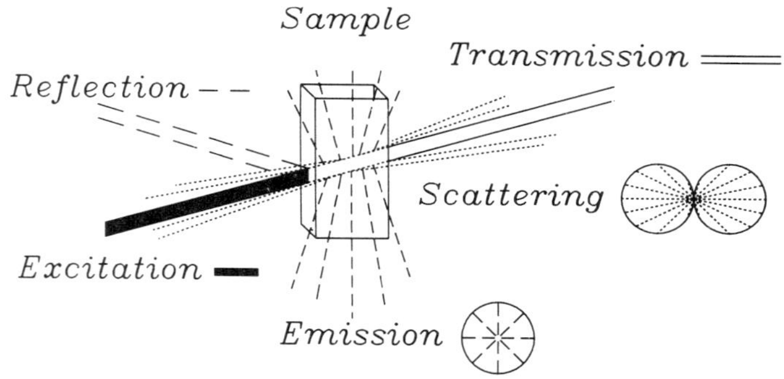

Figure 2.1. Optical-loss processes in a transmitting sample. Incident excitation light can be lost through reflection, scattering, and absorption. Absorption can result in sample heating as well as luminescent emission.

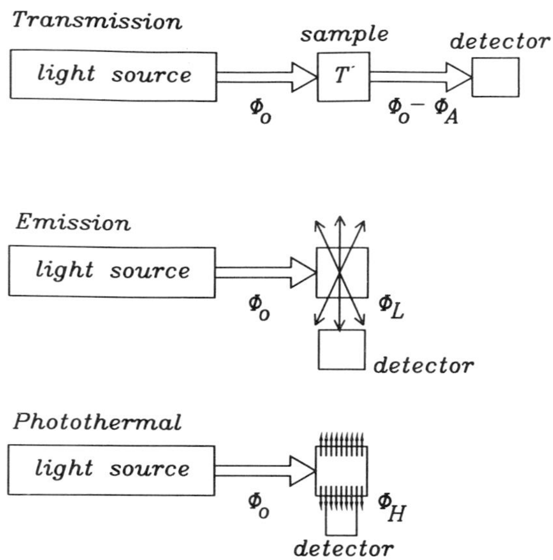

Figure 2.2. Three types of spectroscopy apparatuses. Transmission directly measures the incident excitation source after passing through the sample. In emission, transmission is measured indirectly by monitoring luminescence produced by optical absorption. In photothermal spectroscopy, transmission is measured indirectly by monitoring heat produced by optical absorption.

diffraction, and elastic scattering by themselves do not result in total light power or energy loss. Rather, the light is redistributed in space. This spatial redistribution corresponds to an attenuation of light propagating along the normal ray path. Raman scattering also results in a spatial redistribution. But unlike elastic scattering, there is a power or energy change resulting from the interaction with the sample. Inelastic scattering, in particular Rayleigh, Debye, and Mie, can increase the average interaction pathlength (Patterson et al. 1989, 1991; Zaccanti et al. 1992; Haselgrove et al. 1992). This may result in additional energy or power loss observed in transmission spectroscopy.

Consider the attenuations or power losses along the normal path of the excitation source for the sample shown in Figure 2.1. The main attenuating factors are reflection, sample scattering, and sample absorbance. The excitation source propagates through a window, through the sample, and exits through another window before being detected. We need to know how much power is lost due to each of the attenuation factors. Tracing from the excitation source through the sample, the first loss is that due to reflection. Reflection is an interfacial phenomenon. The amount of power reflected at the entrance window is independent of optical pathlength. The power lost at the entrance window is

$$\Phi_{R1} = R_1 \Phi_0 \tag{1}$$

where Φ_{R1} is the power reflected by the entrance window and R_1 is the window and sample reflectance. Reflectivity of an abrupt interface between two materials with different refractive indices is given by

$$R = \frac{1}{2}\left[\frac{\sin^2(\theta_i - \theta_r)}{\sin^2(\theta_i + \theta_r)} + \frac{\tan^2(\theta_i - \theta_r)}{\tan^2(\theta_i + \theta_r)}\right] \tag{2}$$

where θ_i and θ_r are the angles of the incident and refracted wave measured with respect to the surface normal. The later are related by Snell's law,

$$n_i(\lambda) \sin \theta_i = n_r(\lambda) \sin \theta_r \tag{3}$$

where the $n(\lambda)$ are the refractive indices in the two media. These are given as vacuum-wavelength dependent to illustrate that the reflectivity can change with wavelength. For normal incidence

$$R = \left[\frac{n_i(\lambda) - n_r(\lambda)}{n_i(\lambda) + n_r(\lambda)}\right]^2 \tag{4}$$

In the case of two reflections, one at the air–window interface and one at the window–sample interface, R_1 is the product of the two reflectances. The amount of power reaching the sample is thus $\Phi_0 - \Phi_{R1}$.

Light propagating through the sample will be attenuated both by sample scattering and analyte absorption. The scattered power, Φ_S, depends on the optical pathlength, l(m), and the scattering attenuation factor, $\alpha_S(\mathrm{m}^{-1})$. Similarly, the power lost to the sample by absorption, Φ_A, depends on the optical pathlength and the absorption coefficient, α(m). Attenuation by scattering and optical absorption cannot be separated since both processes result in a decrease in power along the normal path of the excitation beam. Assuming that scattering is minor, attenuation in the sample is due to the combined coefficients, $\alpha_S + \alpha$. The transmission due to scattering and absorption is

$$T_{\alpha_S+\alpha} = T_{\alpha_S} T_\alpha = e^{-(\alpha_S+\alpha)l} \tag{5}$$

By integrating over the sample pathlength one can find the power lost to scattering and absorption. The optical power lost to scattering is

$$\Phi_S = \Phi_0(1 - R_1)\, \frac{\alpha_S}{\alpha_S + \alpha}\, [1 - e^{-(\alpha_S+\alpha)l}] \tag{6}$$

and that due to optical absorption is

$$\Phi_A = \Phi_0(1 - R_1)\, \frac{\alpha}{\alpha_S + \alpha}\, [1 - e^{-(\alpha_S+\alpha)l}] \tag{7}$$

In highly scattered samples, the absorbed power is not a simple function of the sample pathlength and scattering coefficient. The light path is a random walk through the sample. The random walk is closely approximated by diffusion. The diffusive nature of light propagation in scattering media results in a distribution of optical paths through the sample. The distribution of paths is a function of the scattering particle concentration, size and shape distribution, refractive index, and the wavelength of light. All but the first of these factors are parameterized in terms of an asymmetry factor, g, which is also the mean cosine scattering angle of the particles. The power scattered in the sample satisfies the diffusion equation

$$\frac{1}{c}\frac{\partial}{\partial t}\Phi(r, t) - D\nabla^2\Phi(r, t) + \alpha\Phi(r, t) = \Phi_0(t, t) \tag{8}$$

where

$$D = \frac{1}{3\alpha + 3(1 - g)\alpha_S} \tag{9}$$

The impulse point-source solution to this diffusion equation is

$$\Phi(r, t) = \frac{c}{(4\pi Dct)^{3/2}} \exp\left(-\frac{r^2}{4Dct} - \alpha ct\right) \tag{10}$$

Patterson et al. (1989, 1991) have shown that the mean optical pathlength is not l but is given by

$$\langle ct \rangle = \frac{1}{(4D\alpha)^{1/2}} \frac{(l - z_0)\exp(2z_0\sqrt{\alpha/D}) - (l + z_0)}{\exp(2z_0\sqrt{\alpha/D}) - 1} \tag{11}$$

where $\langle ct \rangle$ is the mean pathlength, c being the speed of light, and

$$z_0 = \frac{1}{(1 - g)\alpha_S} \tag{12}$$

is called the scattering depth. For Beer's law optical absorbance, the absorbed power will be proportional to the mean optical pathlength. If the scattering particles are not themselves absorbing,

$$\Phi_A = \Phi_0(1 - R_1)\exp(-\alpha\langle ct \rangle) \tag{13}$$

The increased optical pathlength due to large amounts of scattering is accompanied by an increase in absorbed power. Scattering also causes the absorbed energy to be spread over a larger volume than that illuminated by the excitation source. This can have detrimental effects on photothermal lens, deflection, and diffraction signals since these depend on the absorbed energy being restricted to small volumes. Fortunately, most samples used in chemical analysis will not be highly scattering. Exceptions to this rule occur in biological and medical samples, optical detectors for plate chromatography and electrophoreses, and optical detection in solids and in fluids near critical points.

Reflection also occurs as the transmitted power exits the sample cell. The power loss due to this reflection is

$$\Phi_{R2} = R_2(\Phi_0 - \Phi_{R1} - \Phi_S - \Phi_A) = R_2\Phi_0(1 - R_1)T_{\alpha_S}T_\alpha \tag{14}$$

The power measured in optical transmission spectroscopy is the incident power, less all the power lost through the various loss processes:

$$\Phi_T = \Phi_0 - \Phi_{R1} - \Phi_S - \Phi_A - \Phi_{R2} = \Phi_0(1 - R_1)(1 - R_2)T_{\alpha_S}T_\alpha \tag{15}$$

where Φ_T is the transmitted power that may be monitored with a photodetector. In many cases, optical power loss due to scattering and reflection is small. These losses are not important unless the power loss due to optical absorption is also small. When this is the case, scattering and reflection losses may be measured independently and the transmitted power corrected for these losses is a more accurate indication of analyte absorption.

Under steady-state conditions, the power absorbed by the sample must equal the power lost to the surroundings. The optical power absorbed by the sample may be lost through radiated light. The radiated power is measured in emission spectroscopy. Luminescence processes give rise to a redistribution in the propagation direction of the light used to excite the sample. Since luminescence is not directional. the detector can be placed so as to minimize interference from the excitation source. Figure 2.3 is an energy-level diagram that illustrates excitation and relaxation processes for a typical organic molecule. The three electronic states, the ground and

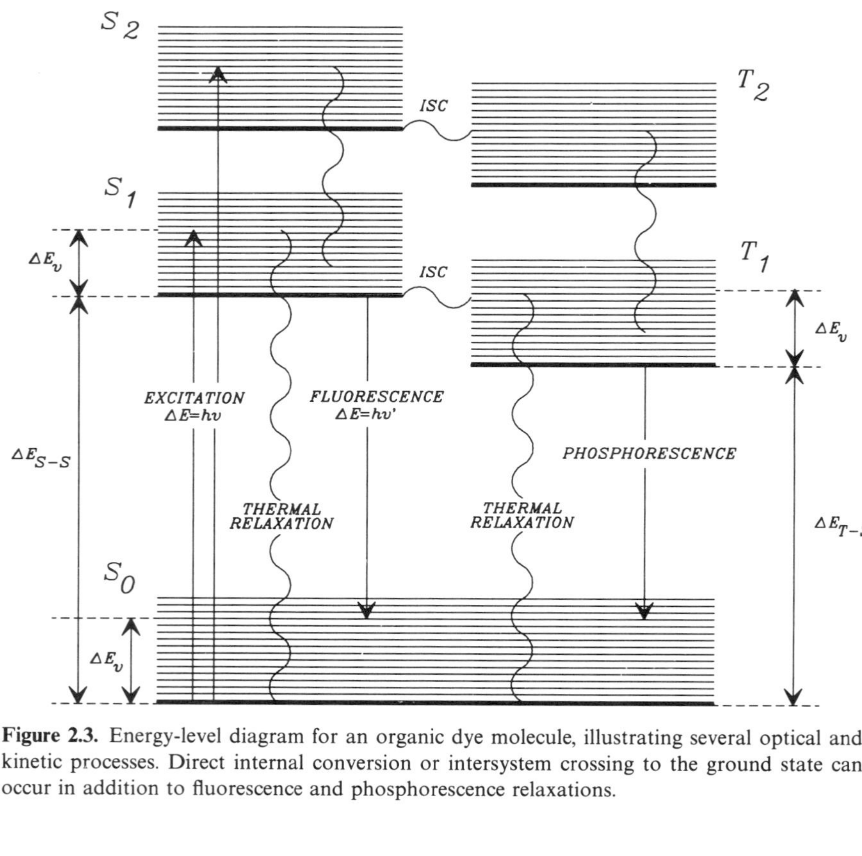

Figure 2.3. Energy-level diagram for an organic dye molecule, illustrating several optical and kinetic processes. Direct internal conversion or intersystem crossing to the ground state can occur in addition to fluorescence and phosphorescence relaxations.

excited singlet states and the triplet state, each have rotational and vibrational states superimposed on the zero point of the electronic state. The molecule absorbs light at a frequency ν and undergoes a transition, resulting in an increase in energy of $h\nu$. The overlap of the rovibrational wavefunctions favor excitation into vibrationally excited states of the excited singlet state. The excited molecule may relax to the ground electronic state through internal conversion (IC), may undergo intersystem crossing (ISC) to the triplet state, or may emit a photon. Spontaneous emission usually results in the production of a vibrationally excited ground electronic state. The light emitted by the sample is usually at lower frequencies than the absorbed light. The vibrationally excited ground electronic state relaxes, resulting in a ground-state molecule that can absorb another photon.

The amount of radiation emitted is related to the power absorbed by the emission efficiency (Ingle and Crouch 1988)

$$\Phi_L = Y_L \Phi_A \tag{16}$$

where Y_L is the luminescence power yield. The yield is related to the quantum efficiency and the shape of the optical transmission by taking into account the energy of each emitted photon emitted at a frequency of ν' by

$$Y_L = \phi_L \int_0^\infty S_{\nu'} \frac{\nu'}{\nu} \, d\nu' \tag{17}$$

where ϕ_L is the luminescence quantum efficiency and $S_{\nu'}$ is the spectral bandshape function. Luminescence is classified as being due to fluorescence and phosphorescence. The luminescence quantum yield is the sum of quantum yields for these two processes. The quantum yields are, in turn, equal to rate of light production relative to the total rate out of the excited state. For example, the fluorescence quantum yield is

$$\phi_F = \frac{k_F}{k_F + k_{NF}} \tag{18}$$

where k_F and k_{NF} are the fluorescence and nonfluorescence relaxation rate constants. Luminescence power is an indirect measurement of optical absorbance. It is related to the optical absorbance through sample absorbance.

Luminescent samples will dissipate excited- and ground-state vibrational energy in the form of heat. The difference between the absorbed power and the power emitted by luminescence processes is dissipated in the form of heat:

$$\Phi_H = \Phi_A - \Phi_L = Y_H \Phi_A \tag{19}$$

where Φ_H is the thermal power or that power which is converted to heat and Y_H is the thermal power yield. This is essentially a statement of the energy conservation principle. In simple systems, energy transferred to the sample by optical absorption must result in luminescence and sample heating. The amount of luminescence and sample heating is proportional to the excitation source power.

Photothermal spectroscopy is illustrated in Figure 2.2. It is based on a measurement of the thermal power. Dissipation of the thermal power is similar to that of the luminescence power. The heat transfer is not directional. There are several means to measure the thermal power in which the excitation source does not interfere with the measurement. Photothermal spectroscopy is similar to luminescence spectroscopy in many respects. It is complementary in that it measures the absorbed power that is not lost through luminescence.

Combining the foregoing expressions for the power absorbed by the sample but not lost via subsequent luminescence, the thermal power in the sample is

$$\Phi_H = \Phi_0 Y_H (1 - R_1) \frac{\alpha}{\alpha_S + \alpha} [1 - e^{-(\alpha_S + \alpha)l}] \tag{20}$$

To obtain an accurate estimate of the thermal power, the incident power, sample cell reflectance, scattering, and luminescence power yield must all be known. No single measurement scheme yields all this information. Emission spectroscopy can be used to determine Y_L and thus Y_H; transmission spectroscopy can be used to determine $\alpha_S + \alpha$ and R by measurement of the sample and blank. In general, only a combination of these three measurements yield enough information for accurate determination of the sample absorption coefficient, α.

Photothermal spectroscopy is normally used when emission spectroscopy does not work. Emission spectroscopy, such as laser-excited fluorescence, is extremely sensitive, owing to the high signal-to-noise ratio. The fluorescence signal can be measured with almost no background, resulting in single-molecule detection limits in certain cases. Nonetheless, photothermal spectroscopy can be used for fluorescent samples. Even species with unit fluorescence quantum will have power yields less than unity because of the excess vibrational energy in the excited state after absorption and in the ground state after emission. Photothermal spectroscopy can be used as a convenient method to determine the luminescent power yields. It is important to recognize that the thermal power yield not only varies from molecule to molecule, in accordance with the fluorescence quantum yield of that species and sample matrix, but also varies with the wavelength of excitation. Shorter excitation wavelengths will result in greater thermal power yields, due to the greater excess vibrational energy in the excited electronic state. An understanding of this and other processes requires knowledge of the kinetics and mechanisms for excited-state relaxation.

2.2 OPTICAL EXCITATION

The electromagnetic radiation spectrum is classified as being composed of separate regions (e.g., microwave, infrared, visible, ultraviolet, etc.). These regions correspond roughly to particular optical excitation mechanisms. Optical excitation mechanisms can be broadly classified by the particular motion or degree of freedom that changes in the absorber. The main types of motions that are excited by electromagnetic radiation are (1) electron or nuclear spin directions, (2) rotations, (3) vibrations and hindered internal rotations of molecules, and (4) electron orbital transitions.

The high-frequency radiowave and microwave portions of the electromagnetic spectrum ($\lambda \leqslant 50\,\mu\text{m}$) correspond to changes in the directions of electron or nuclear spins, to pure rotational transitions of gas-phase molecules, to low-energy vibrations of molecules and phonon vibrations in solids, and to hindered rotations of molecules. The infrared region ($50\,\mu\text{m} \leqslant \lambda \leqslant 2\,\mu\text{m}$) generally corresponds to molecular vibrational transitions, although some electronic transitions can also occur at these relatively low energies. Electronic transitions occur in the near infrared ($2\,\mu\text{m} \leqslant \lambda \leqslant 0.7\,\mu\text{m}$), visible ($0.7\,\mu\text{m} \leqslant \lambda \leqslant 0.4\,\mu\text{m}$), ultraviolet ($0.4\,\mu\text{m} \leqslant \lambda \leqslant 0.2\,\mu\text{m}$), and the far-ultraviolet and x-ray regions ($\lambda < 0.2\,\mu\text{m}$). Because photothermal spectroscopy monitors the optical power absorbed by the sample, excitation into any of these degrees of freedom can result in a photothermal signal. Moreover, the same detection apparatus can be used to monitor absorption at any one or a combination of these regions. Photothermal spectroscopy is thus more versatile than any of the direct spectroscopies associated with a particular type of transition or region of the electromagnetic spectrum. Photothermal measurements are less dependent on the particular type of transition and the particular physics governing the operation of the detector.

2.2.1. Kinetic Treatment of Optical Transitions

Although particular selection rules, transition strengths, and motions all change with the type of transition being excited, some basic equations apply to all forms of optical spectroscopy. Many of the characteristics of spectroscopic transitions can be understood from predictions based on a two-level model using Einstein transition coefficients. The transition is modeled as being composed of a lower state and an upper state. In the absence of nonradiative rates, these states are coupled only by optical transitions

$$\begin{aligned} \frac{dN_l(t)}{dt} &= PB_{ul}N_u(t) - PB_{lu}N_l(t) + A_{ul}N_u(t) \\ \frac{dN_u(t)}{dt} &= PB_{lu}N_l(t) - PB_{ul}N_u(t) - A_{ul}N_u(t) \end{aligned} \tag{21}$$

$N_l(t)$ and $N_u(t)$ are the time-dependent number density populations in the lower and upper states of the transition, respectively, in number per cubic meter, P (photons m^{-3} Hz^{-1}) is the photon density at a particular optical frequency, ν (Hz), B_{lu} and B_{ul} (m^3 s^{-1} Hz) are Einstein coefficients for absorption and stimulated emission at ν, and A_{ul} (s^{-1}) is the Einstein coefficient for optical emission. The B_{ij} are related to absorption and stimulated emission cross sections by $\sigma_{ij}c = B_{ij}$, where c is the speed of light. This simple two-level model may be generalized by including nonradiative transitions and background states. However, the optical process rates will remain the same in any case. Absorption cross sections are also optical frequency dependent and have units of m^2 Hz. Photon densities, cross sections, and Einstein coefficients may also be used in integrated form wherein the frequency dependence is lost in integration over the band. An excellent discussion of the quantitative aspects of this procedure is given in Appendix F of Ingle and Crouch (1988).

The Einstein coefficients are directly related to the quantum mechanical probabilities for absorption and stimulated emission. These relationships are the topic of most books addressing molecular spectroscopy and are outlined only briefly here. The tool used to obtain this relationship is time-dependent perturbation theory. A time-dependent perturbation Hamiltonian, $\hat{H}$, changes the wavefunctions describing the initial state, Ψ_i, so as to allow overlap with a final state. The perturbation Hamiltonian is typically cast in the form of a product of the strength of the perturbing force, with a operator that reflects the action of this force on the wavefunction of the initial state. For an electric or magnetic field given by $A(\nu, t)$, the general perturbation Hamiltonian is

$$\hat{H}' = A_0(\nu)\cos(2\pi\nu t)\cdot\hat{O} \tag{22}$$

where $A_0(\nu)$ is the amplitude of the field at frequency ν and $\hat{O}$ is any transition operator such as the electric or magnetic dipole, quadrapole, and so on. The overlap of the perturbed initial state-wavefunction with the final-state wavefunction, Ψ_f, gives the time rate of change of the transition amplitude. Einstein B_{if} coefficients are essentially second-order rate constants. They are obtained from the quantum mechanical transition rate by recasting the perturbation field strength in terms of the photon density. The first reactant is the optical photon and the second is the species undergoing a transition. Since the transition rate is per species undergoing a transition, the total rate is the single species rate times the number density of species. Kinetics and quantum mechanics are thus related by

$$N_i P B_{if} = N_i A(\nu)^2 \frac{4\pi^2}{h^2} \langle f | \hat{O} | i \rangle^2 \tag{23}$$

The angle-bracket notation is used to indicate the spatial overlap between the perturbed initial state with the final state:

$$\langle f|\hat{O}|i\rangle^2 = \left| \int \Psi_f \hat{O} \Psi_i d\tau \right|^2 \tag{24}$$

This spatial integral is called the transition matrix element for the transition. The photon density term corresponds to the squared electromagnetic field strength and the Einstein coefficient to the time differential. The transition matrix element does not take into account the resonance condition for transition. This resonance condition is found in detailed treatment of time-dependent perturbation theory. Transitions can only occur when the transition matrix element is finite *and* the photon has an energy that will allow the transition to occur (e.g., $\Delta E = h\nu$).

The scale of B and P depends on whether SI or cgs units are used. For any change in scale in the electromagnetic field, there must be a corresponding change of scale in the Einstein coefficient. The photon density in SI units is related to the electric field through

$$P = \frac{\varepsilon_0}{2h\nu} E_0^2 \tag{25}$$

where E_0 (V m^{-1}) is the electric field strength of the electromagnetic radiation, and ε_0 [$10^7/(4\pi c^2)$ C^2 N^{-1} m^{-2}] is the vacuum permittivity. The Einstein coefficient is related to the transition matrix by

$$B_{ij} = \frac{8\pi^2 \nu}{\varepsilon_0 h} \langle f | \hat{O} | i \rangle^2 \tag{26}$$

for transitions resulting from interaction with the electric field. For magnetic dipole or quadrapole transitions, the magnetic induction field, B_0 (T), is related to the electric field through $cB_0 = E_0$, and can similarly be related to the photon density. In general, only electric dipole transitions are observed when allowed. Magnetic dipole and electric quadrapole transitions are approximately 10^{-5} and 10^{-7}, respectively, less than the electric dipole transition strength. These transitions are observed only when the electric dipole transitions is forbidden.

The two-level model may be used for molecules when only one type of transition is of interest. In the Born–Oppenheimer approximation, the total wavefunction is the product of individual wavefunctions describing the different degrees of freedom. A given state has a wavefunction given by

$$\Psi_{\text{total}} = \Psi_e \Psi_v \Psi_r \Psi_s \Psi_t \tag{27}$$

where e, v, r, s, and t are used to indicate electronic, vibration, rotation, spin, and translation wavefunctions. The spin wavefunction can include both electron and nuclear spin terms. For optical transition to occur, the total wavefunction overlap integral

$$\langle f|\hat{O}|i\rangle^2 = \left|\int \Psi_f \hat{O} \Psi_i \, d\tau\right|^2 \tag{28}$$

must be finite. Although the operator can perturb any and all degrees of freedom, the resonance condition usually allows only one perturbed state to undergo a transition. Since the degrees of freedom are orthogonal, the transition matrix element can be decomposed into the overlaps of each individual degree of freedom. For example, the transition matrix element for an optical source resonant with an electronic transition is

$$\left|\int \Psi_f \hat{O} \Psi_i \, d\tau\right|^2 = \langle e_f|\hat{O}|e_i\rangle^2 \langle v_f \,|\, v_i\rangle^2 \langle r_f \,|\, r_i\rangle^2 \langle s_f \,|\, s_i\rangle^2 \langle t_f \,|\, t_i\rangle^2 \tag{29}$$

where

$$\langle x_f \,|\, x_i\rangle^2 = \left|\int \Psi_{f,x} \Psi_{i,x} \, d\tau\right|^2 \tag{30}$$

is the overlap between the unperturbed lower- and upper-state wavefunctions for the x degrees of freedom. This integral serves as the basis for spectroscopy structure analysis and is used extensively in texts on molecular spectroscopy (Flygare 1978, Steinfeld 1985). The transition is allowed only if this overlap is finite. Analysis of the combined overlaps in terms of their being finite or not comprise selection rules. Other features, such as band-shape, come from analysis of the relative magnitudes of the overlap integrals. The overlap between the lower- and upper-state vibration, rotation, and spin all contribute to the spectrum. In electronic spectroscopy, the vibrational overlap is the Franck–Condon factor. The rotational overlap also plays a role in electronic spectroscopy. This terms results in the rotational structure observed in gas-phase electronic transitions. The spin overlap indicates primarily whether or not a transition is electron spin allowed. In small molecules it can give rise to hyperfine structure due to the coupling of nuclear and electron spin and spin and rotation coupling. An excellent treatment of these effects is given by Mizushima's (1975) work on diatomic spectroscopy. Translational overlap is generally disregarded since this term is not important in the molecular or atom fixed frame of reference.

Einstein coefficients can be related to each other through the steady-state solutions of the rate expression. In steady state, the number densities of both upper and lower states are fixed and the time rate of change is zero. The ratio of the population densities found from the rate expression under steady-state conditions is

$$\frac{N_u}{N_l} = \frac{PB_{lu}}{PB_{ul} + A_{ul}} \tag{31}$$

When the photon density is high,

$$\lim_{P\to\infty} \frac{N_u}{N_l} = \frac{B_{lu}}{B_{ul}} \tag{32}$$

The system is still at equilibrium and must follow the Boltzmann distribution law

$$\frac{N_u}{N_l} = e^{-\Delta E/kT} \tag{33}$$

ΔE(J) is the amount of energy exchanged in the interaction, k (J K^{-1}) is the Boltzmann constant, and T (K) is the temperature. To see what the limiting temperature is, we may use the Planck blackbody radiation law:

$$P = \frac{8\pi\nu^3}{c^3} \frac{1}{e^{h\nu/kT} - 1} \tag{34}$$

Since the blackbody radiation photon density increases with temperature, the high photon density limit must correspond to high temperature. In the limit as P approaches infinity, T must also approach infinity. For this to be true,

$$B_{lu} = B_{ul} \tag{35}$$

in the limit as T approaches infinity. The B coefficients are not subject to change with temperature since they are related only to the quantum mechanical wavefunctions. Thus they must be equal at all temperatures.

An optical absorption law optical absorption and stimulated emission can be derived from the rate expressions under steady-state conditions. Photons created through stimulated emission travel in the same direction and are in the same phase as the photon that stimulated the transition. The photon loss is equal to the lower-to-upper state transition rate without spontaneous emission. The latter do not result in photons moving in the same direction as that of the excitation source and do not result in addition or loss of photon from the source. The time-dependent photon density loss in a steady-state system is

$$\frac{dP}{dt} = \frac{dN_l}{dt} = PB_{ul}N_u - PB_{lu}N_l \tag{36}$$

Using the relationships $dx/dt = c$, $B_{ul} = B_{lu}$, and $\sigma c = B_{lu}$, we obtain

$$\frac{dP}{P} = -\sigma(N_l - N_u)\, dx \tag{37}$$

Integration over the pathlength, l, of the sample yields an absorption or Beer's law equation:

$$\frac{P(l)}{P(0)} = e^{-\sigma(N_l - N_u)} \tag{38}$$

Clearly, the absorption coefficient is $\alpha = \sigma(N_l - N_u)$. Thus the absorption coefficient is not necessarily proportional to the total number density of optically active species but rather depends on the population difference between the upper and lower levels. Without nonradiative decay, the number densities can only change through optical interaction. Solving for the population difference under steady-state conditions

$$N_l - N_u = N_{\text{tot}} \frac{A_{ul}}{2PB_{lu} + A_{ul}} \tag{39}$$

where $N_{\text{tot}} = N_l + N_u$. The resulting absorption coefficient is

$$\alpha = \sigma N_{\text{tot}} \frac{A_{ul}}{2E_p\sigma + A_{ul}} \tag{40}$$

where the photon irradiance is defined as $E_P = Pc$ (photon $\text{s}^{-1}\,\text{m}^{-2}$). For photon irradiances $E_P\sigma \ll A_{ul}$, $\alpha = \sigma N_{\text{tot}}$. As the photon irradiance is increased, the absorption coefficient decreases. The decrease in absorption coefficient with increasing irradiance is called optical saturation. The saturation irradiance, E_S, is normally defined as that irradiance that causes the absorption to decrease by a factor of 2. By defining $E_S = A_{ul}/2\sigma$, the absorption coefficient can be put in the convenient form

$$\alpha = \sigma N_{\text{tot}} \frac{1}{1 + E_P/E_S} \tag{41}$$

Saturation conditions are easier to obtain in transitions with small A_{ul} and/or large σ. We can obtain more information regarding the nature of this nonlinear effect because there is a relationship between these two parameters.

The A and B coefficients for a two-level system, in thermal equilibration, result in an equation that is similar to Planck's blackbody radiation law. Combining the steady-state number densities with the Boltzmann distribution law yields

$$\frac{PB_{lu}}{PB_{lu} + A_{ul}} = e^{-\Delta E/kT} \tag{42}$$

Solving for the photon density gives

$$P = \frac{A_{ul}}{B_{lu}} \frac{1}{e^{\Delta E/kT} - 1} \tag{43}$$

Equating this with the blackbody radiation law, we have

$$A_{ul} = \frac{8\pi\nu^3}{c^3} B_{lu} \tag{44}$$

Since the B coefficients are proportional to the absorption cross section, the spontaneous rate is also be related to the absorption cross section:

$$A_{ul} = \frac{8\pi v^3}{c^4} \sigma \tag{45}$$

There is a large amount of empirical data to support this relationship. It has several implications in spectroscopy. First, because of the orthogonal nature of wavefunctions, if there were no perturbation to the upper state, there could be no overlap with the lower state and thus no emission. The thermal relationship between the spontaneous and induced rates are a direct way to account for the vacuum perturbation to the radiative system (notice that by comparing this rate constant to the equivalent for stimulated transitions, the vacuum photon density must be $8\pi v^3/c^3$). Second, it is apparent that the lower the frequency of the transition, the smaller the spontaneous emission rate. Excited states with a small spontaneous emission rates are more likely to be deactivated through nonradiative quenching processes. Nonradiative relaxation is favored for rotational and vibrational transitions, which would radiate in the radiowave and infrared regions of the spectrum. Third, the spontaneous emission relaxation rate is proportional to the absorption cross section. Within a particular wavelength region, spontaneous emission is more probable from species with higher absorption cross sections. Finally, substituting this result into the saturation irradiance definition yields $E_S = 4\pi v^3/c^4$. Thus optical saturation is much more likely to occur for low-frequency transitions.

The direct relationship between σ and A should allow determination of the upper-state fluorescence lifetime though accurate measurements of the absorption coefficient and, conversely, the absorption cross section may be obtained from a measurement of the fluorescence lifetime. The fluorescence lifetime is the time required for a population in the upper state to decay to $1/e$ of the original value. This lifetime can be found by solving the rate expression for the upper state with some initial population.

Without external optical excitation, the optical rate expressions are simply

$$-\frac{dN_l(t)}{dt} = \frac{dN_u(t)}{dt} = -A_{ul}N_u(t) \tag{46}$$

Each species that makes a transition from the upper to lower levels will produce a photon. The photon density is

$$\frac{P(t)}{P(0)} = e^{-A_{ul}t} \tag{47}$$

where $P(0)$ is the initial photon density. The photon density is proportional to the number density of species in the upper level. The upper state will relax producing photons until there is no species remaining, so obviously, the upper state must be prepared in some fashion. The fluorescence lifetime of the upper state is $\tau = A_{ul}^{-1}$.

The finite lifetime of the excited state is related to the homogeneous linewidth of the transition. Any event that shortens the lifetime of a state will result in a broadening of the corresponding transition linewidth. This is a manifestation of the uncertainly principle and broadening due to the finite lifetime is often called uncertainty broadening. The homogeneous linewidth is found from the Fourier transform of the time dependence of the excited state. Transforms of a exponential lifetime functions result in a Lorentzian linewidth of

$$g(\nu) = \frac{1/(2\pi\tau)^2}{\nu^2 + 1/(2\pi\tau)^2} \tag{48}$$

Of course, this function is centered over the transition maximum. The offset to the half height of the homogeneous line is $\Delta\nu\tau = 2\pi$. As predicted from the uncertainty principle, $\nu\tau \geqslant 2\pi$, the linewidth increases with decreasing lifetime. Linewidths that are related directly to the spontaneous emission rate are called naturally broadened lines. Naturally broadened lines are somewhat rare in occurrence. There are usually other mechanisms that increase the homogeneous linewidth of a transition. Collisions can disrupt the oscillation of the electrons or nuclei. This results in a rapid shift in phase of the oscillation. This T_2, or dephasing, process results in increased linewidth. Similarly, since collision is a finite time process, the collision may result in a shift of the line center as well. This is also a consequence of the uncertainty principle. Stimulated emission also broadens the linewidth. Stimulated emission results in a decreased lifetime of the excited state. Since the lifetime is shortened, the effective linewidth is broadened. This type of broadening is referred to as power broadening or saturation broadening. Resonant saturation decreases the absorption cross section as well as changing the number densities in the upper state. This is discussed in more detail in a later section on nonlinear optical effects.

Another implication of the upper-state lifetime is that a two-level system must decay to a state defined by a zero Boltzmann temperature when there is no external excitation source. Any species initially in the upper level will, in time, decay spontaneously to the low level. If we extend the trends of a two-level system to a molecule with several levels and several degrees of freedom, radiative cooling would first decrease the electronic temperature and then the vibrational temperature; finally, the species would be rotationally cooled. Although this may occur for isolated species in black boxes, the isolated system model does not resemble most experiments. We will later find that nonradiative relaxation models predict just the opposite cooling trend (i.e., rotational degrees of freedom cool before vibrational, and electronic temperatures are the slowest to equilibrate).

To summarize, several phenomena are described by the two-level kinetic treatment of optical spectroscopy. Some important phenomena are: (1) the B coefficients are second-order rate constants that describe the probability that a transition will occur upon collision with a photon; (2) these coefficients are directly related to the cross section and to the quantum mechanical transition matrix element; (3) relationship to transition matrix elements can be used to determine the molecular structure through analysis of the overlap integrals; (4) the B coefficients for stimulated emission and absorption are equal; (5) the absorption coefficient depends on number density difference; (6) the A coefficient describes a first-order spontaneous emission process; (7) the spontaneous emission is related to absorption and stimulated processes; (8) spontaneous emission rates are proportional to ν^3; (9) optical saturation occurs when the rate of absorption and stimulated emission become equal to that for spontaneous emission; and (10) the Boltzmann temperature of an isolated system will be zero in the absence of an external optical excitation source.

2.2.2. Nonradiative Transitions

The optically induced kinetic expressions are highly idealized. Every species that absorbs light would have to fluoresce if nonradiative rates did not compete for depopulation of the excited state. The optically coupled model does not allow for nonradiative relaxation. Most experiments are performed with samples that experience collisions. The collisions results in energy transfer between the states. Several new phenomena can be described using nonradiative rates. Unlike the radiative-only transition, including the nonradiative rates allows a finite temperature to be defined in the absence of an externally supplied optical source. The rate expressions can be written as

$$\begin{aligned}\frac{dN_l(t)}{dt} &= E_P\sigma[N_u(t) - N_l(t)] + (A_{ul} + k_{ul})N_u(t) - k_{lu}N_l(t)\\ \frac{dN_u(t)}{dt} &= E_P\sigma[N_l(t) - N_u(t)] - (A_{ul} + k_{ul})N_u(t) + k_{lu}N_l(t)\end{aligned} \tag{49}$$

where k_{ul} and k_{lu} (s^{-1}) are pseudo-first-order nonradiative relaxation rate constants connecting the upper and lower levels. The nonradiative rate constants are actually second order. An energy of ΔE must be gained or lost during the nonradiative transition. The energy is supplied or taken away by the collision partner. Pseudo-first-order rate constants can be used when the collision partner number density does not change.

The optically coupled two-level system must follow the laws governing any kinetic system. Consider first a two-level system in the absence of an external source of radiation. The steady-state condition leads to a relationship between the spontaneous rate constants. In equilibrium steady-state conditions, the populations in the two levels are defined by a Boltzmann temperature. Using the Boltzmann population equation and the steady-state condition results in

$$\frac{N_u}{N_l} = \frac{k_{lu}}{A_{ul} + k_{ul}} = e^{-\Delta E/kT} \tag{50}$$

The relationship between the kinetic rate constants and the temperature, referred to as the principle of detailed balancing, holds at any temperature. It can be used to relate one rate constant to the other(s). A_{ul} does not change with temperature and changes in the number density ratio due to temperature must be due to changes in the nonradiative rate constants. We saw above that without nonradiative transitions, the temperature could be finite only in the presence of an external field. The external field supplied energy to the system. With nonradiative transitions, temperature can be described only by the nonradiative rates or as a combination of nonradiative and radiative rates. Since the ratio of the relaxation rate constants is fixed at a given temperature, two relationships for the number densities can be derived from the rates

$$\begin{aligned} N_l &= \frac{N_{\text{tot}}(A_{ul} + k_{ul})}{A_{ul} + k_{ul} + k_{lu}} \\ N_u &= \frac{N_{\text{tot}} k_{lu}}{A_{ul} + k_{ul} + k_{lu}} \end{aligned} \tag{51}$$

These relative number densities are equal to the total rate into the level divided by the sum of rates both into and out of a level.

In the case where the system is initially perturbed from equilibrium, the return to equilibrium can occur through both radiative and nonradiative mechanisms. In the absence of external optical excitation the time-dependent upper-level population obtained by solving the coupled rate expressions is

$$N_u(t) = \frac{N_{\text{tot}} k_{lu}}{A_{ul} + k_{ul} + k_{lu}} [1 - e^{-(A_{ul} + k_{ul} + k_{lu})t}] + N_u(0) e^{-(A_{ul} + k_{ul} + k_{lu})t} \tag{52}$$

where $N_u(0)$ is the upper-level number density perturbation at $t = 0$. The equilibrium number density is

$$N_u(t = \infty) = \frac{N_{\text{tot}} k_{lu}}{A_{ul} + k_{ul} + k_{lu}} \tag{53}$$

which follows also from the principle of detailed balancing. The time-dependent exponential change in upper-level population is

$$N_u(t) - N_u(t = \infty) = \left[N_u(0) - \frac{N_{\text{tot}} k_{lu}}{A_{ul} + k_{ul} + k_{lu}} \right] e^{-(A_{ul} + k_{ul} + k_{lu})t} \tag{54}$$

It is apparent that the overall relaxation rate constant of the upper level is $A_{ul} + k_{ul} + k_{lu}$. Dynamic equilibrium takes place in a thermally equilibrated system. When the system is perturbed from equilibrium, the population of the upper level continues to be thermally populated but at a rate that is decreased because of the decrease in number density of the lower level. This results in an increase in the overall rate constant for upper-level relaxation. Thus in the approach to equilibrium, rate constants for both population and relaxation of the upper level affect the observed relaxation rate.

The fluorescence quantum efficiency, ϕ_F, accounts for the competition between radiative and nonradiative rates out of the upper level. It is defined as the ratio of rates out of the upper level with and without nonradiative relaxation. In the two-level system without an external source

$$\phi_F = \frac{A_{ul}}{A_{ul} + k_{ul}} \tag{55}$$

It is generally assumed that the fluorescence quantum yield is the ratio of the upper-level lifetime with and without nonradiative relaxation. This is true only for transitions with energies well above kT. For transitions with energies much greater than kT, the upper-state population is small relative to the lower level. (The Boltzmann constant in wavenumbers is $0.6950\,\text{cm}^{-1}\,\text{K}^{-1}$ and at room temperature an energy-level spacing of about $1000\,\text{cm}^{-1}$ will ensure that $<1\%$ of the species are in the upper level.) According to the principle of detailed balance, the rate constant for excitation into the upper level is small relative to that for relaxation into the lower level. Thus k_{lu} is often neglected relative to k_{ul} for visible and ultraviolet fluorescence, and the rate constant for decay of the upper level is taken to be $A_{ul} + k_{ul}$.

The absorption coefficient also changes upon the inclusion of nonradiative relaxation. Upper-level populations have to be considered when

(1) the transition energy is low enough that the upper level can be thermally populated, and (2) the rate of optical excitation exceeds the combined spontaneous relaxation rates. Both of these conditions are more probable at lower transition energies. First, without a strong external excitation source, thermal population of the excited level results in an absorption coefficient of

$$\alpha = N_{\text{tot}}\sigma \frac{A_{ul} + k_{ul} - k_{lu}}{A_{ul} + k_{ul} + k_{lu}} \tag{56}$$

and using a detailed balance,

$$\alpha = N_{\text{tot}}\sigma \frac{1 - e^{-\Delta E/kT}}{1 + e^{-\Delta E/kT}} \tag{57}$$

When $\Delta E \ll kT$, the absorption coefficient is $\alpha \approx \sigma N_{\text{tot}}$.

Second, optical saturation occurs when the rate of excitation exceeds the relaxation rates. The optical saturation condition can be obtained from the rate expression for the upper level. The steady-state lower- and upper-level number densities are

$$\begin{aligned} N_l &= N_{\text{tot}} \frac{E_P\sigma + A_{ul} + k_{ul}}{2E_P\sigma + A_{ul} + k_{ul} + k_{lu}} \\ N_u &= N_{\text{tot}} \frac{E_P\sigma + k_{lu}}{2E_P\sigma + A_{ul} + k_{ul} + k_{lu}} \end{aligned} \tag{58}$$

Since the absorption is proportional to the difference in lower- and upper-level number densities,

$$\alpha = N_{\text{tot}}\sigma \frac{A_{ul} + k_{ul} - k_{lu}}{A_{ul} + k_{ul} + k_{lu}} \frac{1}{1 + E_P/E_S} \tag{59}$$

where the saturation photon irradiance is here defined by

$$E_S = \frac{1}{2\sigma\tau} \qquad \tau = \frac{1}{A_{ul} + k_{ul} + k_{lu}} \tag{60}$$

The absorption coefficient equation is composed of two components. The first is the ratio of rate constants that describe thermal population of the upper level. The second component is the photon irradiance–dependent decrease in the number densities due to optical saturation. Optical saturation limits all types of optical spectroscopy that utilize external sources. It is particularly troublesome in the infrared and radiowave regions of the spectrum, where the spontaneous emission rates are low and the number density in the upper level is already high due to thermal population. What allows spectroscopy to be a feasible tool for the analysis of low-energy transitions are the relatively high rates of nonradiative excited-state relaxation. As discussed below, nonradiative rates typically have Arrhenius-type behavior wherein the rates are greater for low-energy transitions.

Saturation particularly limits absorption spectrophotometry. Because of shot noise, signal-to-noise ratios are enhanced using high-irradiance sources. This enhancement is limited by optical saturation (Winefordner and Rutledge 1985). Photothermal spectroscopy does not monitor the loss of photons from the excitation source directly. Limitations due to optical saturation are less severe. The rate of energy transfer to the collision partner, and thus sample heating, is proportional to $k_{ul}N_u$. Since the upper-level population density increases with irradiance, even beyond the saturation irradiance, the photothermal signal increases with increasing irradiance up to the point where half of the species are in the upper level.

The two-level system can be used to understand the optically induced effects that occur in multiple-level systems. For example, the population difference relationship in the absorption law, optical saturation, and single fluorescence lifetimes are observed in molecules. The major difference in the multiple-level system is that the relationship between the fluorescence lifetime and the absorption cross section must take into account the large finite bandwidth of molecule absorption. There are also limitations to this approach, and several phenomena cannot be described by such a simple picture. For example, optical bleaching and multiple-photon absorption cannot be described by a two-level system even when generalized to include other nearby levels. However, the kinetic basis for understanding these multiple-level effects is the same as that of the two-level system. The latter effects are rather common in photothermal spectroscopy, and some features of these effects will be shown after discussing nonradiative relaxation mechanisms.

2.3. EXCITED-STATE RELAXATION

Optical absorption promotes an absorbing species to an excited state. The internal energy of the absorber is thus increased by an amount equal to the photon energy. In many cases the excited state will relax, causing sample heating with a thermal energy equal to that of the absorbed photon. However, heating is not instantaneous and may consist of several rates.

Heat produced at rates that are slow relative to hydrodynamic relaxation and thermal diffusion may not result in measurable photothermal signals. The energy of the sample is conserved throughout the processes of excited-state relaxation and thermal equilibration. The average energy of the excited species decrease while that of the sample matrix increases. The energy transfer process continues to change the average energies until the thermodynamic temperature of the excited species is equal to that of the sample matrix. Excited-state relaxation is an energy transfer process wherein the excited species transfers energy to the sample matrix. Thus nonradiative energy transfer is a bimolecular second-order rate process. In dilute samples the number density or concentration of energy acceptors is essentially constant and the relaxation processes are pseudo-first order.

Collisional relaxation is similar in many respects to the optical absorption and stimulated emission processes. Just as the photon must have an energy that is compatible with the transition energy, the collision partner must be of an energy that is compatible with the energy transfer event. Large energy transfer events require the collision partner either to (1) have an excited state at an energy that is equal or compatible to the energy to be exchanged with the collision partner, or (2) have a translational energy that is of the same magnitude as the energy to be exchanged. In addition, not every collision will result in a complex that facilitates energy transfer. This results in a collisional energy transfer cross section that is often less than that of the collision rate. In this respect it is similar to the photon absorption cross section.

Several mechanisms may be involved in excited-state relaxation. Each energy transfer generally occurs at a different rate. Excited-state relaxation can take place on time scales ranging from femtoseconds to seconds. The time scales or rates of relaxation are dependent on (1) the nature of the excited state, (2) the intermediate energy levels available to the relaxing species, (3) the nature of the sample matrix, and (4) the concentrations or densities of all species present in the sample matrix.

One goal in the analytical chemistry related to photothermal spectroscopy is to choose sample matrices that enhance the energy transfer. Efficient energy transfer via nonradiative mechanisms allows a maximum amount of energy to be transferred to the sample. The ultimate temperature change that occurs is proportional to the rate at which energy can be transferred to the sample. This, in turn, is proportional to both the excitation irradiance and the rate of excited-state relaxation. Thus enhancing the excited-to-ground state energy transfer will ultimately lead to better photothermal spectroscopy signals. Much of the chemistry of energy transfer has not been exploited in analytical photothermal spectroscopy.

2.3.1. Rotational and Vibrational Relaxation

Flygare (1968, 1978) and Steinfeld, et al. (1989) have discussed mechanisms and time scales involved in rotational and vibrational equilibration and relaxation. The equilibration time is that required to establish an equilibrated canonical distribution within the degree of freedom. Relaxation is the process of transferring energy to translational modes. Rates of rotational and vibrational equilibration and relaxation depend on the collision rate and on the temperature and identity of the collision partner. Gas-phase rovibrational relaxation rate constants are observed to follow Arrhenius behavior, also called the energy gap law, with rare-gas collision partners (Flygare 1968). This empirical behavior has been confirmed using an information or entropy-based approach (Steinfeld, et al. 1989, Chap. 13). Energy transfer on the scale of this description can generally be characterized in terms of the pseudo-first-order rates for different modes of relaxation (Flygare 1968, 1978). For a collection of excited-state species in a sample matrix or solvent that is at equilibrium and described by an equilibrium temperature, the relaxation to an equilibrium state generally follows an Arrhenius-like equation,

$$\frac{1}{\tau} = \eta e^{-\Delta E/kT} \tag{61}$$

where τ (s) is the relaxation time constant and η (s^{-1}) is the frequency of occurrence of conditions resulting in energy transfer (e.g., collision). The relaxation time for individual state-to-state interactions will each be different since (1) the energy gap is different, and (2) the η interaction term depends on the nature of the particular interaction. As a general rule, when averaged over all the possible state-to-state interactions that occur in the approach to equilibrium, the relaxation times will decrease with decreasing quantum energy. For example, rotational relaxation occurs faster than that for vibration.

Another time scale used to characterize excited-state relaxation is that of the lifetime of a level or state. The lifetime of a particular level is different than the relaxation time. The level lifetime is the average time a molecule exists in a particular electronic, vibrational, or rotational level before it makes a transition to another level. Level lifetimes can be obtained from spectroscopic linewidth measurements or by measuring the initial decay out of a state prepared out of equilibrium. Individual level lifetimes apparently depend on the same parameters as do the relaxation rates. However, individual level lifetimes are generally shorter than relaxation times because they involve transitions to all levels, not just those resulting in increased translational energy.

The first type of relaxation is that of translational energy. Translational energy relaxation, or T–T, occurs through collisions. This process is efficient since the translational energy quantum is essentially zero (i.e., there is no energy gap). Equilibration occurs within a few (1 to 10) gas kinetic collisions. An interesting point is that higher-energy species move faster than low-energy species. Higher translational energy species will collide at higher rates and thus relax more rapidly than slow, low-energy species. Rotational to rotation (R–R) and rotation to translation (R-T) relaxation are highly efficient processes due to the small energy-level spacing. Typically, $\Delta E \ll kT$ and the exponential in the Arrhenius-like energy gap expression is essentially 1. Rotational quantum numbers can change by more than 1 upon collision and collision cross sections for R–R relaxation may actually exceed the van der Waals radius. Both of these cause the interaction term, η, to be high for R–R relaxation. Molecules with large permanent dipole moments relax more rapidly than do those of nonpolar species because of long-range dipole interactions. The relatively long-range dipole–dipole interactions are responsible for the greater than unit collision R–R relaxation. Flygare (1978) points out that a consequence of the long-range interactions is that R–R relaxation can actually occur faster than pure translational, or T–T relaxation. T–T relaxation is a purely collisional process and requires physical contact.

On the other hand, R–T relaxation cannot be facilitated through dipole interaction. R–T relaxation is a collision-induced process and relaxation times are about a factor of 100 slower than those for R–R. R–T times are also a factor of about 10 slower than T–T times due to geometric factors. Not all collisions will have the correct geometry for coupling angular momentum to translational momentum. So although the rotational energy gaps are small enough to be neglected, $\Delta E \ll kT$, the collision efficiency factor or interaction term, η, is smaller than that for T–T.

Because the large number of populated rotational levels, R–R and R–T equilibration times may depend on the degree to which the rotational population is perturbed from the equilibrium state. The time required for a given molecule to equilibrate depends on the initial rotational-state distribution. It may take only a few collision times for species to equilibrate to an equilibrium rotational temperature when the only one or two rotational levels are initially out of equilibrium. Distributions that are initially far from equilibrium (e.g., a single rotational level populated by vibrational excitation of a small molecule) may take from 10 to 100 collisions to equilibrate at room temperature. Lifetimes of individual rotational levels can be obtained from spectroscopic linewidths. Small molecules such as OCS, N_2O, and HCN exhibit rotational linewidths indicative of rotational lifetimes varying from 10^{-11} s for polar interactions (e.g., HCN—HCN) to 10^{-10} s for low-mass-collision partner relaxation (e.g., OCS—He) at standard temperature and pressure (Flygare 1978). On the other hand, canonical R–R relaxation times are about 10^{-10} s, while that for R–T relaxation takes about 10^{-8} s under standard gas conditions.

Excited vibrational energy transfer is primarily dependent on collisions, although longer-range dipole interactions can also occur. The types of vibrational relaxation are V–V, V–R, and V–T (Table 2.1). The latter two are similar and often referred to as V–R/T. V–V and V–R/T are distinguished from one another in both the time scales for relaxation and the way that excess energy is distributed among the degrees of freedom. V–V equilibration consists of a set of V–V energy exchange processes. V–V energy exchange can lead to vibrational-level equilibration that is independent of the rotational and translational temperature (Cottrell and McCoubrey 1961). Energy can be distributed among the vibrational levels according to the Boltzmann distribution law. The vibrational temperature of the V–V equilibrated molecule may be higher or lower than the rotational and translational temperature. This can occur only if V–V rates are much faster than V–R/T rates.

Direct evidence of fast V–V equilibration comes mainly from infrared fluorescence and double resonance experiments. In infrared laser-excited infrared fluorescence, vibrational excitation to a low-lying vibrational level often leads to prompt anti-Stokes infrared fluorescence. For example, Yuan

Table 2.1. Relaxation Times for Typical Gas-Phase Species at Standard Temperature and Pressure

Process[a]	Relaxation Time[b] (s)
T–T	10^{-9}
R–T	10^{-8}
R–R	10^{-10}
V–R	10^{-5}
V–V	10^{-8}
V–T	10^{-5}

[a]T = translation, R = rotation, and V = vibration modes.
[b]Estimates are from Flygare (1968) for typical small molecule gases at 300 K, 1 atm pressure.

and Flynn (1972, 1973) found prompt 3- and 5-μm fluorescence from ethylene ν_7 mode excited with a Q-switched CO_2 at 10 μm at low gas pressures. The prompt anti-Stokes fluorescence was attributed to the establishment of a quasi-equilibrium in which vibrational levels above the lowest-lying ν_{10} level at 825 cm^{-1} rapidly equilibrate through V–V transfer. The resulting vibrational energy-state distribution is probably similar to that given by the Boltzmann distribution law, but with a deficit of species in the ground vibrational level. The quasi-equilibrium can occur only if there is rapid V–V energy exchange among the vibrationally excited levels. Hager et al. (1979) later found that the V–V population of unexcited levels takes from 10 to 100 rare-gas collisions to occur. This is on the same order as that required for T–T and R–T relaxation.

Yuan and Flynn (1972) found that the fluorescence decay rate was independent of emission wavelength and thus independent of the identity of the fluorescent level. This fluorescence behavior is consistent with rapid V–V equilibration among the excited vibrational levels. The fluorescence decay is limited by the rate of depopulation of the quasi-equilibrated excited-state ensemble. Depopulation of this state can occur through either fluorescence, which is a very slow process in the infrared, or by V–R/T relaxation to the ground vibrational level. According to the energy gap law, relaxation of the lowest excited level would be the preferred channel, due to relatively low amount of energy to be dissipated in rotational and translational degrees of freedom. Essentially, the excited-state levels all relax through that level or set of levels with the fastest V–R/T rate. Ethylene has a lowest-energy vibrational level 825 cm^{-1} above the ground state, and there are three other vibrational states close in energy. The energy gap law would predict fast V–V relaxation among these levels and slow relaxation to the ground state. Yuan and Flynn (1973) calculate that all four low-lying vibrational levels should relax with about the same efficiency. Excited species will all have to pass through this slow relaxation step or bottleneck to achieve V–R/T equilibration. By measuring the fluorescence decay, it was found that excited-vibrational-state relaxation takes about 2×10^4 collisions with rare gas at room temperature. This low efficiency was confirmed using infrared double resonance (Hager et al. 1979). Similar behavior has been observed for SF_6 (Flygare 1978), CH_3F (Weitz and Flynn 1972), and other species (Weitz and Flynn 1981).

As a general rule, V–R/T relaxation processes take considerably longer than that of V–V. V–R/T relaxation takes from 10^3 to 10^4 collisions to occur, while V–V can take as few as 10 collisions. Energy gap arguments would predict that molecules with low vibrational frequency levels should undergo V–R/T with fewer collisions than those with only high vibrational frequencies. However, the energy gap alone does not account for vibrational relaxation trends. Although the energy gap equation can be used to predict the V–V versus V–R/T rates of a particular species, the prediction accuracy breaks down when comparing one molecule to another. The efficiency factor, η, can vary by an order of magnitude, or perhaps more, from molecule to molecule.

Theories are available for calculating V–R/T energy transfer efficiencies. Unfortunately, most theories use parameters that are difficult to obtain from independent experiments and tend to yield erroneous predictions. Perhaps the most widely used and tested theory is that of Schwartz, Slawsky, and Herzfeld (1952), called the SSH theory. This theory predicts that the efficiency of vibrational energy transfer should decrease with the reduced mass of the collision pair. Yuan and Flynn (1973) tested the SSH theory using vibrational relaxation rate constants determined for ethylene in several different collision partners. The theory was found to be an order of magnitude off for light collision partners (e.g., H_2 through He) and does not predict the correct behavior or trends for heavier partners (e.g., Ne, N_2, Ar, Kr). Xing-Xiao and Zhu-De (1983) have compared vibrational relaxation rate constants of CH_3OH, $C_2F_3Cl_3$, and CF_2Cl_2 obtained using time-resolved photothermal lens spectroscopy to the SSH theory and found it to be in fair qualitative agreement for rare-gas collision partners. It appears that the qualitative agreement between SSH theory and experiment is about as good as what one would predict using the Arrhenius-like energy gap equation.

The vibrational, rotational, and translational relaxation trends allow for prediction of the steps involved in relaxation of a vibrationally excited species. The mechanism for relaxation of vibrationally excited species is illustrated in Figure 2.4. Vibrational excitation using laser sources can excite particular rovibrational state species to higher vibrational levels. Thus a single rovibrational excited state is formed and there may be a hole in the canonical rotational state distribution of the ground state. Immediately following excitation, R–R, R–T, and T–T relaxation begins. R–R relaxation of the excited state results in dispersion of the rovibrational states into a canonical distribution. In the ground state, rotational hole filling takes place. In the excited state, rotational-state dissipation takes place. The combined ground- and excited-state R–R relaxations can either require energy from the matrix or release energy to the matrix. For example, if the excited rotational-state energy is greater than kT, R–R relaxation will produce translational energy. If the state is less than kT, translational energy must be used to produce an equilibrium ro-translational temperature. Using photothermal lens spectroscopy, Grabiner et al. (1972) found that pulsed infrared irradiation of the ν_3 mode of CH_3F results in rapid initial translational cooling in an Ar buffer gas matrix. Translational cooling was

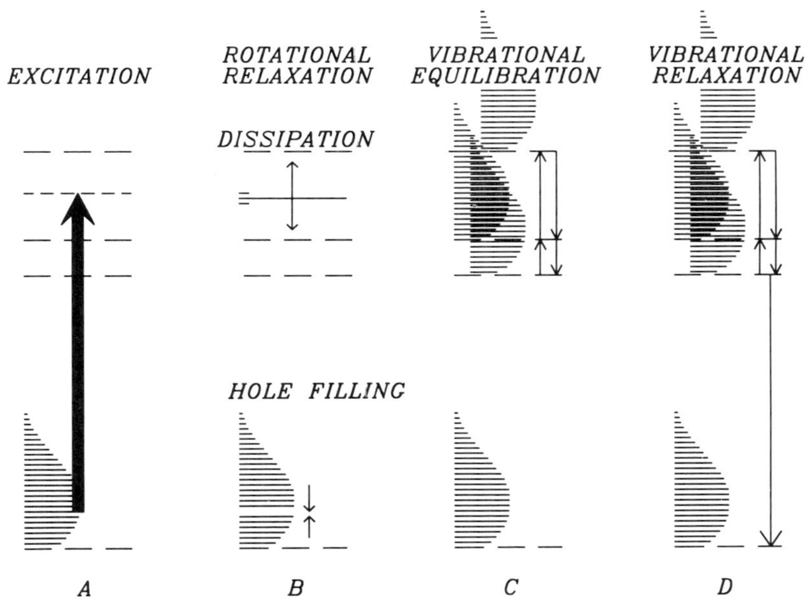

Figure 2.4. Energy level diagram illustrating the relaxation processes for rovibrational excitation. Single rovibrational state transition excitation produces a hole in the rotational population of the lower state and a single rovibrational excited state. Rotational state relaxation rapidly dissipates molecules among the rotational states. Vibrational state relaxation occurs slower. Molecules are often initially equilibrated among excited vibrational states. Equilibrated excited states subsequently relax to the ground state.

attributed to R–T, or rather, T–R processes. Cooling is not unique to CH_3F and has been observed for V–V relaxation in CO_2-N_2 systems as well (Bailey et al. 1985).

Following the more rapid R–R/T and T–T steps, V–V relaxation takes place. V–V relaxation produces a quasi-equilibrium among the *excited* vibrational levels. In a single rovibrational excited state, energy is initially dispersed among the other rotational and excited vibrational states of the molecule. These states are generally not resonant with the excitation source, and further excitation does not occur. The excited vibrational-level populations may attain a canonical distribution prior to the bottleneck relaxation to the ground state. Finally, these excited vibrational states return to the ground state in a time limited by the fastest V–R/T rate. In general, each vibrational excited state decays to the ground state at a rate roughly proportional to the exponential energy gap of the excited state. During the slow V–R/T step, faster V–V steps maintain the excited-state populations in a canonical distribution. It is therefore impossible to distinguish between the V–R/T steps from the individual levels. Instead, the excited vibrational levels decay with a rate constant given by the sum of the individual rates multiplied by the canonical population of the level.

This excited vibrational-state relaxation mechanism has implications for the use of infrared photothermal spectroscopy for gas-phase analysis. If the rate of return of these quasi-equilibrated excited states to the ground state via V–R/T is slower than the rate of excitation, the amount of energy available to be put into the sample is limited. When this is the case, the transition is optically "bleached." The general rule of thumb is that species with low energy vibrational states will have V–R/T relaxation rates that will not allow optical bleaching with common laser sources, while species with lowest excited vibrational energies far in excess of kT will optically bleach with high irradiance sources. Optical bleaching may be remedied to some extent by using a gas matrix that is efficient at promoting V–R/T. An ideal matrix gas would be one with several high-lying vibrational levels that are not resonant with the excitation source and which themselves undergo rapid V–R/T. At the same time this ideal species would not have too many low-lying vibrational levels, since this would increase the heat capacity. The higher heat capacity would mean a lower temperature change for a given amount of energy input to the sample. Another way to enhance the photothermal behavior would be to increase the pressure. Although this also increases the heat capacity, thus lowering the maximum temperature, it turns out that the temperature-dependent refractive index change increases by a proportional amount. Thus high inert gas pressures are favored for enhancing the photothermal signal in systems that have slow relaxation steps.

Vibrational relaxation rates can be very fast in solution phase, where collision frequencies are high. The excited species is surrounded by and coupled intimately to the collision partners in the sample matrix. Vibrational-state lifetimes estimated by examining transition Raman spectrum linewidths indicate that solution-phase lifetimes of relatively low-lying vibrational states range from 1 to 10 ps (Flygare 1978). Visible-wavelength vibrational overtone spectra linewidths indicate that the lifetimes generally decrease with increasing total energy. This decrease may be due to the higher density of vibrational states and a breakdown of the normal coordinates in the highly excited states. There are certain exceptions. High-energy vibrations in molecules possessing mostly low-energy vibrational states may not couple energy efficiently to the solvent. This is apparent in the relatively strong and sharp vibrational overtone spectra of aromatic compounds. These vibrational overtone spectra consist primarily of multiple-quantum C—H stretching excitations. Although the high density of vibrational states at the energy of the excited overtone should favor rapid vibrational relaxation, linewidth measurements indicate that the states exist for several picoseconds. The apparent absence of other vibrational combinations and overtones indicates that their lifetimes are significantly shorter at these high energies.

2.3.2. Electronic States and Transitions

There are two primary methods for labeling electronic states and transitions in spectroscopy. The most accurate method labels states of the molecule, atom, or ion according to the quantum numbers of the species as a whole. In molecules, electrons are promoted between different molecular orbitals. Electronic states are classified by the total electron angular momentum and the total electron spin. The notation used for isolated atom and ion electronic states are capital Roman letters corresponding to atomic orbital angular momentum order (e.g., S, P, D, F, etc.). The electronic states and orbitals of transition metal ion complexes are often labeled with the molecular state and orbital corresponding to the particular symmetry of the complex. Molecular orbitals are also denoted by the orbital angular momentum. The symbols used to indicate the overall orbital angular momentum change depending on the symmetry of the molecule. For example, the electronic states of linear molecules are labeled with the capital Greek letter equivalents of the atomic orbital angular momentum (i.e., Σ, Π, Δ). The electron spin state is noted as a prefix superscript number equal to the total spin quantum number, S, of the state. Selection rules for optical- and collision-induced transitions depend on the state quantum numbers.

The state quantum number labeling is cumbersome and works well only for symmetric species. A scheme noting transitions between individual molecular orbitals is often used instead of the quantum number system. Atom and ion orbitals are labeled according to the atomic orbital scheme (e.g., s, p, d, f). Individual molecular orbitals are labeled according to the linear combination of the atomic orbital–molecular orbital scheme, with lowercase Greek letter equivalents to the atomic orbitals. Orbitals labeled σ, π, and δ correspond to angular momenta of 0, 1, and 2. These molecular orbitals can be either bonding or antibonding. A superscript * notation is used to indicate the antibonding orbitals. Molecules, in particular organic compounds, have a particular molecular orbital filling order. The electron filling order follows the sequence σ, π, π^*, σ^*. σ orbitals arise from the overlap of atomic orbitals with no magnetic angular momentum (e.g., atomic s, p_0, and d_0 orbitals). π molecular orbitals are formed from the overlap of atomic orbitals with magnetic quantum numbers of ± 1, $p_{\pm 1}$ and $d_{\pm 1}$. δ orbitals are rare but can occur in transition metal dimer molecules.

They arise from the overlap of $d_{\pm 2}$ atomic orbitals. This scheme is completed with the assignment of the electrons in nonbonding orbitals. Nonbonding orbitals are given the letter n; n orbitals exist in organic molecules that have N, O, halogens, and so on.

By convention, transitions, either optical or nonradiative, are labeled by the initial and final orbitals. In transition atoms or ions, electrons are promoted between atomic orbitals local to the optically active metal. These are mostly d–d transitions, although f–f atomic orbital transitions occur in lanthanides and actinides. The energy of d–d transitions depends on the ligands associated with the metal. The symmetry of the ligand coordination and the ligand field strength both affect the strength and energy of the atomic electronic transition. On the other hand, lanthanide and actinide transitions are not especially affected by the ligands because the f have a small principal quantum number and therefore do not extend far from the atomic nucleus.

Molecular transitions are also specified by the electron orbitals that change in the transition. Molecular electronic transitions are labeled σ–σ^*, n–σ^*, π–π^*, n – π^*, and so on. π–π^* transitions are of relatively low energy and have very high absorption cross sections, ranging from about 4×10^{-22} to 10^{-20} m^2 (1000 to 30,000 L cm^{-1} mol^{-1}). By comparison, even very strong vibrational transitions in the infrared have maximum cross sections of only 10^{-22} to 10^{-21} m^2. n–π^* transitions are generally of low energy and may occur at energies above or below that for the π–π^* transitions. n–π^* transition have relatively small absorption cross sections, 10^{-24} to 10^{-22} m^2 (10 to 100 L cm^{-1} mol^{-1}). Thus the absorption cross section for π–π^* transitions. n–π^* transition are from 10 to 1000 times greater than those of n–π^* transitions. This can be justified by the transition matrix element component to the Einstein B coefficient. π–π^* transitions will have relatively large electronic transition matrix elements since the electron does not have to move a large distance when making the transition. The small physical displacement corresponds to a large overlap integral. Relative to the π–π^* transition, the electron have to move a greater distance to be in the most probable position for an n–π^* transition.

Solvent polarity is known to affect the absorption maximum energy of the n–π^* and to a lesser extent π–π^* transition. Increased solvent polarity shifts the n–π^* transition maximum to higher energies, thus a blue shift. This is attributed to the stabilization energy associated with the nonbonding orbital. The ground-state energy is lowered in polar solvent due to interaction between the polar solvent and nonbonding electron pair. In the excited n–π^* state, there is one fewer nonbonding orbital electron and thus less stabilizing interaction between the nonbonding orbital electron and the solvent. The absorption maximum of π–π^* transitions often shift to lower energies with increasing solvent polarity. This red shift may be due in part to increased polar solvent stabilization of the more polar $\pi-\pi^*$ state. In some organic molecules the relative ordering of the $\pi-\pi^*$ and $n-\pi^*$ can be changed by choosing solvents that stabilize or destabilize the excited state.

$n-\sigma^*$ transitions occur at relatively high energies with band maxima wavelengths between about 250 and 150 nm. Absorption cross sections are intermediate between the $\pi-\pi^*$ and $n - \pi^*$, ranging from about 4×10^{-23} to 10^{-21} m^2. $n-\sigma^*$ absorptions limit the useful ultraviolet wavelength range of most solvents to above about 200 nm, although some may allow measurements down to about 175 nm. $\sigma-\sigma^*$ have very high transition energies and are often dissociative. These transitions occur in the ultraviolet and vacuum ultraviolet portion of the spectrum ($\lambda \leqslant 135$ nm). They are generally only useful for gas-phase sample analysis since condensed-phase solvents and solids also possess these high-energy transitions.

Optical absorption arising from electron promotion from an atomic orbital to a molecular orbital are also common. Transition metal complexes can exhibit charge transfer transitions wherein the electron is shifted between the atomic orbital of the transition metal ion and a molecular orbital of a ligand. The charge transfer transition involve an oxidation of one species with a reduction in the other. The electron is normally transferred from the ligand to the transition metal ion, although metal ion-to-ligand complexes are known. The cross sections for transfer transitions can be very high ($\sigma \geqslant 10^{-21}$ m^2). Relaxation of the excited charge transfer complex often results in the electron being transferred back to the original donor. However, electron transfer results in two new chemical species, and this complex may decompose prior to relaxation. Charge transfer complexes are very useful for the spectrometric determination of transition metal ions because of their large absorption cross sections. Complexing agents resulting in a charge transfer complex are added to the sample solution containing an unknown amount of transition metal ion. In trace analysis, the absorption cross section of the complexing agents must be very low relative to that of the charge transfer complex to avoid interference (Harris and Williams, 1985). Since the charge transfer complex band develops only in the presence of the transition metal, these reagents are very useful for low-concentration transition metal ion analysis.

The band shape of a molecular electronic absorption is controlled primarily by the Franck–Condon overlap integral component of the transition matrix element. The Franck–Condon principle states that absorption or emission cross sections are maximum for final vibrational states that resemble the initial vibrational state. Since electronic excitation usually changes the bonding, the ground vibrational state of the excited electronic state will have a different nuclear geometry. Excited vibrational states of the final electronic state will generally have equilibrium nuclear positions states which most resemble those of the initial vibronic state. The absorption spectra maxima occur for excitation into higher vibrational levels and emission maxima produce vibrationally excited final states. This principle can also be stated in terms of the potential energy surfaces of the electronic states. Maxima occur into final states that have potential energy surface boundaries that resemble the initial state.

Electronic states are distinctly different from the orbital in which the excited electron resides. The electronic state of the atom or molecule must be labeled according to the optical transition responsible for the formation of that particular state. For example, n–π^* states are different than π–π^* states, although the electron excited is in the π^* state in both cases. These two states will, in general, exist at different energies relative to the ground electronic state and will have different potential energy surfaces. A general electronic state labeling scheme in which the electron spin state is abbreviated as S, for singlet, and T, for triplet, is often used to indicate the particular electronic level of the molecule in molecular spectroscopy. Stable molecules normally have fully occupied bonding orbitals, and thus all electrons are paired. The ground electronic state is thus a singlet. States within a particular spin system are then labeled consecutively starting from the ground singlet state, S_0, to the first excited singlet, S_1, up to the *n*th singlet, S_n. Triplet spin states are associated with the excited singlet states since electronic excitation results in unpaired electrons. The triplet states are labeled starting with the first excited state T_1 and consecutively increasing with the energy of the state, T_2, T_3, and so on. One problem with the S and T state labeling system is that the identity of the excited state is lost unless the state is given explicitly. The identity of the state is important when optical excitation and nonradiative relaxation rates are considered. The state can be specified along with the spin manifold order. For example, $S_1(n-\pi^*)$ can be used to indicate the first excited singlet and is an n–π^* state.

2.3.3. Electronic State Relaxation

Excited electronic-state relaxation is understood well from the body of empirical data, and there are good models describing the trends and the basis for the relaxation mechanisms. Nonradiative relaxation electronic excited states is thought to occur through a series of rovibrational relaxations and electronic-state couplings. Electronic-state couplings are classified as internal conversions or intersystem crossings. Internal conversion is an electronic state coupling within an electron spin system and intersystem crossing is a coupling between electronic states with different spins.

The Franck–Condon factors favor excitation in higher vibrational levels of the excited electronic state. However, excited electronic state vibrational relaxation is extremely fast and the initially excited vibrational state rarely affects the emission spectra. Emission spectra obtained from even low-pressure gas-phase samples have a vibrational structure which indicates that the vibrational temperature of the excited electronic state has equilibrated prior to emission. Given that fluorescence lifetimes are on the order of nanoseconds, excited-electronic state-vibrational energy must occur with collision efficiency a near unity. Vibrational relaxation in condensed-phase samples occur within picoseconds of excitation. Relaxation times for a particular molecule can be as short as the time required to complete several vibrational periods in condensed-phase samples. The rapid vibrational relaxation may be aided by nearby excited electronic states. Internal conversion among electronic states occurs rapidly, allowing coupling between the vibrational and electronic states. Vibrational relaxation of the excited electronic states may occur more rapidly than that of the ground state because of the higher density of states in the excited state.

Guiliano and Hess (1967) give an excellent summary of excited-electronic-state relaxation found for a wide variety of organic molecules. Birks (1989) has rationalized many experimental observations in terms of photochemical and physical processes. Highly vibrationally excited species relax rapidly, owing to the increase in vibrational-state density with energy. The higher the energy above the zero point energy for an electronic state, the higher the density of states. The high density of states allows rapid relaxation since the energy gap is small from each step in the relaxation. Species that undergo internal conversion or intersystem crossing often produce vibrationally excited states on the new electronic state. The density of states is relatively high, so these species rapidly relax. In solution phase, the vibrational relaxation is rapid enough that inverse crossing is unlikely.

Excitation into high-level electronic states (e.g., S_2, S_3, T_2, T_3, etc.) normally undergo rapid internal conversion to the lowest excited state within a given spin multiplicity. For example, excitation into S_3 will result in the rapid production of molecules in the S_1 state. Excited-state internal conversion to the lowest excited state occurs on picosecond time scales, with relaxation rate constants k_{IC} of about 10^{10} to $10^{13}\,s^{-1}$. Empirical evidence for this is common. With the exception of a handful of species such as azulene, most known fluorescent compounds undergo rapid excited-state internal conversion, resulting in fluorescence emanating from lower vibrational states of S_1. Excited-state lifetimes determined from fluorescence indicate that the combined time scales for intersystem crossing and vibrational relaxation are less than nanoseconds.

First-order perturbation theory can be used to describe these results and to quantitate the rate constant for nonradiative relaxation. The result is an equation that is similar to that of the Einstein *B* coefficient:

$$k_{\mathrm{IC}} \propto \rho \langle f|\hat{O}|i\rangle^2 \tag{62}$$

where ρ is the density of states in the final electronic state at the energy of the initial state, and as in the optically induced transition

$$\left|\int \Psi_f \hat{O} \Psi_i \, d\tau\right|^2 = \langle e_f|\hat{O}|e_i\rangle^2 \langle v_f \,|\, v_i\rangle^2 \langle s_f \,|\, s_i\rangle^2 \tag{63}$$

In this case $\hat{O}$ is a collision operator and has the same properties as the electromagnetic transition operator. For internal conversion, the electronic spin overlap integral is large, while for intersystem crossing, the spin overlap small. For internal conversion, the electronic states overlap well with collisional perturbation and the controlling factor is the Franck–Condon vibrational overlap integral. The Franck–Condon principle states that maximum overlap will occur between vibrational states with similar nuclear positions. Places where the potential energy surfaces cross have maximum Franck–Condon overlap. The energy difference between electronic states become consecutively smaller and the possibility of potential curve crossings increases with the number of electronic states. The main reason for rapid excited electronic-state internal conversion is the relatively high overlap of the excited-electronic-state vibrational wavefunctions. The relatively small electronic-state energy gap in the excited electronic states allows results in potential energy curve crossings. The energy gap law thus appears to work for electronic-state relaxation.

Transitions from S_1 to S_0 are much slower than those in the excited electronic states. The empirical evidence for this is that no matter what vibronic state is excited, fluorescence emanates from the ground vibrational level of the first excited singlet. S_1-to-S_0 relaxation is also slow enough that intersystem crossing can compete successfully with the lowest excited-state internal conversion. Molecules with lower π–π^* states tend to fluoresce, while those with n–π^* states of energies lower than π–π^* states tend to relax by nonradiative processes. Justification for this trend is based on the lower absorption cross sections and thus smaller A_{ul} for the n–π^* states. However, Birks (1989) points out that the rates for S_1(n–π^*)-to-T_n(π–π^*) rates are faster than those for either S_1(n–π^*)-to-T_n(n–π^*) or S_1(π–π^*)-to-T_n(π–π^*) transitions (Table 2.2). This would suggest that spin–orbit coupling may enhance the spin–spin transition interactions in ISC. The relative ordering of the S_n(π–π^*) and S_n(n–π^*) may be changed by choosing solvents that stabilize or destabilize the excited state. The S_n(π–π^*)-state lowering in polar solvents can be used to favor fluorescence relaxation over the

Table 2.2. Quantum Yields for Fluorescence and Intersystem Crossing

Molecule (S_1)	ϕ_F	ϕ_{ISC}	$1 - (\phi_F + \phi_{ISC})^a$	$\Delta E_{S_1\text{-}S_0}$ (kcal/mol)
Benzene (π, π^*)	0.05	0.25	0.70	110
1,4-Dimethylbenzene (π, π^*)	0.35	0.65	<0.05	100
Naphthalene (π, π^*)	0.20	0.80	<0.05	92
Anthracene (π, π^*)	0.70	0.30	<0.05	76
Tetracene (π, π^*)	0.15	0.65	0.20	60
Pentacene (π, π^*)	0.10	0.15	0.75	50
Acetone (n, π^*)	0.001	~1.0	<0.05	85
Biacetyl (n, π^*)	0.002	~1.0	<0.05	65
Benzophenone (n, π^*)	0.000	~1.0	<0.05	75
Cyclobutanone (n, π^*)	0.000	0.00	1.0	80
1,3-Pentadiene (n, π^*)	0.000	0.00	1.0	100

[a]Represents an upper limit to ϕ_{IC}.
Source: N. J. Turro, *Modern Molecular Photochemistry*, Benjamin-Cummings, Menlo Park, CA, 1978, p. 181. Adapted from J. W. Birks, in *Chemiluminescence and Photochemical Reaction Detection in Chromatography*, J. W. Birks, ed., VCH, New York, 1989, Chap. 1.

nonradiative relaxation that occurs in nonpolar solvents. On the other hand, photothermal spectroscopy signals may be enhanced in nonpolar solvents where nonradiative relaxation would be favored.

The slow S_1–S_0 relaxation might seem to be an anomaly because both excited state, $S_n \rightarrow S_{n-1}$, and $S_1 \rightarrow S_0$ internal conversions have the same electronic perturbation and spin overlap integrals. However, internal conversion of the first excited singlet is slow because of the lack of potential energy curve crossing. This results in poor Franck–Condon overlap. Again, the relatively large energy difference between the ground and first excited electronic states results in potential energy curves that do not cross. Subsequently, unfavorably large nuclear displacements are required for electronic-state relaxation. S_1 relaxation is often limited by the fluorescence lifetime. Typical fluorescence lifetimes of organic dye compounds are from 1 to 100 ns with corresponding rate constants $k_F \sim 10^7$ to $10^9\ s^{-1}$. Experimentally, fluorescence emission spectra of the vast majority of organic molecules are not dependent on the excitation wavelength, thus confirming that $k_{IC} \gg k_F$ for the lowest excited state.

Because of the violation of spin conservation principles, intersystem crossing rates are smaller than those for excited-state internal conversion (Table 2.3). However, ISC rates are competitive with S_1 to S_0 IC and fluorescence relaxation. Triplet state yields typically constitute 1 to 10% of

Table 2.3. Intersystem Crossing Rate Constants

Molecule	k_{ISC} (s^{-1})	ΔE_{ST} (kcal/mol)	Transition
Napthalene	10^6	30	$S_1(\pi, \pi^*) \rightarrow T_1(\pi, \pi^*)$
1-Bromonaphthalene	10^9	30	$S_1(\pi, \pi^*) \rightarrow T_1(\pi, \pi^*)$
Anthracene	10^8	2–3	$S_1(\pi, \pi^*) \rightarrow T_2(\pi, \pi^*)$
9,10-Dibromoanthracene	$\sim 10^8$	30	$S_1(\pi, \pi^*) \rightarrow T_1(\pi, \pi^*)$
			$S_1(\pi, \pi^*) \rightarrow T_2(\pi, \pi^*)$
Pyrene	10^6	30	$S_1(\pi, \pi^*) \rightarrow T_1(\pi, \pi^*)$
Triphenylene	5×10^7	20	$S_1(\pi, \pi^*) \rightarrow T_1(\pi, \pi^*)$
9-Acetoanthracene	$\sim 10^{10}$	~ 5	$S_1(\pi, \pi^*) \rightarrow T_2(n, \pi^*)$
Acetone	5×10^8	5	$S_1(n, \pi^*) \rightarrow T_1(n, \pi^*)$
Benzophenone	10^{11}	5	$S_1(n, \pi^*) \rightarrow T_2(\pi, \pi^*)$
Benzil	5×10^8	5	$S_1(n, \pi^*) \rightarrow T_1(n, \pi^*)$
Biacetyl	7×10^8	5	$S_1(n, \pi^*) \rightarrow T_1(n, \pi^*)$

Source: N. J. Turro, *Modern Molecular Photochemistry*, Benjamin-Cummings, Menlo Park, CA, 1978. Adapted from J. W. Birks, in *Chemiluminescence and Photochemical Reaction Detection in Chromatography*, J. W. Birks, ed., VCH, New York, 1989, Chap. 1.

the excited-state species in organic molecules but can be as high as 100% in certain cases. These compounds have excited-singlet-state lifetimes of about 1 ns, placing the measurable k_{ISC} rate constants from 10^5 to $10^8\ s^{-1}$. Triplet-to ground-state relaxation rates are generally slow. Triplet-state lifetimes range from microseconds to tens of seconds (k value from 0.1 to $10^5\ s^{-1}$). ISC rates are enhanced by the presence of heavier atoms (e.g., bromides, iodides, and transition metals). Subsequently, ISC and inverse ISC is expected to quench transition metal complexes rapidly down to the ground electronic state. Triplet-state energies are lower than the associated singlet states, and vibrational relaxation in the triplet is fast. Thus once formed, the triplet cannot cross back to the excited singlet except in rare cases (e.g., certain nitrogen heterocycles such as pyrimidine exhibit inverse ISC).

Rates for ISC may also be justified in terms of the Franck–Condon principle. Potential energy surfaces for the triplet are lower but close in energy to the associated singlet states. It is commonly believed that the triplet state is less bound than the associated singlet, and most pictures of the triplet state place the potential energy surface out at slightly larger nuclear displacements. This results in a potential energy surface crossing between the singlet and associated triplet states. If the triplet-state potential energy surfaces crosses near the lowest vibrational level of the singlet, ISC would be most favorable since singlet molecules would relax to this bottleneck state. Subsequently, excited-state ISC is expected to be relatively rapid even though the transition is not allowed based on the spin overlap integral. On the other hand, the vibrational overlap between T_1 and S_0 is expected to be very small. The same reasoning used to justify the slow S_1–S_0 relaxation applies here as well.

For absorbing systems with rapid excited-state relaxation, increasing the excitation power will result in a proportional increase in luminescence and thermal powers. But the excitation process is dynamic and there may be "traps" to this seemingly basic principle. Triplet states are of lower energy than the associated excited singlet states. Triplet-state vibrational relaxation will occur after intersystem crossing. After relaxation, the triplet state molecule does not have enough energy to cross back into the excited singlet state. Reverse intersystem crossing to the ground singlet or phosphorescence can occur. Of course, triplet-state phosphorescence emission lifetimes are long and molecules that do not cross efficiently to the ground singlet state are metastable.

Metastable states such as the triplet are one of the ways that power absorbed from the optical excitation source can be trapped. This power is not immediately available as either emission or thermal power. Molecules trapped in metastable states may not absorb more light. Thus metastable state production will result in a nonlinear relationship between excitation and emission or thermal powers. If the triplet metastable state lifetime is long compared to the times required for rovibrational equilibration and relaxation, the triplet will exist at the same thermodynamic temperature as that of the sample matrix. In this case the metastable state is essentially a different species.

Time scales for thermal relaxation will depend on the nature of the excited and intermediate states and as well as the sample matrix. A collection of excited-state species may each relax by a different route. Each relaxation step can be thought of as being reversible or at least microscopically reversible. Microscopic reversibility implies that each step of the relaxation cascade occurs under equilibrium conditions. The collection of reversible pseudo-first-order steps have apparent first-order kinetic behavior. Excited-state relaxation rates are very important in pulsed excitation experiments. The rates of energy transfer will control the signal rise times. In continuous excitation conditions, relaxation rates will affect the amount of power that the sample can absorb. Rate-limiting steps (e.g., those describing the relaxation of metastable states) will affect the number of species in the ground state and thus able to absorb radiation.

The addition of an excited-state quencher to the sample matrix can result in enhanced energy transfer when optical bleaching limits the rate of optical energy deposition in the sample. Direct electronic-state quenching is not common but does occur when the sample contains a species that can accept the relatively large electronic-state energy. When added to solutions containing a fluorescent component, quencher species can accept the excited-state energy by a direct intermolecular energy transfer mechanism. Criteria for quenching via this direct mechanism are that the quencher (1) have an excited electronic state near the excited state of the fluorescent compound, (2) be nonfluorescent, and (3) have a short-lived excited state. Excited-state quenching species must be used with caution since their addition may interfere with analyte detection since the energy transfer step must be isoenergetic. Triplet-state quenching can also occur. Several efficient triplet-state quenchers are known and are often used to enhance the power output from continuous dye lasers.

Perhaps the most common excited-state quencher is molecular oxygen. Unless purposefully removed, O_2 will be found in all solvents. Dissolved O_2 has a $^3\Sigma_g^-$ ground electronic state. Triplet quenching produces one of two excited singlets through

$$T_1 + O_2(^3\Sigma_g^-) \rightarrow S_0 + O_2(^1\Sigma_g^+) \text{ or } O_2(^1\Delta_g) \quad (64)$$

T_1 and S_0 are used to indicate the absorbing species. The $^1\Sigma_g^+$ state is about 13,000 cm^{-1} above the ground state and the $^1\Delta_g$ is about 7800 cm^{-1} above the ground state. When formed, the O_2 $^1\Sigma_g^+$ state will undergo rapid internal conversion to the metastable $^1\Delta_g$ state. The singlet oxygen lifetime varies by about two orders of magnitude, depending on the solvent (Birks 1989, Chap. 7). Since the return to the ground state is required to recover the energy transferred to the oxygen, photothermal spectroscopy may best be performed in samples with shorter O_2 $^1\Delta_g$ lifetimes. O_2 $^3\Sigma_g^-$ can also quench singlet states. In the latter, O_2 is promoted to the $^1\Delta_g$ while facilitating internal conversion to S_0 of the excited analyte.

O_2 $^1\Delta_g$ is more reactive than the ground state and may oxidize the analyte or solvent. Birks (1989, Chap. 7) has discussed several high-performance liquid chromatography (HPLC) detection schemes based on singlet oxygen photochemistry. The concentration of O_2 $^1\Delta_g$ can be high in solutions where it is produced by energy transfer from the excited analyte (Table 2.4). Since quenching of the analyte allows further optical excitation, each analyte can produce many of the metastable O_2 $^1\Delta_g$ species. This buildup in concentration will increase the rate of analyte oxidation. Using only moderate laser excitation sources (e.g., ~1 mW), we have observed analyte (dityrosine) oxidation resulting in a 50% decrease in signal after only 2 minutes of irradiation of aqueous solutions. To obtain accurate quantitative results, sample deoxygenation must be performed prior to measurement in cases where oxidation takes place.

Table 2.4. Lifetimes of Singlet Oxygen in Various Solvents

Solvent	τ (μs)	Solvent	τ (μs)
H_2O	2	CH_3CN	30
CH_3OH	7	$CHCl_3$	60
C_2H_5OH	12	CS_2	200
C_6H_{12}	17	$CDCl_3$	300
D_2O	20	C_6F_6	600
C_6H_6	24	CCl_4	700
$(CH_3)_2CO$	26	$CFCl_3$	1000

Source: D. R. Kearns, in *Singlet Oxygen*, H. H. Wasserman and R. W. Murray, eds., Academic Press, New York, 1979, Chap. 4, and K. Gollnick and H. J. Kuhn, in *Singlet Oxygen*, Chap. 8. Adapted from J. W. Birks, in *Chemiluminescence and Photochemical Reaction Detection in Chromatography*, J. W. Birks, ed., VCH, New York, 1989, Chap. 7.

2.4. RELAXATION KINETICS

There are two main approaches to finding the overall relaxation kinetics resulting in heat production. The first is the density matrix approach wherein the relative population in each intermediate state is given along the diagonals of the density matrix (Steinfeld and Houston 1978, Steinfeld et al. 1989). Evolution of the density matrix is found by solving the equation (Stone et al. 1980)

$$\frac{d\rho}{dt} = \left(\frac{d\rho}{dt}\right)_{\text{optical}} + \frac{1}{\tau}(\mathbf{K} - 1)\rho \tag{65}$$

The first term on the right-hand side describes the excitation and emission rates, and the second term gives the kinetic relaxation rates in terms of the mean collision time, τ, and the relaxation matrix, K. The elements of the **K** matrix are the probabilities of relaxation into a state

$$(\mathbf{K}\rho)_{ij} = 0 \qquad i \neq j \tag{66}$$

$$(\mathbf{K}\rho)_{ii} = \sum_j P_{ij}\rho_{jj} \tag{67}$$

where P_{ij} is the probability of a transition from state j to state i. The average energy in an absorber can be obtained from the product of the energy matrix and the density matrix, ρ:

$$\langle E \rangle = \mathrm{Tr}\,\mathbf{W}\rho \tag{68}$$

where $\mathbf{W}$ is a matrix giving the energies of the individual states along the diagonal (e.g., $W_{ii} = E_i$). Similarly, the thermal power can be obtained from the time rate of change of the density matrix. Under steady-state conditions, $d\rho/dt = 0$, and the thermal power is proportional to

$$\Phi_H \propto \frac{1}{\tau}\,\mathrm{Tr}\,\mathbf{W}(\mathbf{K} - 1)\rho \tag{69}$$

Stone and Goodman (1978) and Stone et al. (1980) have used this approach to determine the heat and thermal reaction product production in pulsed infrared excited gas-phase species using energy gap law–based transition probabilities. A similar approach was used by Wood et al. (1969) and by Burak et al. (1969) to describe optical saturation effects in continuously excited gaseous species. Stettler and Witrol (1977) described a similar matrix-based method for describing photoacoustic signals obtained from systems with multiple-level absorption and relaxation.

Although the density matrix approach is appealing in its all-inclusive correctness, it is difficult to implement. The density matrix approach to modeling excited-state relaxation requires either extensive knowledge pertaining to the individual rate steps, or approximations that are difficult to check, owing to the complexity of the model. Exact rate information is rarely known well enough to allow its use. The model is probably best used as a check for validity of the energy gap law and the Fermi golden rule for excited vibrational-state excitation. More information and practical insights can be gained using approximate models that account for only the rate-limiting steps.

The second method used to model the rates of excited-state relaxation is a discrete-state kinetic model of selected states. It is essentially the same as the density matrix model, with the exception that only selected states are accounted for. In this model all but a few rate-limiting relaxation processes are assumed to be very fast compared to the rate-limiting steps. The kinetics of the rate-limiting steps will control the overall rates of heat production by excited-state relaxation. This approach is applicable in condensed phases where vibrational relaxation is extremely fast and metastable state lifetimes are long compared to vibrational states. It apparently works well for electronic transition modeling since excited-state internal conversion and vibrational relaxation steps are extremely fast. The model may also be applied to gas-phase samples if the rates of high rovibrational-level relaxation are fast compared to a few bottleneck transitions. In particular, the discrete-state model should adequately describe relaxation of the low-energy vibrational excited state when the density of states is low (e.g., for small molecules).

In the selected state model, the power liberated at each of the fast and slow rate steps must be accounted for carefully. The relative fraction of molecules relaxing by a particular route is given by a quantum efficiency or power yield. Each relaxation route will have associated rates and heat releases. In the organic molecule example, there may be three separate relaxation routes. That fraction of the excited singlet states that relax directly to the ground singlet state via intersystem crossing will release an amount of heat equal to the photon energy. The quantum yield for direct relaxation can be defined as ϕ_H. That fraction fluorescing is given by the fluorescence quantum efficiency, ϕ_F, and will release heat equal to the difference between the absorbed and emitted photons. That fraction crossing to the triplet state is given by the quantum efficiency for intersystem crossing is ϕ_{ISC}. Triplet-state relaxation will be slow and the heat will be liberated in a series of fast and slow steps. Each major relaxation route has an associated characteristic relaxation-time constant and an average energy transferred to the sample. A series of rate expressions describing the relative populations in the rate-limiting or bottleneck states are set up and solved under the irradiation condition. Guiliano and Hess (1967) and deVries and Wiersma (1980) have used this approach to model optical saturation due to trap states in organic dyes under pulsed and continuous conditions. The rate equations can generally be solved for continuous excitation because of the steady-state condition. With pulsed excitation the steady-state approximation may not be valid and time-dependent solutions may have to be found.

Using either method results in an equation that will describe the time-dependent characteristics of the heating. If we consider a system that is in an excited state or collection of excited states with a total excess energy of Q (J), the thermal energy deposited in the sample upon relaxation will be given by a kinetic relaxation function

$$Q_H(t) = QR(t) \tag{70}$$

$Q_H(t)$ is the kinetic-controlled time-dependent heat and the function $R(t)$ is a relaxation function that specifies the time dependence of the heat production. The later is obtained from the time-dependent concentrations. In the case of a series of first-order steps, the heat produced is

$$Q_H(t) = Q \sum_{s=1}^{S} Y_s(1 - e^{-k_s t}) \tag{71}$$

The heat is produced by a collection of relaxation steps, Y_s is the yield of relaxation step s, and k_s is the relaxation rate constant for that particular step. The rate of heat production is $R'(t) = -dR(t)/dt$ which we will call the kinetic relaxation function. In this case $R'(t)$ is given by

$$R'(t) = \sum_{s=1}^{S} Y_s k_s e^{-k_s t} \tag{72}$$

The first-order kinetic relaxation function is a system response function, and systems response theory can be used to determine the time and frequency response of a first-order system (Poularikas and Seely 1985). This theory is useful for determining both the time-dependent evolution of heat in the case of pulsed excitation and the frequency response of the heat production using modulated excitation sources. The relaxation function gives the power response to an impulse excitation, resulting in an amount of energy Q in the excited states. For the pulsed excitation source the time-dependent thermal power is simply

$$\Phi_H(t) = QR'(t) \tag{73}$$

Time-dependent excitation will produce a time-dependent power of $\Phi(t)$ in the excited state. Assuming that the response is linear (i.e., there is a linear relationship between excitation and response), the time-dependent power is then the convolution of the $\Phi(t)$ with $R'(t)$:

$$\Phi_H(t) = \int_0^t \Phi(t')R'(t - t')\,dt' = \Phi(t) * R'(t) \tag{74}$$

Here $*$ is used to indicate convolution. For continuous excitation, the time-dependent power is given by

$$\Phi_H(t) = \Phi \int_0^t R'(t')\,dt' = \Phi R(t) \tag{75}$$

where ϕ is the power of the continuous excitation source. For times much longer than required for the slowest relaxation,

$$\Phi_H(t) = \Phi \sum_{s=1}^{S} Y_s = \Phi Y_H \tag{76}$$

In the frequency domain, the response of the sample heating is obtained from the product of the frequency-dependent sample heat with the Fourier transform of the kinetic relaxation function

$$\Phi_H(\omega) = \Phi(\omega)R'(\omega) \tag{77}$$

where ω is the angular frequency, $2\pi f$. For practical purposes, $R'(\omega)$ may be obtained from the Laplace transform

$$R'(\omega) = e^{i\omega t}|\mathscr{L}\{R'(t)\}|_{\zeta=i\omega} \tag{78}$$

where $\mathscr{L}$ is used to indicate the Laplace transform with a transform variable of ζ. Other techniques for determining the periodic kinetic response based on matrix methods are given by Stettler and Witrol (1977) and Steinfeld and Houston (1978).

Although useful in many instances, the system response approach must be used with caution. This approach works only if the relaxation is linear and first-order. This is true in most cases. However, there are several cases where the sample response is not a linear function of excitation power. When this is the case, rate expressions have to be "coupled" to the excitation in a more complex manor. Examples of these situations are discussed below.

2.5. NONLINEAR ABSORPTION

Nonlinear absorption spectroscopy can yield information about atomic and molecular structure and dynamics not able to be obtained using conventional spectroscopic techniques. But in addition to the beneficial aspects, nonlinear absorptions measurements can be plagued with experimental artifacts that are difficult to describe and characterize (Burak et al. 1969, Wood et al. 1969). There are several causes of experimental artifacts. First, optical power delivered to the sample lasers can result in significant sample heating. This heating results in a change in energy-level populations and can also cause photothermal lensing. Also, the attenuation varies along the sample pathlength for highly absorbing samples with nonlinear absorption. This gives rise to complicated nonlinear irradiance-dependent absorption behavior. This problem is further complicated if the excitation source has a spatially varying beam profile.

There are several nonlinear absorption processes now known to occur in photothermal spectroscopy when high-irradiance excitation lasers are used to irradiate the sample. Nonlinear absorption is a phenomenon wherein the amount the energy or power absorbed by a sample depends on the irradiance of the source used to excite the sample. Nonlinear absorption can complicate signal interpretation for quantitative analysis. However, these phenomena are of themselves interesting and can yield information of use to the analyst.

Nonlinear optical absorption is common in pulsed laser–excited photothermal spectroscopy. Twarowski and Kliger (1977a,b) realized the importance of pulsed laser–excited photothermal lens spectroscopy for measuring multiphoton transitions. The result of their study showed that multiphoton absorption can enhance the photothermal lens signal by narrowing spatial temperature change profile. Long and Bialkowski (1984, 1985) and Bialkowski and Long (1987) have examined some of the implications of optical saturation in gas-phase infrared photothermal spectroscopy. Poston and Harris (1990) estimated saturation irradiances, energy transfer kinetics, and enthalpies of triplet benzophenone in condensed-phase samples by examining both time-dependent photothermal deflection signals over a range of excitation irradiances. Quantitative information on condensed-phase samples has also come from photothermal-wave mixing techniques. McGraw et al. (1987) and Zhu and Harris (1988, 1898) have examined nonlinear absorption in condensed-phase species using thermal gratings.

There are two main types of nonlinear absorption (Figure 2.5). In the first type, the total amount of energy or power absorbed by the sample increases

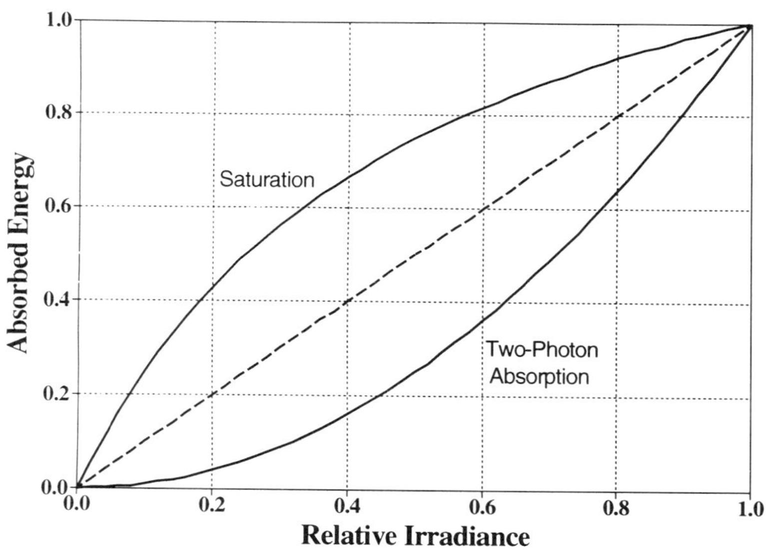

Figure 2.5. Nonlinear irradiance-dependent effects on absorbed energy. In optical saturation the absorption coefficient decreases with increasing irradiance. With multiple-photon absorption, the absorption increases with irradiance. The dashed line shows linear absorption behavior.

with increasing excitation irradiance. This type of behavior is caused by either multiphoton or consecutive multiple photon absorption by the sample. In effect, the absorption coefficient increases with increasing irradiance. The second type of nonlinear absorption effect causes the relative energy or power absorbed by the sample to decrease with increasing irradiance. The absorption coefficient decreases relative to the low-irradiance, or small-signal, absorption coefficient in this case. Decreased absorption can be due either to stimulated emission of the excited state of the optical transition or by depletion of species from states that participate in light-induced transitions.

The first nonlinear absorption effect exploited in photothermal spectroscopy was that of the multiphoton absorption effect. Twarowski and Kliger (1977a,b) developed a general theory for pulsed laser–excited photothermal lens spectroscopy that includes multiphoton absorption effects. This was the first comprehensive theory describing pulsed laser–excited photothermal lens spectroscopy. In fact, this theory and the multiphoton absorption application preceded analytical application by several years. In multiphoton absorption, more than one photon is absorbed per optical transition. A related effect, multiple-photon absorption, results in absorption of more than one photon per molecule, but the absorptions occur consecutively instead of simultaneously as in the case of multiphoton absorption. In either case, the amount of energy or power absorbed by the sample increases with increasing excitation irradiance.

Optical saturation, or optical bleaching, was discovered shortly after the advent of analytical applications of photothermal spectroscopy using pulsed laser excitation sources to measure concentrations of single-photon absorbers. The theory for optical saturation and bleaching is more complicated than that of multiphoton absorption in that it is sensitive to the excitation and relaxation rates of the excited species. Optical saturation and bleaching limit the amount or energy or power absorbed by the sample by either "saturating" the upper state of the optically induced transition or by removing molecules from the optically induced transition states by metastable state formation. In either case the amount of energy or power absorbed by the sample decreases with increasing excitation irradiance.

2.5.1. Multiphoton Absorption

Multiphoton absorption occurs when two or more quanta of electromagnetic radiation are absorbed from the source simultaneously. This process is often pictured as one where the species passes through an intermediate "virtual" state. The virtual state exists at energies resonant with the excitation source wavelength. The virtual state is really a manifestation of the uncertainly principle. This state is constructed, at least in theory, as a superposition of all real states. The multiphoton transition can be found through high-order perturbation theory. First-order perturbation theory gives one-photon interaction, second-order perturbation yields terms that describe two-photon effects, and n-order theory describes n-photon effects.

The general n-photon transition operator can be expressed as

$$a_{ge}^{n}(t) = \left(\frac{1}{\hbar}\right)^{n} \sum_{j_1, j_2 \ldots j_n}^{\infty} \frac{\langle g|H'|j_1\rangle\langle j_1|H'|j_2\rangle}{(\omega_{gj_1} - \omega)(\omega_{gj_2} - \omega)} \cdots \frac{\langle j_{n-2}|H'|j_{n-1}\rangle\langle j_{n-1}|H'|e\rangle}{\omega_{gj_{n-1}} - (n-1)\omega} \frac{\exp[i(\omega_{\gtrless} - n\omega)t] - 1}{\omega_{ge} - n\omega} \tag{79}$$

where $a_{ge}^{n}(t)$ is the time-dependent probability amplitude for the transition, the ω_{j1j2} are the angular frequencies associated with the energy differences between levels, $\hbar$ is Planck's constant $h/2\pi$, and H' is the perturbation Hamiltonian. For dipole selection rules, $H' = -\frac{1}{2}\mathbf{dE}$, $\mathbf{d}$ being the dipole transition operator and $\mathbf{E}$ the electric field strength. The only real states are the ground (g) and excited (e) states. The summation over the j_m indices includes all states, including the ground and excited states of the transition. Summations over the j_m of terms such as $|j_m\rangle\langle j_m|$ constitute the virtual states. Although the probability of transition to any of the real states in the summation is very low owing to the off-resonance character of the transition, the sum over all of these low-probability transitions results in a finite probability of making a transition through the virtual state.

For single-wavelength excitation, the transition rate is proportional to the square of the transition probability and the n-photon absorption coefficient is the rate divided by the excitation irradiance raised to a power equal to the number of photons absorbed. For simplicity, the summation terms and states can be cast in terms of a n-photon tensor operator, O^n. The n-photon absorption cross section is proportional to

$$\sigma_{ge}^{n} \propto |\langle e|O^n|g\rangle|^2 \tag{80}$$

The tensor operator can be analyzed for the selection rules associated with the multiphoton transition (McClain and Harris 1978, Chen and Yeung 1978, Halpern et al. 1980, Bialkowski and Guillory 1981). As a general rule, the n-photon selection rules can be obtained from n repeated applications of the single-photon dipole transition rules. However, if one particular state is near a virtual state, the selection rules will be dominated by transitions into and out of that particular state (McClain and Harris 1978, Halpern et al. 1980). The power absorbed per unit volume by the multiphoton absorber is

$$\delta P = N\sigma_{ge}^{n} E_0^{n} \tag{81}$$

where N is the number density of absorbing species, $\delta P\,(\mathrm{W\,m^{-3}})$ is the power density change, and E_0 $(\mathrm{W\ m^{-2}})$ is the irradiance. Since the absorbed energy is proportional to the irradiance raised to the number of photons required for the transition, multiphoton transitions are enhanced by either (1) increasing the energy of the excitation source, (2) decreasing the excitation pulse width at constant average power, and (3) decreasing the area of sample excitation by focusing the source into the sample.

Multiphoton absorption coefficients are many orders of magnitude smaller than those of direct, one-photon absorption because of the low transition probability through virtual states. Twarowski and Kliger (1977b) report an absorption cross section for benzene of only $10^{-59}\,\mathrm{m^4\,molec^{-1}\,s\,photon^{-1}}$. Continuous laser excitation sources generally do not have irradiances high enough to induce multiphoton absorptions. However, the high pulse energy and spatial coherence of pulsed dye and infrared molecular lasers allows the irradiance to be increased to levels high enough to induce multiphoton transitions.

2.5.2. Optical Saturation of Two-Level Transitions

Optical saturation occurs when the excited state is populated by a significant amount. When this occurs, stimulated emission competes with absorption. Stimulated emission results in light propagating in the same direction as the stimulating light. In transmission spectroscopy, optical saturation results in transmission that increased with irradiance. In photothermal spectroscopy, saturation limits the amount of energy able to be absorbed by the sample.

Optical saturation is best understood by examining the rate expressions for a two-level system. The two levels can be broadened by both homogeneous and inhomogeneous processes, resulting in generalization to a multilevel system with two main levels. Thus the two-level system can be used to model molecular electronic transitions where the electronic states are the two levels and rovibrational structure is considered to be an inhomogeneous line broadening. It can also be used to describe vibrational excitation in which the two levels are vibrational states and the rotational-translational states comprise the inhomogeneous line.

A two-level system with nondegenerate states has the kinetic rate expressions

$$\frac{dN_l(t)}{dt} = E_P\sigma(\nu)(N_u(t) - N_l(t)) + kN_u(t)$$
$$\frac{dN_u(t)}{dt} = E_p\sigma(\nu)(N_l(t) - N_u(t)) - kN_u(t) \tag{82}$$

where N_l and N_u (m^{-3}) are the number densities in the lower and upper states, E_p (photons m^{-2}) the photon irradiance at frequency ν (s^{-1}), $\sigma(\nu)$ (m^{-2}) the absorption cross section, and k (s^{-1}) the kinetic rate constant for decay out of the upper level. The lower-state rate expression consists of three terms: the loss due to absorption, $-E_p\sigma N_l$; the gain due to stimulated emission, $E_p\sigma N_u$; and the gain due to fluorescence or other first-order mechanisms returning the upper-state species to the lower-state kN_u.

Optical saturation occurs when the stimulated emission rate competes with that for relaxation. In the limit where the dephasing time (T_2) is much shorter than either optical pumping or relaxation times (T_1), the incoherent system can be treated by coupled rate expressions (Steinfeld 1985). The net result is that the optical absorption coefficient can be recast in a form that is dependent on the strength of the excitation source. The absorption does not occur only at the frequency, $h\nu_0$, which is equal to the energy difference between the two states. The optical frequency-dependent absorption coefficient is given by the convolution of the homogeneous linewidth with the inhomogeneous bandshape

$$\alpha(\nu) = \alpha_0(\nu_0) \int_0^{\infty} \rho(\nu)g(\nu_0 - \nu)\, d\nu \tag{83}$$

where $\alpha_0(\nu_0)$ (m^{-1}) is the small-signal absorption coefficient at the homogeneous line band center, $\rho(\nu)$ the normalized inhomogeneous line shape function, and $g(\nu_0 - \nu)$ the Lorentzian line shape function:

$$g(\nu_0 - \nu) = \frac{1/(2\pi\tau)^2}{(\nu_0 - \nu)^2 + (1 + E/E_S)/(2\pi\tau)^2} \tag{84}$$

The homogeneous linewidth has contributions from both relaxation time, τ (s), and power broadening, $(1 + E/E_S)$, where E (W m^{-2}), is the irradiance and E_S ($W\,m^{-2}$), the saturation irradiance, is defined by

$$E_S = \frac{h\nu_0}{2\sigma(\nu_0)\tau} \tag{85}$$

The absorption cross section is related to the small-signal absorption coefficient by

$$\alpha_0(\nu) = (N_l - N_u)\sigma(\nu) \tag{86}$$

where $N_l - N_u$ (m^{-3}) is the number density difference between the lower and upper levels of the transition. Relaxation times used to define the saturation irradiance generally include both T_1 and T_2 processes. However, the rate equations are strictly valid only when $T_2 \leqslant T_1$.

The absorption coefficient equation can be integrated for two important limits. For gas-phase molecules with rovibrational transition spacing wider than the homogeneous linewidth, the inhomogeneous line shape may be approximated as a delta function at band center, ν_0. The exponential absorption coefficient for the homogeneous line is just $\alpha(\nu) = \alpha_0(\nu_0)g(\nu - \nu_0)$ and

$$\alpha(\nu) = \frac{\alpha_0(\nu)}{1 + E/E_S(\nu)} \tag{87}$$

where $\alpha_0(\nu) = \alpha_0(\nu_0)/[1 + \delta(\nu)]$, $E_S(\nu) = E_S[1 + \delta(\nu)]$, and $\delta(\nu) = [2\pi\tau(\nu - \nu_0)]^2$.

Large molecules have rovibrational transition spacings which are narrower than the homogeneous linewidth and the absorption is inhomogeneously broadened. By definition, inhomogeneous line broadening means that the molecules exist in several, unconnected states. If the time scale for nonresonant state relaxation into the resonant state is shorter than that of excitation to the excited state, every state in the inhomogeneously broadened transition is in effect coupled. This results in depletion of the entire population of lower states by excitation through the resonant state. In other words, the transition is effectively homogeneously broadened since all states are connected on a time scale shorter than that of the excitation. Thus even though the absorption linewidth is much greater than the excitation line, the vibrational relaxation into states depleted by the saturating excitation source is so rapid that saturation appears to be homogeneous. This is a common situation in solution-phase samples (Giuliano and Hess 1967) and probably occurs in gas-phase samples of moderate pressure (Wood et al. 1969).

On the other hand, if the time scales for relaxation into the levels excited by the laser are longer than that of excitation, the effects of the inhomogeneously broadened transition on the effective saturation absorption coefficient must be taken into account. Thus the second saturation limit is of inhomogeneously broadened transitions with relatively long backgroundstate relaxation times. For large molecules with a wide inhomogeneous absorption, the line shape function is essentially constant over the homogeneous linewidth excited by the narrowband laser. In this case the irradiance-dependent absorption coefficient is

$$\alpha(\nu) = \frac{\alpha_0(\nu)}{(1 + E/E_S)^{1/2}} \tag{88}$$

where the inhomogeneous band center absorption coefficient is $\alpha_0(\nu) = \alpha_0(\nu_0)\rho(\nu)/4\tau$. This type of saturation, called inhomogeneous saturation, has been found to occur in semiconductors (Keilmann 1976).

The inhomogeneous absorption coefficient does not decrease as rapidly with irradiance as the homogeneous one because of the square-root power dependence in the denominator. This effect is due to the increased number of species that can be excited at high irradiances because of power broadening. Thus, although saturation may occur between states with energy differences at or near the excitation energy (i.e., homogeneous saturation), the total number of species excited increases as a result of power broadening of the excitation source linewidth. Inhomogeneous broadening is more likely at lower gas pressures since rotational relaxation is limited by collision rates.

2.5.3. Optical Bleaching

Optical bleaching is similar to optical saturation in that the absorption coefficient decreases with increasing excitation irradiance. In optical saturation, the population in the excited state increases to the point and the rate of stimulated emission becomes competitive with the rate of absorption. In optical bleaching, the decrease in absorption is due to a decrease in the number density of both the ground state and the state being excited. When the ground state is depleted, the amount of energy capable of being absorbed by the sample diminishes. Ground- and excited-state depletion normally occurs by metastable state production. Bleaching occurs when the kinetic rate of return of metastable states to the ground state is slower than the rate of metastable state production. Another mechanism of bleaching is photolysis. Photolysis of the absorbing species results in a decrease in the number density of the absorbing species, and thus the absorption decreases.

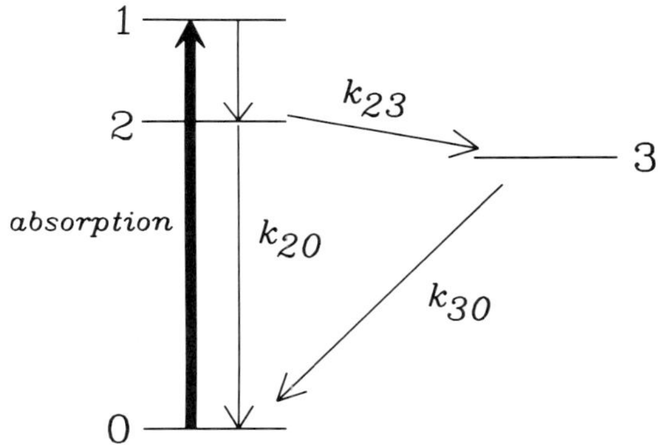

Figure 2.6. Three-level energy diagram used to describe kinetic effects of optical bleaching.

Optical bleaching can be seen in a simple three-level model. A three-level system is illustrated in Figure 2.6. This model has been found to be useful for optical bleaching in infrared laser–excited gas-phase species and for organic species that have near-unity triplet quantum yield. In the former the levels correspond to vibrational levels in the molecule, and in the latter they correspond to electronic states. Levels 0 and 1 are the ground and optically excited states. The optically excited state relaxes reversibly only to level 2. Level 2 is a metastable state if the lifetime is longer than that of the photoexcited ground state. The rate expressions for this system are

$$\begin{aligned}\frac{dN_0(t)}{dt} &= -E_P\sigma_{01}N_0(t) + E_P\sigma_{10}N_1(t) + k_{20}N_2(t)\\ \frac{dN_1(t)}{dt} &= E_P\sigma_{01}N_0(t) - E_P\sigma_{10}N_1(t) - k_{12}N_1(t) + k_{21}N_2(t) \quad (89)\\ \frac{dN_2(t)}{dt} &= k_{12}N_1(t) - (k_{20} + k_{21})N_2(t)\end{aligned}$$

The N_i are the number densities of absorbing species in level i, k_{ij} the first-order rate constants for transitions from level i to j, σ_{ij} the absorption cross sections, $E_p = E/h\nu$ (photons m^{-1}) the excitation photon irradiance, and $h\nu$ the excitation energy.

In vibrational excitation experiments, level 0 is the ground state, level 1 the vibrational state being excited, and level 2 represents a collection of excited vibrational-state levels. These levels collectively relax to the ground state. In the unit ISC quantum yield molecule, level 1 is any excited singlet state and level 2 represents the metastable triplet state. This case may be modeled better by not allowing stimulated emission because of the rapid vibrational relaxation in the excited triplet state. This system of equations can be solved for either the steady-state response or the dynamic system response. The steady-state solution is obtained by setting the time rate of number density changes equal to zero. The steady-state approximation is valid for continuous excitation and when ation rates are faster than the rates of excitation and stimulated emission. The transition cross sections are related to each other through the level degeneracies, $g_0\sigma_{01} = g_1\sigma_{10}$. For equal degeneracies, $\sigma_{01} = \sigma_{10}$. In this case the steady-state populations are

$$
\begin{aligned}
N_0 &= N_{\text{tot}} \frac{k_{12}k_{20} + (k_{21} + k_{20})E_P\sigma_{01}}{E_p\sigma_{01}(k_{12} + 2k_{20} + 2k_{21}) + k_{12}k_{20}} \\
N_1 &= N_{\text{tot}} \frac{(k_{20} + k_{21})E_P\sigma_{01}}{E_P\sigma_{01}(k_{21} + 2k_{20} + 2k_{21}) + k_{12}k_{20}} \\
N_2 &= N_{\text{tot}} \frac{k_{12}E_P\sigma_{01}}{E_p\sigma_{01}(k_{12} + 2k_{20} + 2k_{21}) + k_{12}k_{20}}
\end{aligned}
\tag{90}
$$

where N_{tot} is the total number density of the absorber. The absorption coefficient is found from

$$\alpha(\nu) = \sigma_{01}(N_0 - N_1) \tag{91}$$

By defining the effective saturation irradiance, E_S, due to optical bleaching as

$$E_S = \frac{h\nu}{2\sigma_{01}\tau_{\text{effective}}} \qquad \tau_{\text{effective}} = \frac{k_{12} + 2k_{20} + 2k_{21}}{2k_{12}k_{20}} \tag{92}$$

an absorption coefficient similar to that for optical saturation can be defined:

$$\alpha = \frac{\alpha_0}{1 + E/E_S} \tag{93}$$

Compared to the result obtained for two-level saturation, it is clear that optical bleaching exhibits irradiance-dependent behavior that is equivalent to that of optical saturation using this model.

The optical bleaching solution was obtained without regard for the linewidth of the transition. The implicit assumption is that the T_2 lifetime is very much shorter than the kinetic or stimulated emission lifetime of the states. This assumption should hold in all but the most extreme cases, or high-irradiance and long dephasing times. Thus E_S is dependent on ν only through absorption cross-section dependence. In this case the homogeneous linewidth is

$$g(\nu_0 - \nu) = \frac{1/(2\pi\tau)^2}{(\nu_0 - \nu)^2 + 1/(2\pi\tau)^2} \tag{94}$$

where the τ value is that of the optically excited state. This τ is generally much shorter than that due to kinetic decay of the excited state; in fact, it is on the order of the T_2 time. Since optical bleaching has no effect on the homogeneous linewidth, convolution with the inhomogeneous linewidth will also be irradiance independent. Since optical bleaching results in a depletion of states, the hole burned in the inhomogeneous line will be

$$\frac{\rho(\nu)g\,(\nu_0 - \nu)}{1 + g(\nu_0 - \nu)E/E_S} \tag{95}$$

For a sufficiently broad inhomogeneous width, integration results in the irradiance-dependent absorption coefficient

$$\alpha(\nu) = \frac{\alpha_0(\nu_0)\rho(\nu_0)}{4\pi\tau} \frac{1}{\sqrt{1 + E/E_S}} \tag{96}$$

where the τ is that of the optically excited state and E_S has been redefined as $E_S/4\pi^2\tau^2$. As with the case of optical saturation, inhomogeneous optical bleaching results in an inverse square-root irradiance dependence. Thus although the processes are entirely different, the effect of optical bleaching is equivalent to that of optical saturation in terms of the phenomena observed.

2.5.4. Response Times During Optical Bleaching

The steady-state solution is valid for continuous excitation, after the time required to reach steady state. The time required to reach steady state is a function of the relaxation rate constants and the irradiance. To determine the steady-state time, the time-dependent solution for the rate expressions must be obtained. A time-dependent solution for the rate equations can be obtained for continuous excitation. Excitation starts at $t = 0$ and is continuous up to any time t. This is called step excitation. Solutions to the rate equations for initial conditions such that populations in all levels other than the ground state are zero are

$$N_0(t) = N_{\text{tot}} \frac{E_P\sigma_{01}(k_{20} + k_{21}) + k_{12}k_{20} + E_P\sigma_{01}(k_{12} + k_{20} + k_{12})e^{-\kappa t/2}}{k_{12}k_{20}(1 + E/E_S)}$$

$$N_1(t) = N_{\text{tot}} k_{12} E_P \sigma_{01} \frac{1 - e^{-\kappa t/2}}{k_{12}k_{20}(1 + E/E_S)} \tag{97}$$

$$N_2(t) = N_{\text{tot}}(k_{21} + k_{20})E_P\sigma_{01} \frac{1 - e^{-\kappa t/2}}{k_{12}k_{20}(1 + E/E_S)}$$

where

$$\kappa = -\sqrt{[2E_P\sigma_{01} - (k_{21} + k_{20})]^2 + k_{12}(k_{12} - 2k_{20} + 2k_{21}} + 2E_P\sigma_{01} + k_{12} + k_{20} + k_{21} \tag{98}$$

The solution for the absorption coefficient is

$$\alpha = \frac{\alpha_0}{1 + E/E_S}\left(1 + \frac{E}{E_S}\, e^{-\kappa t/2}\right) \tag{99}$$

Optical bleaching offers an interesting way to probe relaxation kinetics. Since E_S is a function of the relaxation rate constants, the rate constants may be determined from the irradiance-dependent absorption coefficient. In fact, more rate information can be obtained from optical bleaching experiments than from the more conventional excited relaxation experiments. This is because the absorbing species is driven by the excitation source. The driven system may be limited by fast relaxation steps as well as by slow ones, whereas the response in impulse relaxation experiments is dominated by the slowest relaxation step.

The steady-state condition is of some consequence when using pulsed lasers for sample excitation. It is convenient to use the steady-state approximation solutions since they yield easily interpretable irradiance-dependent results. The general rule is that the steady-state approximation can be used if the time-dependent excitation irradiance varies on a time scale that is much greater than the time constant of the system. A pulsed excitation source can be thought of as being composed of sequential portions. Each portion is at a constant irradiance. Thus the system will respond to a portion as if it were a step excitation. To be valid, the system response time for the irradiance of each portion has to be shorter than the duration of that portion for the steady-state approximation solution.

For step excitation of the three-level system discussed above, steady state will be reached at times much greater than $2/\kappa$. Although the expression is complicated by the square root, the two limiting response times can be found. For irradiances much less than that required for optical bleaching, the system response can be found from

$$\lim_{E_P \to 0} \kappa = k_{12} + k_{20} + k_{21} - \sqrt{(-k_{12} + k_{20} + k_{21})^2 + 4k_{12}k_{21}} \quad (100)$$

This expression can be further reduced to when there is a rate-limiting step or rate bottleneck in the relaxation. For example, if the rate-limiting step is given by the rate of level 2 to level 0 relaxation, k_{20} can be neglected in the square term of the square root and

$$\lim_{k_{20} \to 0} \kappa = k_{20} \quad (101)$$

Any relaxation (e.g., downward), rate-limiting step will most strongly influence κ. As the irradiance increases, the rate of excitation becomes competitive with that due to relaxation. Increasing the irradiance lengthens the response time of the dynamic system. In fact, at very high irradiances, the κ is zero and the system never reaches steady state. The minimum response time of the system is at least that of the longest kinetic relaxation time. For example, it was shown above that κ reduces to the rate-limiting rate constant if the difference between it and all other rate constants is large. In addition, the response time *increases* with excitation irradiance.

2.5.5. Optical Bleaching of Organic Dyes

A three-level model may also be used to describe optical bleaching and time-dependent heat generation in organic molecules. This model was described by Guiliano and Hess (1967) and has been tested quite extensively by Harris and co-workers (McGraw et al. 1987; Zhu and Harris 1988, 1989; Poston and Harris 1990; Zhu et al. 1992). This model is illustrated in Figure 2.7, where the ground state, labeled 0, is excited to a vibrationally excited singlet state, labeled level 1. This can be the first excited singlet or a higher singlet state. The optically excited state is extremely short lived. It relaxes rapidly to lower vibrational states of the first excited singlet, level 2. The excited singlet can relax to the ground state or can undergo intersystem

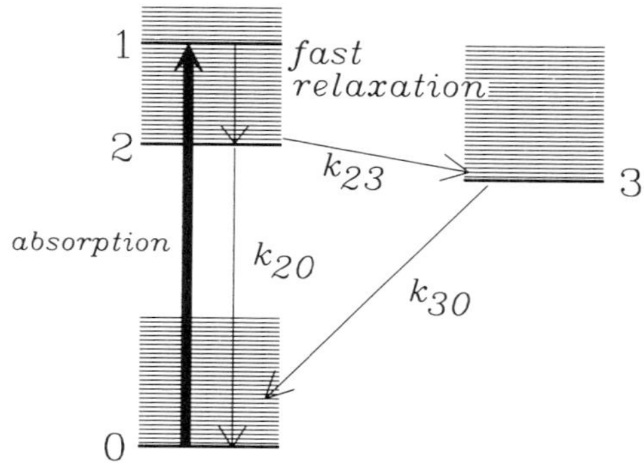

Figure 2.7. Three-level energy diagram used for a kinetic description of many organic dye molecules. Level 0 is the ground singlet state; level 2 is the ground vibrational level of the first excited singlet; level 1 is the state excited by optical irradiation, which may be an excited singlet or an excited vibrational state of the lowest excited singlet; and level 3 is the lowest triplet state of the species.

crossing to form the triplet state, level 3. The triplet then relaxes to the ground state. In this model all the relaxations are nonreversible owing to large energy gaps. There are two rate-limiting steps: the excited singlet state and the triplet-state relaxation to the ground state. Guiliano and Hess give convincing arguments for the validity of this model and claim that the kinetic equations should model most organic dye compounds with only minor modifications.

The rate expressions that describe this model are simply

$$\frac{dN_0(t)}{dt} = -E_P\sigma_{01}N_0(t) + k_{20}N_2(t) + k_{30}N_3(t)$$

$$\frac{dN_2(t)}{dt} = E_P\sigma_{01}N_0(t) - (k_{20} + k_{23})N_2(t) \qquad (102)$$

$$\frac{dN_3(t)}{dt} = k_{23}N_2(t) - k_{30}N_3(t)$$

where k_{20} is the sum of fluorescent, k_f, and nonradiative, k_{nr}, relaxation rate constants, and k_{30} is the sum of the phosphorescence, k_p, and nonradiative triplet state relaxation, k_{nr}, rate constants. The extremely rapid optically excited singlet relaxation does not allow for stimulated emission. Assuming that populations in all excited states are initially zero, the time-dependent solutions for step excitation are

$$N_0(t) = N_{\text{tot}} - N_2(t) - N_3(t)$$

$$N_2(t) = N_{\text{tot}} E_P \sigma_{01} \times \frac{k_{30}R(2 - e^{-at} - e^{-bt}) - [k_{30}(k_{20} + k_{23} - k_{30} + E_P\sigma_{01}) + 2k_{23}E_P\sigma_{01}](e^{-at} - e^{-bt})}{2R[(k_{23} + k_{30})E_P\sigma_{01} + (k_{20} + k_{23})k_{30}]}$$

$$N_3(t) = N_{\text{tot}} k_{23} E_P \sigma_{01} \frac{R(2 - e^{-at} - e^{-bt}) + (k_{20} + k_{23} + k_{30} + E_P\sigma_{01})(e^{-at} - e^{-bt})}{2R[(k_{23} + k_{30})E_P\sigma_{01} + (k_{20} + k_{23})k_{30}]} \tag{103}$$

where

$$R = \sqrt{E_P^2\sigma_{01}^2 + 2E_P\sigma_{01}(k_{20} - k_{23} - k_{30}) - 2k_{30}(k_{20} + k_{23}) + (k_{20} + k_{23})^2}$$

$$a = \tfrac{1}{2}(E_P\sigma_{01} + R + k_{20} + k_{23} + k_{30}) \tag{104}$$

$$b = \tfrac{1}{2}(E_P\sigma_{01} - R + k_{20} + k_{23} + k_{30})$$

These time-dependent solutions show optical bleaching behavior. The denominator can be rearranged:

$$N_2(t) = N_{\text{tot}} E_P \sigma_{01} k_{30}(k_{20} + k_{23}) \times \frac{k_{30}R(2 - e^{-at} - e^{-bt}) - [k_{30}(k_{20} + k_{23} + k_{30} + E_P\sigma_{01}) + 2k_{23}E_P\sigma_{01}](e^{-at} - e^{-bt})}{2R(1 + E/E_S)}$$

$$N_3(t) = N_{\text{tot}} E_P \sigma_{01} k_{23} k_{30}(k_{20} + k_{23}) \times \frac{R(2 - e^{-at} - e^{-bt}) + (k_{20} + k_{23} + k_{30} + E_P\sigma_{01})(e^{-at} - e^{-bt})}{2R(1 + E/E_S)} \tag{105}$$

$$E_S = \frac{h\nu}{2\sigma\tau_{\text{effective}}} \qquad \tau_{\text{effective}} = \frac{k_{23} + k_{30}}{2k_{30}(k_{20} + k_{23})}$$

The system has two response times. The slower of the two is 1/b. The steady-state solution can be used if the excitation irradiance does not change significantly over times greater than the slower response time (Guiliano and Hess 1967). Organic dye molecules may have absorption cross sections of $\sigma_{01} \sim 10^{-20}\,\text{m}^2$, excited singlet relaxation rate constants of $k_{20} \sim 10^9\,\text{s}^{-1}$, intersystem crossing rates of $k_{23} \sim 10^8\,s^{-1}$ (for $\phi_{\text{ISC}} = 0.1$), and triplet relaxation constants of $k_{30} \sim 10^6\,\text{s}^{-1}$ or less. Using these typical constants, $\tau \sim 46$ ns and $E_S \approx 5 \times 10^8\,\text{W m}^{-2}$. This might seem like a high irradiance values, but it is easily reached with conventional pulsed lasers. A typical N_2

laser pumped dye laser delivers a 10-μJ pulse in 5 ns. If this laser is focused to a diameter of 100 μm, the irradiance in on the order of 10^{11} W m^{-2}. This excitation will clearly deplete the ground state. Moreover, the steady-state approximation, obtained from the $t \to \infty$ behavior of the expressions given above, cannot be used since the effective relaxation time is on the order of the pulse duration, and the overall rate of relaxation, $1/\tau_{\text{effective}}$, is slower than the rate of excitation, $E_P\sigma_{01}$.

A worse case situation may be typified by benzophenone. This molecule has a unit ISC quantum yield and the rate constant for triplet relaxation to the ground state can be rather small in certain solvents, $k_{30} \sim 25\ \text{s}^{-1}$ in CCl_4 (Poston and Harris 1990). The absorption cross section is about 4×10^{-23} at 322 nm. For this species the effective saturation irradiance is $E_S \sim 4 \times 10^5$ W m^{-2}. This irradiance can be exceded by a continuous 10-mW ultraviolet HeCd laser focused to a beam waist of 100 μm.

2.5.6. Relaxation for Impulse Excitation

This organic dye model can be used to illustrate the heating rate discussed in Section 2.3. Relaxation of the excited species is easily described. For impulse excitation, with a pulse duration short enough to prevent inter-system crossing, the decay of excited singlet and triplet states is described by

$$N_2(t) = N_2(0)e^{-(k_{20}+k_{23})t}$$
$$N_3(t) = N_2(0)k_{23}\frac{e^{-k_{30}t} - e^{-(k_{20}+k_{23})t}}{k_{20}+k_{23}-k_{30}} \tag{106}$$

Sample heating will occur by three primary means. First, rapid excited-singlet-state relaxation will produce effectively instantaneous heating. The amount of heat in this step is proportional to the difference between the photon energy and ground-to-excited singlet state zero-point energy. Second, the excited-state singlet, N_2, will relax with a rate of $(k_{20} + k_{23})N_2(t)$. Assuming negligible phosphorescence, the amount of energy in this step is

$$\Delta E = (1 - Y_F)[(1 - \phi_{\text{ISC}})\,\Delta E_{20} + \phi_{\text{ISC}}\,\Delta E_{23}] = (1 - Y_F)\,\frac{k_{20}\,\Delta E_{20} + k_{23}\,\Delta E_{23}}{k_{20} + k_{23}} \tag{107}$$

where Y_F is the fluorescence power yield. The last means of heat generation is that produced by triplet-state relaxation. The rate of heat released in this step is $k_{30}N_3(t)$ and the energy is the triplet-to-ground state energy difference. In total, the rate of energy release to the sample is

$$R'(t) = \Delta E_{12}\delta(t) + (1 - Y_F)(k_{20}\Delta E_{20} + k_{23}\Delta E_{23})e^{-(k_{20}+k_{23})t}$$
$$+ (1 - Y_F)\,\Delta E_{30}k_{23}k_{30}\,\frac{e^{-k_{30}t} - e^{-(k_{20}+k_{23})t}}{k_{20}+k_{23}-k_{30}} \qquad (108)$$

The rate of heat release is comprised of an instantaneous portion and a complex exponential portion. The complex exponential contains information regarding both the singlet and triplet relaxation times. In theory, these exponential constants can be obtained from photothermal signal rise times. In practice it is difficult to determine accurate rate constants from multiple exponential decays. Exponential lifetime determinations are complicated further by the acoustic limited signal rise times observed in photothermal spectroscopy discussed later in the book.

2.5.7. Multiple-Photon Absorption

Multiple-photon absorption is similar to multiphoton absorption in that more than one photon is absorbed. However, the absorbing species is excited through real intermediate states. In electronic-state spectroscopy, multiple-photon absorption can occur when an excited singlet or triplet state absorbs a photon and the molecule makes a transition, for example, from S_1 to S_2 or from T_1 to T_2. Multiple-photon absorption also occurs in vibrational transitions excited with infrared sources. In this case absorbed photons result in consecutive excitations to higher vibrational levels.

Multiple-photon absorption can occur either through consecutive excitation of a species, called double resonance (Steinfeld and Houston 1978), or via a multistep process where the lower levels of the subsequent transitions are connected to the initially excited state by nonradiative relaxation routes. Excitation of the initially excited state will compete with excited-state relaxation. Unlike multiphoton absorption, the absorbed energy is first linearly dependent on excitation irradiance, but becomes nonlinear at higher excitation irradiances. The higher-order irradiance dependence begins when the rate of secondary excitation of the initially excited-state becomes competitive with the rate of excited-state relaxation. Thus multiple-photon absorption is a kinetic effect wherein the rate of excitation competes with that for relaxation.

One can model the effects of multiple absorption on the absorbed energy using the rate expressions for an absorbing system. Consider the general four-level system shown in Figure 2.8. This energy-level diagram is typical

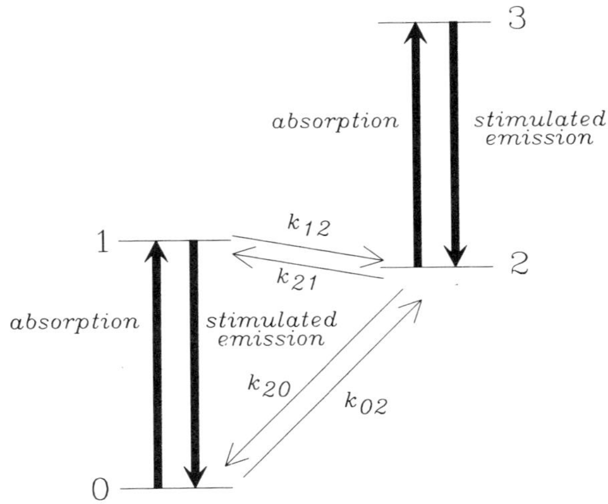

Figure 2.8. Kinetic description of multiple-photon absorption, including excited-state absorption and stimulated emission.

of both electronic and vibrational excitation/relaxation schemes. The ground state, level 0, is excited by a one-photon process to the first excited state, level 1. This state reversibly relaxes to level 2, which may then either absorb a second photon or relax to the ground state. Disconnecting levels 1 and 2 circumvents any coherent effects in the multiple-photon absorption. The rate expressions for the four levels of this system are

$$\begin{aligned}
\frac{dN_0(t)}{dt} &= -(E_p\sigma_{01} + k_{02})N_0(t) + E_p\sigma_{10}N_1(t) + k_{20}N_2(t) \\
\frac{dN_1(t)}{dt} &= E_p\sigma_{01}N_0(t) - (E_p\sigma_{10} + k_{12})N_1(t) + k_{21}N_2(t) \\
\frac{dN_2(t)}{dt} &= k_{02}N_0(t) + k_{12}N_1(t) - (E_p\sigma_{23} + k_{20} + k_{21})N_2(t) + (E_p\sigma_{32} + k_{32})N_3(t) \\
\frac{dN_2(t)}{dt} &= E_p\sigma_{23}N_2(t) - (E_p\sigma_{32} + k_{32})N_3(t)
\end{aligned} \tag{109}$$

The relaxation rate constants could include emission rates; optical emission is not explicitly taken into account in this scheme.

Steady-state solutions for the N_i are found and the absorption coefficient is calculated from

$$\alpha = \sigma_{01}N_0 - \sigma_{10}N_1 + \sigma_{23}N_2 - \sigma_{32}N_3 \tag{110}$$

The power absorbed per unit volume by the system is then determined by the relationship αE. A steady-state solution for this set of equations is easily obtained. However, the results are long and difficult to interpret. The solution is greatly simplified by assuming that the upward relaxation rate constants are very slow relative to the downward rate constant in all transitions except those connecting levels 1 and 2. Also, it can be assumed that the absorbing species is not perturbed significantly in the number-density populations, so the stimulated emission rates can be neglected relative to the "dark" relaxation rates. Thus $\sigma_{10} = 0$ and $\sigma_{32} = 0$. The latter assumption is generally valid for condensed-phase organic dye compounds since internal conversion of upper singlet and triplet states occurs in picoseconds. Setting the excited transition stimulated emission rate to zero is probably valid for vibrational transitions because the high density of states will quickly scramble the identity of the highly excited vibrational state. With these assumptions

$$\alpha = \sigma_{01}N_0 + \sigma_{23}N \tag{111}$$

where

$$\begin{aligned} N_0 &= N_{\text{tot}}k_{32}\frac{k_{20}k_{12}}{E_p^2\sigma_{01}\sigma_{23}k_{12} + k_{32}[E_p\sigma_{01}(k_{12} + k_{20} + k_{21}) + k_{20}k_{12}]} \\ N_2 &= N_{\text{tot}}k_{32}\frac{E_p\sigma_{01}k_{12}}{E_p^2\sigma_{01}\sigma_{23}k_{12} + k_{32}[E_p\sigma_{01}(k_{12} + k_{20} + k_{21}) + k_{20}k_{12}]} \end{aligned} \tag{112}$$

The power per unit volume absorbed by the sample, found for the case where the rate of level 3 relaxation is much faster than that for excitation from level 2, is

$$U = N_{\text{tot}}E_p\sigma_{01}\frac{k_{20}k_{12} + E_p\sigma_{23}k_{12}}{E_p\sigma_{01}(k_{12} + k_{20} + k_{21}) + k_{20}k_{12}} \tag{113}$$

Optical bleaching can occur. By defining

$$E_S = \frac{h\nu}{2\sigma_{01}\tau_{\text{effective}}} \qquad \tau_{\text{effective}} = \frac{k_{12} + k_{20} + k_{21}}{2k_{12}k_{20}} \tag{114}$$

the power absorbed per unit volume is

$$U = N_{\text{tot}} \frac{E_P\sigma_{01} + E_P^2\sigma_{01}\sigma_{23}k_{20}^{-1}}{1 + E/E_S} \tag{115}$$

Two-photon absorption behavior (e.g., $\alpha \propto E_P$) , can occur when $E_P\sigma_{23} > k_{20}$ only if $E < E_S$. The later inequality is equivalent to

$$E_P\sigma_{01} < \frac{k_{12}}{k_{12} + k_{20} + k_{21}} k_{20} \tag{116}$$

Together, these relationships imply that σ_{01} must be less than σ_{23} for two-photon absorption behavior to occur.

Poston and Harris (1990) have examined the excitation irradiance-dependent absorption of benzophenone solutions using photothermal deflection spectroscopy. Benzophenone is an interesting molecule in that it has a triplet quantum yield of unity. The four-level model described above with $k_{21} = 0$ is appropriate for this case. Benzophenone was reported to exhibit multiple-photon absorption, although the data were not precise enough to allow quantitative determination of the excited-state lifetime. Zhu et al. (1992) have recently reviewed the use of photothermal diffraction spectroscopy to measure both optical bleaching and multiple-photon excitation. These data are interpreted using an amended three-level organic dye model similar to that of Guiliano and Hess (1967). The three-level organic dye models are amended to include an excited-state absorption. Since the rate of relaxation of the state produced by excited-state absorption is very fast, the irradiance-dependent absorption coefficient is

$$\alpha = N_0\sigma_{01} + N_2\sigma_2 + N_3\sigma_3 \tag{117}$$

where σ_2 is the absorption cross section for S_1–S_2 transitions, σ_3 is that for T_1–T_2 transitions, and the number densities are those found using the time/irradiance-dependent organic dye molecule model.

2.6. Absorbed Energy

The absorbed energy ultimately results in the photothermal signal. In this section the energy absorbed by a sample excited with a pulsed, TEM_{00}, or Gaussian profile laser will be derived. A Gaussian laser beam is chosen to illustrate the methods because of its prevalence in photothermal spectroscopy. It is assumed that the radial profile does not change along the axis of propagation of the excitation beam (i.e., the cylindrical symmetry approximation). This assumption is implicit in most derivations of photothermal signals. For this assumption to be valid the sample must be optically thin and the sample cell length along the axis of propagation must be such that the excitation beam waist does not vary significantly through the cell. The cylindrical symmetry approximation is quite good if the cell length is shorter than the Rayleigh range of the focused beam.

An excitation source propagating in the fundamental TEM_{00} mode has an instantaneous irradiance at radius r from the beam center of

$$E_0(r, t) = \frac{2\Phi(t)}{\pi w^2} e^{-2r^2/w^2} \tag{118}$$

where $\Phi(t)$ (W) is the instantaneous radiant power and w (m) is the electric field beam waist radius parameter. The absorbance law must be modified to include the irradiance-dependent absorption coefficient.

For multiphoton absorption, the absorption coefficient is proportional to the irradiance raised to a power. The absorbed energy is highly dependent on the temporal and spatial characteristics of the excitation laser. Consider a pulsed laser operating in the fundamental, TEM_{00}, spatial mode, and with an exponential time dependence

$$\Phi(t) = \frac{Q}{\tau_p} e^{-t/\tau_p} \tag{119}$$

where Q is the total or integrated pulse energy and τ_p is the exponential pulse width. For a collimated beam, the amount of energy absorbed by an n-photon absorber in a section close to the sample cell entrance window will be

$$Q_A = \alpha_{ge}^n N_a \int_0^\infty E(r, t)^n \, dt \tag{120}$$

where N_A is the number density of absorbing species in the sample. Because of the nonlinear dependence, the absorbed energy will be

$$Q_A(r) = \frac{2^n Q^n}{n\tau_p^{(n-1)} \pi^n w^{2n}} e^{-2nr^2/w^2} \tag{121}$$

The amount of energy absorbed depends inversely on the pulse duration and is a complex function of the number of photons absorbed per molecule,

n. The spatial distribution of the absorbed energy also differs from that of the excitation source. The beam waist radius has been reduced by a factor of $1/\sqrt{n}$.

The signal dependence on the excitation source beam waist radius and temporal characteristics make quantitative measurements of multiphoton absorption spectra difficult. As pulsed dye lasers are scanned, the wavelength, confocal parameter, and temporal characteristics of the excitation radiation may all change. The change in excitation beam waist radius affects not only the total energy absorbed but also the photothermal signal through the signal dependence on absorbed energy spatial distribution. In the case of optical saturation, the differential expression for the irradiance along the sample path, z, is dependent on the irradiance-dependent absorption coefficient

$$-\frac{dE_z(r, t)}{dz} = E_z(r, t)\alpha(v) = \frac{E_z(r, t)\alpha_0(v)}{[1 + E_z(r, t)/E_S]^f} \tag{122}$$

The exponential factor, f, is 1 for homogeneous and $\frac{1}{2}$ for inhomogeneously broadened transitions. For an optically thin sample (i.e., sample transmission over 99%), the irradiance is essentially constant along the path dz and the $[1 + E_z(r, t)/E_S]^f$ term can be ignored in the integration. This, together with the fact that for small α, $\exp[-\alpha(v)l] \approx 1 - \alpha(v)l$, , results in

$$E_l(r, t) - E_0(r, t) \approx -E_0(r, t)\alpha(v)l \tag{123}$$

where $E_0(r, t)$ is the instantaneous irradiance of the pulsed laser entering the sample and $E_l(r, t)$ is the exiting irradiance after passing through a sample of pathlength l. The $\alpha(v)$ is either the homogeneous or inhomogeneous saturation absorption coefficient. The amount of heat deposited in the sample per unit volume per pulse is the integrated irradiance, $U(r)$ (J m^{-3}). This is found by integrating the irradiance difference per unit length

$$U(r) = \int_0^\infty \alpha(v)E_0(r, t)\,dt \tag{124}$$

Because of the nonlinear irradiance dependence, the result must be obtained for specific temporal pulse shapes. Constant $E_0(r, t)$ from $t = 0$ to $t = \tau_p$ is a rectangular pulse. This pulse duration may be obtained in electrooptically chopped laser pulses. The energy deposited in the sample over the pulse is

$$U_r(r) = \frac{\alpha_0(v)He^{-2r^2/w^2}}{[1 + (E/E_S)e^{-2r^2/w^2}]^f} \quad (125)$$

where the subscript r is used to indicate a rectangular pulse. An average irradiance of $E = 2\Phi(0)/\pi w^2$ (W/m^2) and a time-integrated irradiance of $H = 2\tau_p\Phi(0)/\pi w^2$ (J/m^2) are defined for a rectangular pulse duration of τ_p and a constant radiant power of $\Phi(0)$.

Signal generation in photothermal spectroscopy is based on changes in the refractive index that occur upon sample heating. Signals are dependent on the spatial distribution as well as the magnitude of the energy absorbed by the sample. The above shows that the spatial distribution can change quite dramatically when optical saturation occurs. The shape of the energy density for saturation of a homogeneously broadened transition is shown in Figure 2.9. Optical saturation results in the energy being distributed over a wider area (or larger volume) and the curvature at the top is diminished over that of the undistorted Gaussian. But while both of these effects may diminish the photothermal spectroscopy signal response, the signal will still be finite. In contrast, the transmission spectroscopy signal will decrease to zero for irradiances beyond that required for optical saturation.

Another important case is an exponential temporal profile which may be used to resemble more closely the output of transverse discharged lasers such as the TEA–CO_2, N_2, and excimer lasers. In this case

$$\Phi(t) = \Phi(0)e^{-t/\tau_p} \quad (126)$$

The exponentially averaged power is $\Phi(0) = Q/\tau_p$ where Q (J) is the total pulse energy. The energies absorbed by the sample are

$$U_e(r) = \alpha_0(v)E_S\tau_p \ln[1 + (E/E_S)e^{-2r^2/w^2}] \quad (127)$$

for homogeneously broadened transitions and

$$U_e(r) = 2\alpha_0(v)E_S\tau_p[\sqrt{1 + (E/E_S)e^{-2r^2/w^2}} - 1] \quad (128)$$

for inhomogeneously broadened transitions. The average irradiance is defined here as $E = 2Q/\tau_p\pi w^2$.

Because of the similarity in irradiance dependencies, the methods can also be used for optical bleaching. If the steady-state assumption holds, integration over square or exponential pulses can be performed analytically. If

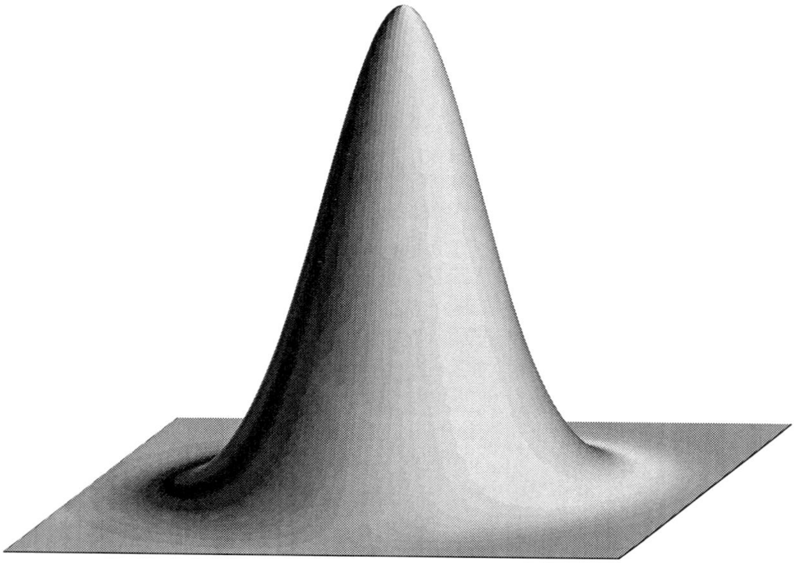

(a)

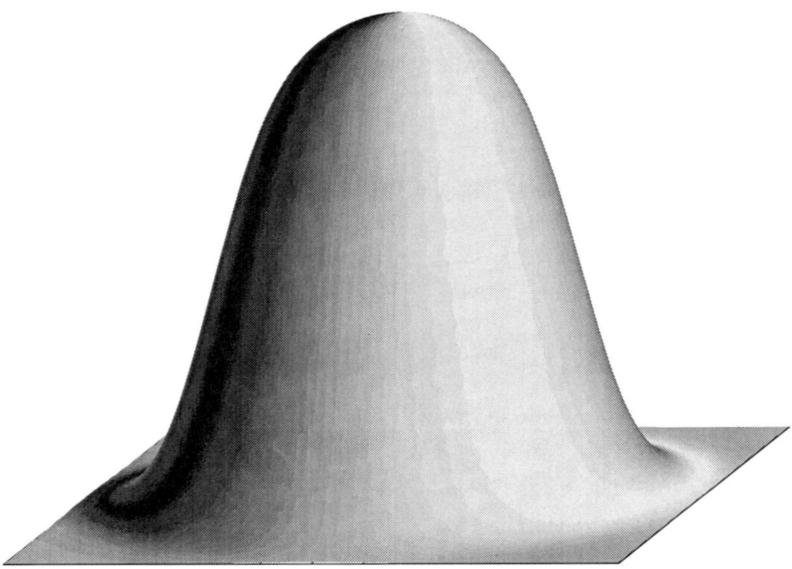

(b)

Figure 2.9. Effects of optical saturation on the Gaussian profile: (*a*) energy density produced by a Gaussian laser beam without optical saturation; (*b*) distortion that occurs as a result of optical saturation. The latter, E/E_S, was 10.

the steady-state approximation results are not valid for the excitation conditions of the experiment, then, in general, analytical results can be obtained only for square-pulse lasers.

Integration over time-variant excitation is difficult because of the exponential irradiance dependence found for optical bleaching and multiple-photon absorption models. However, the exponential dependence does not present a problem for constant irradiance excitation. If the excitation pulse can be approximated by a square pulse, analytical results for optical bleaching and multiple-photon excitation can be obtained.

REFERENCES

Bailey, R. T.; Cruickshank, F. R.; Pugh, D.; and Middleton, K. M. *J. Chem. Soc. Faraday Trans.* 2 **81** 255 (1985).

Bialkowski, S. E. and Guillory, W. A. *Chem. Phys.* **55** 229 (1981).

Bialkowski, S. E. and Long, G. R. *Anal. Chem.* **59** 873 (1987).

Birks, J. W. In *Chemiluminescence and Photochemical Reaction Detection in Chromatography*, Birks, J. W., ed., VCH, New York (1989).

Burak, I.; Steinfeld, J. I.; and Sutton, D. G. *J. Quant. Spectrosc. Radiat. Transfer* **9** 959 (1969).

Chen, K.-M. and Yeung, E. S. *J. Chem. Phys.* **69** 43 (1978).

Cottrell, T. L. and McCoubrey, J. C. *Molecular Energy Transfer in Gases*, Butterworth, London (1961).

deVries, H. and Wiersma, D. A. *J. Chem. Phys.* **72** 1851 (1980).

Flygare, W. H. *Acc. Chem. Res.* **1** 121 (1968).

Flygare, W. H. *Molecular Structure and Dynamics*, Prentice Hall, Englewood Cliffs, NJ (1978).

Grabiner, F. R.; Siebert, D. R.; and Flynn, G. R. *Chem. Phys. Lett.* **17** 189 (1972).

Guiliano, C. R. and Hess, L. D. *IEEE J. Quant. Electron.* **3** 358 (1967).

Hager, J.; Krieger, W.; Ruegg, T.; and Walther, H. *J. Chem. Phys.* **70** 2859 (1979).

Halpern, J. B.; Zacharias, H.; and Wallenstein, R. *J. Mol. Spectrosc.* **79** 1 (1980).

Harris, T. D. and Williams, A. W. *Appl. Spectrosc.* **39** 28 (1985).

Haselgrove, J. C.; Schotland, J. C.; and Leigh, J. S. *Appl. Opt.* **31** 2678 (1992).

Ingle, J. D., Jr. and Crouch, S. R. *Spectrochemical Analysis*, Prentice Hall, Englewood Cliffs, NJ (1988).

Keilmann, F. *IEEE J. Quant. Electron.* **QE-12** 592 (1976).

Long, G. R. and Bialkowski, S. E. *Anal. Chem.* **56** 2806 (1984).

Long, G. R. and Bialkowski, S. E. *Anal. Chem.* **57** 1079 (1985).

McClain, W. M. and Harris, R. A. In *Excited States*, Vol. 3, Lim, E. C., ed., Academic Press, New York (1978).

McGraw, D. J.; Michaelson, J.; and Harris, J. M. *J. Chem. Phys.* **86** 2536 (1987).

Mizushima, M. *The Theory of Rotating Diatomic Molecules*, Wiley, New York (1975).

Patterson, M. S.; Chance, B.; and Wilson, B. C. *Appl. Opt.* **28** 2331 (1989).

Patterson, M. S.; Moulton, J. D.; Wilson, B. C.; Berndt, K. W.; and Lakowicz, J. R. *Appl. Opt.* **30** 4474 (1991).

Poston, P. E. and Harris, J. M. *J. Amer. Chem. Soc.* **112** 644 (1990).

Poularikas, A. D. and Seely, S. *Signals and Systems*, PWS, Boston (1985).

Schwartz, R. N.; Slawsky, Z. I.; and Herzfeld, K. F. *J. Chem. Phys.* **20** 1591 (1952).

Steinfeld, J. I. *Molecules and Radiation*, MIT Press, Cambridge, MA (1985).

Steinfeld, J. I. and Houston, P. L. In *Laser and Coherence Spectroscopy*, Steinfeld, J. I., ed., Plenum, New York (1978).

Steinfeld, J. I.; Francisco, J. S.; and Hases, W. L. *Chemical Kinetics and Dynamics*, Prentice Hall, Englewood Cliffs, NJ (1989).

Stettler, J. D. and Witriol, N. M. In *Optoacoustic Spectroscopy and Detection*, Pao, Y.-H., ed., Academic Press, New York (1977).

Stone, J. and Goodman, M. F. *Phys Rev. A* **18** 2618 (1978).

Stone, J.; Thiele, E.; Goodman, M. J.; Stephenson, J. C.; and King, D. S. *J. Chem. Phys.* **73** 2259 (1980).

Twarowski, A. J. and Kliger, D. S. *Chem. Phys.* **20** 253 (1977a).

Twarowski, A. J. and Kliger, D. S. *Chem. Phys.* **20** 259 (1977b).

Weitz, E. and Flynn, G. *J. Chem. Phys.* **56** 6060 (1972).

Weitz, E. and Flynn, G. In *Photoselective Chemistry*, Part 2, Jortner, J., ed., Wiley, New York (1981).

Winefordner, J. D. and Rutledge, M. *Appl. Spectrosc.* **39** 377 (1985).

Wood, O. R.; Gordon, P. L.; and Schwarz, S. E. *IEEE J. Quant. Electron.* **QE5** 502 (1969).

Xing-Xiao, M. and Zhu-De, X. *Chem. Phys. Lett.* **98** 563 (1983).

Yuan, R. C. L. and Flynn G. W. *J. Chem. Phys.* **57** 1316 (1972).

Yuan, R. C. L. and Flynn G. W. *J. Chem. Phys.* **58** 649 (1973).

Zaccanti, G.; Bruscaglioni, P.; Ismaelli, A.; Carraresi, L.; Gurioli, M.; and Wei, Q. *Appl. Opt.* **31** 2141 (1992).

Zhu, X. R. and Harris, J. M. *Chem. Phys.* **124** 321 (1988).

Zhu, X. R. and Harris, J. M. *J. Chem. Phys.* **93** 75 (1989).

Zhu, X. R.; McGraw, D. J.; and Harris, J. M. *Anal. Chem.* **64** 710A (1992).

CHAPTER

3

HYDRODYNAMIC RELAXATION: HEAT TRANSFER AND ACOUSTICS

3.1. LOCAL EQUILIBRIUM

An equilibrium state is one wherein the temperature, pressure, and energy density are equal throughout the sample. Irradiated samples are generally not in an equilibrium state. In the irradiated volume, the fractional amount of species in excited states is higher than that given by the Boltzmann distribution for the average temperature, so the energy density is higher than that of the rest of the sample. Excited-state species relax, causing subsequent sample temperature increases with corresponding changes in pressure and density. Subsequently, the temperature, pressure, and energy density of the irradiated volume differ from those of the rest of the sample.

Two different disequilibrium states can be identified in the irradiated sample: one the internal energy of the absorber versus that of the sample matrix, the other the spatial disequilibrium in the thermodynamic parameters. Excited-state relaxation follows the path of least action, wherein the energy density does not deviate too far from the current value. For the excited-state species, the approach to equilibrium occurs through one or a series of relaxation rate steps. There may be fast and slow steps in the total process, but the net result is the same. Each step produces a state that is closer to equilibrium than the preceding state. Relaxation of a spatial disequilibrium state occurs by a similar process. For a sample that is spatially heterogeneous in terms of the temperature, pressure, and energy density, the approach to the state at the boundaries of the sample also follows a path of least action. Each successive volume element along a line connecting the irradiated volume with the sample cell boundary will be closer than the preceding one to the boundary state.

If the minimum volume element contains many molecules or atoms, the sample can be thought of as a continuous fluid medium. The continuous fluid or continuum model is much easier to describe since the result of individual particle collisions does not have to be accounted for. Mass, momentum, and heat are quantized on the particle level but are modeled as continuous in the continuum model. The condition for the continuum approximation is that the mean free path be much less than the length of interest. The ratio between the mean free path, Λ, and the length of interest, l, is the Knudson number:

$$Kn = \frac{\Lambda}{l} \tag{1}$$

The condition for valid application of the continuum fluid model is $Kn \ll 1$. The minimum length of interest, or more precisely the length monitored, associated with optical spectroscopy is on the order of the wavelength of light used to interrogate the sample. The mean free path for atmospheric-pressure samples is about 10^{-8} m. Condensed samples have intermolecular spacings on the order of the molecular radius, about 10^{-9} m. These are to be compared with typical optical wavelengths, which are on the order of 10^{-6} to 10^{-7} m. Clearly, the minimum volume element probed or excited in optical spectroscopy is large compared to the intermolecular spacings, and the continuum fluid model should be valid for all but very low pressure gas samples.

The minimum volume element contains from 10^3 to 10^9 molecules or atoms of both the matrix and the absorbing species. Considering the fact that photothermal spectroscopy is generally used as the trace analysis technique, measuring number density concentrations on the order of ppm to ppt, there are generally few excited species at any given time within each minimum volume element. Consequently, excited-state populations constitute only a small departure from a thermodynamic equilibrium condition. A 1-ppm sample that is far from optical saturation will have roughly 1 ppt of the absorbers in an excited state at any given moment. Subsequently, only one molecule in 10^9 is out of equilibrium. This small departure is generally negligible compared to the total excess energy of the volume element. Subsequently, each volume element may be modeled as being in a local equilibrium state. In the local equilibrium state, collective properties of the large number of absorber and matrix species can be described in terms of macroscopic thermodynamic and hydrodynamic parameters. The sample can be thought of as being in local equilibrium within each of these volume elements. The local equilibrium state of the sample is characterized by a thermodynamic temperature and pressure within the volume element. These in turn are related to the mass or number density, and the energy density at that particular position within the sample.

Spatial disequilibrium approaches equilibrium through a homogeneous series of steps wherein the thermodynamic excess in one volume element is transferred to the next. The thermodynamic exchange rate is first order in time, although second order in spatial difference. The driving force for this relaxation is entropy or the tendency to equilibrate the temperature of the system. Energy is transferred in space by means of the hydrodynamic relaxation modes. There are two principal types of hydrodynamic modes: diffusive modes, which equilibrate energy through diffusive flux, and propagating modes, which transfer energy through waves. In terms of thermal or kinetic energy, these modes correspond to thermal diffusion and propagating or acoustic relaxation.

3.2. THERMODYNAMIC AND OPTICAL PARAMETERS IN PHOTOTHERMAL SPECTROSCOPY

Local and global equilibrium can both be described in terms of the thermodynamic temperature and pressure. For global equilibrium to occur, the temperature and pressure of all elements within the sample must be the same. A temperature variation will cause heat transfer in a direction that will act to decrease the temperature variation. Similarly, pressure variation will cause mass transfer in a direction that decreases the pressure change. In fluids, temperature, pressure, and density are related through the state equation. In this case the state equation describes the density that arises due to a given equilibrium temperature and pressure in the sample. In the case of local equilibrium, the temperature and pressure of all elements within a given volume within the sample must be nearly the same. The temperature and pressure can, in turn, be used to determine the density in a given volume element.

3.2.1. Enthalpy and Temperature

Photothermal spectroscopy deals with energetics. Optical energy is absorbed by the sample, resulting in an increase in potential energy. Energy is transfered to the sample as a whole through various kinetic processes. In this fashion, the potential energy is converted to kinetic energy. Assuming that the kinetic energy is equilibrated (e.g., the energy is canonically partitioned among the various degrees of freedom), the excess energy results in a temperature increase. The increase in temperature is not restricted to the optically excited region. Thermal diffusion transfers heat throughout the sample and eventually to the surroundings.

Photothermal spectroscopy is normally performed under constant-pressure conditions. The conventional approach to describing photothermal effects is to model the temperature change as a function of the excess enthalpy. The experimental observables (e.g., the change in refractive index) are then related to the temperature change. In a given volume element of the sample, the sample enthalpy is the sum of enthalpies of the analyte plus that of the solvent or sample matrix. The enthalpy of each component, in turn, is the sum of the average populations in each energy level times the enthalpy of that level:

$$H_c = \sum_j H_{c,j} n_{c,j} \tag{2}$$

where $H_{c,j}$(J) is the enthalpy per molecule of component c in energy level j in a given volume element and $n_{c,j}$ is the number of molecules in the j energy level in that volume element. The total enthalpy density of volume element is

$$h_{\text{total}} = \sum_c H_c N_c \tag{3}$$

where N_c (m^{-3}) is the number density of each species.

Equilibrium is defined as a state of constant temperature. The temperature is related to the heat through the heat capacity. The heat is the excess enthalpy, or enthalpy in the form of kinetic energy. The excess enthalpy density is

$$h = \sum_c (H_c - H_{c,0}) N_c \tag{4}$$

where $H_{c,0}$ is the ground or zero-point state enthalpy of component c. This relationship can be used to account for the difference between potential and kinetic enthalpies. For cases where the absorbed energy is in the form of potential energy, $H_{c,0}$ is the ground or zero-point enthalpy of the excited state.

The enthalpy density is related to the temperature by the specific heat capacity, $C_{P,a}$ (J kg^{-1} K^{-1}), and the absorber density, $\rho_a = n_a m_a$, where m_a (kg) is the mass of the absorbing species and N_a (m^{-3}) is the number density:

$$T_a = \frac{h_a}{\rho_a C_{P,a}} \tag{5}$$

T_a (K) is used to indicate the absorber temperature. The sample matrix also has excess enthalpy and temperature related by

$$T_m = \frac{h_m}{\rho_m C_{P,m}} \tag{6}$$

where the subscript m is used to indicate the matrix or solvent. Under local equilibrium conditions, the absorber and sample matrix are at equilibrium with each other, and the absorber and matrix temperatures are equal. The total excess enthalpy density of the sample, h (J m^{-3}), is the sum of contributions from the absorber and the matrix:

$$h = h_a + h_m \tag{7}$$

The absorber and matrix temperatures are the same, and equal to that of the sample. Thus

$$T = \frac{h}{\rho_a C_{P,a} + \rho_m C_{P,m}} \tag{8}$$

where T has been used with the understanding that absorber and matrix temperatures are the same. The sample heat capacity per unit volume is $\rho C_P = \rho_a C_{P,a} + \rho_m C_{P,m}$. In photothermal spectroscopy of trace levels of optical absorbers, the density of the absorbing species is very small and may be neglected relative to that of the sample and $\rho C_P \approx \rho_m C_{P,m}$. If the temperature is changed by a small amount relative to the equilibrium sample temperature, $T = T_0 + \delta T$, the temperature change is related to the enthalpy density change through

$$\delta T = \frac{\delta h}{\rho C_P} \tag{9}$$

The relationship between temperature and the heat absorbed by the sample serves as a means to model the photothermal effect. The relationship between enthalpy and temperature is valid when the total excess energy is distributed canonically and when the pressure is equilibrated in each volume element. It is generally assumed that the amount of energy absorbed constitutes a small departure from the equilibrium sample enthalpy. This in turn implies small temperature, density, and heat capacity changes. It is also assumed that the pressure remains constant under all conditions. Thus the densities and heat capacities may be taken as their equilibrium values, and these values are used to relate the temperature change to that of the enthalpy.

The temperature change is used to determine other physical properties of the sample. Several temperature-dependent properties that constitute the photothermal effects are given in Table 3.1. These macroscopic properties change when the sample is heated. In addition to the thermodynamic

Table 3.1. Physical Properties That Change by the Photothermal Effect

Physical Property	Thermodynamic Relationship
Temperature, T (K)	$\delta T = \delta h/\rho C_P$ or $\delta T = \delta U/\rho C_V$
Density, ρ (kg m^{-3})	$(\partial\rho/\partial T)_P\,\delta T = -\rho\beta\,\delta T$
Pressure, P (Pa)	$(\partial P/\partial T)_V\,\delta T = \beta K_T^{-1}\,\delta T$
Refractive index, n	$(\partial n/\partial T)\,\delta T$
Emittance, M (W m^{-2})	Stefan–Boltzmann law
Reflectivity, r	Related to δn
Absorbance, A	Boltzmann population equation

parameters, optical parameters of the sample may change upon excitation. The refractive index of a transparent sample can change with temperature. This temperature-dependent refractive index change serves as the basis for most of the photothermal spectroscopy methods. The other physical properties listed in Table 3.1 are related to the photothermally induced temperature change. Many of these properties are related. For example, the refractive index change is due primarily to the temperature-dependent density. The temperature-dependent volume change is also related to density. The volume change is responsible for surface deformation of solid materials and for pressure wave generation in the photoacoustic effect. Emittance and absorbance are related to the temperature through the Boltzmann distribution equation relating excited quantum state populations to thermodynamic temperature. Reflectance is related to the refractive index change.

3.2.2. Energy and Dynamic Change

Problems in the temperature-based approach arise when the sample is perturbed far from equilibrium and when energy is absorbed by the sample in times shorter than that required to equilibrate the pressure. If the pressure is not equilibrated, the excess enthalpy density is no longer a useful parameter. In this case it may be easier to account for the photothermal effects through the excess energy density. The energy density change, U (J m^{-3}), is related to temperature through the differential relationship

$$\left(\frac{\partial U}{\partial T}\right)_P = \rho C_P - \frac{P}{V}\left(\frac{\partial V}{\partial T}\right)_P \qquad (10)$$

The change in energy density is thus

$$\delta U = \rho C_P \delta T - \beta P \delta T \tag{11}$$

where β (K^{-1}), the volume expansion coefficient, has been used in place of the differential. As discussed above, the $\rho C_P \delta T$ term is the kinetic energy at constant pressure, or enthalpy. The $\beta P \delta T$ term is the loss of energy density due to volume-change ($P \Delta V$) work.

Although difficult to interpret in this equilibrium form, this relationship has important consequences in the dynamics of photothermal spectroscopy. In general, δU will be static, fixed by the amount of energy absorbed by the sample. On the other hand, δT, ρ, and P are all dynamic parameters. The evolution of these three parameters must be understood in order to understand the signals observed in photothermal spectroscopy. Photothermal spectroscopy uses an energy source (e.g., light), to excite the sample. Enthalpy density changes that occur in the sample must therefore be inferred from the energy density equation. For example, with fast sample heating, the temperature change is initially that due to the increase in energy at constant volume. This is because the density may not change rapidly enough to ensure isobaric conditions. The temperature is initially high and will decrease during volume expansion. The decreased temperature is due to the energy required to do $P \Delta V$ work.

It is important to model changes in density, temperature, and pressure that occur during the return to equilibrium in order to model signals observed in photothermal spectroscopy. The density change is perhaps most important since it is primarily responsible for the refractive index changes that are the basis of photothermal spectroscopy. The density is found directly from the hydrodynamic equations discussed below. In most cases the density is all that will be necessary to model the photothermal signals. However, a description of the temperature change is necessary in photothermal radiometry. The temperature can be found through relationships between the total energy density and the density. The pressure may then be related to temperature and density. The difference between these and the conventional approach is that no assumptions have to be made regarding the pressure or heat capacity of the sample during the initial expansion.

3.3. CONSERVATION EQUATIONS

The conservation equations are central to the idea of local equilibrium. Mass, momentum, and energy conservation laws must all be satisfied during nonequilibrium relaxation. Transport coefficients used in the conservation equations are based on phenomenological laws or are based on properties of ideal Newtonian fluids. This continuum fluid level of detail is adequate for a description of most samples encountered in photothermal spectroscopy. The conservation equations and the molecular parameters used to find the transport coefficients of gases and liquids have been discussed in the classic work by Hirschfelder, Curtiss, and Bird (1954). The equations described below are adapted from Hirschfelder et al. but include the modern mathematical treatments illustrated by Berne and Pecora (1976) and McLennan (1989). The interested reader is encouraged to consult the latter and the text by Temkin (1981) with regard to the basis of fluid mechanics.

Relaxation of spatially anisotropic conserved quantities is driven by entropy. Relaxation is a nonreversible process in which the entropy increases. During relaxation, the amounts of certain quantities, called conserved quantities, cannot change overall unless there is input from or output to an external source or sink. Any conserved intensive quantity, A (m^{-3}), is expressed by the general law

$$\frac{\partial A}{\partial t} + \nabla \cdot \mathbf{j}_A = \sigma_A \tag{12}$$

$\mathbf{j}_A$ ($\mathrm{m}^{-2}\,\mathrm{s}^{-1}$) is the flux of the conserved quantity relative to the volume surrounding the element, and σ_A ($\mathrm{m}^{-3}\,\mathrm{s}^{-1}$) is the rate at which this quantity is put into or removed from a volume element. A positive flux indicates the loss of A from the volume element. A positive σ_A indicates that the quantity is going into the volume.

This conservation equation follows from simple considerations. The flux is the amount of A crossing a unit cross-sectional area, dS, surrounding the volume element per unit time. Because the flux is a vector quantity, having a magnitude and a direction, the amount of A leaving the volume will be proportional to $dS \cdot \mathbf{j}_A$. For any volume element, the time rate of change in A must be equal to the sum of fluxes across the surface of the volume,

$$\int_V dV \frac{\partial A}{\partial t} = -\int_S dS \cdot \mathbf{j}_A \tag{13}$$

From Gauss's theorem,

$$\int_S dS \cdot \mathbf{j}_A = \int_V dV \, \nabla \cdot \mathbf{j}_A \tag{14}$$

and thus

$$\int_V dV \frac{\partial A}{\partial t} = -\int_V dV\ \nabla \cdot \mathbf{j}_A \tag{15}$$

The conservation equation with no external source or sink follows by differentiation with respect to volume:

$$\frac{\partial A}{\partial t} + \nabla \cdot \mathbf{j}_A = 0 \tag{16}$$

Conservation equations are cast in the form of flux across a unit area, surrounding an infinitesimal volume, placed anywhere within the system (or even at the boundary of the system). Including the σ_A source term allows the use of these equations in open systems (i.e., systems with sources and/or sinks).

During the course of relaxation, mass, momentum, and energy must be conserved overall. For a homogeneous fluid, the mass conservation or continuity equation is

$$\frac{\partial \rho}{\partial t} + \nabla \cdot \mathbf{j}_m = 0 \tag{17}$$

where ρ (kg m^{-3}) is the mass density and $\mathbf{j}_m$ (kg m^{-2} s^{-1}) is the mass flux vector (Table 3.2). The mass conservation equation is equal to zero since no mass is entering the system. Any mass flux must result in a proportional change in the rate of density change. In a nonhomogeneous fluid, masses of each individual species must be maintained separately. In this case the mass conservation law is the mass diffusion equation.

Table 3.2. Vector and Density Quantities Used in the Conservation Equations

Parameter	Symbol	Units
Mass flux	$\mathbf{j}_m$	kg m^{-2} s^{-1}
Pressure tensor	$\mathbf{t}$	kg m^{-1} s^{-2}
Energy density	U	J m^{-3}
Energy flux	$\mathbf{j}_E$	J m^{-2} s^{-1}
Flow velocity	$\mathbf{u}$	m s^{-1}
Enthalpy density	h	J m^{-3}

Conservation of momentum is given by

$$\frac{\partial \mathbf{j}_m}{\partial t} + \nabla \cdot \mathbf{t} = 0 \tag{18}$$

where $\mathbf{t}$ (kg m^{-1} s^{-2}) is the pressure tensor or momentum flux. The pressure tensor is made up of elements, t_{ij}, that represents the jth component of momentum flux across the ith direction. Terms with equivalent indices (e.g., t_{xx}) represent momentum flux parallel to the direction; those with different indices (e.g., t_{yx}), represent shear momentum flux contributions. Temkin (1981) describes the origin of the pressure tensor and presents an excellent discussion of its relationship to viscosities and pressures in terms of molecular degrees of freedom. For the purpose of catagorization, it is sufficient to say that *ideal fluids* are represented by a pressure tensor that accounts only for isotropic pressure, $t_{ij} = \delta_{ij}P$, P being pressure, *Newtonian fluids* are those that also have nonisotropic stress tensor components. The components are related to the viscosities of the fluid.

The energy conservation law is

$$\frac{\partial U}{\partial t} + \nabla \cdot \mathbf{j}_E = 0 \tag{19}$$

where U (J m^{-3}) is the energy density and $\mathbf{j}_E$ (J m^{-2} s^{-1}) is the energy flux vector. Energy density is a function of temperature and pressure. Any spatially varying temperature or pressure will produce a time-dependent energy density change. These three basic conservation equations are related to each other by casting the fluxes in terms of the linear flow velocity. Defining a flow velocity $\mathbf{u}$ (m s^{-1}), the $\mathbf{j}_m = \rho\mathbf{u}$ and mass conservation law becomes

$$\frac{\partial \rho}{\partial t} + \nabla \cdot \rho\mathbf{u} = 0 \tag{20}$$

and since

$$\frac{\partial \mathbf{j}_m}{\partial t} = \mathbf{u}\,\frac{\partial \rho}{\partial t} + \rho\,\frac{\partial \mathbf{u}}{\partial t} \tag{21}$$

the conservation of momentum equation defined in terms of the flow velocity becomes

$$\rho\,\frac{\partial \mathbf{u}}{\partial t} = \mathbf{u} \cdot \nabla\rho\mathbf{u} - \nabla \cdot \mathbf{t} \tag{22}$$

Analysis of the pressure tensor for an isotropic fluid results in the fluid flow equation (Temkin 1981, Pugh 1989)

$$\rho \frac{\partial \mathbf{u}}{\partial t} = -\nabla P + (\tfrac{1}{3}\eta_s + \eta_b)\nabla(\nabla \cdot \mathbf{u}) + \eta_s \nabla^2 \mathbf{u} \tag{23}$$

where P (Pa) is pressure and η_s and η_b are the shear and bulk viscosities (Table 3.3). Shear viscosity gives rise to the pressure increase due to shear flow (e.g., the motion of planes relative to each other). The bulk viscosity gives rise to a pressure lag in expansion or contraction of a fluid due to the different rates of translational and rovibrational equilibration. It should be negligible in monatomic gases and liquids. Including this term allows the Newtonian fluid model to account accurately for the hydrodynamic properties of liquids and gases.

The energy conservation equation can also be written in terms of energy density flow and pressure changes, plus an excess energy flux due to mass diffusive flux of various chemical species and heat flux due to external heating and thermal diffusion. The last two energy fluxes are cast as an excess flux, and when the pressure tensor is broken up into isotropic and anisotropic terms,

$$\frac{\partial U}{\partial t} + \mathbf{u} \cdot \nabla U = -h \nabla \mathbf{u} - \mathbf{t} : \mathbf{u} - \nabla \cdot \mathbf{j}_E^* \tag{24}$$

Table 3.3. Bulk Parameters Used in Hydrodynamic Equations

Parameter	Symbol	Units
Shear viscosity	η_s	$\mathrm{kg\,m^{-1}\,s^{-1}}$
Bulk viscosity	η_b	$\mathrm{kg\,m^{-1}\,s^{-1}}$
Isothermal compressibility	$K_T = -V^{-1}(\partial V/\partial P)_T = \rho^{-1}(\partial \rho/\partial P)_T$	$\mathrm{Pa^{-1}}$
Volume expansion coefficient	$\beta = V^{-1}(\partial V/\partial T)_P = -\rho^{-1}(\partial \rho/\partial T)_P$	$\mathrm{K^{-1}}$
Thermal conductivity	κ	$\mathrm{J\,m^{-1}\,s^{-1}\,K^{-1}}$
Thermal diffusivity	$D_T = \kappa/\rho C_P$	$\mathrm{m^2\,s^{-1}}$
Specific heat at constant volume	$C_V = \rho^{-1}(\partial U/\partial T)_\rho$	$\mathrm{J\,kg^{-1}\,K^{-1}}$
Specific heat at constant pressure	$C_P = \rho^{-1}(\partial h/\partial T)_P$	$\mathrm{J\,kg^{-1}\,K^{-1}}$
Heat capacity ratio	$\gamma = C_P/C_V$	unitless
Isentropic sound speed	$c = \sqrt{\gamma \rho^{-1} K_T^{-1}}$	$\mathrm{m\,s^{-1}}$

where $h = U + P$ (J m^{-3}) is the enthalpy per unit volume, $\mathbf{j}_E^*$ is the excess flux, and the double dot is the tensor product, $t_{ij}(\partial \mathbf{u}_i/\partial x_j)$, x_j being a spatial coordinate. Energy is converted to temperature using the heat capacity and the equation of state, $\delta\rho = \rho(K_T \delta P - \beta\,\delta T)$:

$$\frac{\partial T}{\partial t} + \mathbf{u}\cdot\nabla T = -\frac{\beta T}{\rho C_V K_T}\nabla\cdot\mathbf{u} - \frac{1}{\rho C_V}(\mathbf{t}:\nabla\cdot\mathbf{u} + \nabla\cdot\mathbf{j}_E^*) \tag{25}$$

where C_V (J kg^{-1} K^{-1}) is the specific heat at constant volume and K_T (Pa^{-1}) is the isothermal compressibility. It is assumed that these parameters are constant over the range of the temperature change. In the absence of external input, the excess energy flux is the heat flux due to thermal diffusion:

$$-\nabla\cdot\mathbf{j}_E^* = \kappa\nabla^2 T \tag{26}$$

where κ (J m^{-1} s^{-1} K^{-1}) is the thermal conductivity. Thus

$$\frac{\partial T}{\partial t} + \mathbf{u}\cdot\nabla T + \frac{\beta T}{\rho C_V K_T}\nabla\cdot\mathbf{u} + \frac{1}{\rho C_V}(\mathbf{t}:\mathbf{u} - \kappa\nabla^2 T) = 0 \tag{27}$$

In cases where there is an external input into the system, the energy conservation equation is

$$\frac{\partial T}{\partial t} + \mathbf{u}\cdot\nabla T + \frac{\beta T}{\rho C_V K_T}\nabla\cdot\mathbf{u} + \frac{1}{\rho C_V}(\mathbf{t}:\nabla\cdot\mathbf{u} - \kappa\nabla^2 T) = \frac{1}{\rho C_V}\sigma_E \tag{28}$$

Here σ_E (W m^{-3}) is the rate of excess energy input per volume element. The input energy is scaled by the heat capacity to give the rate of temperature change per unit volume.

In photothermal spectroscopy, energy gained by the sample enters into the energy balance equation through the heating rate:

$$\sigma_E = q_H \tag{29}$$

where q_H (W m^{-3}) is the rate of heat production per unit volume in the sample due to the photothermal effect. In this case

$$\frac{\partial T}{\partial t} + \mathbf{u}\cdot\nabla T + \frac{\beta T}{\rho C_V K_T}\nabla\cdot\mathbf{u} + \frac{1}{\rho C_V}(\mathbf{t}:\nabla\cdot\mathbf{u} - \kappa\nabla^2 T) = \frac{q_H}{\rho C_V} \tag{30}$$

The rate of heat production can be limited by the excited-state relaxation rate. In this case mass diffusion can spatially redistribute excited-state species prior to relaxation. To account for these effects, the external heating rate is given by

$$\sigma_E = -\sum_{r=1}^{R} \left(\frac{\partial U_{\text{excess}}}{\partial n_r}\right) \frac{\partial}{\partial t} n_r \tag{31}$$

where n_r (m^{-3}) is the number density of excited-state species r and U_{excess} is the excess energy (i.e., the energy in excess of the ground-state energy). The sign convention used is such that a loss in excited-state number density increases the energy. The r index is used to account for the particular energy state of the absorbing species. Optical absorption is accounted for by the populations of particular excited-state species. The time rate of change of the spatially dependent excited-state number densities are found from the mass diffusion equation. The number density of each diffusing excited-state species relaxing with a first-order relaxation constant of k_r, is found from the mass diffusion equation:

$$\frac{\partial n_r}{\partial t} = D_r \nabla^2 n_r - k_r n_r \tag{32}$$

where D_r ($m^2\ s^{-1}$) is the mass diffusion coefficient of species r. Generally, all excited-state species have the same diffusion coefficient, D. The excess energy input per unit volume is thus

$$\sigma_E = -\sum_{r=1}^{R} \frac{\partial U_{\text{excess}}}{\partial n_r} (D\,\nabla^2 n_r - k_r n_r) \tag{33}$$

Excited-state mass diffusion is negligible except for rarified gases and in cases where long-lived metastable states are formed after excitation. When mass diffusion is slow compared to excited-state relaxation, there is no spatial redistribution of the energy, and mass diffusion may be neglected. In this case the excess energy is simply that restricted by the rate of excited-state relaxation:

$$\sigma_E = \sum_{r=1}^{R} \frac{\partial U_{\text{excess}}}{\partial n_r} k_r n_r \tag{34}$$

3.4. HYDRODYNAMIC EQUATIONS

The three conservation equations must be solved simultaneously in order to describe changes in pressure, density, and temperature that follow sample irradiation. Most derivations of the changes start with the three conservation equations and through a variety of approximations, solve for a particular parameter. In this section a general solution to the conservation equations is given. The conservation equations are first cast in the form of the Navier–Stokes equations by considering perturbations to be small relative to the bulk thermodynamic parameters. Sample perturbation by irradiation constitutes an external input of energy noted as q_H, the rate of heat production. Mass diffusion of excited-state or metastable species is not considered but can be added to the general solutions.

Berne and Pecora (1976) and McLennan (1989) illustrate the various modes of hydrodynamic relaxation available to a relaxing system by solving all the conservation equations simultaneously. Berne and Pecora (1976) solve these equations without distinguishing between the perpendicular and parallel flow velocities. McLennan (1989) approximates the solution using an eigenvector expansion. The treatment shown below has elements of both derivations but is consolidated in order to describe the hydrodynamic relaxation of a sample that has an initial, or impulse, temperature perturbation. The latter is consistent with the type of perturbation expected in photothermal spectroscopy.

In photothermal spectroscopy, it is normally assumed that the power delivered to the sample results in only small perturbations to the thermodynamic parameters. In this case the temperature, density, pressure, and linear velocity are defined in terms of small perturbations from equilibrium values:

$$\begin{aligned} T(x_i, t) &= T + \delta T(x_i, t) \qquad & \rho(x_i, t) &= \rho + \delta\rho(x_i, t) \\ P(x_i, t) &= P + \delta P(x_i, t) \qquad & \mathbf{u}(x_i, t) &= \delta\mathbf{u}(x_i, t) \end{aligned} \tag{35}$$

The equilibrium values of the temperature, density, and pressure are finite, while that of the flow velocity is zero. The zero flow velocity greatly simplifies the density and energy conservation equations since they contain terms proportional to **u**. The pressure tensor term may also be dropped from the energy conservation equation since it is second order in $\delta\mathbf{u}$. Without the pressure tensor term, changes are adiabatic. Momentum conservation is maintained in the form of viscous transport. Thus substitution of these into the conservation equations results in the Navier–Stokes equations:

$$\frac{\partial \delta\rho}{\partial t} + \rho\nabla\cdot\delta\mathbf{u} = 0$$

$$\rho\frac{\partial \delta\mathbf{u}}{\partial t} + \nabla\delta P - \eta_s\nabla^2\delta\mathbf{u} - (\tfrac{1}{3}\eta_s + \eta_b)\nabla(\nabla\cdot\delta\mathbf{u}) = 0$$

$$\frac{\partial \delta T}{\partial t} + \frac{\beta T}{\rho C_V K_T}\nabla\cdot\delta\mathbf{u} - \frac{\kappa}{\rho C_V}\nabla^2\delta T = \frac{q_H}{\rho C_V} \tag{36}$$

Here the parameters P, T, and ρ are equilibrium values of the bulk, and δP, δT, $\delta\rho$, and $\delta\mathbf{u}$ are the perturbations. The excess energy flux terms due to external source heat production and mass diffusion have been dropped for simplicity.

The linear flow can be broken up into two components. The perpendicular component is flow across the surface of the volume element and the parallel component is normal to the surface. The gradient, $\nabla\cdot$, terms are finite only for the parallel component. The Laplacian, ∇^2, term is finite for both flow directions. When the flow velocity is broken up into parallel, $\|$, and perpendicular, $\perp$, components, the linear flow conservation equation is broken into two orthogonal components

$$\frac{\partial \delta\rho}{\partial t} + \rho\nabla\cdot\delta\mathbf{u}_{\|} = 0$$

$$\frac{\partial \delta\mathbf{u}_{\|}}{\partial t} + \nabla\delta P - \nu_l\nabla^2\delta\mathbf{u}_{\|} = 0$$

$$\frac{\partial \delta\mathbf{u}}{\partial t} - \nu\nabla^2\delta\mathbf{u}_{\perp} = 0$$

$$\frac{\partial \delta T}{\partial t} + \frac{\beta T}{\rho C_V K_T}\nabla\cdot\delta\mathbf{u}_{\|} - \frac{\kappa}{\rho C_V}\nabla^2\delta T = \frac{q_H}{\rho C_V} \tag{37}$$

where $\nu = \eta_s/\rho$ ($m^2\ s^{-1}$), the kinematic viscosity, and $\nu_l = (4\eta_s/3 + \eta_b)/\rho$ ($m^2\ s^{-1}$), the longitudinal kinematic viscosity. These are diffusive terms, having the same units as thermal or mass diffusion, and correspond to diffusion of the perpendicular and parallel linear flow or momentum. The third equation describes the hydrodynamic modes for $\delta u_{\perp}$ (there are two components to $\delta u_{\perp}$). This equation is independent of the other three equations. $\delta u_{\perp}$ dispersion, called a viscous shear mode, is perpendicular to the heat or density motion. Viscous shear modes occur at constant density and temperature and therefore do not dissipate energy. This mode of hydrodynamic relaxation does not have to be considered further since it does not couple with the temperature or density.

The remaining three equations can be simplified further. The gradient flow velocity term in the temperature-change equation is reduced to a density derivative using the first equation. The pressure is a function of temperature and density:

$$\delta P = \left(\frac{\partial P}{\partial T}\right)_\rho \delta T + \left(\frac{\partial P}{\partial \rho}\right)_T \delta \rho \tag{38}$$

The pressure change term in the flow velocity equation can be eliminated with

$$\delta P = \frac{\beta}{K_T} \delta T + \frac{1}{\rho K_T} \delta \rho \tag{39}$$

Performing the grad, $\nabla\cdot$, operation on the flow equation and defining the linear flow gradient, $\psi = \nabla \cdot \mathbf{u}_{||}$, the equations can be written

$$\frac{\partial \delta\rho}{\partial t} + \rho\psi = 0$$

$$\frac{\partial \delta\psi}{\partial t} + \frac{1}{\rho^2 K_T} \nabla^2 \delta\rho + \frac{\beta}{\rho K_T} \nabla^2 \delta T - \nu_l \nabla^2 \delta\psi = 0$$

$$\frac{\partial \delta T}{\partial t} - \frac{\beta T}{\rho^2 C_V K_T} \frac{\partial \rho}{\partial t} - \frac{\kappa}{\rho C_V} \nabla^2 \delta T = \frac{q_H}{\rho C_V} \tag{40}$$

These three differential equations may be solved using the Fourier–Laplace transform. A Fourier transform is taken over the spatial coordinates and a Laplace transform is taken over time. The Fourier transform of the Laplacian is

$$\mathscr{F}\{\nabla^2 f(x, y, z, t)\} = -\mathbf{k}^2 f(\mathbf{k}_x, \mathbf{k}_y, \mathbf{k}_z, t) \tag{41}$$

where the Fourier transform is sybolized by $\mathscr{F}$ and $i\mathbf{k}$ is the Fourier transform variable. The Laplace transform of the time derivative is

$$\mathscr{L}\left\{\frac{\partial}{\partial t} f(x, y, z, t)\right\} = \zeta f(x, y, z, \zeta) - f(x, y, z, t = 0) \tag{42}$$

where the Laplace transform is given by $\mathscr{L}$ and ζ is the Laplace transform variable. Performing these transforms results in

$$\zeta\delta\rho(\mathbf{k}, \zeta) + \rho\psi(\mathbf{k}, \zeta) = \delta\rho(\mathbf{k}, 0)$$

$$\zeta\psi(\mathbf{k}, \zeta) - \frac{1}{\rho^2 K_T}\mathbf{k}^2\delta\rho(\mathbf{k}, \zeta) - \frac{\beta}{\rho K_T}\mathbf{k}^2\delta T(\mathbf{k}, \zeta) + v_l\mathbf{k}^2\psi(\mathbf{k}, \zeta) = \psi(\mathbf{k}, 0)$$

$$\zeta\delta T(\mathbf{k}, \zeta) - \frac{\beta T}{\rho^2 C_V K_T}\zeta\delta\rho(\mathbf{k}, \zeta) + \frac{\kappa}{\rho C_V}\mathbf{k}^2\delta T(\mathbf{k}, \zeta) = -\frac{\beta T}{\rho^2 C_V K_T}\delta\rho(\mathbf{k}, 0) + \delta T(\mathbf{k}, 0) + \frac{q_H(\mathbf{k}, \zeta)}{\rho C_V} \tag{43}$$

the boundary conditions, $\delta\rho(\mathbf{k}, 0)$, $\psi(\mathbf{k}, 0)$, and $\delta T(\mathbf{k}, 0)$, being the density, flow gradient, and temperature change at $t = 0$. The equations can be abbreviated using the thermodynamic relationship for the sound speed, $c^2 = \gamma/\rho K_T$ (m s^{-1}), $\gamma = C_P/C_V$, and the thermal diffusion coefficient $D_T = \kappa/\rho C_P$ (m^2 s^{-1}). Of course, C_P (J kg^{-1} K^{-1}) is specific heat at constant pressure.

The resulting set of equations can be written in the form of a matrix equation:

$$\begin{bmatrix} \zeta & \rho & 0 \\ -c^2\mathbf{k}^2/\gamma\rho & \zeta + v_l\mathbf{k}^2 & -\beta c^2\mathbf{k}^2/\gamma \\ -\zeta\beta T/\rho^2 C_V K_T & 0 & \zeta + \gamma D_T\mathbf{k}^2 \end{bmatrix} \begin{bmatrix} \delta\rho(\mathbf{k}, \zeta) \\ \psi(\mathbf{k}, \zeta) \\ \delta T(\mathbf{k}, \zeta) \end{bmatrix} = \begin{bmatrix} 1 & 0 & 0 \\ 0 & 1 & 0 \\ -\beta T/\rho^2 C_V K_T & 0 & 1 \end{bmatrix} \begin{bmatrix} \delta\rho(\mathbf{k}, 0) \\ \psi(\mathbf{k}, 0) \\ \delta T(\mathbf{k}, 0) + q_H(\mathbf{k}, \zeta)/\rho C_V \end{bmatrix} \tag{44}$$

The hydrodynamic relaxation modes are found by solving the matrix equation

$$\mathbf{A}(\mathbf{k}, \zeta)\xi(\mathbf{k}, \zeta) = \mathbf{B}\xi(\mathbf{k}, 0) \tag{45}$$

The solution,

$$\xi(\mathbf{k}, \zeta) = \mathbf{A}(\mathbf{k}, \zeta)^{-1}\mathbf{B}\xi(\mathbf{k}, 0) \tag{46}$$

is found using standard matrix procedures. The inverse **A** matrix is the cofactor matrix divided by the determinant:

$$\mathbf{A}(\mathbf{k},\zeta)^{-1}$$

$$=\frac{1}{|\mathbf{A}(\mathbf{k},\zeta)|}\begin{bmatrix} (\zeta+v_l\mathbf{k}^2)(\zeta+\gamma D_T\mathbf{k}^2) & -\rho(\zeta+\gamma D_T\mathbf{k}^2) & -\beta\rho c^2\mathbf{k}^2/\gamma \\ c^2\mathbf{k}^2(\zeta+D_T\mathbf{k}^2)/\rho & \zeta(\zeta+\gamma D_T\mathbf{k}^2) & \zeta\beta c^2\mathbf{k}^2/\gamma \\ \zeta(\zeta+v_l\mathbf{k}^2)\beta T/\rho^2 C_V K_T & -\zeta\beta T/\rho^2 C_V K_T & \zeta(\zeta+v_l\mathbf{k}^2)+c^2\mathbf{k}^2/\gamma \end{bmatrix} \tag{47}$$

where use was made of the relationship $(\gamma - 1) = \beta^2 T/\rho C_V K_T$ to obtain this result. The matrix on the right is obtained from the cofactors of $\mathbf{A}(\mathbf{k}, \zeta)$ and $|\mathbf{A}(\mathbf{k}, \zeta)$ is the determinate:

$$|\mathbf{A}(\mathbf{k},\zeta)| = \zeta^3 + \zeta^3\mathbf{k}^2(\gamma D_T + v_l) + \zeta\mathbf{k}^2(c^2 + \gamma v_l D_T \mathbf{k}^2) + D_T c^2 \mathbf{k}^4 \tag{48}$$

The hydrodynamic relaxation equations are those given by the inverse Fourier–Laplace transform of

$$|\mathbf{A}(\mathbf{k},\zeta)|\xi(\mathbf{k},\zeta) = \text{cofac}[\mathbf{A}(\mathbf{k},\zeta)]\mathbf{B}\xi(\mathbf{k},0) \tag{49}$$

3.5. HYDRODYNAMIC RESPONSE TO PHOTOTHERMAL EXCITATION

To determine the changes in thermodynamic parameters that occur upon sample excitation, it is assumed that the initial density change, $\delta\rho(\mathbf{k}, 0)$, flow, $\psi(\mathbf{k}, 0)$, and temperature change, $\delta T(\mathbf{k}, 0)$, are all minor compared to the energy input from photothermal heating. These initial values are usually taken to be due to statistical fluctuations in the thermodynamic parameters. Assuming that the fluctuation values are negligible is equivalent to the assumption that thermodynamic parameter perturbations due to photothermal heating are much greater than the statistical fluctuations. In this case, q_H is the only term that need be considered and

$$\xi(\mathbf{k},0) = \begin{bmatrix} 0 \\ 0 \\ q_H(\mathbf{k},\zeta)/\rho C_V \end{bmatrix} \tag{50}$$

It follows that the vector equation that describes the hydrodynamic relaxation is

$$|\mathbf{A}(\mathbf{k}, \zeta)|\xi(\mathbf{k}, \zeta) = \begin{bmatrix} -\rho\beta c^2\mathbf{k}^2/\gamma \\ \zeta\beta c^2\mathbf{k}^2/\gamma \\ \zeta(\zeta + v_l\mathbf{k}^2) + c^2\mathbf{k}^2/\gamma \end{bmatrix} \frac{q_H(\mathbf{k}, \zeta)}{\rho C_V} \tag{51}$$

Solutions for any one of the thermodynamic parameters must satisfy the same equation, but with different source-term dependencies. The individual equations are

$$\begin{aligned} &\zeta^3\delta\rho(\mathbf{k}, \zeta) + \zeta^2\mathbf{k}^2(\gamma D_T + v_l)\delta\rho(\mathbf{k}, \zeta) + \zeta\mathbf{k}^2(c^2 + \gamma v_l D_T\mathbf{k}^2)\delta\rho(\mathbf{k}, \zeta) \\ &\qquad + D_T c^2\mathbf{k}^4\delta\rho(\mathbf{k}, \zeta) = -\frac{\beta c^2\mathbf{k}^2}{C_P}\, q_H(\mathbf{k}, \zeta) \\ &\zeta^3\psi(\mathbf{k}, \zeta) + \zeta^2\mathbf{k}^2(\gamma D_T + v_l)\psi(\mathbf{k}, \zeta) + \zeta\mathbf{k}^2(c^2 + \gamma v_l D_T\mathbf{k}^2)\psi(\mathbf{k}, \zeta) \\ &\qquad + D_T c^2\mathbf{k}^4\psi(\mathbf{k}, \zeta) = \frac{\beta c^2}{\rho C_P}\, \zeta\mathbf{k}^2 q_H(\mathbf{k}, \zeta) \\ &\zeta^3\delta T(\mathbf{k}, \zeta) + \zeta^2\mathbf{k}^2(\gamma D_T + v_l)\delta T(\mathbf{k}, \zeta) + \zeta\mathbf{k}^2(c^2 + \gamma v_l D_T\mathbf{k}^2)\delta T(\mathbf{k}, \zeta) \\ &\qquad + D_T c^2\mathbf{k}^4\delta T(\mathbf{k}, \zeta) = \left[\zeta(\zeta + v_l\mathbf{k}^2) + \frac{c^2\mathbf{k}^2}{\gamma}\right]\frac{q_H(\mathbf{k}, \zeta)}{\rho C_V} \end{aligned} \tag{52}$$

3.5.1. Solving the Hydrodynamic Equations

We only need to solve one of these equations to find the photothermal effects. The density change equation has a simple source term and is easier to interpret than ψ. Density change is also related directly to the refractive index change and thus to the photothermal signal for most samples. This equation will be used to examine the hydrodynamic relaxation modes. Diffusive and propagating mode solutions for the density change can be obtained from the inverse transform of the density change equation.

The density change is found by first solving for the transformed density change, $\delta\rho(\mathbf{k}, \zeta)$,

$$\delta\rho(\mathbf{k}, \zeta) = -\frac{\beta c^2}{C_P}\frac{\mathbf{k}^2}{|\mathbf{A}(\mathbf{k}, \zeta)|}\, q_H(\mathbf{k}, \zeta) \tag{53}$$

and then taking the inverse transforms. The algebraic equation includes the energy source term. The transform of this term is generally a function of space and time. To keep the density change solution general, the energy source term is replaced with a delta function of both space and time. The density change found in this case is the impulse response of the system. After the impulse response is found, the response to any source function can be found by convolution.

The inverse Laplace transform is taken first. The inverse Fourier transform is treated after manipulation of the inverse Laplace transform result. The inverse Laplace transform of the density change is found from

$$\delta\rho(\mathbf{k},\, t) = -\frac{\beta c^2 \mathbf{k}^2}{C_P} \sum_{j=1}^{3} \lim_{\zeta \to \zeta_j} \frac{\zeta - \zeta_j}{|\mathbf{A}(\mathbf{k},\, \zeta)|} \exp(\zeta_j t) \tag{54}$$

where the ζ_i are the roots of the determinant equation. Being third order in ζ, this determinant equation results in three roots:

$$|\mathbf{A}(\mathbf{k},\, \zeta)| = (\zeta - \zeta_1)(\zeta - \zeta_2)(\zeta - \zeta_3) \tag{55}$$

The resulting density change is the sum of three terms. Each term will represent a hydrodynamic relaxation mode. For a source with finite time duration, the Laplace transform of $q_H(\mathbf{k}, \zeta)$ is convoluted with the hydrodynamic relaxation modes to give the solution

$$\delta\rho(\mathbf{k},\, t) = \sum_{j=1}^{3} \int_0^t \delta\rho_i(\mathbf{k},\, t') q_H(\mathbf{k},\, t - t')\, dt' \tag{56}$$

where $\delta\rho_i(\mathbf{k}, t)$ represent the impulse response of the three terms in the Laplace transform.

Exact roots of the determinant equation may be obtained using the general cubic equation solution. However, the exact roots have complicated **k** dependence and cannot be inverted. An approximate solution for the roots can be obtained with terms through second order in **k**, which allows subsequent Fourier inversion (McClennan 1989). The approximate roots are given by the three terms

$$\begin{aligned} \zeta_1 + \mathbf{k}^2 D_T &= 0 \\ \zeta_{2,3} \pm i\mathbf{k}c + \Gamma\mathbf{k}^2 &= 0 \end{aligned} \tag{57}$$

The diffusive constant Γ ($\mathrm{m}^2\ \mathrm{s}^{-1}$) is defined by

$$\Gamma = \tfrac{1}{2}[\nu_l + (\gamma - 1)D_T] \tag{58}$$

Thus $|\mathbf{A}(\mathbf{k}, \zeta)$ can be factored into the approximate equation

$$|\mathbf{A}(\mathbf{k}, \zeta)| = (\zeta + D_T\mathbf{k}^2)(\zeta + i\mathbf{k}c + \Gamma\mathbf{k}^2)(\zeta - i\mathbf{k}c + \Gamma\mathbf{k}^2) \tag{59}$$

These three approximate root terms give the hydrodynamic relaxation modes of a sample excited with an impulse source. The inverse Fourier–Laplace transform of the roots results in the differential equations that, to this order of approximation, describe the hydrodynamic relaxation modes of a fluid are

$$\frac{\partial \xi}{\partial t} - D_T\nabla^2\xi = 0$$

$$\frac{\partial^2 \xi}{\partial t^2} - 2\Gamma\frac{\partial}{\partial t}\nabla^2\xi - (c^2\nabla^2 - \Gamma^2\nabla^2\nabla^2)\xi = 0 \tag{60}$$

where ξ is used to indicate any of the thermodynamic parameters. The first expression is the usual heat diffusion law, and the second, obtained from the product of the second two roots, is the acoustic-wave equation. One would expect that the dissipation of any fluctuation will occur by these two main mechanisms or modes. By examining the eigenvectors of the matrix equation, McLennan shows that the thermal diffusion mode pertains to conditions when $\delta P = 0$ (i.e., for heat dissipation at constant pressure). The complex frequency, $i\zeta = \omega$, is $\omega = \pm c\mathbf{k} + i\Gamma\mathbf{k}^2$. The first term corresponds to a wave propagation at a velocity, c, and the second corresponds to an attenuation. Acoustic-wave attenuation is due to thermal conduction and viscosity. McLennan shows that sound propagation is adiabatic in this approximation.

3.5.2. Thermal Diffusion Mode

The first term in the Laplace transform is due to thermal diffusion. Inversion with the first root results in

$$\delta\rho_1(\mathbf{k}, t) = -\frac{\beta}{C_P}\frac{c^2}{(\Gamma - D_T)^2\mathbf{k}^2 + c^2}e^{-\mathbf{k}^2 D_T t} \tag{61}$$

The attenuation term is not important for short times or for short distances of wave propagation. For most cases of practical importance in photothermal spectroscopy, this term can be set to zero:

$$\delta\rho_1(\mathbf{k}, t) = -\frac{\beta}{C_P}\frac{c^2}{D_T^2\mathbf{k}^2 + c^2}e^{-\mathbf{k}^2 D_T t} \tag{62}$$

Table 3.4. Approximate Hydrodynamic Parameters for Selected Substances near 25°C

Substance[a]	ρ (kg/m^3)	c (m/s)	ν^b (m^2 s^{-1})	D_T^c (m^2 s^{-1})	γ^d	Γ (m^2 s^{-1})
Air	1.18	331	1.5×10^{-5}	2.1×10^{-5}	1.40	1.2×10^{-5}
Argon	1.78	319	1.2×10^{-5}	2.2×10^{-5}	1.58	1.3×10^{-5}
Nitrogen	1.25	334	1.4×10^{-5}	2.2×10^{-5}	1.31	1.1×10^{-5}
Ethyl ether	713	985	3.1×10^{-7}	8.2×10^{-8}	1.68	1.8×10^{-7}
Methanol	790	1103	6.9×10^{-7}	1.0×10^{-7}	1.21	3.6×10^{-7}
Ethanol	800	1207	1.4×10^{-6}	8.9×10^{-8}	1.21	7.0×10^{-7}
Acetone	800	1174	4.0×10^{-7}	1.1×10^{-7}	1.44	2.2×10^{-7}
Water	997	1497	8.9×10^{-7}	1.4×10^{-7}	1.01	4.5×10^{-7}
Chloroform	1520	987	3.6×10^{-7}	8.1×10^{-8}	1.63	2.0×10^{-7}
Carbon tetrachloride	1600	926	5.7×10^{-7}	7.5×10^{-8}	1.46	3.0×10^{-7}

[a]Gases are at 101.3 kPa.
[b]From *Handbook of Chemistry and Physics* (1980).
[c]From Dovichi (1987).
[d]γ estimated from β, K_T, ρ, and C_P values.

Second, the value of the thermal diffusion coefficient is generally much smaller than the acoustic velocity and can be neglected in the denominator. The condition for neglecting the thermal diffusion term is $c \gg D_T\mathbf{k}$. Using the data in Table 3.4, this condition is apparently met when $\mathbf{k} \ll 10^7\ \text{m}^{-1}$ in gases, $\mathbf{k} \ll 10^{10}\ \text{m}^{-1}$ in liquids. This corresponds to distances greater than 10^{-7} to 10^{-10} m. This condition is met in optical experiments where the minimum distance is on the order of the wavelength of light or greater. With these approximations, the first root term is

$$\delta\rho_1(\mathbf{k}, t) = -\frac{\beta}{C_P}\, e^{-\mathbf{k}^2 D_T t} \tag{63}$$

The inverse Fourier transform of this equation gives the impulse response for a point source:

$$\delta\rho_{\text{point}}(x, y, z, t) = -\frac{\beta}{C_P}\frac{1}{(4\pi D_T t)^{3/2}}\, e^{-(x^2+y^2+z^2)/4D_T t} \tag{64}$$

An additional factor of $(2\pi)^{-3/2}$ occurs because the impulse response is a convolution with the source function. The density change solution to any source function is the convolution

$$\delta\rho(x, y, z, t) = \int_0^t \int_{-\infty}^{\infty} \int_{-\infty}^{\infty} \int_{-\infty}^{\infty} q_H(x', y', z', t')\,\delta\rho_{\text{point}} \times (x - x', y - y', z - z', t - t')\,dx'\,dy'\,dz'\,dt' \tag{65}$$

Solutions for one-, two-, and three-dimensional diffusion can be obtained by integrating over the constant dimension(s). A one-dimensional solution corresponds to a heat source, equally distributed in the y–z plane. Thus $q_H(x, t)$ is the power density per unit area of the plane. The two-dimensional case is for a line source, and $q_H(x, y, t)$ is the power density per unit length of the line. The three-dimensional case is for an arbitrary distribution in (x, y, z). In this case $q_H(x, y, z, t)$ is the power density.

The thermal diffusion equation can also be cast in terms of temperature change. Although not strictly valid for all cases, the connection will be made to show the equivalence. When the relationship $\delta\rho = -\rho\beta\,\delta T$, is valid, the temperature change is directly proportional to the density change and

$$dT_{\text{point}}(x, y, z, t) = \frac{1}{\rho C_P}\frac{1}{(4\pi D_T t)^{3/2}}\,e^{-(x^2+y^2+z^2)/4D_T t} \tag{66}$$

The usual differential equation for the temperature change may also be obtained. To see this, the inverse transforms are taken. Including the inverse Laplace of the source term, we have

$$\frac{\partial}{\partial t}\,\delta T - D_T \nabla^2 \delta_T = \frac{q_H}{\rho C_P} \tag{67}$$

which is the form the thermal diffusion equation is normally given in.

3.5.3. Fourier–Laplace Solutions for the Thermal Diffusion Equation

A solution to the diffusion equation can be obtained using the Fourier–Laplace transform in cases where there is no boundary (e.g., a homogeneous medium of infinite or semi-infinite extent). Carslaw and Jaeger (1986) discuss solutions to the thermal diffusion equation for several other boundary conditions. The temperature change and the heat source function both vary in space and time:

$$\frac{\partial \delta T(x, y, z, t)}{\partial t} - D_T \nabla^2 \delta T(x, y, z, t) = \frac{q_H(x, y, z, t)}{\rho C_P} \tag{68}$$

where q_H (J s^{-1} m^{-3}) is the power per unit volume generated by the source. Taking the Fourier transform in space and the Laplace transform in time yields

$$[\zeta + D_T(\mathbf{k}_x^2 + \mathbf{k}_y^2 + \mathbf{k}_z^2)]\,\delta T(\mathbf{k}_x, \mathbf{k}_y, \mathbf{k}_z, \zeta) = \frac{1}{\rho C_P}\,\mathscr{L}\{\mathscr{F}\{q_H(x, y, z, t)\}\} \tag{69}$$

When q_H can be broken up into space- and time-dependent components,

$$q_H(x, y, z, t) = \Phi(t)S(x, y, z) \tag{70}$$

then

$$[\zeta + D_T(\mathbf{k}_x^2 + \mathbf{k}_y^2 + \mathbf{k}_z^2)]\,\delta T(\mathbf{k}_x, \mathbf{k}_y, \mathbf{k}_z, \zeta) = \frac{1}{\rho C_P}\,\mathscr{L}\{\Phi(t)\}\mathscr{F}\{S(x, y, z)\} \tag{71}$$

Solving for $\delta T(\mathbf{k}_x, \mathbf{k}_y, \mathbf{k}_z, \zeta)$ yields

$$\delta T(\mathbf{k}_x, \mathbf{k}_y, \mathbf{k}_z, \zeta) = \frac{1}{\rho C_P}\,\frac{\mathscr{L}\{\Phi(t)\}\mathscr{F}\{S(x, y, z)\}}{\zeta + D_T(\mathbf{k}_x^2 + \mathbf{k}_y^2 + \mathbf{k}_z^2)} \tag{72}$$

The solution is obtained by first taking the inverse Laplace transform,

$$\delta T(\mathbf{k}_x, \mathbf{k}_y, \mathbf{k}_z, t) = \frac{1}{\rho C_P}\,\Phi(t) * [e^{-D_T(\mathbf{k}_x^2 + \mathbf{k}_y^2 + \mathbf{k}_z^2)t}\,\mathscr{F}\{S(x, y, z)\}] \tag{73}$$

The spatial convolution for a balanced Fourier transform is defined by

$$\mathscr{F}^{-1}\{F_1(\mathbf{k}_{x,i})F_2(\mathbf{k}_{x,i})\} = \frac{1}{(2\pi)^{1/2}}\int_{-\infty}^{\infty} f_1(x_i')\,f_2(x_i - x_i')\,dx_i' = \frac{1}{(2\pi)^{1/2}}\,f_1(x_i)\otimes f_2(x_i) \tag{74}$$

for the x, y, and z spatial coordinates. The inverse Fourier transform of the diffusion equation is

$$\delta T(x, y, z, t) = \frac{1}{\rho C_P}\,\Phi(t) * \left[\frac{1}{(4\pi D_T t)^{3/2}}\,e^{-(x^2 + y^2 + z^2)/4D_T t}\right] \otimes S(x, y, z) \tag{75}$$

The rightmost convolution, $\otimes$, represents the spatial convolution over all coordinates associated with the Fourier transform, and $*$ represents the time convolution associated with the Laplace transform. Expanding the time and space convolutions results in

$$\delta T(x, y, z, t) = \frac{1}{\rho C_P} \int_0^t \frac{\Phi(t')}{[4\pi D_T(t-t')]^{3/2}} \int_{-\infty}^{\infty} \int_{-\infty}^{\infty} \int_{-\infty}^{\infty} \times S(x', y', z')\, e^{-[(x-x')^2+(y-y')^2+(z-z')^2]/4D_T(t-t')}\, dx'\, dy'\, dz'\, dt' \tag{76}$$

The convolution is an integral over space and time of the function describing heat production with the characteristic equation describing diffusion. Diffusion is a linear, additive transport phenomenon. As such, the systems approach can be used to solve many diffusion problems. The convolution shows that the total temperature change is the superposition of individual point sources. The diffusion of each individual point source of power is the same. For an unit energy instantaneous source power of $\Phi(t) = \delta(t)$ (s^{-1}), and a spatial point source, $S(x, y, z) = \delta(x)\delta(y)\delta(z)$ (m^{-3}) per unit volume, the characteristic equation describing diffusion is

$$\delta T_{\text{point}}(x, y, z, t) = \frac{1}{\rho C_P} \frac{1}{(4\pi D_T t)^{3/2}}\, e^{-(x^2+y^2+z^2)/4D_T t} \qquad 0 \leqslant t \tag{77}$$

The temperature change solution to any source function is the convolution

$$\delta T(x, y, z, t) = \int_0^t \int_{-\infty}^{\infty} \int_{-\infty}^{\infty} \int_{-\infty}^{\infty} q_H(x', y', z', t')\, \delta T_{\text{point}} \times (x-x', y-y', z-z', t-t')\, dx'\, dy'\, dz'\, dt' \tag{78}$$

The integrand is often given the symbol for the Green's function, $G(x, y, z, x', y', z', t)$. Solutions for one-, two-, and three-dimensional diffusion can be obtained by integrating over the constant dimension(s). Solutions are given for infinite, $-\infty \leqslant x_i \leqslant \infty$ boundary conditions are tabulated in Table 3.5. The one-dimensional solution corresponds to a heat source, equally distributed in the y–z plane. Thus $q_H(x, t)$ is the power density per unit area of the plane. The two-dimensional case is for a line source, and $q_H(x, y, t)$ is the power density per unit length of the line. The three-dimensional case is for an arbitrary distribution in (x, y, z). In this case, $q_H(x, y, z, t)$ is the power density. Of course, the temperature change solutions are equivalent to those for the density within a factor of $-\beta/\rho$ when the system is at constant pressure.

Table 3.5. Point Source Solutions for One-, Two-, and Three-Dimensional Thermal Diffusion Equation

$$\delta T_{\text{point}}(x, t) = \frac{1}{\rho C_P} \frac{1}{(4\pi D_T t)^{1/2}} e^{-x^2/4D_T t} \tag{79}$$

$$\delta T_{\text{point}}(x, y, t) = \frac{1}{\rho C_P} \frac{1}{4\pi D_T t} e^{-(x^2+y^2)/4D_T t} \tag{80}$$

$$\delta T_{\text{point}}(x, y, z, t) = \frac{1}{\rho C_P} \frac{1}{(4\pi D_T t)^{3/2}} e^{-(x^2+y^2+z^2)/4D_T t} \tag{81}$$

3.5.4. Propagating Mode

The second two roots give the propagating hydrodynamic mode. Laplace inversion using the second two roots results in

$$\delta\rho_{2,3}(\mathbf{k}, t) = \frac{\beta}{C_P} \frac{(\Gamma - D_T)\mathbf{k}c \; \sin(c\mathbf{k}t) + c^2 \; \cos(c\mathbf{k}t)}{(\Gamma - D_T)^2\mathbf{k}^2 + c^2} e^{-\Gamma\mathbf{k}^2 t} \tag{82}$$

With the assumption that the acoustic attenuation and thermal diffusion coefficients are much less than the acoustic velocity for the volume being analyzed, this equation reduces to

$$\delta\rho_{2,3}(\mathbf{k}, t) = \frac{\beta}{C_P} \cos(c\mathbf{k}t) e^{-\Gamma\mathbf{k}^2 t} \tag{83}$$

The general space- and time-dependent impulse-response solution is difficult to obtain because $\mathbf{k} = (k_x^2 + k_y^2 + k_z^2)^{1/2}$ in the cosine term. The inverse Fourier transform will yield different impulse solutions for different geometries and symmetries. The one-dimensional solution is

$$\delta\rho_{\text{impulse}}(x, t) = \frac{\beta}{2C_P} \frac{1}{(4\pi\Gamma t)^{1/2}} \left[e^{-(x-ct)^2/4\Gamma t} + e^{-(x+ct)^2/4\Gamma t}\right] \tag{84}$$

The differential equation for the propagating acoustic mode is found by taking the Fourier–Laplace transform:

$$\left(\frac{\partial}{\partial t} - \Gamma\nabla^2\right)^2 \delta\rho_{2,3} - c^2\nabla^2 \delta\rho_{2,3} = \frac{\beta}{C_P}\left(\frac{\partial}{\partial t} - \Gamma\nabla^2\right) q_H \tag{85}$$

Acoustic attenuation is usually neglected. This approximation is valid in gases where the viscosity is low. Acoustic attenuations are on the order of a fraction of a percent per kilometer in air. Attenuation can always be reincorporated into the wave propagation equation as an exponentially decreasing function of distance (Temkin 1981). Assuming that the acoustic attenuation term is not important for short times or for short distances of propagation yields

$$\frac{\partial^2}{\partial t^2}\,\delta\rho_{2,3} - c^2\nabla^2\,\delta\rho_{2,3} = \frac{\beta}{C_p}\frac{\partial}{\partial t}\,q_H \tag{86}$$

The propagating wave equation is solved in terms of pressure. Including the appropriate scale, $\delta P = c^2\delta\rho$ (Temkin 1981), the usual form of the pressure wave equation can be obtained:

$$\frac{\partial^2}{\partial t^2}\,\delta P - c^2\nabla^2\,\delta P = \frac{\beta c^2}{C_p}\frac{\partial}{\partial t}\,q_H \tag{87}$$

The right-hand side of this equation shows that the pressure wave is generated by a periodic source. The generating pressure fluctuation is proportional to the time rate of change of the thermal power. Continuous excitation does not result in acoustic-wave generation since the heat is dissipated slowly, and presumably under isobaric conditions.

However, the attenuated propagating equation shows that there may be some density perturbation due to the propagation mode even for continuous source excitation. This can be shown with the convolution theorem. If the source is continuous, the density change attributed to the acoustic roots is

$$\delta\rho_{2,3}(\mathbf{k}, t) = \lim_{t\to\infty}\frac{\beta q_H}{C_p}\int_0^t \cos(c\mathbf{k}t')\,e^{-\Gamma\mathbf{k}^2 t'}\,dt' = \frac{\beta q_H}{C_p}\frac{\Gamma^{-1}}{c^2/\Gamma^2 + \mathbf{k}^2} \tag{88}$$

For a one-dimensional case, the inverse Fourier transform gives the spatially dependent density change

$$\delta\rho_{2,3}(x, t) = \frac{\pi^{1/2}\beta q_H}{2^{1/2}cC_p}\,e^{-c|x|/\Gamma} \tag{89}$$

This density perturbation will fall off rapidly with distance from the irradiated volume. c/Γ values range from about $3\times10^7\ \mathrm{m}^{-1}$ in air to $3\times10^9\ \mathrm{m}^{-1}$ in methanol. Thus the perturbation distances are on the order of, or less than, the wavelengths of light. The maximum density perturbation is a factor of $1/c$ less than that due to the thermal diffusion mode. The density perturbation is only about 10^{-3} of that due to thermal diffusion. In cases where the acoustic attenuation is neglected, $\Gamma = 0$, and the density perturbation due to the propagation acoustic mode is identically zero.

3.5.5. Summary of Hydrodynamic Mode Solutions

It is clear that solutions for the hydrodynamic response to photothermal heating can be classified as being due to thermal diffusion and due to the propagating or acoustic mode. Purely thermal diffusive solutions are the most used in photothermal spectroscopy and are used extensively in subsequent chapters. These solutions may be accurate enough to allow quantitative prediction for continuous excitation. Since the propagating acoustic mode is proportional to the time rate of change of the energy input to the sample, it should not have to be accounted for in continuous source excitation.

Purely acoustic mode solutions are used in photothermal spectroscopy. Photoacoustic spectroscopy uses pressure transducers to measure the pressure wave generated by rapid optical heating. The pressure transducer may also respond to temperature. The pressure transducer is normally placed relatively far from the optically excited region to avoid interference from scattered radiation and from sample heating. In this case, the propagating acoustic solution can be used without considering thermal diffusion. Thermal diffusion is relatively slow compared to acoustic-wave propagation. Using argon as an example, the time required for heat to diffuse a distance of $l = 1$ cm is $\tau_c = l^2/4D_T \approx 1$ s, and the time required for the acoustic wave to traverse the same path is $\tau_a = l/c \approx 3 \times 10^{-5}$ s. It is clear that the acoustic wave can be measured without interference from the thermal perturbation with detectors placed relatively far from the region of sample excitation.

Although thermal diffusion is not an important component in photoacoustic spectroscopy, the opposite is not always true. Propagating mode solutions are important in photothermal spectroscopy using pulsed or high-frequency, chopped or oscillating sample excitation. Propagating acoustic solutions are particularly valuable when the short time response to pulse excitation is sought. The acoustic-limited density change response is important to analyze when excited-state relaxation rates compete with the acoustic-limited rate of density change. Grabiner et al. (1972), used purely acoustic mode solutions to describe the rates of sample density decrease found in their photothermal lens studies of vibrational relaxation rate constants.

Since the density or pressure change associated with the propagating wave solution is proportional to the time rate of change of the heat source, accounting for either thermal diffusion or acoustic propagation alone gives only an approximate indication of the sample density changes when the sample is irradiated with pulsed or rapidly chopped light. In quantitative prediction using pulsed or chopped light sources, the response to both the thermal diffusive and propagating density changes must be accounted for.

The magnitude of the propagating density wave perturbation is the same as for the thermal diffusion component. Both solutions are proportional to the rate and magnitude of the thermal perturbation, q_H, and both scale the temperature change to the density change by a factor of β/C_P. The thermal diffusion perturbation lowers the density of the sample while the propagating perturbation corresponds to an increase in density. Thus density, and therefore mass, is conserved overall. The propagating mode density perturbation moves at the speed of sound while the diffusive component is stationary in the sense that the maximum density change does not change position. Because of these two different time–space responses, the temporal development of the density perturbation will be a function of both diffusive and propagating modes.

3.6. DENSITY RESPONSE TO IMPULSE EXCITATION

Accounting for the total density change that occurs in time-dependent photothermal spectroscopy can be accomplished by two means. First, solutions to both propagating and thermal diffusion equations can be obtained independently and combined to give the total density change result. This method can be quite accurate and yields results that are more facile to interpretation. Second, a solution to the differential equation describing the hydrodynamic behavior of the sample can be obtained. This can be accomplished using numerical techniques to find the roots to the determinate. This coupled solution is the most general and applies to all samples and all excitation schemes. However, solutions are difficult to interpret since they do not result in analytical equations.

Using either method, it is generally easiest to obtain the impulse excitation solution. The impulse excitation solution may then be convoluted with any time dependences due to excited-state relaxation and/or the excitation source. In cases where the approximations used to obtain separate thermal diffusion and propagation equations hold, the total density change is the sum of the two components convoluted with the excitation source:

$$\delta\rho(\mathbf{k}, t) = \frac{\beta}{C_P} \int_0^t [e^{-\Gamma \mathbf{k}^2 t'} \cos(c\mathbf{k}t') - e^{-\mathbf{k}^2 D_T t'}] q_H(\mathbf{k}, t - t')\, dt' \tag{90}$$

When sample excitation and excited-state relaxation are both fast relative to the propagating and diffusive relaxation, the convolution is the impulse response of the sample and

$$\delta\rho_{\text{impulse}}(\mathbf{k}, t) = \frac{\beta q_H(\mathbf{k})}{C_P} \left[e^{-\Gamma\mathbf{k}^2 t}\cos(c\mathbf{k}t) - e^{-\mathbf{k}^2 D_T t}\right] \tag{91}$$

For any finite $\mathbf{k}$, the density change is initially zero. Since $c \gg D_T$, the density will decrease with time in a cosine fashion. The rate of this density change is limited by the acoustic propagation of the density, and the acoustic rise time is approximately, $\tau_a \sim \pi/2\mathbf{k}c$ or $\tau_a \sim l/c$, where l is the length the wave has to travel. The oscillating propagating term is exponentially damped with a rate of $\Gamma\mathbf{k}^2$. For times long after impulse heating, the density will decrease in an exponential fashion. The decrease in density change is limited by thermal diffusion.

3.6.1. One-Dimensional Case

The inverse spatial Fourier transform of the impulse-response solution gives the temporal and spatial density change. Consider the one-dimensional problem of impulse excitation of a thin sample in a homogeneous medium. This is a common problem in the photothermal spectroscopy of surfaces. Optical absorption occurs only in a thin membrane of infinite extent in y and z dimensions, and the density change is that which occurs along the x axis.

The spatial Fourier transform of the impulse response is

$$\begin{aligned}\delta\rho_{\text{impulse}}(x, t) &= \frac{\beta}{2C_P}\frac{1}{(4\pi\Gamma t)^{1/2}}\left[e^{-(x-ct)^2/4\Gamma t} + e^{-(x+ct)^2/4\Gamma t}\right] \\ &\quad - \frac{\beta q_H}{C_P}\frac{1}{(4\pi D_T t)^{1/2}}\, e^{-x^2/4D_T t}\end{aligned} \tag{92}$$

This equation gives the density change of a infinitesimally thin absorbing region in a sample. The first exponential term is due to the right positive-x propagating acoustic wave, the second to the left propagating wave. The last exponential term is that due to thermal diffusion. The sample response is somewhat difficult to interpret because of the inverse square-root time dependence of both propagating and diffusive terms. To remedy this, the solution can be convoluted with a finite width spatial function. A useful function for this purpose is the Gaussian. A spatially dependent Gaussian energy source is defined as

$$q_H(x, t) = \frac{Q\delta(t)}{w}\left(\frac{2}{\pi}\right)^{1/2} e^{-2x^2/w^2} \tag{93}$$

to facilitate interpretation. Q is the total energy deposited in the sample.

Spatial convolution with a Gaussian function results in

$$\delta\rho(x, t) = \frac{\beta Q}{(2\pi)^{1/2} w C_P} \frac{e^{-2(x-ct)^2/w^2(1+2t/\tau_\Gamma)} + e^{-2(x+ct)^2/w^2(1+2t/\tau_\Gamma)}}{(1 + 2t/\tau_\Gamma)^{1/2}} - \frac{2^{1/2}\beta Q}{\pi^{1/2} w C_P} \frac{e^{-2x^2/w^2(1+2t/t_c)}}{(1 + 2t/t_c)^{1/2}} \tag{94}$$

$\tau_\Gamma = w^2/4\Gamma$ is the acoustic attenuation time constant and $t_c = w^2/4D_T$ is the characteristic thermal decay time constant. Figure 3.1 illustrates the time-dependent behavior of this density perturbation. At zero time, the density perturbation is zero, although the energy has been deposited into the sample. As time progresses the density in the region of the thermal perturbation decreases. There is a corresponding increase in density surrounding this region. This is a manifestation of the conservation of mass principle. At consecutively longer times, two density waves can be seen. These waves propagate away from the initially perturbed region, one traveling in the positive, right, the other in the negative, left, directions. The density wave broadens in time. The half-width of the wave is given by $w(1 + 2t/\tau_\Gamma)^{1/2}$. Γ and D_T are similar and the width of the density wave will be about the same as that of the thermal density perturbation that remains in the excitation region. At long times, the propagating density wave is out of the picture, and the density change produced by the initial excitation broadens and decreases in magnitude as heat diffuses out of the perturbed region. The half-width of the Gaussian density change profile is $w(1 + 2t/t_c)^{1/2}$. This time-dependent broadening is illustrated in Figure 3.2. In this figure the time increases from rear to front and the density decrease is plotted for clarity.

There are a few interesting features worth noting. First, the propagating density perturbation is not a wave, but rather, is a Gaussian-shaped bump. "Waves" normally have both positive and negative excursions about the mean value. Apparently, one-dimensional propagating modes do not have this wavelike feature. Although the density waves are positive only, they are not technically solitons since they do broaden in time. Second, the propagating mode is Gaussian with a width that is initially equal to that of the spatial source. The width of this and the thermal diffusive density change

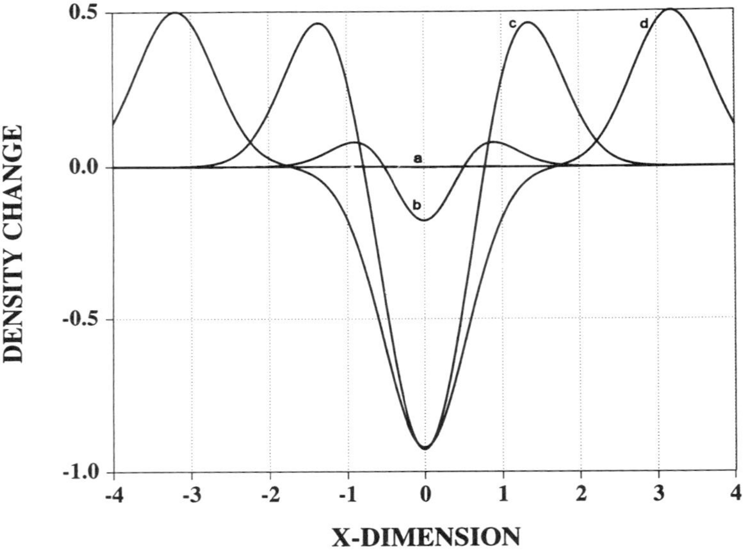

(a)

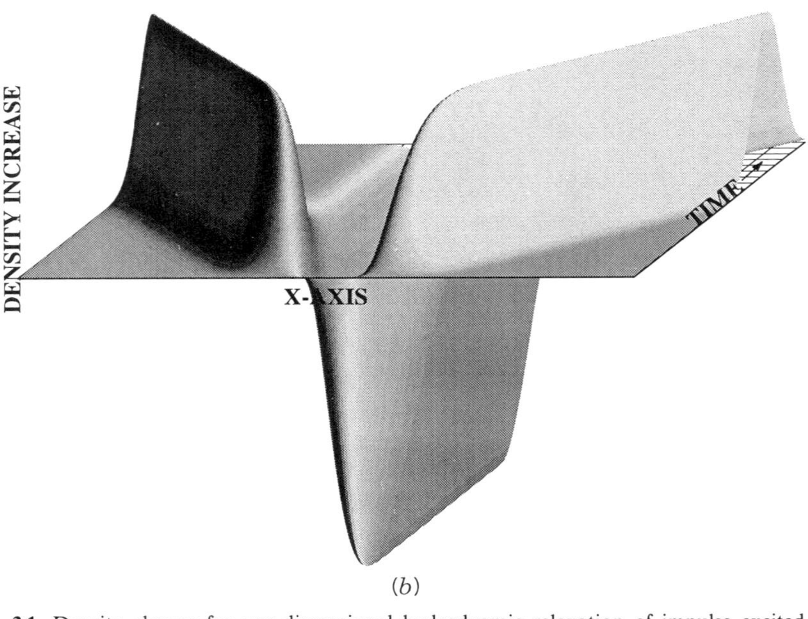

(b)

Figure 3.1. Density change for one-dimensional hydrodyamic relaxation of impulse excited sample. The parameters use for these calculations are $w = 10\ \mu\text{m}$, $D_T = \Gamma = 2.2 \times 10^{-5}\ \text{m}^2\ \text{s}^{-1}$ $c = 300\ \text{m}\ \text{s}^{-1}$. (*a*) Spatially dependent density change plotted at (a) 10^{-9}, (b) 10^{-8}, (c) 4×10^{-8}, and (d) 10^{-7} s; (*b*) three-dimensional plot of the density versus time and spatial coordinate. The density decrease in the middle is the diffusive mode and the "wings" are the propagating acoustic mode.

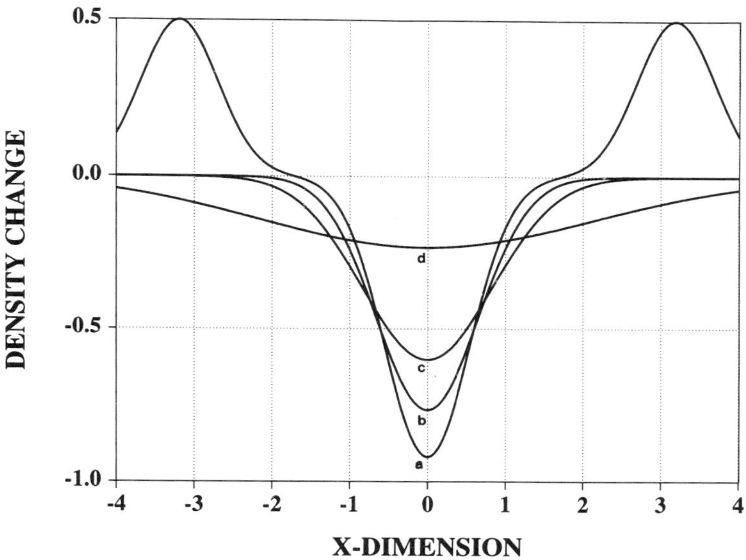

(*a*)

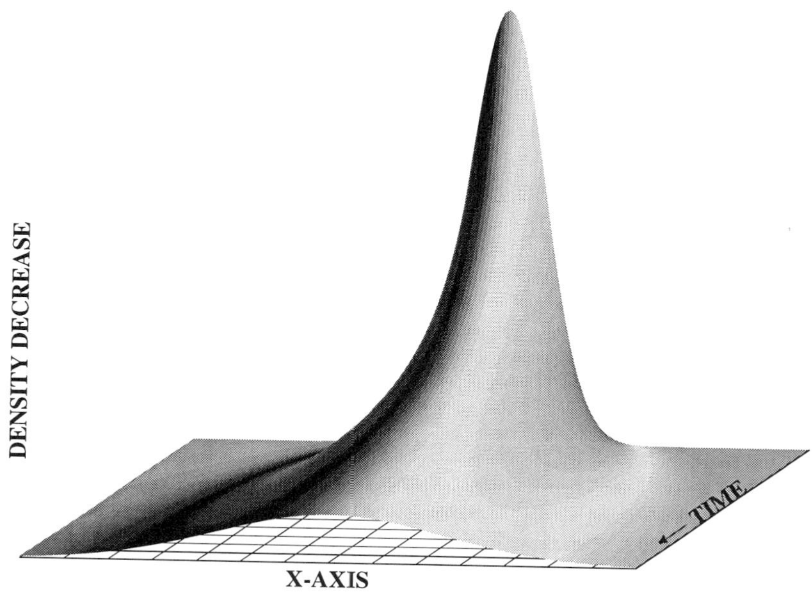

(*b*)

Figure 3.2. Density change for one-dimensional hydrodynamic relaxation for times longer than these required for propagating model dissipation. Same conditions as Figure 3.1. (*a*) Spatial dependence at (a) 10^{-7}, (b) 4×10^{-7}, (c) 10^{-6}, and (d) 10^{-5} s; (*b*) time-dependent spatal evolution. Notice that the density and time axes are reversed from those of part (*A*).

component both broaden in time. It may be surmized that acoustic attenuation is a diffusive broadening effect wherein the "wavelength" is increased with a corresponding decrease in magnitude. Third, for the infinitesimally thin profile, the widths of the propagating and diffusive density changes are initially infinitesimally thin. They broaden into a Gaussian form as time progresses. Diffusion and acoustic attenuation are linear Gaussian processes. The linearity means that impulse solutions can be convoluted with any source. The Gaussian character occurs because the time evolution of a spatial impulse results in a Gaussian-shaped profile.

Finally, the density perturbation takes a finite time to develop. The time required for the density to change in the region of the perturbation can be found by setting $x = 0$ in the density equation. Assuming that the acoustic attenuation and thermal time constants are small compared to the density change development time yields

$$\delta\rho(x = 0, t \ll t_c) \approx \frac{2^{1/2}\beta Q}{\pi^{1/2} w C_p} [e^{-2(ct)^2/w^2} - 1] \tag{95}$$

The acoustic-limited density perturbation occurs on a time scale of about $t_a = w/c$. This time constant is often referred to as the acoustic-limited rise time for sample perturbation. The acoustic-limited rise time places a lower limit on times that can be observed in photothermal spectroscopy. Pulsed excitation or excited-state relaxation that occurs in times much shorter than the acoustic-limited rise time do not influence the rate of density change.

3.6.2. Two-Dimensional Cylindrically Symmetric Example

Propagating acoustic and thermal density changes are important to determine in the case of Gaussian-profile laser-excited homogeneous samples. This experimental configuration is often used to study excited-state relaxation rates (Grabiner et al. 1972, Siebert et al. 1974, Barker and Rothem 1982). It is important to be able to model the effect that the propagating mode has on the signal rise time when relaxation times are on the order of the acoustic-limited rise time. However, equations describing the propagation mode do not yield simple solutions for the two-dimensional radially symmetric geometry of this experimental configuration.

Grabiner et al. (1972) and Siebert et al. (1974) used solutions for density change without thermal diffusion. Barker and Rothem (1982), Barker and Toselli (1989), and Pugh (1989) consider solutions for the space dependent density change that correspond to obtaining solutions to the coupled hydrodynamic mode equation. Barker and Rothem solved the conservation equations for a nonviscous gas, including mass diffusion of the excited-state. After examining the model for several test cases, Barker and Toselli concluded that mass diffusion was not needed for most practical samples. In all cases to date, hydrodynamic equations are solved assuming that a single excited-state relaxes through first-order kinetics.

Shown below is a solution to the density change equation, including viscous damping of the acoustic wave. The equation is solved for impulse excitation. Solutions in cases where the rate of heat deposited in the sample is limited by either excited-state relaxation, or the excitation irradiance can be found by subsequent convolution of the time-dependent heat generation terms with the impulse response solution. The solution may be obtained as the sum of two components:

$$\delta\rho_{\text{impulse}}(r, t) = \delta\rho_{\text{diffusion}}(r, t) + \delta\rho_{\text{acoustic}}(r, t) \tag{96}$$

The first component is the impulse-response density change due to thermal diffusion. This component is found under the assumption that the cell radius is much greater than the thermal diffusion length at any time of practical importance. This assumption allows the simpler infinite media solution to be used. The second component is the impulse response of the propagating mode density change. The integral equation for the impulse-response propagating mode density change for infinite boundary condition is given in Appendix 3.B. However, the infinite media solution does not correctly model the experimental observations for an enclosed sample. The density wave propagates much more rapidly than the spread of heat due to thermal diffusion. The acoustic wave may reflect off the sample cell wall and return to the center of the sample several times before the heat has diffused to the cell wall. The infinite media approximation cannot be used when this is the case. The general solution is found by using a series of Bessel's functions.

The sample is optically thin and homogeneous. The sample is contained in a cylindrical cell, of radius b, and of a length sufficient to merit neglecting the longitudinal hydrodynamic terms. The cell radius is large enough that the cell wall boundary does not affect thermal diffusion. After pulsed irradiation, there is a semi-infinite line of heat deposited in the center of the cell given by

$$q_H(r, t) = \frac{2Q\delta(t)}{\pi w^2} e^{-2r^2/w^2} \tag{97}$$

The density change due to thermal diffusion is found by convolution with the impulse-response function given in Section 3.5:

$$\delta\rho_{\text{diffusion}}(r, t) = \int_0^t \int_{-\infty}^{\infty} \int_{-\infty}^{\infty} \int_{-\infty}^{\infty} q_H(\sqrt{x'^2 + y'^2}, t')\delta\rho_{\text{impulse}} \times (x - x', y - y', z - z', t - t')\, dx'\, dy'\, dz'\, dt' \tag{98}$$

where the radius is defined in the x–y plane as $r^2 = x^2 + y^2$ and the z-axis is the longitudinal axis. Convolution yields the diffusion component

$$\delta\rho_{\text{diffusion}}(r, t) = -\frac{2\beta Q}{\pi C_P w^2(1 + 2t/t_c)} e^{-2r^2/w^2(1+2t/t_c)} \tag{99}$$

The characteristic thermal decay time, $t_c = w^2/4D_T$, is the same as that for the one-dimensional problem, and the time-dependent width of the Gaussian density change is $w(1 + 2t/t_c)^{1/2}$.

The propagating component is found by solving the differential equation

$$\left(\frac{\partial}{\partial t} - \Gamma\nabla^2\right)^2 \delta\rho_{\text{acoustic}} - c^2\nabla^2\,\delta\rho_{\text{coustic}} = \frac{\beta}{C_P}\left(\frac{\partial}{\partial t} - \Gamma\nabla^2\right) q_H \tag{100}$$

with the finite cylindrical cell boundary conditions. The differential equation is similar to the Bessel equation of order zero, so it is assumed that the solution can be obtained as a series expansion of zero-order Bessel's functions, $J_0(r)$. Sums over the series of $J_0(X_j r)$ span the finite radial space between 0 and 1. X_j is a zero of the first-order Bessel function, $J_1(X_j) = 0$. Bessel's functions also have two practical advantages: (1) the radial Laplacian operation in cylindrical symmetry has the simple result, $\nabla^2 J_0(ar) = -a^2 J_0(ar)$, and (2) the $J_0(X_i r)$ possess an orthogonality relationship,

$$\int_0^1 rJ_0(X_i r)J_0(X_j r)\, dr = \delta_{i,j}\,\frac{J_0^2(X_i)}{2} \tag{101}$$

(Abramowitz and Stegun 1965) where the $\delta_{i,j}$ is used to indicate that the integral is zero if $i \neq j$.

Bessel's functions span the space of r from 0 to 1 and the solution will be of the general form

$$\delta\rho_{\text{acoustic}}(r, t) = \sum_{j=1}^{\infty} c_j(t)J_0(v_j r) \tag{102}$$

where $v_j = X_j/b$. The zeros of $J_1(r)$ are used to satisfy the boundary condition $d\delta\rho(b, t)/dr = 0$, which ensures that the acoustic wave possesses the correct phase (Bates et al. 1970). A similar orthogonality relationship exists for the zeros of $J_0(r)$, and density expansions satisfy the condition $\rho(b, t) = 0$ (Barker and Toselli 1989). However, using these zeros does not result in the correct phase relationship for times greater that b/c. The time-dependent expansion coefficients, $c_j(t)$, are determined from the differential equation. To facilitate the solution, the heat source is also expanded as a series zero-order Bessel function:

$$q_H = \sum_{i=1}^{\infty} G_i J_0(v_i, t) \tag{103}$$

The G_i are found by using the orthogonality integral relationship. Multiplying both sides of the equation by sums over $J_0(v_j, r)$ and integrating over r results in

$$G_j = \frac{2}{b^2[J_1(v_j r)]^2} \int_0^b r J_0(v_j r) q_H(r)\, dr \tag{104}$$

By assuming that the cell radius is much greater than the beam waist radius, the integration can be carried out from 0 to ∞. Using the definite integral

$$\int_0^{\infty} r e^{-ar^2} J_0(br)\, dr = \frac{1}{2a} e^{-b^2/4a} \tag{105}$$

the G_j obtained with this approximation are

$$G_j = \frac{Q}{\pi b^2[J_0(v_j)]^2} e^{-v_j^2 w^2/8} \tag{106}$$

$\delta\rho$ and q_H are both sums over identical Bessel's functions and the differential equation can be solved for each term in the series independently. The differential equation is solved as follows. The Bessel's function expansions for $\delta\rho$ and q_H are for a given j substituted into the differential equation. The Laplacian of the spatially dependent Bessel's functions are found. Then a Laplace transform of the resulting expression is taken. Performing these steps for one j term is equivalent to doing so for each term in the series. The result for a single term is

$$(\zeta + \Gamma v_j^2)^2 c_j(\zeta) + c^2 v_j^2 c_j(\zeta) = \frac{\beta}{C_P}(\zeta + \Gamma v_j^2) G_j \tag{107}$$

The $J_0(v_j r)$ have canceled on both sides of the equation since both $\delta\rho$ and q_H are expansions of J_0. The Laplace transform of $c_j(t)$ is simplified since $c_j(t = 0)$ and its derivatives are equal to zero (Barker and Rothem 1982). The solution for the $c_j(\zeta)$ is found in terms of ζ, Γ, and v_j:

$$c_j(\zeta) = \frac{\beta}{C_P} \frac{G_j(\zeta + \Gamma v_j^2)}{(\zeta + \Gamma v_j^2)^2 + c^2 v_j^2} \tag{108}$$

The inverse Laplace transform is

$$c_j(t) = \frac{\beta}{C_P} G_j e^{-v_j^2 \Gamma t} \cos ctv_j \tag{109}$$

The density change due to the propagating mode found by substitution of the $c_j(t)$ into the general density solution formula is

$$\delta\rho_{\text{acoustic}}(r, t) = \frac{\beta Q}{\pi b^2 C_P} \sum_{j=1}^{\infty} \frac{e^{-v_j^2 w^2(1+2t/\tau_\Gamma)/8} \cos ctv_j}{[J_0(v_j)]^2} \tag{110}$$

Combining this result with that obtained for the diffusion density change, and using a unit Q, results in the impulse response of the sample to Gaussian profile laser excitation:

$$\delta\rho_{\text{impulse}}(r, t) = \frac{\beta}{\pi b^2 C_P} \sum_{j=1}^{\infty} \frac{e^{-v_j^2 w^2(1+2t/\tau_\Gamma)/8} \cos ctv_j}{[J_0(v_j)]^2} J_0(v_j r) - \frac{2\beta}{\pi C_P w^2(1 + 2t/t_c)} e^{-2r^2/w^2(1+2t/t_c)} \tag{111}$$

The acoustic-limited rise time can be found by examining the behavior at $r = 0$. This time dependence is given by

$$\delta\rho_{\text{impulse}}(0, t) = \frac{\beta}{\pi b^2 C_P} \sum_{j=1}^{\infty} \frac{e^{-v_j^2 w^2(1+2t/\tau_\Gamma)/8} \cos ctv_j}{[J_0(v_j)]^2} - \frac{2\beta}{\pi C_P w^2(1 + 2t/t_c)} \tag{112}$$

Figure 3.3 shows the time-dependent density change due to the first, propagating mode, term at $r = 0$. The time axis is given in units of w/c. The early time-dependent behavior is independent of Γ. The density change due to the propagating mode takes w/c to change from the initial value of 1 to the minimum value. This is the acoustic-limited rise time. To see what effect this component has on the total density requires the diffusion mode component. A combined propagating and diffusion mode density change solution for argon with a beam waist of $100\,\mu\text{m}$ is shown in Figure 3.4. The relative density change attributed to the diffusion mode is equal and opposite to that of the propagating mode at zero time. The density decreases to a minimum at the acoustic-limited rise time of w/c and then increases slowly. The slow increase is due to thermal diffusion. The time required to reach the minimum density change in this case is about 320 ns.

Radial behavior of the density change is illustrated in Figure 3.5. In this example the Gaussian beam radius is $10\,\mu\text{m}$ and the time is $2\,\mu\text{s}$ after impulse irradiation. The propagating acoustic wave exhibits both positive and negative density changes relative to the average. This is more typical of the wavelike behavior that one would expect from a sound wave. The leading form of the wave is a compression, density increase, while the tailing

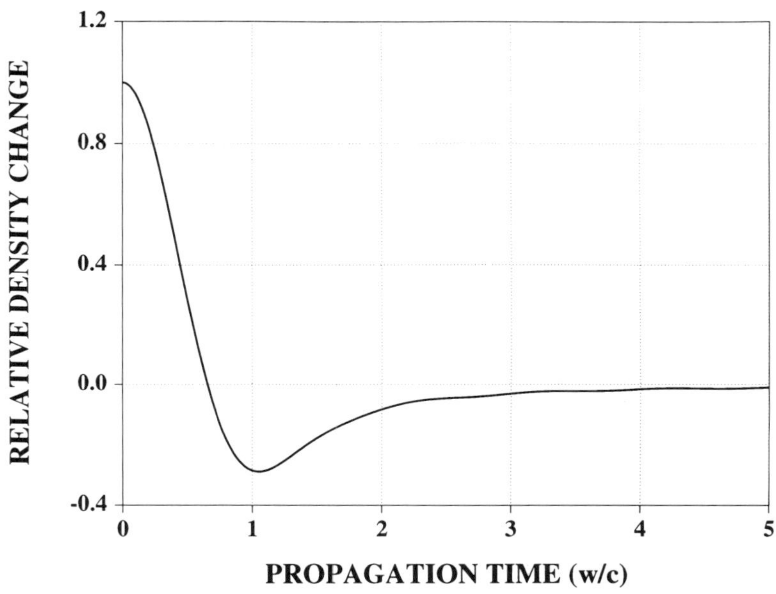

Figure 3.3. Propagating or acoustic mode contribution to the on-axis density change for the radially symmetric impulse excitation case. The time axis units are w/c.

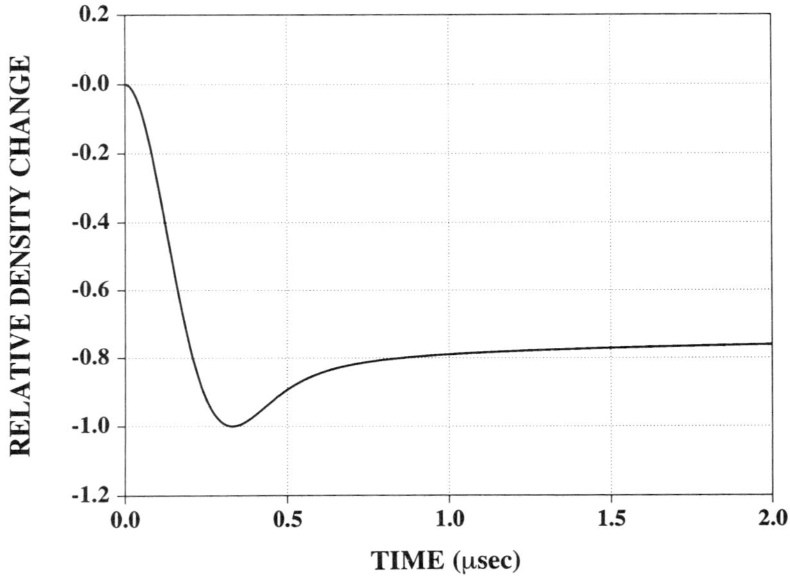

Figure 3.4. Combined diffusion and propagating mode density changes for the on-axis radially symmetric case. Hydrodynamic parameters for argon gas at standard temperature and pressure were used in the calculations. The Gaussian beam waist radius is 100 μm.

part of the wave is a rarefaction. This wave behavior is found in both cylindrical and spherical symmetric acoustic waves (Temkin 1981) but not in the planar waves discussed earlier. The long-time behavior is illustrated in Figure 3.6. As the wave propagates, it broadens and decreases with a characteristic time constant of τ_Γ. The magnitude of the wave also decreases because it occupies larger volume at larger radiuses. There are two mechanisms for diminishing the size of the density perturbation with time and, consequently, distance: one due to viscous damping and the other due to the volume, which falls of as $1/r = 1/ct$.

At very long times the propagating mode will reflect off the cell walls. The reflected wave is focused back into the center of the cell. The zero radius density change illustrating this effect is shown in Figure 3.7. The density change perturbation at 63 μs is due to the reflected acoustic wave. The reflected wave has the correct phase (high-density leading) and is comparable to that found by Jacobs (1989) using a numerical approach to solve the hydrodynamic equations. The returning propagating wave will be significant only in cylindrical sample cells. For rectangular sample cells, the propagating wave continues to diverge after reflection off of the flat surface. This case may be approximated either by using an infinite cylindrical cell radius or by

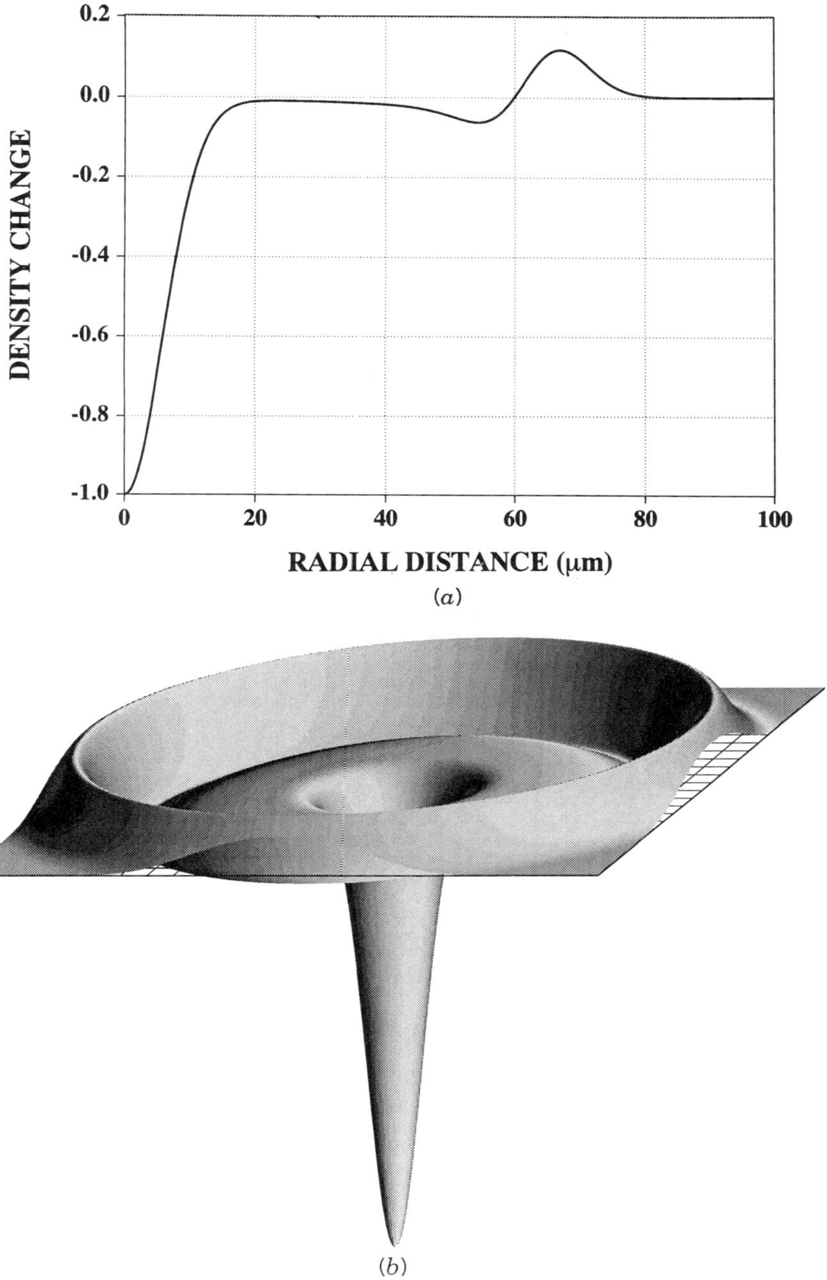

Figure 3.5. Combined diffusion and propagating mode radially dependent density changes for Gaussian impulse excitation. The hydrodynamic parameters are the same as those in Figure 3.4. The time is 200 ns after irradiation. The Gaussian beam waist radius is 10 μm. The acoustic wave can be seen at about 65 μm. (*a*) Radially dependence; (*b*) three-dimensional projection.

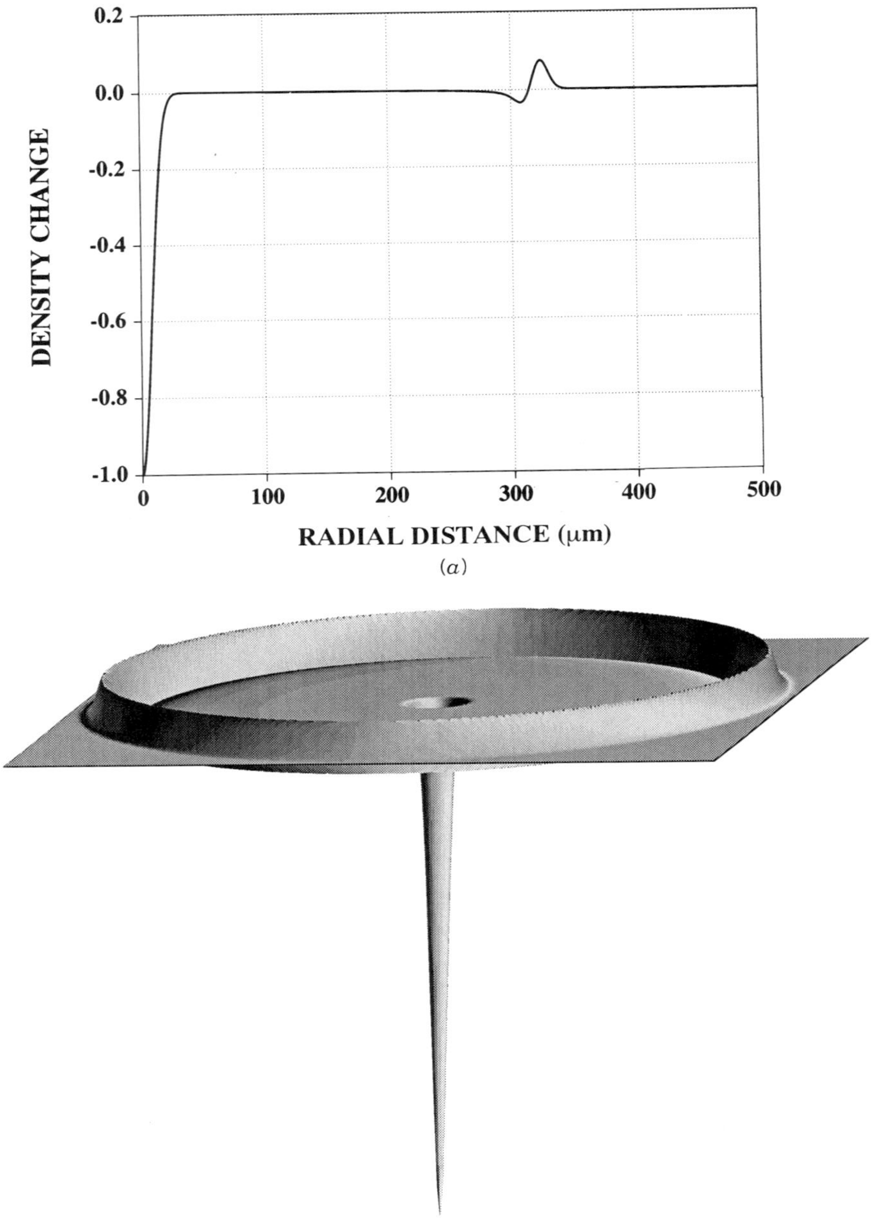

Figure 3.6. Radially depentent density changes for Gaussian impulse excitation. The hydrodynamic parameters are the same as those in Figure 3.5. The time is 1 μs. (*a*) Radially dependence; (*b*) three-dimensional projection.

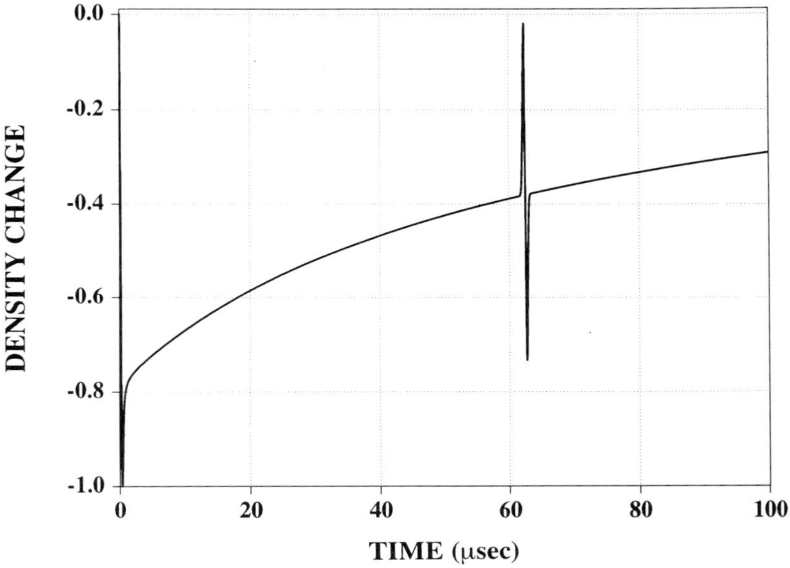

Figure 3.7. Time-dependent density changes for Gaussian impulse excitation in a 1-cm-diameter cylindrical cell. The hydrodynamic parameters are the same as those in Figure 3.4. The acoustic perturbation at $\sim 64\,\mu s$ is the reflected acoustic wave.

solving for the impulse response for an infinite medium. Infinite-medium solutions for the propagating component are given in Appendix 3B.

3.6.3 Coupled solutions

The implicit assumption in obtaining the solution discussed above was that the differential equation describing the density change could be factored into the three approximate roots. However, this assumption does not have to be made. The mathematical tools used to obtain the propagating density change solution can be applied to the differential equation resulting from the solution to the conservation equations. In this case there is no apparent distinction between thermal diffusion and propagating modes. Using the same expansion for the density change, the coupled hydrodynamic mode impulse-response solution is

$$c_j(\zeta) = -\frac{\beta c^2}{C_P}\frac{v_j^2 G_j}{\zeta^3 + \zeta^2 v_j^2(\gamma D_T + v_l) + \zeta v_j^2(c^2 + \gamma v_l D_T v_j^2) + D_T c^2 v_j^4} \tag{113}$$

The cubic equation in the demoninator has three roots that can be determined using either numerical techniques or the cubic equation solution.

These three roots consist of one real negative root and two roots that are complex conjugates. The real root corresponds to the diffusion mode, and the complex-conjugate roots correspond to the propagating mode. The inverse Laplace transform of the product of the real root with the two complex-conjugate roots can be obtained. The cubic equation roots can be separated as

$$c_j(\xi) = -\frac{C_j}{(\zeta + R_j)(\zeta + A_j + iB_j)(\zeta + A_j - iB_j)} \tag{114}$$

The inverse transform is

$$c_j(t) = -C_j \frac{e^{-A_j t}(R_j - A_j)\sin B_j t - e^{-(A_j + R_j)t} B_j \cos B_j t + B_j e^{-R_j t}}{B_j(R_j^2 - 2A_j R_j + A_j^2 + B_j^2)} \tag{115}$$

The latter is similar to that obtained using separate diffusion and propagating modes for $R_j \approx v_j^2 D_T$, $A_j + R_j \approx v_j^2 \Gamma$, and $B_j = v_j c$. In fact, assuming that $R_j - A_j \approx 0$ results in

$$c_j(t) = \frac{\beta v_j^2 G_j}{C_P} [e^{-(A_j + R_j)t} \cos B_j t - e^{-R_j t}] \tag{116}$$

With $R_j = v_j^2 D_T$ and $A_j + R_j = v_j^2 \Gamma$, this is exactly the solution obtained using the Bessel's function expansion for the separate diffusive and propagating modes.

Predictions of a coupled solution have been compared to those of the separate diffusive and propagating mode solutions discussed above for a number of test cases by this author. The real exponential and trigonometric parts of the model predictions are examined separately. In most cases the real part is equivalent to the thermal diffusion term in the approximate solution, and the trigonometric part is equivalent to the propagating mode solution obtained with the Bessel's function series. One exception to this generalization occurs when the sample cell diameter is on the order of the thermal diffusion length. In this case the infinite-media approximation fails and the Bessel's function series solution presumably yields more accurate predictions.

3.7. SOLUTIONS INCLUDING MASS DIFFUSION

Mass diffusion can affect the radial distribution of the heat source in photothermal spectroscopy. Mass diffusion may have to be taken into account when the lifetime of the excited-state is on the order of the mass diffusion time, $t_D = w^2/4D$, where D is the binary mass diffusion coefficient for the analyte in a particular solvent. The effects of mass diffusion have been considered for gas-phase samples. Barker and Rothem (1982) and Barker and Toselli (1989) considered the effect on the rise time of the temperature change, and Bialkowski (1984) has considered the effects on the magnitude of analytical signals. Although the problem was addressed for gas-phase samples, the results should hold equally well for liquid samples. When mass diffusion competes with relaxation, energy deposited in the absorbing species spreads out over a volume greater than the irradiated volume. This results in a temperature change that is less than that expected without mass diffusion. Relatively long excited-state lifetimes may affect the relative accuracy of absorption coefficient quantitation for pulsed and continuously excited samples. The inaccuracy is compounded in photothermal lens spectroscopy since signals are proportional to the curvature of the temperature change.

Impulse excitation with a laser pulse of a duration much shorter than the time scale for excited-state relaxation will result in an excited-state population that can diffuse in space prior to relaxation and heat production. For impulse, Gaussian profile laser excitation, the number density of species in the excited-state will be

$$N^*(r, 0) = \frac{2N_{\text{tot}}\sigma Q}{\pi w^2 h\nu} e^{-2r^2/w^2} \tag{117}$$

where N^* (m^{-3}) is the number density in the excited-state, σ the absorption cross section at frequency ν, and $Q(J)$ is the pulse energy. It is assumed that the pulse energy was much less than that required for optical bleaching and that no relaxation has occurred during the excitation time. In the absence of mass diffusion, an excited-state that relaxes by a first-order process will deposit energy in the sample with a time-dependent power per unit area of

$$q_H(r, t) = h\nu N^*(r, 0)ke^{-kt} \tag{118}$$

where k is the rate constant for excited-state relaxation. Assumed here is the energy loss of $h\nu$ per relaxation event.

On the other hand, mass diffusion causes species to disperse. Mass diffusion occurs independent of any distribution. It occurs in isotropic and anisotropic distributions. Excited-state species are effectively labeled by their energy level, so diffusion results in a dispersion or redistribution in space. The differential mass diffusion equation describing this process for the two-dimensional geometry of collimated laser excitation is

$$\frac{\partial N^*(r, t)}{\partial t} - D\nabla^2 N^*(r, t) = N^*(r, 0) \tag{119}$$

where $N^*(r, 0)$ is the source of diffusion species at zero time. The solution of the mass diffusion problem is essentially the same as that for thermal diffusion. For an excited-state that does not relax, the mass diffusion solution for times greater than or equal to zero is

$$N^*(r, t) = \frac{2N^*(r, 0)}{\pi w^2(1 + 2t/t_D))} e^{-2r^2/w^2(1 + 2t/t_D)} \tag{120}$$

The characteristic mass diffusion time, $t_D = w^2/4D$, has been used. The similarity of this definition to that for the characteristic thermal diffusion time is due to the fact that the diffusion equations are equivalent in form. Assuming that the mass diffusion coefficient does not change with the energy state of the absorbing species, a reasonable assumption, the power produced by excited-state relaxation is simply the product of the excited-state mass diffusion distribution with the relaxation rate

$$q_H(r, t) = \frac{2N_{\mathrm{tot}}\sigma Qk}{\pi w^2(1 + 2t/t_D)} e^{-kt} e^{-2r^2/w^2(1 + 2t/t_D)} \tag{121}$$

The resulting density change can be found by convolution of the mass diffusive source with the impulse response for thermal diffusion. The diffusion term has inseparable space and time dependencies. Performing the spatial convolution results in

$$\delta\rho(r, t) = \frac{2\beta N_{\mathrm{tot}}\sigma Qk}{\pi w^2 C_P} \int_0^t \frac{e^{-k\tau} e^{-2r^2/w^2(1 + 2\tau/t_D + 2(t-\tau)/t_c)}}{1 + 2\tau/t_D + 2(t - \tau)/t_c} d\tau \tag{122}$$

This result can only be expressed in terms of the time convolution of the exponential source with the diffusion term, but the limiting behavior may be understood by examining the time constants for the various processes.

The three competitive time constants are that for thermal diffusion, t_c, that for mass diffusion, t_D, and that for excited-state relaxation, $t_e = 1/k$. It is apparent that when t_e is very short compared to t_c and t_D, the exponential decay term approaches the delta function, $\delta(\tau)$. Evaluating the time convolution integral results in the impulse response for thermal excitation. In the gas phase, the mass and thermal diffusion coefficients are related. Hard-sphere collision theory yields the theoretical transport coefficients,

$$D = \frac{3}{8}\left(\frac{k_B T}{\pi M}\right)^{1/2} \frac{1}{d^2 n} \tag{123}$$

$$D_T = \frac{25}{32\gamma}\left(\frac{k_B T}{\pi M}\right)^{1/2} \frac{1}{d^2 n}$$

where D is the self-mass diffusion coefficient, k_B the Boltzmann constant, M the molecular weight, d the collision diameter, and n the molar density. The molar density is related to temperature and pressure by $n = P/k_B T$. D and D_T are both inversely proportional to pressure and proportional to $T^{3/2}$. Changing the pressure or temperature will thus affect t_D and t_c in the same fashion. Since an excited-state lifetime much shorter than that for either mass or thermal diffusion results in the thermal diffusion impulse response, it can be said that if the time constant for excited-state relaxation is much shorter than that for diffusion, mass diffusion does not have to be considered.

The implication that $t_D \approx t_c$ is of some consequence to gas-phase excited-state relaxation rate studies. In these studies a short pulsed laser is used to prepare the excited state, and the rise time of the temperature change or photothermal signal is used to determine the rate of excited-state relaxation. To obtain an accurate estimate of the rise time, t_e must be longer than the acoustic rise time, $t_a = w/c$. When this is the case, the time convolution may be approximated by

$$\delta\rho(r, t) \approx \frac{2\beta N_{\text{tot}}\sigma Q}{\pi w^2(1 + 2t/t_c)C_P}(1 - e^{-kt})e^{-2r^2/w^2(1 + 2t/t_c)} \tag{124}$$

since the τ mass and thermal diffusion dependencies cancel. Interestingly enough, when $t_D = t_c$, mass and thermal diffusion contributions cancel over the time of the convolution, resulting in a simple analytical expression for the time-dependent temperature change that occurs in gas-phase photothermal spectroscopy. This result can be cast in a form using the generalized relaxation term

$$\delta\rho(r, t) \approx R(t)\, \delta\rho_{\text{impulse}}(r, t) \tag{125}$$

Of course, the density change magnitude and temporal form will vary with the excited-state relaxation time. The mass and thermal diffusion time constants are proportional to the squared beam waist radius. On the other hand, the excited-state lifetime is independent of excitation geometry. Subsequently, the relative effects of mass diffusion and excited-state relaxation can change with experimental geometry. In addition, the excited-state lifetime is usually inversely proportional to P since relaxation is a collision-induced energy transfer process. Subsequently, decreasing the pressure will increase t_E while decreasing both t_D and t_c. Mass diffusion may be important in gas-phase samples when (1) the excitation laser beam waist radius is small, and (2) the total gas pressure is low.

The hard-sphere collision transport coefficients are only approximations. The assumptions used in obtaining the simplified equation should be tested. Binary mass diffusion coefficients have been tabulated for a number of different mixtures for gases, liquids, and solids (Cussler 1984). In addition, there are several fairly accurate theories relating molecular parameters to gas-phase binary diffusion coefficients. Binary gas-phase mass diffusion coefficients range from about $10^{-5}\,m^2\,s^{-1}$ for n-octane/N_2 to $10^{-4}\,m^2\,s^{-1}$ for H_2/He atmospheric pressure gas mixtures. Although the mass diffusion coefficients do vary from species to species, for small molecules the mass diffusion coefficient is on the order of, and often very close to, the thermal diffusion coefficent for the matrix gas. This, in conjunction with III-125 might explain the success of gas-phase excited-state relaxation studies based on measurement of the initial rate of signal production.

Determination of the liquid mass diffusion coefficients based on the size of the diffusing solute and viscosity of the solvent is tricky because the solute may form a solvation complex; the coefficients should be obtained experimentally. Liquid diffusion coefficients do not exhibit strong temperature and pressure dependencies and the diffusion coefficients are about the same for most midsized species. Typical binary mass diffusion coefficients are on the order of $10^{-9}\,m^2\,s^{-1}$ (e.g., benzoic acid in water). The mass diffusion coefficient is much smaller than that for thermal diffusion in condensed-phase samples. The latter are on the order of $D_T \sim 10^{-7}\,m^2\,s^{-1}$.

A simple calculation can be used to see if mass diffusion need be considered in condensed-phase samples. A well-focused visible laser can produce a TEM_{00} beam with a beam waist radius of 10 μm or less. The mass diffusion time in a liquid would be on the order of $t_D \sim 25$ ms. This is about the time scale for relaxation of organic molecule triplet states. So mass diffusion may have to be considered in liquid samples when long-lived metastable states are formed with high quantum yield. On the other hand, mass diffusion coefficients are two or more orders of magnitude smaller than the thermal diffusion coefficients and may therefore be neglected in the time convolution integral. In this case the time convolution reduces to

$$\delta\rho(r, t) \approx \frac{2\beta N_{\text{tot}}\sigma Q k}{\pi w^2 C_P} \int_0^t \frac{e^{-k\tau} e^{-2r/w^2(1 + 2(r - \tau)/t_c)}}{1 + 2(t - \tau)/t_c}\, d\tau \tag{126}$$

which is just

$$\delta\rho(r, t) \approx R'(r) * \delta\rho_{\text{impulse}}(r, t) \tag{127}$$

The effects of mass diffusion on the density that occurs in photothermal spectroscopy performed with any excitation geometry should be similar to those for TEM_{00} laser excitation since convolution can be used to model any experimental geometry. In gas-phase samples with $t_D \approx t_c$, the effect of mass diffusion cancels the effect of thermal diffusion in the convolution. This results in an analytical expression that can be used to model the competative effects of diffusion and first-order excited-state relaxation on the temperature change magnitude and rise time. In homogeneous solution-phase samples, the time scales for thermal diffusion are much shorter than those for mass diffusion. Subsequently, mass diffusion may be neglected relative to thermal diffusion. The temperature change that occurs is a convolution of the relaxation rate with thermal diffusion for the impulse response.

When the time scale for excited-state relaxation is much shorter than that required for thermal diffusion, the convolution describing the time-dependent density change can be simplified. In this case

$$\delta\rho(r, t) \approx \frac{2\beta N_{\text{tot}}\sigma Q k}{\pi w^2(1 + 2t/t_c)C_P}(1 - e^{-k\tau})e^{-2r^2/w^2(1 + 2t/t_c)} \tag{128}$$

and as is the case of the gas-phase sample,

$$\delta\rho(r, t) \approx R(t)\,\delta\rho_{\text{impulse}}(r, t) \tag{129}$$

This case is valid only in the limit of $t_E \ll t_c$.

The equations derived by Barker and Rothem (1982) and later Barker and Toselli (1989) take into account both thermal diffusion and propagation mode energy dissipation. This solution also assumes that impulse excitation first produces excited-state species with a given number density. The excited-state species simultaneously diffuse, through mass diffusion, and relax to produce an energy increase local to the region bound by the mass diffusion process. Although no distiction is made between thermal diffusion and acoustic modes of energy dissipation, the solution does result in stationary and periodic density change components that can be interpreted as the thermal diffusion and propagating mode components, respectively.

A solution for the excited-state mass diffusion problem is first obtained by solving the governing differential equation in the absence of the energy-dissipative hydrodynamic relaxation modes. For impulse Gaussian laser excitation, the mass diffusion/excited-state relaxation equation solved for a cylindrical sample cell yields

$$N^*(r, t) = \sum_{i=1}^{\infty} G_i J_0(v_i r)\exp(\alpha_i t) \tag{130}$$

where N^* is the number density of excited species, G_i the expansion coefficients given above, v_i the relative Bessel's function zero, x_i/b, (b is again the cylinrical sample cell radius), and α_i is given by

$$\alpha_i = -(Dv_i^2 + k) \tag{131}$$

where D is the binary mass diffusion coefficient and k is the first-order excited-state relaxation rate constant. Including the hydrodynamic equations with this mass diffusive source results in the governing equation of motion for the density change:

$$\begin{aligned}\frac{\partial^3 \delta\rho}{\partial t^3} - \gamma D_T \nabla^2 \frac{\partial^2 \delta\rho}{\partial t^2} - c^2\nabla^2 \frac{\partial \delta\rho}{\partial t} + c^2 D_T \nabla^2\nabla^2\delta\rho \\ = \frac{H_1}{\rho C_V(\partial T/\partial P)_\rho}(D\nabla^2 - k)\nabla^2 N_1(r, t)\end{aligned} \tag{132}$$

where H_1 is the energy of the excited-state able to be released to kinetic energy in the sample matrix. This equation is solved using the Laplace transform techniques discussed above. The expansion coefficients are

$$c_j(\zeta) = -\frac{H_1}{\rho C_V(\partial T/\partial P)_\rho} \frac{v_j^2 G_j(Dv_j^2 + k)}{\zeta^3 + \zeta^2 v_j^2 \gamma D_T + \zeta v_j^2 c^2 + v_j^4 D_T c^2} \tag{133}$$

This equation has a cubic equation in the denominator which is determined using numerical techniques. The three roots consist of one real, negative root, and two roots that are complex conjugates. As with the coupled solution without mass diffusion, the real root corresponds to the diffusion mode and the complex roots correspond to the propagating mode. The inverse Laplace transform of the product of the real root with the two complex-conjugate roots can be obtained. The solutions will be similar to those given in equation 116, but will include the mass diffusion and excited-state relaxation terms,

$$c_j(t) \approx \frac{\beta H_1 v_j^2 G_j(Dv_j^2 + k)}{C_P}[e^{-(A_j + R_j)t}\cos B_j t - e^{-R_j t}] \tag{134}$$

Barker and Rothem have run simulations under a number of conditions. They conclude that mass diffusion can affect the apparent rise time of the photothermal signal when the mass diffusion time is not well separated from the times for thermal diffusion and excited-state relaxation. In these cases the full coupled hydrodynamic solution is used to model the density change or photothermal signal, and experimental data are compared to the theoretical results to estimate the relaxation rate constant.

3.8. EFFECT OF HYDRODYNAMIC RELAXATION ON TEMPERATURE

The temperature change may be approximated from the density change due to both the diffusive and propagating modes. Temperature is related to pressure and density through the equation of state,

$$\delta T = \frac{K_T}{\beta}\,\delta P - \frac{1}{\rho\beta}\,\delta\rho \tag{135}$$

The results of the hydrodynamic relaxation equation allow one to estimate both the density change and the pressure change. The pressure change is related to the propagating mode density change through the scale factor, c^2, at constant temperature. The temperature change can thus be estimated from the two components of density change. For the cylindrically symmetric example, the radially dependent temperature change is

$$\delta P(r, t) = c^2 \delta\rho_{\text{acoustic}}(r, t) \tag{136}$$

Using the total density change and this pressure change in the equation of state results in the temperature estimate

$$\delta T(r, t) = \frac{K_T c^2}{\beta}\,\delta\rho_{\text{acoustic}}(r, t) - \frac{1}{\rho\beta}\,[\delta\rho_{\text{acoustic}}(r, t) + \delta\rho_{\text{diffusion}}(r, t)] \tag{137}$$

This can be further simplified using the isentropic sound velocity

$$\delta T(r, t) = \frac{1}{\rho\beta}\,[(\gamma - 1)\,\delta\rho_{\text{acoustic}}(r, t) - \delta\rho_{\text{diffusion}}(r, t)] \tag{138}$$

The diffusion component of the density change decreases the density while the propagating acoustic component increases the density. Thus both terms contribute to a temperature change increase. For impulse excitation, one would expect the initial temperature change to be

$$\delta T(r, 0) = \frac{q_H(r, 0)}{\rho C_V} \tag{139}$$

since the sample expansion would not have occurred. The temperature change formula based on acoustic and diffusion contributions to the density change gives this result. Since

$$\begin{aligned} \delta\rho_{\text{acoustic}}(r, 0) &= \frac{\beta q_H(r, 0)}{C_P} \\ \delta\rho_{\text{diffusion}}(r, 0) &= -\frac{\beta q_h(r, 0)}{C_P} \end{aligned} \tag{140}$$

the temperature change is

$$\delta T(r, t) = \frac{1}{\rho\beta}\left[(\gamma - 1)\frac{\beta q_H(r, 0)}{C_P} + \frac{\beta q_H(r, 0)}{C_P}\right] = \frac{q_H(r, 0)}{\rho C_V} \tag{141}$$

The propagating acoustic component to the density change is relatively short lived. The temperature change should evolve from the initial value to a value of

$$\delta T(r, t_c \gg t \gg \tau_a) \approx \frac{q_H(r, 0)}{\rho C_P} \tag{142}$$

At times (or in volumes) where the propagating mode contribution to the density change is negligible,

$$\delta T(r, t \gg \tau_a) \approx -\frac{1}{\rho\beta}\,\delta\rho_{\text{diffusion}}(r, t) \tag{143}$$

Thus over the course of propagating mode relaxation, the temperature change will evolve from $q_H/\rho C_V$ to $q_H/\rho C_P$. The on-axis temperature change evolution for the pulsed Gaussian-shaped laser excitation of a homogeneous sample is illustrated in Figure 3.8.

3.9. THERMODYNAMIC FLUCTUATION

The hydrodynamic equations can be used to find the spatial and temporal density fluctuations that occur in finite temperature samples. Fluctuating thermodynamic parameters dissipate by the same hydrodynamic modes that dissipate the external source perturbations. Density dissipation is found from the inverse Fourier–Laplace transform of the autocorrelation function. Assuming that there is no external source–induced perturbation (e.g., $q_H = 0$), the autocorrelation is obtained by multiplying the matrix solution

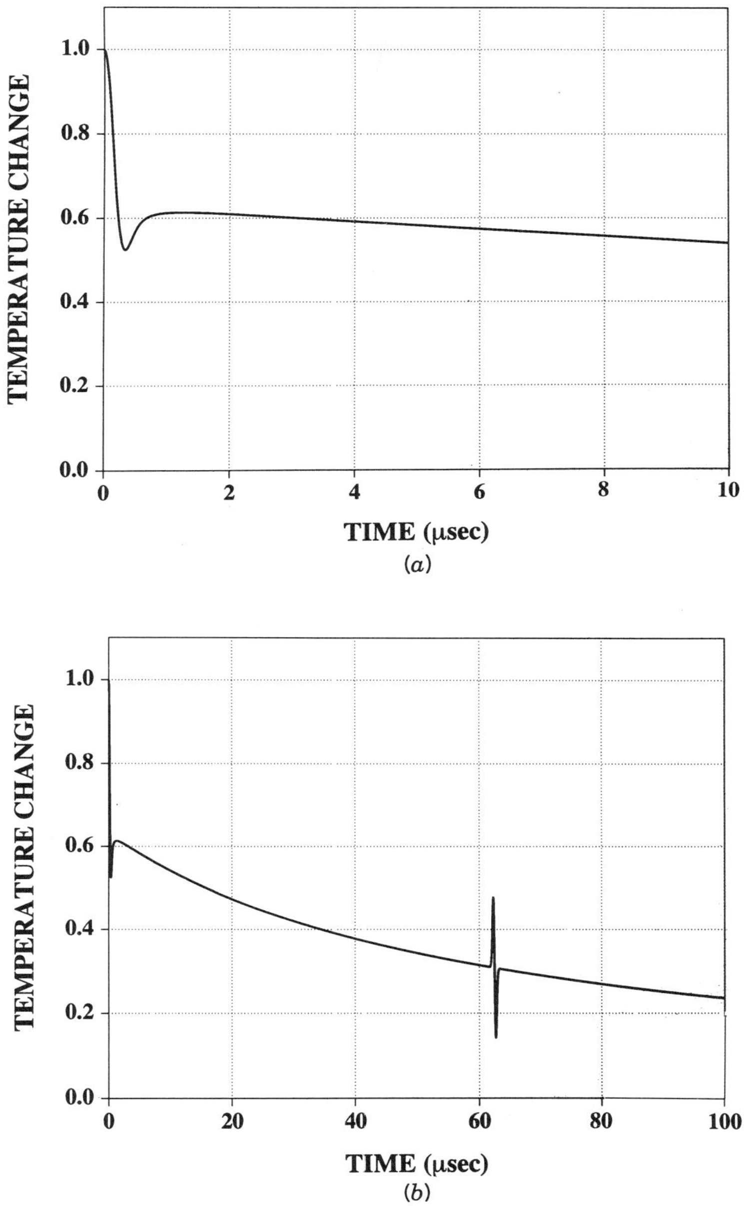

Figure 3.8. Time-dependent temperature changes for Gaussian impulse excitation in a 1-cm-diameter cylindrical cell. The hydrodynamic parameters are the same as those in Figure 3.7. (*a*) Early time behavior. The temperature is initially a maximum and decreases rapidly as the pressure equilbrates due to propagating mode dissipation. (*b*) Long-term temperature change. The temperature change slowly decreases due to thermal diffusion. The temperature perturbation about 64 μs is due to the reflected acoustic wave.

to the conservation equations by $\delta\rho^*(\mathbf{k}, 0)$ and then ensemble averaging

$$\frac{\langle \delta\rho^*(\mathbf{k}, 0)\, \delta\rho(\mathbf{k}, \zeta)\rangle}{\langle \delta\rho^*(\mathbf{k}, 0)\, \delta\rho(\mathbf{k}, 0)\rangle} = \frac{(\zeta + \gamma D_T \mathbf{k})(\zeta + v_l \mathbf{k}) + (1 - 1/\gamma)c^2\mathbf{k}^2}{|A(\mathbf{k}, \zeta)|} \tag{144}$$

where the angular brackets signify the time average or expectation. The time dependence of the density autocorrelation is found from the Laplace inversion,

$$\frac{\langle \delta\rho^*(\mathbf{k}, 0)\, \delta\rho(\mathbf{k}, t)\rangle}{\langle \delta\rho^*(\mathbf{k}, 0)\, \delta\rho(\mathbf{k}, 0)\rangle} = \sum_{j=1}^{3} \lim_{\zeta \to \xi_j} \frac{[(\zeta_j + \gamma D_T \mathbf{k})(\zeta_j + v_l \mathbf{k}) + (1 - 1/\gamma)c^2\mathbf{k}^2](\zeta - \zeta_j)}{|A(\mathbf{k}, \zeta)|} \exp(\zeta_j t) \tag{145}$$

The inverse Laplace transform is obtained using the same determinate roots as were used to obtain the thermal diffusion and propagating modes. Berne and Pecora (1976) find that the Laplace transform of the density autocorrelation function can be approximated by

$$\frac{\langle \delta\rho^*(\mathbf{k}, 0)\, \delta\rho(\mathbf{k}, t)\rangle}{\langle \delta\rho^*(\mathbf{k}, 0)\, \delta\rho(\mathbf{k}, 0)\rangle} \approx \frac{\gamma - 1}{\gamma} e^{-\mathbf{k}^2 D_T t} + \frac{1}{\gamma}\left(\cos c\mathbf{k}t + \frac{3\Gamma - v_l}{\gamma c}\mathbf{k} \sin c\mathbf{k}t\right) e^{-\mathbf{k}^2 \Gamma t} \tag{146}$$

The first term on the right-hand side is clearly dissipation due to thermal diffusion. The second term is dissipation due to the damped sound wave. This density autocorrelation function is used to describe dynamic light-scattering phenomena. Rayleigh scattering is due to long-lived density fluctuation correlations, or the diffusive term in the autocorrelation above. The Brillouin scattering component is due to light scattering off the propagating acoustic wave components of the sample. The Brillouin component is described by the second term in the autocorrelation function.

There are many similarities between dynamic light scattering and photothermal spectroscopy. In fact, photothermal spectroscopy has been called forced Rayleigh scattering by some authors. The density fluctuations responsible for light scattering will result in a false photothermal signal, or more technically, an intrinsic noise similar to the thermal noise in electronic circuits. To see the similarity, the dynamic light-scattering signal is proportional to correlated fluctuations in the dielectric constant:

$$\Phi_{\mathrm{DLS}}(\mathbf{k}) \propto \langle \delta^* \varepsilon(\mathbf{k}, 0)\, \delta\varepsilon(\mathbf{k}, t)\rangle \tag{147}$$

where ε is the dielectric constant. Since $n = \sqrt{\varepsilon}$,

$$\Phi_{\mathrm{DLS}}(k) \propto \left(\frac{\partial \varepsilon}{\partial \rho}\right)^2 \langle \delta^* \rho(\mathbf{k}, 0)\, \delta\rho(\mathbf{k}, t)\rangle = \frac{1}{n^2}\left(\frac{\partial n}{\partial \rho}\right)^2 \langle \delta^* \rho(\mathbf{k}, 0)\, \delta\rho(\mathbf{k}, t)\rangle \quad (148)$$

Photothermal signals are proportional to the spatial derivatives of the refractive index change. For refractive indexes proportional to density, the **k**-space photothermal spectroscopy signal is proportional to

$$\Phi_{\mathrm{PTS}}(k, t) \propto \frac{1}{n}\left(\frac{\partial n}{\partial \rho}\right)(i\mathbf{k})^p\, \delta\rho(\mathbf{k}, t) \quad (149)$$

where the factor $(i\mathbf{k})^p$ is equivalent to the spatial derivatives used to calculate the interferometric, $p = 0$, deflection, $p = 1$, and photothermal lens, $p = 2$, optical elements. Since the two signal powers are related to density and the density-dependent refractive index, they may be directly compared to each other.

3.10. NOISE-EQUIVALENT DENSITY FLUCTUATION

The intrinsic noise can be found from the density fluctuation autocorrelation function and compared to the photothermal-induced density autocorrelation to obtain the signal-to-noise power of the photothermal signal. The Brillouin component is a high-frequency component ($\sim$3 GHz in argon gas). The photothermal-induced density perturbation relaxation time is on the order of the thermal diffusion time constant and subsequently does not contain the high-frequency components (except during the impulse excited rise time). Long ensemble averaging will favor the diffusive term relative to the propagation mode. Trigonometric terms will average to zero and the density fluctuation autocorrelation function may be approximated by

$$\langle \delta\rho^*(\mathbf{k}, 0)\, \delta\rho(\mathbf{k}, t)\rangle_{\mathrm{fluctuation}} \approx \langle \delta\rho^2\rangle \frac{\gamma - 1}{\gamma} e^{-\mathbf{k}^2 D_T t} \quad (150)$$

The autocorrelation of a fluctuation from the average equilibrium value can be interpreted as the time- and/or space-dependent variance of that parameter. The $\langle \delta\rho^2\rangle$ term is the autocorrelation with zero time displacement. This term is the density fluctuation variance. A well-known relationship, obtained from thermodynamic fluctuation considerations for the density variance is

$$\langle \delta\rho^2 \rangle = \sigma_\rho^2 = \frac{\rho^2 k_B T K_T}{V} \tag{151}$$

where k_B is the Boltzmann constant and V is the volume under consideration.

The density fluctuation autocorrelation may be used to estimate the theoretical limits of detection using photothermal spectroscopy. The noise equivalent power (NEP) is that used to excite the sample to obtain a signal equal to the noise. Photothermal spectroscopy measurements are actually refractive index measurements. The refractive index is in turn related to the density. The measured signal and noise variance are thus both proportional to density.

The signal is the density change that can be induced by the photothermal effect. The signal power is the autocorrelation of the time-dependent signal

$$\langle \delta\rho(\mathbf{k},0)\,\delta\rho(\mathbf{k},t)\rangle_{\text{photothermal}} = \lim_{t\to\infty} \frac{1}{t} \int_0^t \delta\rho_{\text{impulse}}(\mathbf{k},t')\,\delta\rho_{\text{impulse}}(\mathbf{k},t-t')\,dt' \tag{152}$$

where the density changes are those for an impulse excitation. The integral may be approximated by assuming the cosine terms average to zero in the long integration. In this case

$$\langle \delta\rho(\mathbf{k},0)\,\delta\rho(\mathbf{k},t)\rangle \approx \left[\frac{\beta q_H(k)}{C_P}\right]^2 e^{-\mathbf{k}^2 D_T t} \tag{153}$$

The NEP is found from the relationship

$$\langle \delta\rho(\mathbf{k},0)\,\delta\rho(\mathbf{k},t)\rangle_{\text{photothermal}} = \langle \delta\rho(\mathbf{k},0)\,\delta\rho(\mathbf{k},t)\rangle_{\text{fluctuation}} \tag{154}$$

Using the results from above yields

$$q_H^2(\mathbf{k})\,\frac{\beta^2}{C_P^2} = \langle \delta\rho^2 \rangle\,\frac{\gamma - 1}{\gamma} \tag{155}$$

which, with statistical thermodynamic result for $\langle \delta\rho^2 \rangle$, is

$$q_H^2(\mathbf{k})\,\frac{\beta^2}{\rho^2 C_P^2} = \frac{k_B T K_T}{V}\,\frac{\gamma - 1}{\gamma} \tag{156}$$

The γ term can be determined from the thermodynamic relationship

$$\frac{\beta^2 T}{\rho C_P K_T} = \frac{\gamma - 1}{\gamma} \tag{157}$$

This results in the simpler expression

$$q_H^2(\mathbf{k})\,\frac{1}{\rho C_P} = \frac{k_B T^2}{V} \tag{158}$$

An important point to recognize here is that since the density fluctuations disperse in the same manner as do the photothermal-induced density changes, the time and spatial dependence of the hydrodynamic relaxation modes of the signal cancels that of the noise. Moreover, any photothermal signal measurement (e.g., interferometric, deflection, or lens) will have the same limiting NEP. Multiplication of the signal by any factor of $i\mathbf{k}$ to obtain the spatial derivative must be matched with an equal multiplication to obtain the appropriate noise power.

The right-hand side of the NEP equation is related to the temperature change. Since

$$\delta T_H^2(\mathbf{k}) = \frac{q_H^2(\mathbf{k})}{\rho^2 C_P^2} \tag{159}$$

the resulting NEP temperature is

$$\delta T_H^2(\mathbf{k}) = \frac{k_B T^2}{\rho C_P V} \tag{160}$$

The photothermal heating term is a function of space and thus of the particular geometry of the sample and the light used to excite the sample. The NEP temperature change given above can be used to determine whether or not a particular optically induced temperature change can be discriminated over the statistical fluctuations.

To illustrate the use of the density fluctuation to determine detection limits in photothermal spectroscopy, consider the case where a pulsed

Gaussian laser beam is used to excite a homogeneous sample. A pulsed laser with an integrated irradiance of

$$H(x, y) = \frac{2Q}{\pi w^2} e^{-2(x^2 + y^2)/w^2} \tag{161}$$

where H ($\mathrm{J\,m^{-2}}$) is the integrated irradiance, Q (J) the pulse energy, and w the beam waist radius, will produce an instantaneous temperature change of

$$\delta T_H(x, y) = \frac{2\alpha Q}{\pi w^2 \rho C_P} e^{-2(x^2 + y^2)/w^2} \tag{162}$$

in a sample with a small absorption coefficient, $\alpha\,(\mathrm{m^{-1}})$. The temperature change will occur over the sample pathlength, l, directed along the z-axis.

The Fourier transform of the temperature change is

$$\delta T_H(\mathbf{k}_x, \mathbf{k}_y) = \frac{\alpha Q}{2\pi \rho C_P} e^{-(\mathbf{k}_x^2 + \mathbf{k}_y^2)w^2/8} \tag{163}$$

and the inverse transform of the square is

$$\delta T_H^2(x, y) = \left(\frac{\alpha Q}{\pi \rho C_P}\right)^2 \frac{1}{2w^2} e^{-(x^2 + y^2)/w^2} \tag{164}$$

The inverse transform of the NEP temperature equation yields

$$\delta T_H^2(x, y) = 2\pi\delta(x)\delta(y) \frac{k_B T^2}{\rho C_P V} \tag{165}$$

Taking the volume element to be that of the excitation laser, $V = \pi w^2 l$, the noise equivalent signal is

$$\alpha_{\mathrm{NEP}} = \frac{2\pi T}{Q} \sqrt{\frac{k_B \rho C_P}{l}} \tag{166}$$

Gas-phase samples may be treated as ideal. A monatomic gas has $\rho C_P \approx 3R/2V_m$, where R is the gas law constant and V_m is the molar volume. The α_{NEP} for a given excitation energy and optical path length at standard conditions [e.g., 101 kPa (1 atm) and 298 K] is

$$\alpha_{\mathrm{NEP}} \approx 2.2 \times 10^{-7} \frac{1}{Ql^{1/2}} \tag{167}$$

The α_{NEP} has units of $\mathrm{J}^{-1}\,\mathrm{m}^{-1/2}$ pulse$^{-1/2}$. The inverse-square-root pulse dependence is a consequence of signal-averaging considerations. With signal averaging, the signal estimate precision increases proportional to the number of averages, and the standard deviation due to random noise increases as the square root of the number of averages.

Calculation of α_{NEP} for condensed samples requires the empirical thermodynamic parameters. The data given in Table 3.4 can be used for these calculations. The α_{NEP} are given in Table 3.6. With the exception of water, these data show that the α_{NEP} values are about the same for all the liquids. Even water with its high heat capacity differs by only a factor of 1.5 to 2. The α_{NEP} value for liquids is apparently

$$\alpha_{\mathrm{NEP}} \approx 10^{-5} \frac{1}{Ql^{1/2}} \tag{168}$$

for the liquids examined. Using the conventional 1-cm-pathlength cuvette, a detection limit of $\alpha_{\mathrm{NEP}} \approx 2 \times 10^{-4}$ will be obtained. Of course, this is a single-pulse detection limit. Using ensemble averaging, the signal estimate will increase proportionally to the number of measurement repetitions averaged, while the noise variance will increase as the square root of the number.

The foregoing estimates assume that the excitation laser can be collimated to give a Gaussian profile that does not change dimensions over the length of the sample cell. This is not always the case. Gaussian laser beams have a beam waist radius that depends on the distance from the focus,

$$w^2(z) = w_0^2\left(1 + \frac{z^2}{z_0^2}\right) \tag{169}$$

where w_0^2 is the minimum beam waist radius at the focus and $z_0 = \pi w_0^2 n/\lambda$ is the confocal distance. The length over which the photothermal interaction takes place is known to be limited by the confocal distance (Fang and Swofford 1979, Bialkowski 1986). The interaction length is limited to about $\pm 2z_0$ around the focus for photothermal lens spectroscopy. For tightly focused beams, the effective interaction volume is

$$V \approx \frac{2}{\pi} \int_{-2z_0}^{2z_0} \int_{-\infty}^{\infty} \int_{-\infty}^{\infty} \frac{e^{-2(x^2+y^2)/w^2(z)}}{w^2(z)}\, dx\, dy\, dz \tag{170}$$

or $V \approx 4z_0$.

Table 3.6. Theoretical Pulsed Laser Photothermal Spectroscopy Absorption Coefficient Detection Limits

Substance	Unfocused[a] $10^6 \times \alpha_{NEP}$ ($J^{-1} m^{-1/2} pulse^{-1/2}$)	Focused[b] $10^9 \times \alpha_{NEP}$ ($J^{-1} m^{-1/2} pulse^{-1/2}$)
n-Hexane	8.4	3.0
Ethyl ether	8.9	3.1
Cyclohexane	8.2	2.9
Methanol	9.7	3.4
Ethanol	9.5	3.4
Acetone	9.2	3.2
Water	14	5.0
Methylene chloride	8.3	2.9
Chloroform	8.3	2.9
Carbon tetrachloride	8.1	2.8

[a]For $V = \pi w^2 l$.
[b]Tightly focused excitation at $\lambda = 400$ nm.

Replacing the volume term in the α_{NEP} equation results in the estimate

$$\alpha_{NEP} \approx \frac{\pi T}{Q} \sqrt{\frac{k_B \lambda}{n \rho C_P}} \tag{171}$$

This NEP detection limit is applicable to photothermal interferometry and lens spectroscopies where the effective interaction lengths are probed. The α_{NEP} value for the ideal monatomic gas is found to be

$$\alpha_{NEP} \approx 5 \times 10^{-7} \frac{\lambda^{1/2}}{Q} \tag{172}$$

For example $\alpha_{NEP} \approx 10^{-9}\,J^{-1}$ pulse$^{-1/2}$ in the infrared ($\lambda = 10\,\mu$m) and $\alpha_{NEP} \approx 10^{-10}\,J^{-1}$ pulse$^{-1/2}$ in the visible ($\lambda = 400$ nm). Using the data in Table 3.4, and for a tightly focused laser operating in the visible region, $\lambda = 400$ nm, results in the detection limits given in Table 3.6. The results are all within a factor of 2 with $\alpha_{NEP} \approx 3 \times 10^{-9}\,J^{-1}$ pulse$^{-1/2}$.

3.11. SUMMARY

A solution to the hydrodynamic relaxation was found that allowed separation of diffusive and propagating modes. By separating these two modes, the usual thermal diffusion and propagating mode equations can be obtained. Relaxation of an optically excited sample occurs by these two modes. Relaxation is approximated well by the sum of separate diffusion and propagating acoustic modes. The hydrodynamic equations can also be solved without the assumptions required to separate the modes. The results obtained for the coupled solution are similar in form to those obtained using separate diffusive and propagating modes.

Hydrodynamic relaxation also occurs for thermally induced fluctuations in thermodynamic parameters. Fluctuation dissipation results in a time- and space-dependent variance in sample density. The sample density variance can be compared to the density perturbation induced by the photothermal effect. This comparison results in an estimate of the noise equivalent absorption coefficient. For pulsed, tightly focused beams, the noise equivalent absorption coefficient was found to be 10^{-9} to $10^{-10}\,J^{-1}$ pulse$^{-1/2}$. The theoretical values are slightly lower than those obtained experimentally.

The results of this section have been based on there being a local equilibrium in the sample. This assumption may not always hold. If the absorbing species concentration is high, there can be a significant departure from equilibrium. With long-lived metastable states, several state-specific species may need to be accounted for in the hydrodynamic equations. An exact treatment of photothermal signal that occurs in this case can be obtained by solving the hydrodynamic equations coupled with the rate equations describing the relaxation. The methods shown in this chapter can be applied to this more complex problem.

APPENDIX 3A: THERMODYNAMIC PARAMETER CALCULATION

The thermodynamic parameters given in Table 3.4 were obtained from various sources and calculations. Density, heat capacity, refractive index, and isothermal compressibility data are for 101 kPa pressure, 25°C (*Handbook of Chemistry and Physics*, 1980). The n values are for sodium D ($\lambda = 589.3$ nm); the β values for liquids are from Perry and Chilton (1973) or calculated based on the refractive index temperature-dependence data compiled by Dovichi (1987). Density and refractive index are related by the molar refractivity. Using the Lorentz–Lorenz relationship and assuming that dn/dT depends only on the density, β can be found from

$$\beta = -\frac{1}{\rho}\left(\frac{\partial \rho}{\rho T}\right)_P = \frac{6n}{(n^2-1)(n^2+2)}\left(\frac{\partial n}{\partial T}\right)_P \tag{173}$$

The β and K_T values are given in Table 3.7 together with γ values calculated using the standard thermodynamic relationship.

Table 3.7. Thermodynamic and Optical Data for Common Solvents and Gases at 25°C

Substance	n	$10^4 \times dn/dT$ (K^{-1})	$10^4 \times \beta$ (K^{-1})	γ	$10^{10} \times K_T$ (m^2/N)	ρ (kg/m^3)	C_P (J/kg K)
Air	1.0003	−0.0098	[a]	1.40	[a]	1.18	1010
Argon	1.0003	−0.0094	[a]	1.58	[a]	1.78	520
Nitrogen	1.0003	−0.0100	[a]	1.31	[a]	1.25	1140
Ethyl ether	1.352	−6.1	−1.6	1.68	10.7	713	2310
Methanol	1.326	−3.9	−1.2	1.21	12.6	790	2460
Ethanol	1.359	−4.0	−1.1	1.21	11.5	800	2360
Acetone	1.357	−5.4	−1.5	1.44	12.4	800	2180
Water	1.333	−0.81	−0.26	1.01	4.57	997	4180
Chloroform	1.444	−6.0	−1.4	1.63	9.74	1520	950
Carbon tetrachloride	1.459	−6.1	−1.2	1.45	10.7	1600	850

[a]Assuming ideal behavior for gases, $\beta = T^{-1}$ and $K_T = P^{-1}$.

For gases, the temperature-dependent refractive index can be related to β and n

$$\left(\frac{\partial n}{\partial T}\right)_P = \beta \frac{(n^2 - 1)(n^2 + 2)}{6n} \tag{174}$$

The data in Table 3.7 are based on this formula and the refractive index of the gases at 0°C.

APPENDIX 3B: PROPAGATING MODE IMPULSE RESPONSE FOR POLAR COORDINATES IN INFINITE MEDIA

The inverse Fourier transform in two-dimensional Cartesian coordinates is

$$f(x, y) = \frac{1}{2\pi} \int_{-\infty}^{\infty} \int_{-\infty}^{\infty} e^{i\mathbf{k}_x x + i\mathbf{k}_y y} f(\mathbf{k}_x, \mathbf{k}_y)\, d\mathbf{k}_x\, d\mathbf{k}_y \tag{175}$$

Transformation into polar or cylindrical coordinates, where the radius, r, and the angle, θ, locate the points, gives the transform integral

$$f(r, \theta) = \frac{1}{2\pi} \int_0^{2\pi} \int_0^{\infty} e^{i\mathbf{k}_r r \cos(\theta - \mathbf{k}_\theta)} f(\mathbf{k}_r, \mathbf{k}_\theta) k_r \, d\mathbf{k}_r \, d\mathbf{k}_\theta \tag{176}$$

For a symmetric function (i.e., one that is not a function of θ), $f(\mathbf{k}_r, \mathbf{k}_\theta) = f(k_r)$ and the integration over θ can be performed without accounting for the transformed function. θ integration yields

$$f(r) = \int_0^{\infty} J_0(\mathbf{k}_r r) f(\mathbf{k}_r) \mathbf{k}_r \, d\mathbf{k}_r \tag{177}$$

where $J_0(k_r r)$ is the Bessel's function of order zero. The transform given by this integral is sometimes called a Hankel transform. It has many properties similar to those of the Fourier transform.

One can find the propagating mode impulse response of a Gaussian profile source function

$$f(r, t) = \frac{2Q_a \delta(t)}{\pi w^2} e^{-2r^2/w^2} \tag{178}$$

The spatial Fourier transform of this source is

$$f(\mathbf{k}_r, t) = \frac{Q_a \delta(t)}{2\pi} e^{-\mathbf{k}_r^2 w^2/8} \tag{179}$$

The product with the Fourier transformed solution for the propagating mode,

$$\delta\rho(\mathbf{k}_r, t) = \frac{\beta}{C_P} e^{-\Gamma \mathbf{k}_r^2 t} \cos \mathbf{k}_r ct \tag{180}$$

gives the impulse-response

$$f_{\text{impulse}}(\mathbf{k}_r, t) = \frac{\beta Q_a}{2\pi C_P} e^{-\mathbf{k}_r^2 (w^2/8 + \Gamma t)} \cos \mathbf{k}_r ct \tag{181}$$

The inverse Fourier transform is the complicated integral

$$\delta\rho_{\text{impulse}}(r, t) = \frac{\beta Q_a}{2\pi C_P} \int_0^{\infty} J_0(\mathbf{k}_r r) e^{-\mathbf{k}_r^2 (w^2/8 + \Gamma t)} \cos(\mathbf{k}_r ct) \mathbf{k}_r \, d\mathbf{k}_r \tag{182}$$

Integration can be performed by Taylor expansion of either the cosine or Bessel's function. Either expansion results in a series of degenerate or confluent hypergeometric functions, Φ (Abramowitz and Stegun 1965). For example, the cosine expansion results in

$$\delta\rho_{\text{impulse}}(r, t) = \frac{2\beta Q_a}{\pi C_P w^2(1 + 2t/\tau_\Gamma)} \sum_{j=0}^{\infty} \frac{(-1)^j j!}{(2j)!} \alpha(t)^j \Phi(j + 1, 1; -z(r, t))$$
$$\alpha(t) = \frac{8c^2t^2}{w^2(1 + 2t/\tau_\Gamma)} \qquad z(r, t) = \frac{2r^2}{w^2(1 + 2t/\tau_\Gamma)} \tag{183}$$

The confluent hypergeometric series can be evaluated with the aid of recursion relationships. These particular confluent hypergeometric functions can be reduced to a power series and an exponential. The confluent hypergeometric function, $\Phi(j + 1, 1; -z)$, can be expressed as the product of the exponential with the familiar Laguerre polynomial (Abramowitz and Stegun 1965):

$$\Phi(j + 1, 1; -z) = e^{-z} L_j^{(0)}(z)$$
$$L_j^{(0)}(z) = \sum_{m=0}^{j} (-1)^m \binom{j}{j - m} \frac{z^m}{m!} \tag{184}$$

Thus

$$\delta\rho_{\text{impulse}}(r, t) = \frac{2\beta Q_a}{\pi C_P w^2(1 + 2t/\tau_\Gamma)} e^{-2r^2/w^2(1 + 2t/\tau_\Gamma)} \sum_{j=0}^{\infty} \frac{(-1)^j j!}{(2j)!} \alpha(t)^j L_j^{(0)}[z(r, t)] \tag{185}$$

Usually, only a few cosine expansion terms are needed for a reasonably accurate prediction. The first five terms of the series are

$$\sum_{j=0}^{5} \frac{(-1)^j j!}{(2j)!} \alpha^j L_j^{(0)}(z)$$
$$= 1 + \frac{z - 1}{2} \alpha + \frac{z^2 - 4z + 2}{24} \alpha^2 + \frac{z^3 - 9z^2 + 18z - 6}{720} \alpha^3 \tag{186}$$
$$+ \frac{z^4 - 16z^3 + 72z^2 - 96z + 24}{40{,}320} \alpha^4$$

The inverse transform can also be given as an integral expression. This is done by taking the transform of two parts and convoluting. The first part is that without acoustic attenuation. This results in

$$f_1(r) = \int_0^\infty J_0(\mathbf{k}_r r)\cos(\mathbf{k}_r ct)\mathbf{k}_r\, d\mathbf{k}_r = \begin{cases} -\dfrac{ct}{(c^2t^2 - r^2)^{3/2}} & \text{for } c^2t^2 > r^2 \\ 0 & \text{for } c^2t^2 < r^2 \end{cases} \tag{187}$$

For the spatially Gaussian terms, the inverse spatial Fourier transform results in

$$f_2(r) = \frac{2Q_a}{\pi w^2(1 + 2t/\tau_\Gamma)}\, e^{-2r^2/w^2(1 + 2t/\tau_\Gamma)} \tag{188}$$

The spatial convolution of this term with the transformed cosine term yields the propagating mode density impulse response. Using the same method of transforming from Cartesian to polar cylindrical coordinates as was shown for the Fourier transform, convolution in radial coordinates can be shown to be given by the integral

$$\int_0^\infty r' I_0(rr') f_1(r - r') f_2(r')\, dr' \tag{189}$$

where I_0 is the modified Bessel's function of zero order. Substitution of the cosine and Gaussian terms in the convolution integral results in the integral equation for the propagating mode impulse response

$$\delta\rho_{\text{impulse}}(r) = -\frac{\beta Q_a ct}{\pi^2 C_P w^2(1 + 2t/\tau_\Gamma)} \int_0^{ct} r' I_0(rr')\, \frac{e^{-2(r - r')^2/w^2(1 + 2t/\tau_\Gamma)}}{(c^2t^2 - r'^2)^{3/2}} \tag{190}$$

REFERENCES

Abramowitz, M. and Stegun, I. A., eds. *Handbook of Mathematical Functions*, Dover, New York (1965).

Barker, J. R. and Rothem, T. *Chem. Phys.* **68** 331 (1982).

Barker, J. R. and Toselli, B. M. In *Photothermal Investigations in Solids and Fluids*, Sell, J. A., ed., Academic Press, New York (1989).

Bates, R. D.; Jr.; Flynn, G. W.; Knudtson, J. T.; and Ronn, A. M. *J. Chem. Phys.* **53** 3621 (1970).

Berne, B. J. and Pecora, R. *Dynamic Light Scattering*, Wiley, New York (1976).

Bialkowski, S. E. *Chem. Phys. Lett.* **104** 448 (1984).

Bialkowski, S. E. *Anal. Chem.* **58** 1706 (1986).

Carslaw, H. S. and Jaegar, J. C. *Conduction of Heat in Solids*, 2nd ed., Clarendon, Oxford (1986).

Cussler, E. L. *Diffusion*, Cambridge University Press, New York (1984).

Dovichi, N. J. *CRC Crit. Rev. Anal. Chem.* **17** 357 (1987).

Grabiner, F. R.; Siebert, D. R.; and Flynn, G. W. *Chem. Phys. Lett.* **17** 189 (1972).

Handbook of Chemistry and Physics, 61st ed., Weast, R. C., ed., CRC Press, Boca Raton, FL (1980).

Hirschfelder, J. O.; Curtiss, C. F.; and Bird, R. B. *Molecular Theory of Gases and Liquids*, Wiley, New York (1954).

Jacobs, S. J. *Chem. Phys.* **132** 71 (1989).

McLennan, J. A. *Introduction to Nonequilibrium Statistical Mechanics*, Prentice Hall, Englewood Cliffs, NJ (1989).

Perry, R. H. and Chilton, C. H., eds. *Chemicals Engineer's Handbook*, McGraw-Hill, New York (1973).

Pugh, D. In *Photoacoustic, Photothermal and Photochemical Process in Gases*, Hess, P., ed., Springer-Verlag, New York (1989).

Siebert, D. R.; Grabiner, F. R.; and Flynn, G. W. *J. Chem. Phys.* **60** 1564 (1974).

Temkin, S. *Elements of Acoustics*, Wiley, New York (1981).

CHAPTER

4

OPTICAL PRINCIPLES FOR PHOTOTHERMAL SPECTROSCOPY

To model the photothermal spectroscopy signal, signal generation is divided into several processes, each of which influences the signal used to determine sample absorption. In this chapter we address some of the methods used to calculate the propagation behavior of optical radiation through a homogeneous medium. The optical propagation behavior of light in general, and laser light in particular, is central to several of the processes resulting in photothermal signal generation. The principal processes are optical excitation and excited-state relaxation, refractive index change, optical element generation, optical element monitoring, optical signal development, and detection. With the exception of the photophysics of optical excitation and excited-state relaxation, an understanding of these processes depends heavily on an understanding of optics and electromagnetic wave propagation.

The chapter begins with a brief description of Maxwell's equations and shows simple plane wave solutions to these equations for electromagnetic wave propagation in a vacuum and in dielectric media. Gaussian beam propagation mode solutions, applicable to laser sources, are given. The Gaussian beam equations are also solutions of Maxwell's differential equations but are more flexible in that they allow modeling of light that is not of infinite extent. The different transverse modes of the Gaussian beam are also given.

Several methods for determining light propagation characteristics and beam transformation with optics are outlined. Techniques for paraxial ray tracing and Gaussian beam propagation are discussed. Basic equations for finding photothermal deflection angles and lens strengths are found from the ray tracing equations. Gaussian beam focusing equations are derived. A major part of these discussions is devoted to diffraction optics. Diffraction theory is probably the most quantitative method for determining the effects of an optical system on a paraxial beam. This theory has not been used extensively in photothermal spectroscopy. It is included here to serve as a basis for understanding the few papers that have used this theory for more accurate predictions of the photothermal signals.

Because of the number of topics covered, only brief explanations can be given. The topics are introduced to allow (1) better understanding of the optical designs used in photothermal spectroscopy, (2) a basis for calculating the measurable photothermal signals, (3) a means of calculating the effects of the photothermal signals on the propagation of light, and (4) a basis from which other apparatus can be designed and optimized. More detailed discussion of optics appears in the references listed.

4.1. LIGHT PROPAGATION

The propagation of electromagnetic radiation is described by Maxwell's equations. Maxwell's equations are semiclassical in that they do not account for the quantum nature of matter. The media in which the light propagates are considered to be continuous and to possess bulk properties. Just as in the case of the semiclassical fluid mechanics or hydrodynamic equations, these bulk properties can be related to atomic and/or molecular properties. Maxwell's equations have been solved for a number of conditions and for a number of different modes of propagation. The term *condition* is used here to mean the electrical and dielectric properties of the media. A propagation mode is a specific solution for electromagnetic propagation that satisfies Maxwell's equations.

4.1.1. Maxwell's Equations

Maxwell's equations in SI units can be written as (Flygare 1978)

$$\nabla \times \mathbf{E} + \frac{\partial \mathbf{B}}{\partial t} = 0 \qquad \nabla \times \mathbf{H} - \frac{\partial \mathbf{D}}{\partial t} = \mathbf{J}$$
$$\nabla \cdot \mathbf{D} = \rho \qquad \nabla \cdot \mathbf{B} = 0$$
$$\mathbf{D} = \varepsilon_0 \boldsymbol{\varepsilon} \cdot \mathbf{E} \qquad \mathbf{B} = \mu_0 \boldsymbol{\mu} \cdot \mathbf{H} \tag{1}$$

E (V m^{-1}) and **H** (T m H^{-1}) are the electric and magnetic field vectors associated with the electromagnetic wave. **D** (V m^{-1}) and **B** (T) are the electric and magnetic induction field vectors associated with **E** and **H**. The other quantities are **J** (C $s^{-1} m^{-2}$) , the current density; ρ (C m^{-3}), the charge density; $\boldsymbol{\mu}$ the relative magnetic permittivity; and $\boldsymbol{\varepsilon}$, the relative electric permittivity, or dielectric constant, of the medium. μ_0 (H m^{-1}) and ε_0 (F m^{-3}) are the vacuum magnetic permittivity and vacuum permeability constants, respectively. $\boldsymbol{\mu}$ and $\boldsymbol{\varepsilon}$ are, in general, unitless matrix or diadic parameters. This form of the equations is convenient to relate to refractive index since the magnetic permittivity and dielectric constant are relative constants.

These equations are the embodiment of classical laws of electronics. The first equation is Faraday's law wherein a time-dependent magnetic flux density gives rise to an electric field. The second is the Ampère–Oersted law, showing that magnetic fields exist when there is electrical current. The third is Coulomb's law. The last equation is similar to the Coulomb law for magnetic density. The divergence of **B** is zero as a consequence of there being no magnetic analog to electrical charge. The last two equations relate the electric and magnetic fields to the fields induced in the medium. These equations are completed by defining the proportionality between the current density and the electric field:

$$\mathbf{J} = \sigma \cdot \mathbf{E} \tag{2}$$

where σ is the conductivity dyadic. This is the equivalent to Ohm's law.

4.1.2. Plane Waves in Vacuum

Simultaneous solution to these equations results in laws and insights into the propagation of electromagnetic radiation. The equations can be solved for any material. Solutions for nonconducting transparent materials are most important for light propagation. The simplest solution is for light propagation in a vacuum. In a vacuum there are no free or bound charges, and the current, **J**, and charge density, ρ, are zero. In a vacuum, the relative magnetic permittivity and dielectric constants are both equal to 1. The resulting Maxwell's differential equations for electromagnetic wave propagation in a vacuum are

$$\begin{aligned} \nabla \times \mathbf{E} &= -\frac{\partial \mathbf{B}}{\partial t} \qquad \nabla \times \mathbf{B} = \mu_0 \varepsilon_0 \frac{\partial \mathbf{E}}{\partial t} \\ \nabla \cdot \mathbf{E} &= 0 \qquad \nabla \cdot \mathbf{B} = 0 \end{aligned} \tag{3}$$

A solution is found by taking the curl ($\nabla \times$) of the first two equations and substituting the last two equations. After some manipulation, the vacuum wave equations that result are

$$\nabla^2 \mathbf{E} - \mu_0 \varepsilon_0 \frac{\partial^3 \mathbf{E}}{\partial t^2} = 0 \qquad \nabla^2 \mathbf{B} - \mu_0 \varepsilon_0 \frac{\partial^2 \mathbf{B}}{\partial t^2} = 0 \tag{4}$$

It is apparent that **E** and **B** both obey the general wave equation given by

$$\nabla^2 \mathbf{A} - \frac{1}{c^2}\frac{\partial^2 \mathbf{A}}{\partial t^2} = 0 \tag{5}$$

where **A** is any vector quantity and c (m s^{-1}) is the speed of wave propagation. The general wave equation can be used to relate the speed of light to properties of the media. Comparing the wave equation to those for the **E** and **B** vectors given the speed of light in a vacuum yields

$$c_0 = \frac{1}{\sqrt{\mu_0 \varepsilon_0}} \tag{6}$$

In SI units, μ_0 is $4\pi \times 10^{-7}$ H m^{-1}, and ε_0 is $(4\pi c^2)^{-1} \times 10^7$ C V^{-1} m^{-3}. Magnetic permittivities and dielectric constants generally depend on the frequency of the electromagnetic radiation. This is not so in a vacuum. Subsequently, there is no dispersion in a vacuum and all electromagnetic waves travel at the same speed.

Plane wave solutions for **E** and **B** can be cast in the form

$$\mathbf{E} = \mathbf{E}_0\, e^{i(\omega t - \mathbf{k}\cdot\mathbf{r} + \phi)} \qquad \mathbf{B} = \mathbf{B}_0\, e^{i(\omega t - \mathbf{k}\cdot\mathbf{r} + \phi)} \tag{7}$$

where $\omega = 2\pi v$ (rad s^{-1}) is the angular frequency, **k** (m^{-1}) is the wave vector, and **r** (m) is the radial distance vector. The radial distance is measured relative to an arbitrary location in space. The relative phase shift at this position is accounted for by the constant phase shift, ϕ (rad). The wave vector, **k**, points in the direction of propagation and has a magnitude

$$|\mathbf{k}| = \frac{\omega}{c_0} = \frac{2\pi}{\lambda_0} \tag{8}$$

where λ_0 (m) and c_0 are the vacuum wavelength and speed, respectively. The exponential solutions have the divergence

$$\nabla \cdot \mathbf{E} = -i\mathbf{k} \cdot \mathbf{E} \qquad \nabla \cdot \mathbf{B} = -i\mathbf{k} \cdot \mathbf{B} \tag{9}$$

Since the last two Maxwell's equations state that the divergence is zero, $\mathbf{E}_0$ – and $\mathbf{B}_0$ must both point in directions perpendicular to **k**. Also, from

$$\nabla \times \mathbf{E} = -i\mathbf{k} \times \mathbf{E} \tag{10}$$

and

$$\frac{\partial \mathbf{B}}{\partial t} = i\omega\mathbf{B} \tag{11}$$

it follows from the first of the Maxwell's equations in a vacuum that

$$\mathbf{B}_0 = \frac{\mathbf{k} \times \mathbf{E}_0}{c_0|\mathbf{k}|} \tag{12}$$

This shows that the **E** and **B** vectors are perpendicular to each other and that the two are in phase with each other. The three vectors, $\mathbf{E}_0$, $\mathbf{B}_0$, and **k**, form a right-handed Cartesian coordinate system. The electromagnetic wave propagates in the direction of **k** at a velocity c_0. This is the familiar model of electromagnetic radiation where the **E** and **B** vectors are perpendicular to the direction of propagation. The magnitude of the B field is related to the E field by $E_0^2/B_0^2 = c_0^2$ in the SI system. Finally, the average irradiance of an electromagnetic field is proportional to the squared electric field amplitude. This relationship is given by

$$E = \tfrac{1}{2}c\varepsilon_0\mathbf{E}^*\mathbf{E} \tag{13}$$

4.1.3. Polarization

Polarization is not accounted for in this idealized solution. The description for plane wave propagation constitutes only one of two orthogonal electric field polarizations that can exist simultaneously. An electromagnetic wave polarized in a plane including the x-axis does not interact in a vacuum with one polarized in a plane containing the y axis. For z-axis propagation, the two polarizations are orthogonal and can exist simultaneously without interaction. The relative phase and magnitude of the waves propagating in two polarizations result in the characteristic polarization of composite wave. If the phases of the orthogonal waves are equal, the radiation is plane polarized since the fields combine to point in a particular direction. If the phase of the two orthogonal waves are different, the composite electric and magnetic fields rotate in space and time. Orthogonal waves that are $\pm\pi$ out of phase from one another and of equal magnitude result in circularly polarized radiation. In this case the magnitude of the electric and magnetic fields do not change, but the direction of these fields rotates in the x–y plane. If the phase difference is different than π, the radiation will be elliptically polarized. If the amplitudes of the fields are different, the polarization can be described by a linear combination of elliptical and linear polarized components.

4.1.4. Plane Wave Propagation in Dielectric Media

Maxwell's equations for propagation in dielectric media are

$$\nabla^2 \mathbf{E} - \frac{\mu\varepsilon}{c_0^2}\frac{\partial^2 \mathbf{E}}{\partial t^2} = 0 \qquad \nabla^2 \mathbf{B} - \frac{\mu\varepsilon}{c_0^2}\frac{\partial^2 \mathbf{B}}{\partial t^2} = 0 \tag{14}$$

The plane wave solution is essentially that shown for vacuum propagation, with the exception that the speed of light is now

$$c = \frac{c_0}{\sqrt{\mu\varepsilon}} \tag{15}$$

Refractive index, n, is a unitless quantity that describes the relative speed of light in a particular medium. It is defined as the ratio of the speed of light in a vacuum, c_0, to that in a particular medium, c:

$$n = \frac{c_0}{c} \tag{16}$$

The real part of the refractive index is always positive and greater than or equal to 1.

The electric field plane wave solution for electromagnetic radiation propagation in an isotropic medium is

$$\mathbf{E} = \mathbf{E}_0\, e^{i(\omega t - \mathbf{k}\cdot\mathbf{r} + \phi)} \tag{17}$$

The wave vector magnitude is now

$$|\mathbf{k}| = \frac{\omega}{c} = \frac{2\pi n}{\lambda_0} \tag{18}$$

The physical wavelength is increased by an amount proportional to the refractive index:

$$\lambda = n\lambda_0 \tag{19}$$

The dielectric coefficient is generally complex, having real and imaginary components. Subsequently, the refractive index will also have real and imaginary components since $n = \sqrt{\mu\varepsilon}$. What is commonly referred to as the refactive index is the real part of the complex refractive index. The imaginary part is related to the optical absorbance or gain in the dielectric medium. This can be seen in the plane wave propagation equation. Since $\mathbf{k}$ is proportional to n, and $n = n_r + in_i$, where n_r is the real part and n_i the imaginary part, the imaginary part of $\mathbf{k}$ will result in a real exponential term. This real exponential corresponds to either an amplification, if the exponent is positive, or a diminishing, if the exponent is negative, of the electric field as a function of distance traveled through the medium. The relationship between the refractive index, the dielectric coefficient, and to a much lesser extent, the relative magnetic permittivity, are important in photothermal spectroscopy. Any optically induced change in the refractive index must be reflected in a change in one or both of the latter two parameters.

4.1.5. Refractive Index

According to the Maxwell equations, the refractive index of a substance is related to the dielectric constant, $\boldsymbol{\varepsilon}$, and relative magnetic permittivity, $\boldsymbol{\mu}$, through

$$n = \sqrt{\boldsymbol{\varepsilon\mu}} \tag{20}$$

In anisotropic media, the dielectric constant and relative magnetic permittivity are dyadic quantities. Thus refractive index can vary with electric and magnetic field directions, or the polarization, in anisotropic media. In isotropic media they are scalar. Like the refractive index, the dielectric constant and relative magnetic permittivity are unitless quantities. $\boldsymbol{\mu}$ is nearly equal to 1 in most substances and generally can be neglected in terms of its effect on the refractive index of a substance. It can have values that range from slightly less than 1, for diamagnetic materials to slightly greater than 1 for paramagnetic materials.

The dielectric constant varies more widely than $\boldsymbol{\mu}$ and is mostly responsible for slowing the propagation of light. The dielectric constant is a function of the frequency of the electromagnetic radiation and thus may result in dispersion. Dielectric constants vary with frequency of the electromagnetic radiation, generally decreasing with increasing frequency. The smallest dielectric constant is for a vacuum, with a value of 1. The highest static values are for polar substances in a constant electric field or at least very low frequencies. Water has a dielectric constant of 80 for static electric fields. In the visible and infrared portions of the electromagnetic spectrum, some of the larger dielectric constants are 2.7 for liquid CS_2 in the visible to about 17 for germanium in the near infrared. The dielectric constant may also vary with the strength of the electric field. This results in what is broadly classified as nonlinear optical phenomena (Flytzanis 1975).

4.1.6. Relating Macroscopic Dielectric Constant to Molecular Parameters

Consider the effect of an electric field on the electron distribution about an atom or molecule. Being composed of electrons and nuclei, the electric field will perturb or polarize the charge distribution, resulting in an induced electric dipole moment. Since all species in the media will be polarized, the field seen by any given atom or molecule will be different than the field posed across the medium. When viewed as a whole, the individual induced dipoles will add up to a macroscopic dipole. The macroscopic dipole is a field parameter called the polarization. The polarization in a medium can be defined on both a microscopic and a macroscopic level. On the microscopic level, a local electric field induces a dipole moment in atom molecule. For linear response, the induced moment is

$$\mathbf{d}_{\text{ind}} = \alpha \mathbf{E}_{\text{local}} \tag{21}$$

where α (C m^2 V^{-1}) is defined as the electric polarizability and $\mathbf{E}_{\text{local}}$ is the electric field local to the atom or molecule. The local electric field polarizes the charges in the medium to induce an electric dipole moment of $\mathbf{d}_{\text{ind}}$ (C m). α is a tensor quantity or a 3×3 matrix of values $\alpha_{i,j}$ which specifies how the local field component j affects the i component of the induced dipole. $\mathbf{E}_{\text{local}}$ and $\mathbf{d}_{\text{ind}}$ are both vector quantities describing the x-, y-, and z-axis components of these fields. The $\alpha_{i,j}$ vary with the frequency of the local electric field. In general, the $\alpha_{i,j}$ vary inversely proportional to frequency. The higher the local field frequency, the smaller the $\alpha_{i,j}$. In isotropic media, α is an electric field-frequency-dependent scalar.

The polarization of the medium, $\mathbf{P}$ (C m^{-2}), is defined as the induced moment per unit volume:

$$\mathbf{P} = N\mathbf{d}_{\text{ind}} \tag{22}$$

where N is the number density of atoms or molecules. In terms of the local electric field,

$$\mathbf{P} = N\alpha\mathbf{E}_{\text{local}} \tag{23}$$

The local field is made up of the three separate components: the external field, the field due to the polarization induced by this field, and a field due to any static charge separation in the medium. The third component only has to be included for solid materials with static polarization (e.g., ferroelectric crystals). For isotropic fluids, only the first two fields need to be considered.

The average polarization around each particular atom or molecule results in a second local field. Lorentz's method for polarization averaging assumes a sphere large enough to be modeled using bulk dielectric properties of the fluid, yet small relative to the extent of the macroscopic sample (Fröhlich 1949). This method is apparently accurate and results in relationships that model empirically derived dielectric coefficients quite well, even before quantum theory was used. For the isotropic fluid, the sum of the first two fields is given by the relationship

$$\mathbf{E}_{\text{local}} = \mathbf{E} + \frac{\mathbf{P}}{3\varepsilon_0} \tag{24}$$

The second term is the result of the Lorentz average.

On the macroscopic level, the polarization is defined in terms of the difference between electric induction fields with and without the dielectric medium present. Consider an electric field applied across a sample composed of two sample compartments. The compartments are arranged such that the electric field must pass through one compartment, then the next. In one compartment a vacuum exists. The other compartment contains the dielectric medium. In the vacuum the electric induction field is $\mathbf{D}_{\text{vacuum}} = \varepsilon_0\mathbf{E}$. The field lines must pass through the medium wherein $\mathbf{D}_{\text{dielectric}} = \boldsymbol{\varepsilon}\varepsilon_0\mathbf{E}$. Since the field is not dissipated in the dielectric medium, there must be a field in the medium that opposes the external field. This is the polarization. So the induction field in the dielectric medium can also be written as

$$\mathbf{D}_{\text{dielectric}} = \mathbf{D}_{\text{vacuum}} + \mathbf{P} \tag{25}$$

Substitution of the two induction fields yields

$$\boldsymbol{\varepsilon}\varepsilon_0\mathbf{E} = \varepsilon_0\mathbf{E} + \mathbf{P} \tag{26}$$

The polarization can be found by rearranging this expression as

$$\mathbf{P} = (\boldsymbol{\varepsilon} - 1)\varepsilon_0\mathbf{E} \tag{27}$$

The microscopic and macroscopic equations for the polarization are used to relate the dielectric coefficient to the atomic or molecular electric polarizability. Substitution of the local electric field equation into the polarization based on polarizability, solving for **P**, and equating to the polarization based on the dielectric constant results in the relationship given by the Clausius–Mossotti formula,

$$\frac{\boldsymbol{\varepsilon} - 1}{\boldsymbol{\varepsilon} + 2} = \frac{N\alpha}{3\varepsilon_0} \tag{28}$$

A temperature-dependent term arises for molecules that have a permanent dipole moment. The external field will tend to align the dipole while kinetic energy will tend to disperse the molecular alignment. For gases and dilute solutions of polar species in nonpolar solvents, molecules with permanent dipole moments, μ_0, will interact with the local electric field, producing a polarization

$$\mathbf{P}_{\text{dipole}} = N\,\frac{\mu_0^2 \mathbf{E}_{\text{local}}}{3k_B T} \tag{29}$$

In classical mechanics, the temperature dependence arises because while the local electric field tends to align the individual dipoles, the kinetic energy tends to redirect the oriented molecules. The total polarization is the sum of terms due to the electric polarizability and the dipole orientation. In quantum mechanics, the temperature dependence arises due to the finite population in the ground rotational state (Flygare 1978). The classical and quantum mechanical calculations both yield the same result for the dipole effect on molecular polarizability. Combining these leads to the Debye equation for the dielectric constant:

$$\frac{\boldsymbol{\varepsilon} - 1}{\boldsymbol{\varepsilon} + 2} = \frac{N}{3\varepsilon_0}\left(\alpha + \frac{\mu_0^2}{3k_B T}\right) \tag{30}$$

Similar equations describing the dielectric constant for pure or concentrated solutions of polar substances are known (Fröhlich 1949).

4.1.7. Refractive Index at Optical Frequencies

Terms corresponding to molecular alignment in the electric field do not influence the dielectric coefficient at optical frequencies (Guillory 1977). Molecular alignment in an electric field corresponds to rotational motion of the molecule. Frequencies corresponding to rotational motion are in the microwave to far-infrared regions of the electromagnetic spectrum. In the liquid state, rotational orientation times are about 10^{-12} s even for small molecules. Thus the permanent dipole moment contribution to the dielectric constant for visible radiation is negligible relative to the polarizability. This is very apparent for water. The dielectric constant for water in a static electric field is about 80. On the other hand, the dielectric constant obtained from the refractive index for 589.3 sodium light is about 1.77 at room temperature. For infrared and higher-frequency radiation, the dielectric coefficient can be replaced directly by the squared refractive index, resulting in

$$\frac{n^2 - 1}{n^2 + 2} = \frac{N\alpha}{3\varepsilon_0} \tag{31}$$

This relationship, referred to as the Lorenz–Lorentz equation, is often found in a form relating the refractive index to the molar refractivity:

$$\frac{n^2 - 1}{n^2 + 2} = \frac{R_m \rho}{M} \tag{32}$$

This equation relates the refractive index to the molar refractivity, R_m ($m^3\ mol^{-1}$), of a substance and the density of the material. M ($kg\ mol^{-1}$), the molecular weight, is used to relate the weight density to the molar density. The molar refractivity is related to the molecular polarizibility through $R_m = N_A\alpha/3\varepsilon_0$, where N_A is Avogadro's number. Tables of molar refractivities can be found in standard handbooks for chemical and physical information (e.g., *Handbook of Chemistry and Physics* 1980), and molar refractivities based on functional groups can often be used if the molar refractivity of a given substance is not found. The molar refractivity can also be calculated from refractive index and density information. It is known from the vast amounts of data that the refractive index ratio given above is generally a linear function of density but also depends on wavelength and to a minor extent, temperature. The only term that can be wavelength and temperature dependent is the molar refractivity, which in turn is related to the atomic or molecular polarizability.

The Lorenz–Lorentz equation can be used to determine the relationship between refractive index and sample composition. Since the refractive index is a function of density, the total refractive index of a mixture is

$$\frac{n^2 - 1}{n^2 + 2} = \sum_j \frac{R_{m,j} \rho_j}{M_j} \tag{33}$$

In low-density materials, in particular, gases, the refractive index is nearly 1 and

$$\frac{n^2 - 1}{n^2 + 2} = \frac{(n - 1)(n + 1)}{n^2 + 2} \approx \frac{2}{3}(n - 1) \tag{34}$$

and the total refractive index is linearly dependent on the densities of the components of the gas mixture:

$$\begin{aligned} n &\approx 1 + \frac{3}{2}\sum_j \frac{R_{m,j}\rho_j}{M_j} \\ &\approx 1 + \frac{1}{2\varepsilon_0}\sum_j N_j\alpha_j \end{aligned} \tag{35}$$

Refractive index and molar refractivity values are normally listed for 589.3 nm, corresponding to the atomic sodium line. Of course, the refractive index will be a function of wavelength. Wavelength dependence is due to the wavelength-dependent polarizability of the species. This interdependence can be described using both classical and quantum mechanical descriptions of matter.

4.1.8. Polarizability

The relationship between the refractive index and wavelength can be explored using the semiclassical interpretation of the Lorentz electron theory of matter. In the classical mechanics picture of optical interactions, an absorbing species is a damped oscillating dipole. The damped dipole can absorb electromagnetic radiation through resonant coupling with the electromagnetic radiation. The oscillating dipole is polarizable, so a collection of damped dipole oscillators result in a net polarization at optical frequencies. Although the relationships were derived for classical oscillators, the results may still be related to the quantum mechanical description of matter through the oscillator strength. The detailed description of the semiclassical approach is extensive and can only be summarized here. Many of the details may be found in Flygare (1978) and Verdeyen (1981).

The classical Lorentz electron oscillator is governed by the equation

$$\frac{\partial^2}{\partial t^2}x(t) + \gamma\frac{\partial}{\partial t}x(t) + \omega_0^2 x(t) = e\mathbf{E}(t) \tag{36}$$

where $x(t)$ is the displacement of the electron from the rest position, ω_0 the resonant frequency of the oscillator, γ a damping term, e the charge on the electron, and $\mathbf{E}(t)$ the driving electric field. The resonant frequency of this oscillator is $\omega_0 = k/m$, where k is the Hooke's law force constant and m is the mass of the electron. When driven by an external electric field oscillating with an angular frequency, ω,

$$\mathbf{E}(t) = \mathbf{E}_0 e^{-i\omega t} \tag{37}$$

The solution for the electron's displacement is

$$x(t) = \frac{l}{m} \frac{\mathbf{E}(t)}{\omega_0^2 - \omega^2 - i\omega/\tau} \tag{38}$$

where τ is the relaxation time defined by m/γ. The induced dipole of this oscillator is simply the displacement times the charge on the electron,

$$\mathbf{d}_{\text{ind}} = ex(t) = \alpha \mathbf{E}(t) \tag{39}$$

and the polarizability for the classical oscillator is

$$\alpha = \frac{e^2}{m} \frac{1}{\omega_0^2 - \omega^2 - i\omega/\tau} \tag{40}$$

The α can be related to the dielectric coefficient through the Clausius–Mossotti equation. However, since the polarizability has real and imaginary components, the real and imaginary components of the dielectric constant are difficult to find using this relationship. We subsequently resort to using the low-density limit for the dielectric constant,

$$\frac{\varepsilon - 1}{\varepsilon + 2} \approx \frac{1}{3}(\varepsilon - 1) \tag{41}$$

Thus at low density

$$\varepsilon - 1 = \frac{e^2 N}{m\varepsilon_0} \frac{1}{\omega_0^2 - \omega^2 - i\omega/\tau} \tag{42}$$

This is the classical Lorentz oscillator result for the dielectric constant.

The semiclassical result is obtained by relating the classical oscillator to the quantum mechanical result using the oscillator strength. Most atoms and molecules have several electrons and are thus composed of several damped oscillators. In addition, each electron has several resonant frequencies, corresponding to transitions to different excited states. The total dielectric coefficient is given as a sum of those due to the individual oscillators,

$$\varepsilon - 1 = \frac{N}{\varepsilon_0} \sum_j f_j \alpha_j \tag{43}$$

The weighing factor is the oscillator strength, f_j, of the jth oscillator and the α_j are the electronic polarizabilities:

$$\alpha_j = \frac{e^2}{m} \frac{1}{\omega_j^2 - \omega^2 + i\omega/\tau_j} \tag{44}$$

The dielectric coefficient in a frequency region near a particular resonant transition that is far removed from the other transitions is

$$\varepsilon = 1 + \frac{e^2 N f}{m\varepsilon_0} \frac{1}{\omega_0^2 - \omega^2 - i\omega'\tau} \tag{45}$$

This can be broken up into the real and imaginary components:

$$\begin{aligned} \varepsilon_r &= 1 + \frac{e^2 N f}{m\varepsilon_0} \frac{\omega_0^2 - \omega^2}{(\omega_0^2 - \omega^2)^2 + \omega^2/\tau^2}, \\ i\varepsilon_i &= \frac{e^2 N f}{m\varepsilon_0} \frac{i\omega/\tau}{(\omega_0^2 - \omega^2)^2 + \omega^2/\tau^2} \end{aligned} \tag{46}$$

Near the resonance frequency, $\omega_0 \approx \omega$, and $\omega_0^2 - \omega^2 = (\omega_0 - \omega)(\omega_0 + \omega) \approx 2\omega_0(\omega_0 - \omega)$. Using this approximation and the substitutions $2\pi\nu = \omega$, $2\pi\,\Delta\nu = \tau^{-1}$, the real part can be reduced to

$$\varepsilon_r = 1 + \frac{e^2 N f}{4mc\varepsilon_0} \frac{c}{2\pi\nu_0} \frac{\nu_0 - \nu}{\Delta\nu} \frac{\Delta\nu/2\pi}{(\nu_0 - \nu)^2 + (\Delta\nu/2)^2} \tag{47}$$

This formula is written in a form to make the following substitutions more apparent. First, it is apparent that the dielectric coefficient has Lorentzian line shape. Using the normalized Lorentzian line shape function

$$g(\nu) = \frac{\Delta\nu/2\pi}{(\nu_0 - \nu)^2 + (\Delta\nu/2)^2} \tag{48}$$

results in

$$\varepsilon_r = 1 + \frac{e^2 N f}{4mc\varepsilon_0} \frac{c}{2\pi\nu_0} \frac{\nu_0 - \nu}{\Delta\nu} g(\nu) \tag{49}$$

Second, oscillator strength of a transition is the ratio of the quantum mechanical to the classical transition probability. The relationship between the oscillator strength and the exponential absorption coefficient is (e.g., Ingle and Crouch 1988)

$$\alpha(\nu) = \alpha(\nu_0)g(\nu) = \frac{e^2Nf}{4mc\varepsilon_0}\, g(\nu) \tag{50}$$

where $\alpha(\nu)$ is the absorption coefficient for a Lorentzian line transition given by

$$\alpha(\nu) = \sigma(\nu)(N_1 - N_2) \tag{51}$$

for a nondegenerate two-level absorber with an absorption cross section, $\sigma(\nu)$, where N_1 is the population in the lower level and N_2 the population in the excited state; $\alpha(\nu_0)$ is the absorption coefficient at the band center. Using these relationships, the classical polarizibility can be related to the absorption coefficient in the quantum mechanical description. The dielectric coefficient is

$$\varepsilon_r = 1 + \frac{\lambda_0}{2\pi}\,\alpha(\nu_0)\,\frac{\nu_0 - \nu}{\Delta\nu} \tag{52}$$

This equation is just one of several results that can be derived from the Kramers–Kronig relationships. The latter have been used to relate the real and imaginary parts of any dispersive physical parameter. The main steps in this derivation do not depend on the isolation of the atom or molecule. Thus α may be used in the Clausius–Mossotti equation.

4.1.9. Quantum Mechanical Polarizability

The results of the semiclassical polarizability calculations have been applied to many systems although the theory leading to this result is known to be incorrect. Similar results can be obtained using a complete quantum mechanical approach. Flygare (1978) shows that the molecular or atomic polarizibility is related directly to the dipole transition operator through

$$\alpha = \sum_j \alpha_j = \frac{4\pi}{h} \sum_j \frac{|D_j|^2\omega_j}{\omega_j^2 - \omega^2} \tag{53}$$

where α is total polarizibility in cgs units and α_j is the polarizability contribution corresponding to a particular oscillator or transition. D_j is the transition dipole matrix element moment operator for the jth transition from the ground state with a resonant angular frequency ω_j:

$$D_j = \langle j|\hat{O}|0\rangle \tag{54}$$

where $\hat{O}$ is the dipole moment operator. For frequencies near a particular resonance transition, $\omega \approx \omega_j$ and the contribution due to the resonant transition dominates. In this case the contribution to the polarizibility is dominated by

$$\alpha_j \approx \frac{2\pi}{h} |D_j|^2 \frac{\omega_j - \omega}{(\omega_j - \omega)^2} \tag{55}$$

The quantum mechanical absorption cross section for the jth transition in terms of this dipole operator is

$$\sigma_j(\omega) = \frac{8\pi^2 |D_j|^2 \omega_j^3}{hc\omega^2} \frac{1/\tau_j}{(\omega_j - \omega)^2 + (1/\tau_j)^2} \tag{56}$$

again in cgs units. For frequencies near resonance, yet sufficiently removed that the linewidth term in the denominator of the Lorentzian is much less that $(\omega_j - \omega)^2$,

$$\sigma_j(\omega) \approx \frac{8\pi^2 |D_j|^2 \omega_j^3}{hc\omega^2} \frac{1/\tau_j}{(\omega_j - \omega)^2} \tag{57}$$

Substituting the absorption cross section into the polarizibility, changing angular frequency to frequency, and defining the transition linewidth, $2\pi\Delta\nu = \tau^{-1}$, results in

$$\alpha_j \approx \frac{\sigma_j(\nu_j)\lambda_j(\nu_j - \nu)}{8\pi^2 \Delta\nu_j} \tag{58}$$

where ν_j is the band center of the transition. In cgs units, the dielectric coefficient is related to the polarizibility through $\varepsilon - 1 = 4\pi N\alpha$. The contribution to the dielectric coefficient for the jth transition is

$$\varepsilon - 1 = \frac{\lambda_j}{2\pi} \alpha_j(\nu_j) \frac{\nu_j - \nu}{\Delta\nu_j} \tag{59}$$

where $\alpha_j(\nu_j)$ is the band center absorption coefficient. This is, of course, the same as the value obtained using the semiclassical approach.

The polarizibility and refractive index do not arise due to a single transition. Far from any resonant absorptions, the total polarizibility will be a sum of polarizibilities due to the individual resonances in the damped oscillator. In this case

$$\varepsilon - 1 = \frac{1}{2\pi} \sum_j \lambda_j \alpha_j(\nu_j) \frac{\nu_j - \nu}{\Delta\nu_j} \tag{60}$$

This result can be used to relate absorption coefficients to refractive index. For example, one may need to estimate what the effect of changing the number density of states involved in a near-resonant transition has on the dielectric coefficient or refractive index. The resonant and nonresonant contributions are

$$\varepsilon - 1 = \frac{1}{2\pi} \sum_{j>0} \lambda_j \alpha_j(\nu) \frac{\nu_j - \nu}{\Delta\nu_j} + \frac{\lambda_0}{2\pi} \alpha_0(\nu_0) \frac{\nu_0 - \nu}{\Delta\nu_0} \tag{61}$$

The summation is effectively constant since each absorption contributes to the dielectric coefficient only in a nonresonant fashion. This equation can subsequently be rewritten as

$$\varepsilon - 1 = \varepsilon' + \frac{\lambda_0}{2\pi} \alpha_0(\nu) \frac{\nu_0 - \nu}{\Delta\nu_0} \tag{62}$$

where ε' is the dielectric coefficient due to the nonresonant terms in the summation. ε' is always large enough that the resonance term never results in a dielectric coefficient of less than 1. The refractive index may be estimated using the square-root approximation $(1 + \delta)^{1/2} \approx 1 + \delta/2$:

$$n = n' + \frac{c}{4\pi} \alpha_0(\nu_0) \frac{1 - \nu/\nu_0}{\Delta\nu} \tag{63}$$

where n' is the index contribution from the nonresonant transitions. This equation gives the explicit frequency dependence of refractive index around a resonant transition.

4.2. FUNDAMENTAL MODES IN HOMOGENEOUS MEDIA

The plane wave is not the most fundamental mode of propagation. One cannot, for example, describe an arbitrary image by superposition of plane waves. One definition of a fundamental mode is that it is one that can describe propagation of the smallest segments of electromagnetic radiation. Larger segments can then be constructed by a linear superposition of these smaller, fundamental segments. The strength of the optical elements formed by the photothermal effect are inversely proportional to the dimension of the thermal perturbation. Thus the fundamental mode of electromagnetic radiation is not without consequence to these experiments. Using a fundamental mode, the smallest possible area will be illuminated. This, in turn, will result in the greatest possible optical element strength. In this respect, there is a great advantage to using laser source for sample excitation. Well-behaved laser sources operate in what is known as the fundamental transverse electromagnetic mode (TEM_{00}). This fundamental mode is one solution to Maxwell's equations for the propagation of light. This particular propagation mode solution has the smallest volume of any of the mode solutions. Because it has the smallest possible volume, optical elements formed in a photothermal experiment using this mode for excitation will have the largest possible optical element strength. In addition, any optical source can be mathematically constructed from a superposition of fundamental modes. Thus an understanding of the propagation characteristics of the fundamental mode is important to an understanding of the propagation of all forms of light, including the plane wave.

4.2.1. TEM_{00} Mode

The fundamental TEM_{00} mode is found by solving Maxwell's equations under the constraint that the resulting wave be radially symmetric. The details of this solution can be found in most textbooks on laser radiation. The solution has a spatial electric field profile that can be expressed as the product of three exponential terms (Verdeyen 1981, Yariv 1985):

$$\mathbf{E}(x, y, z) = \mathbf{E}_0 \frac{w_0}{w(z)} \exp\left[\frac{-(x^2 + y^2)}{w^2(z)}\right] \times \exp(-i[\mathbf{k}z - \eta(z)]) \times \exp\left[-i\mathbf{k}\frac{x^2 + y^2}{2R(z)}\right] \tag{64}$$

$\mathbf{E}(x, y, z)$ is the electric field magnitude, $\mathbf{E}_0$ the maximum electric field at $(x, y, z) = 0$, $w(z)$ (m) the electric field beam radius, $\eta(z)$ an additional phase-shift term, and $R(z)$ the radial phase factor. The x and y are the transverse dimensions and z is often called the longitudinal dimension. The laser beam propagates along the longitudinal dimension or the positive z

axis. This description does not include electric field polarization. The x and y dependencies describe only the amplitude of the electric field. Electric field polarization is accounted for in $\mathbf{E}_0$. The $\mathbf{B}$ field can be found by the relationship to $\mathbf{E}$. As with the plane wave, the $\mathbf{B}$ field is perpendicular to $\mathbf{E}$ and $\mathbf{k}$.

Only the first exponential term is real. The second two exponential terms are complex and must therefore describe phase terms. In combination with the beam waist and origin electric field, the real term is the amplitude of the electric field of the wave,

$$|\mathbf{E}(r, z)| = \mathbf{E}_0 \frac{w_0}{w(z)} e^{-r^2/w^2(z)} \tag{65}$$

It is apparent that the amplitude is radial symmetric Gaussian since $x^2 + y^2 = r^2$. The electric field beam waist radius has a longitudinal functionality

$$w^2(z) = w_0^2 \left(1 + \frac{z^2}{z_0^2}\right) \qquad z_0 = \frac{\pi w_0^2 n}{\lambda} \tag{66}$$

The z-axis origin is at the position of the minimum electric field beam waist radius. A Gaussian beam at and near the origin is illustrated in Figure 4-1. z_0 (m) is the confocal distance or Rayleigh range. This is the distance over which the beam radius varies by less than $\sqrt{2}$. The minimum beam waist radius is w_0 (m) and is often called the minimum spot size. Within the confocal distance, the beam is reasonably collimated. In other words, the

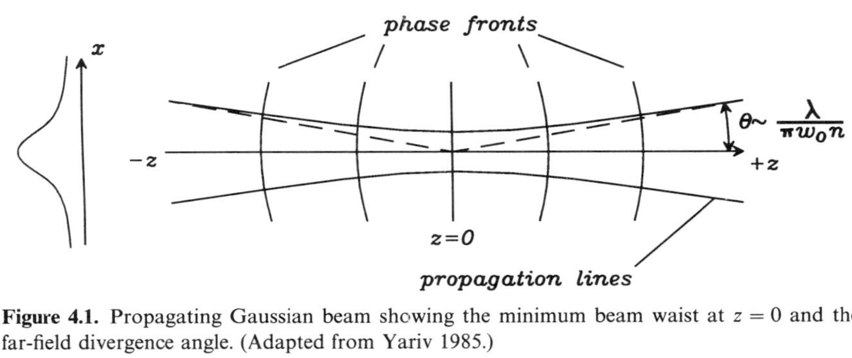

Figure 4.1. Propagating Gaussian beam showing the minimum beam waist at $z = 0$ and the far-field divergence angle. (Adapted from Yariv 1985.)

beam waist radius does not change significantly over the range $\pm z_0$. The width of the Gaussian field increases with increasing distance from the origin. At distances far from the origin, the beam waist radius increases in a linear fashion with z:

$$w(z \gg z_0) = \frac{\lambda z}{\pi n w_0} \tag{67}$$

Locations along the z axis where this condition is met are called the far field. The beam divergence angle in the far field is

$$\theta = \frac{2\lambda}{\pi n w_0} \tag{68}$$

Clearly, the smaller the minimum beam waist radius, the greater the beam divergence in the far field. This relationship can be attributed to the uncertainty principle.

The two exponential phase-shift terms have different dimension dependencies. The longitudinal phase shift is given by

$$\phi = kz - \eta(z) \tag{69}$$

where

$$\eta(z) = \tan^{-1}\frac{z}{z_0} \tag{70}$$

The phase velocity is close to, but slightly greater than, that for the plane wave. Near the focus, the tangent term is negligibly small, and the phase velocity is equivalent to the plane wave. In other words, Gaussian beams resemble plane waves near their focus. In another extreme, the phase velocity also approaches that for a plane wave at large w_0, hence z_0.

The radial phase-shift term is

$$R(z) = z\left(1 + \frac{z_0^2}{z^2}\right) \tag{71}$$

This is the radius of curvature of the wave. At the minimum electric field beam waist origin, the radius of curvature is infinite. Thus the wavefront is planar at $z = 0$. At z-axis positions far away from zero and definitely beyond the confocal distance, $R(z) \sim z$. At these distances the wave looks like a spherical wave with an origin at $z = 0$. In combination, the two phase-shift terms show that TEM_{00} modes have a spherical, single-phased surface. This is what is meant by the term *spatial coherence*. The phases of the electric fields at any radius from the z axis are equal or coherent.

By inspecting the components to the fundamental mode equation, it is easy to see that as the minimum beam waist radius approaches infinity, the TEM_{00} electric field equation become that of the plane wave. A plane wave is just a specialized form of the Gaussian propagation mode. For physically realizable waves, the distance over which the electric field resembles a plane wave is the confocal distance. The combined effects of the confocal distance, the beam waist radius equation, and the wavefront radius of curvature equation indicate that beams with larger w_0 will have longer confocal distances, smaller beam divergence angles, and will behave as plane waves over longer distances.

4.2.2. Irradiance of Gaussian Beams

The irradiance of the Gaussian TEM_{00} beam is found from the complex square of the electric field, **E*E**. The imaginary terms cancel in the complex square and the irradiance is thus

$$E(r,\ t) = \frac{c\varepsilon_0 \mathbf{E}_0^2(t) w_0^2}{w^2(z)} e^{-2r^2/w^2(z)} \tag{72}$$

where the electric field, $\mathbf{E}_0(t)$, has been generalized to allow for a time-dependent irradiance, $E(r, t)$. It is often more convenient to cast the irradiance in terms of the time-dependent radiometric power, $\Phi(t)$ (W):

$$E(r,\ t) = \frac{2\Phi(t)}{\pi w^2(z)} e^{-2r^2/w^2(z)} \tag{73}$$

since power is most commonly measured quantity. The relationship between power and electric field is easily obtained from these two irradiance equations.

4.2.3. Higher-Order Gaussian Modes

The TEM_{00} mode is only one of an infinite number of $TEM_{n,m}$ modes that satisfy Maxwell's equations for propagation in homogeneous media. Higher-order modes (e.g., those with finite n and m) can be specified in cylindrical or Cartesian coordinates. The most common convention is to use Cartesian coordinates. The electric field for the $TEM_{n,m}$ mode is given by

$$\mathbf{E}(x, y, z) = H_m\left[\frac{2^{1/2}x}{w(z)}\right] H_n\left[\frac{2^{1/2}y}{w(z)}\right] \exp[i(m+n)\eta(z)] \times \mathbf{E}_{00}(x, y, z) \quad (74)$$

where the symbols are the same as those used to describe the TEM_{00} mode. $\mathbf{E}_{00}$ is the TEM_{00} electric field given above. $R(z)$ and $w(z)$ are the same as those for the TEM_{00} mode and $\eta(z)$ is now

$$\eta(z) = (n + m + 1)\tan^{-1}\frac{z}{z_0} \quad (75)$$

The $H_n(\xi)$ are Hermite polynomials of order n of the variable ξ. In the $TEM_{n,m}$ mode, they are explicit functions of x and y but depend on z only through the electric field beam waist radius term. Propagation modes described by this equation are called Gaussian–Hermite modes. The $TEM_{n,m}$ modes have a spatial profile with n x-axis and m y-axis nodes or lines where the electric field is zero. Several $TEM_{n,m}$ modes are illustrated in Figure 4-2.

The high-order TEM electric field mode equation is similar, though more general than that for TEM_{00}. The major differences are in the dependence of the longitudinal phase shift on transverse mode number and the transverse electric field strength dependence induced by the Hermite polynomial functions. The longitudinal phase factor will affect the phase-velocity phase coherence to some degree. Higher-order $TEM_{n,m}$ modes will have faster phase velocities. In laser operation, the higher phase velocity affects the frequency of the laser radiation. Single-frequency lasers must operate with a single Gaussian–Hermite TEM mode.

The first few Hermite polynomials are

$$\begin{aligned} H_0(\xi) &= 1 \\ H_1(\xi) &= 2\xi \\ H_2(\xi) &= 4\xi^2 - 2 \\ H_3(\xi) &= 8\xi^3 - 12\xi \end{aligned} \quad (76)$$

The first Hermite polynomial is symmetrical about the axis origin, the second is asymmetric, the third symmetric, and so on. The electric field changes sign when crossing the propagation axis in the asymmetric modes. Subsequently, there is a node in the electric field strength at the propagation axis for odd-numbered nodes.

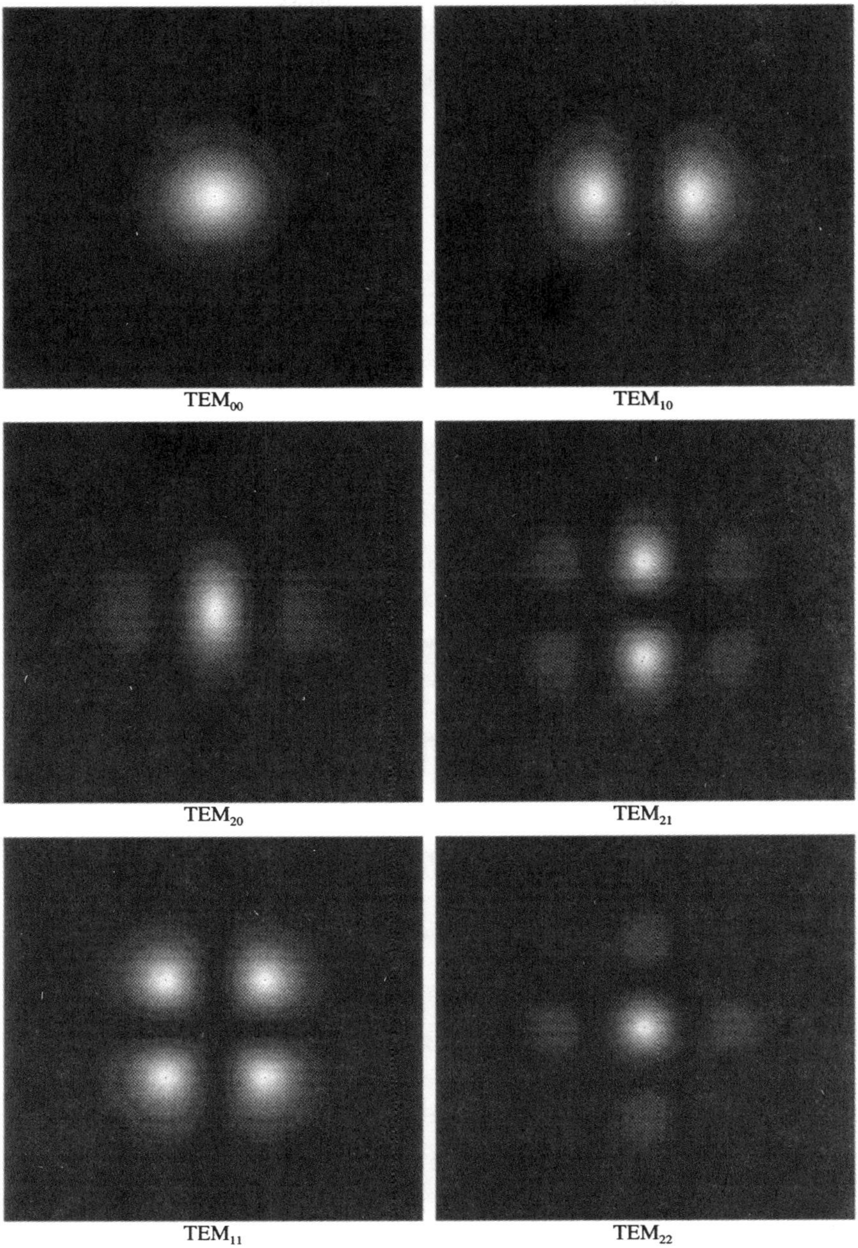

Figure 4.2. Spatial profile of several low-order TEM$_{n,m}$ modes. The TEM$_{0,0}$ mode has the lowest mode volume.

The beam divergence of all $TEM_{n,m}$ modes is the same. The minimum beam waist radius is defined as that for the Gaussian TEM_{00} mode. However, the area illuminated by the higher-order Gaussian–Hermite modes is greater than that of the TEM_{00} mode. This can be seen from the integral

$$\int_{-\infty}^{\infty} H_n^2(\xi)e^{-\xi^2}\,d\xi = \pi^{1/2}2^n n! \tag{77}$$

The square is used because the irradiance is proportional to the square of the electric field strength. The area irradiated by the $TEM_{n,m}$ mode is

$$A_{n,m} = \frac{\pi w_0^2}{2}\, n!\, m!\, 2^{n+m} \tag{78}$$

Subsequently, the smallest irradiation area occurs for the TEM_{00} mode and the area increases with increasing mode index. Persons working with laser sources know the value of the TEM_{00} mode. Not only does it result in the smallest possible spot size, but lasers restricted to TEM_{00} mode operation also give more stable output energies or powers. Lasers that operate multimode produce output in a number of different TEM modes. The distribution of power in these modes very often changes with time, in a chaotic fashion, resulting in light that is very difficult to characterize and control, so that while the average output power remains nearly constant, the power in a particular mode will vary. This variation is even more apparent when the laser beam is transformed with lenses.

The higher-order modes in cylindrical coordinates are given by similar equations; only the Hermite polynomials are replaced with generalized Laguerre polynomials, resulting in Gaussian–Laguerre modes (Kogelnik and Li 1966). The electric field is given by

$$\mathbf{E}(x, y, z) = \left[\frac{\sqrt{2}r}{w(z)}\right]^l L_p^l\left[2\frac{r^2}{w^2(z)}\right]\exp[i(2p+l)\eta(z)] \times \mathbf{E}_{00}(r, z) \tag{79}$$

where L_p^l is the generalized Laguerre polynomial and p and l are the radial and angular mode numbers. The $R(z)$ and $w(z)$ are the same as those for the TEM_{00} and Gaussian–Hermite modes and $\eta(z)$ is

$$\eta(z) = (2p + l + 1)\tan^{-1}\frac{z}{z_0} \tag{80}$$

The angular mode number corresponds to nodes across the beam profile, while the radial mode number corresponds to radial nodes that are similar to the radial nodes found in the solution to the wave equation for the hydrogen atom.

The first few generalized Laguerre polynomials are

$$\begin{aligned} L_0^l(\xi) &= 1 \\ L_1^l(\xi) &= l + 1 - \xi \\ L_2^l(\xi) &= \tfrac{1}{2}(l+1)(l+2) - (l+2)\xi + \tfrac{1}{2}\xi^2 \end{aligned} \tag{81}$$

Lasers with circular mirrors normally operate with $l = 0$. The generalized Laguerre are closely related to the Hermite polynomials (Abramowitz and Stegun 1965) and the mode volume increases with mode number in a fashion similar to that of the Gaussian–Hermite modes.

4.3. SPATIAL TRANSFORMATION OF RAYS BY OPTICAL ELEMENTS

4.3.1. Paraxial Ray Tracing

The propagation characteristics of light are transformed on passing through media with varying refractive indexes. The method for describing these transformations is ray tracing. Rays that have direction vectors that make only small angles with the main axis of propagation are said to be paraxial. The paraxial approximation is technically valid when $\theta \approx \sin\theta$. In this case, transformations that occur at interfaces between materials of different refractive indexes can be performed through linear operations instead of through Snell's law. The linear mathematics used to determine the effect that a refractive index change has on the ray is called the ray-matrix technique. The ray-matrix technique is described in most modern textbooks covering optics and will only be summarized here.

In paraxial ray tracing, a ray is defined by a vector that includes both position and direction information. The position is relative to an axis of symmetry. Optical elements are thought to exist along this symmetry axis. Optical elements (e.g., mirrors, lenses, prisms, diffraction gratings, polarizers, etc.) are defined by transformation matrixes that describe how the characteristics of the ray are transformed at or passing through the element. For beams and optical devices that are cylindrically symmetric, the ray coordinates and angles reduce to a set of two parameters: the radial offset from the axis of symmetry, r, and the angle relative to the axis of symmetry, θ. Any given ray is represented by a vector, $\mathbf{s} = [r, \theta]^{\mathrm{T}}$. The optical element transformation matrixes, **T**, each consist of four elements. To determine the ray characteristics after passing through an element, $\mathbf{s}_{out}$, one need only find the vector-matrix product

$$\mathbf{s}_{out} = \mathbf{T}\mathbf{s}_{in} \tag{82}$$

where $\mathbf{s}_{in}$ is the initial ray vector.

When the optical system contains several individual elements, a single ray transfer matrix can be found that describes the overall characteristics of the system. The ray transfer matrix is the product of the individual ray matrixes comprising the optical system being defined:

$$\mathbf{T}_{total} = \mathbf{T}_k \mathbf{T}_{k-1} \cdots \mathbf{T}_2 \mathbf{T}_1 \tag{83}$$

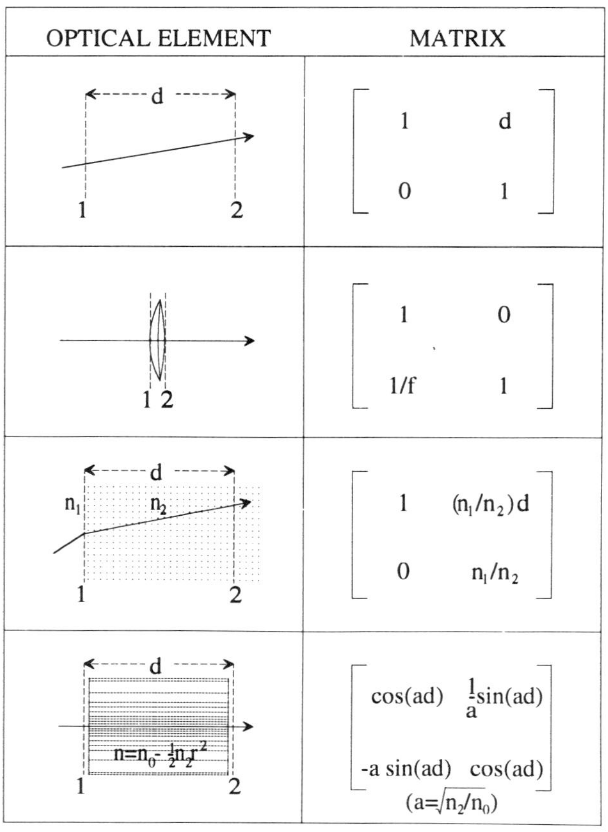

Figure 4.3. Several useful ray transfer matrixes. The positions marked 1 and 2 are input and output positions, respectively. The term n_2 corresponds to d^2n/dr^2 in the text.

The arrangement of the transformation matrixes is such that the ray passes through each element in sequence. Matrix multiplication starts with the first element and procedes sequentially through the kth element. The composite ray transfer matrix is constructed to reflect this sequence. The last element that the ray passes through is the first in the list.

For purposes of determining the propagation of rays through transparent media, only four transformation matrixes are important. These are the matrixes for free-space propagation, for changes that occur at refractive index boundaries, the effect of an ideal thin lens, and for propagation through lenslike media. These ray transformation matrices are shown in Figure 4.3.

4.3.2. Ray Propagation in Isotropic Media

The matrix for propagation over a distance, d, is

$$\mathbf{T} = \begin{bmatrix} 1 & d \\ 0 & 1 \end{bmatrix} \tag{84}$$

Consider the simple case of a ray propagating in free space with initial radial offset r_{in} and at an initial angle θ_{in}. Multiplication of the ray transfer matrix by the input ray vector results in

$$\begin{bmatrix} r_{\text{out}} \\ \theta_{\text{out}} \end{bmatrix} = \begin{bmatrix} 1 & d \\ 0 & 1 \end{bmatrix} \begin{bmatrix} r_{\text{in}} \\ \theta_{\text{in}} \end{bmatrix} = \begin{bmatrix} r_{\text{in}} + \theta_{\text{in}} d \\ \theta_{\text{in}} \end{bmatrix} \tag{85}$$

The propagating ray deviates from the axis by an amount proportional to the distance and the ray angle. Of course, the ray propagation angle does not change with this transformation.

Ray angles change in passing an interface between materials with different refractive indexes. According to Snell's law, the angle change that occurs passing from a medium with refractive index n_1 to one of n_2 is

$$n_1 \sin \theta_{\text{in}} = n_2 \sin \theta_{\text{out}} \tag{86}$$

We need to determine the output angle (in the n_2 medium) as a function of the refractive indexes and the input angle. For small angles, Snell's law can be written

$$\theta_{\text{out}} = \frac{n_1}{n_2} \theta_{\text{in}} \tag{87}$$

At a flat interface, the ray angle changes but the radial offset remains fixed. The transformation matrix for the interface is

$$\mathbf{T} = \begin{bmatrix} 1 & 0 \\ 0 & n_1/n_2 \end{bmatrix} \tag{88}$$

The change in ray angle at the interface accounts for the change in refractive index. Thus the optical pathlength or propagation distance through the media with a different refractive index is the physical distance. For example, a ray entering a glass block with a refractive index of n_2 and then propagating over a physical distance of d will have a ray vector of

$$\begin{bmatrix} r_{\text{out}} \\ \theta_{\text{out}} \end{bmatrix} = \begin{bmatrix} 1 & d \\ 0 & 1 \end{bmatrix} \begin{bmatrix} 1 & 0 \\ 0 & n_1/n_2 \end{bmatrix} \begin{bmatrix} r_{\text{in}} \\ \theta_{\text{in}} \end{bmatrix} = \begin{bmatrix} r_{\text{in}} + d\theta_{\text{in}} n_1/n_2 \\ \theta_{\text{in}} n_1/n_2 \end{bmatrix} \tag{89}$$

4.3.3. Ray Propagation Through a Lens and Lenslike Media

Two different types of elements can focus or defocus rays passing through an element. The first is a thin lens, which is one that does not occupy any distance along the propagation path. Rays entering the lens at some position along the z axis will exit the lens at the same position on the axis. Like the refractive index interface, a lens changes the direction angle of a ray but not the offset from the z axis. Real lenses can introduce beam offset. The differences between real and ideal thin lenses are small for long-focal-length lenses, and by defining the focal length relative to a point on the z axis that is not necessarily at the midpoint of the real lens, the thin-lens approximation can be quite accurate for short-focal-length devices as well. The ray transfer matrix for a thin lens is

$$\mathbf{T} = \begin{bmatrix} 1 & 0 \\ -1/f & 1 \end{bmatrix} \tag{90}$$

Note that this transfer matrix is for a lens with a positive focal length of f. Diverging lenses are represented by negative f. Consider for the moment the effect of an ideal thin lens on the wavefront of a plane wave passing through the lens. When the plane wave enters the lens, all rays are in phase. After passing through the lens, the wavefront of the on-axis rays are retarded relative to those that passed though the lens near the perimeter. The exiting wave has a spherical wavefront with a radius of curvature equal to the focal length of the lens.

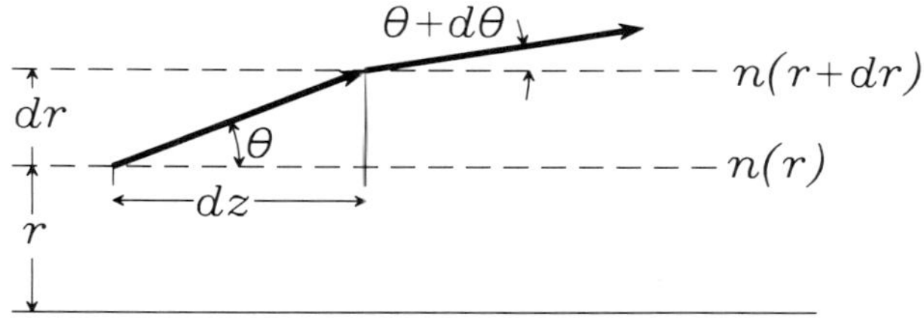

Figure 4.4. Ray path in a medium with inhomogeneous refractive index. (Adapted from Verdeyan 1981.)

The second type of element that can focus or defocus a bundle of rays is based on a refractive index gradient. Lenses that are constructed based on this principle are called graded refractive index (GRIN) lenses. A ray transfer matrix can be constructed for propagation in the inhomogeneous index medium. The inhomogeneous index medium of interest in photothermal spectroscopy is one where the refractive index has a squared radial dependence. This is the same refractive index dependence as that used in the GRIN lens devices.

For purposes of illustration, the basic equation governing the propagation of a ray through a medium with a radial-dependent refractive index will be derived and then applied to determine the ray transfer matrix for a GRIN lens. A paraxial ray traveling through a medium with a radially dependent refractive index, $n(r)$, is illustrated in Figure 4-4. The ray passes through thin sections or shells with different refractive indexes. Snell's law applies at each of the infinitely thin shell boundaries. In this case

$$n(r)\cos\theta_1 = n(r + \delta r)\cos(\theta_1 + \delta\theta) \tag{91}$$

Series expansion of the second refractive index results in

$$n(r + \delta r) = n(r) + \frac{dn(r)}{dr}\,\delta r + \frac{1}{2!}\,\frac{d^2 n(r)}{dr^2}\,\delta r^2 + \cdots \tag{92}$$

Retaining only the first two terms and expanding the cosine sum yields

$$n(r)\cos\theta_1 = \left[n(r) + \frac{dn(r)}{dr}\,\delta r\right](\cos\theta_1\cos\delta\theta - \sin\theta_1\sin\delta\theta) \tag{93}$$

For small $\delta\theta$, this equation can be rearranged as

$$\frac{dn(r)}{dr} = n(r)\,\frac{\delta\theta}{\delta r}\tan\theta_1 \tag{94}$$

Since $\tan\theta_1 = \delta r/\delta z$,

$$\frac{1}{n(r)}\frac{dn(r)}{dr} = \frac{\delta\theta}{\delta z} \tag{95}$$

This equation is important in the context of photothermal deflection spectroscopy. The angle at which a ray exits a medium with a refractive index gradient is found by integrating the above over the optical path

$$\theta = \int_{\text{path}} \frac{1}{n(r)}\frac{dn(r)}{dr}\,dz \approx \frac{1}{n_0}\int_{\text{path}} \left[\frac{dn(r)}{dr}\right]_{r_0} dz \tag{96}$$

where n_0 is the bulk refractive index of the medium and r_0 is the offset of the ray entering the medium.

In the paraxial approximation $\theta \approx \delta r/\delta z$; thus the Snell's law equation yields

$$\frac{d^2 r}{dz^2} = \frac{1}{n(r)}\frac{dn(r)}{dr} \tag{97}$$

This is the basic equation for ray propagation in an inhomogeneous medium. In the most general interpretation, r is the radial distance from the main axis of propagation, but the refractive index variation does not have to be a cylindrical symmetric. For the cases considered below, the refractive index change will be cylindrically symmetric.

Any cylindrically symmetric radial dependent refractive index can be expanded in a Maclaurin series as a function of r:

$$n(r) = n_0 + \left(\frac{dn}{dr}\right)_{r=0} r + \frac{1}{2!}\left(\frac{d^2 n}{dr^2}\right)_{r=0} r^2 + \cdots \tag{98}$$

The first term of this expansion is the constant refractive index of the medium, the second is the gradient of the refractive index, the third the curvature, and so on. When the refractive index changes proportional to r^2, the resulting optical element is similar to a lens. In the case where the rays are propagating near the axis, the odd derivatives vanish and only the first and third terms of the Maclaurin series are significant:

$$n(r) = n_0 + \frac{r^2}{2!}\left(\frac{d^2 n}{dr^2}\right)_{r=0} \tag{99}$$

Using the basic equation for beam propagation, we obtain

$$\frac{d^2r}{dz^2} = \frac{1}{n(r)}\frac{dn(r)}{dr} = \frac{1}{n_0 + (r^2/2)(d^2n/dr^2)_{r=0}}\left(\frac{d^2n}{dr^2}\right)_{r=0} r \tag{100}$$

Restricting the radial offset to positions where $(d^2n/dr^2)r^2 \ll 1$ yields

$$\frac{d^2r}{dz^2} = -\frac{r}{n_0}\frac{d^2n}{d^2r} \tag{101}$$

The solution to this differential equation for an input ray at radial offset r_{in} and angle θ_{in} is

$$r(z) = r_{\text{in}} \cos az + \frac{\theta_{\text{in}}}{a} \sin az \tag{102}$$

where the a term is defined by

$$a = \sqrt{\frac{-1}{n_0}\left(\frac{d^2n}{dr^2}\right)_{r=0}} \tag{103}$$

The propagation angle is found by differentiating with respect to z:

$$\theta(z) = -ar_{\text{in}} \sin az + \theta_{\text{in}} \cos az \tag{104}$$

The ray transfer matrix for this optical element is

$$\mathbf{T} = \begin{bmatrix} \cos ad & 1/a \sin ad \\ -a \sin ad & \cos ad \end{bmatrix} \tag{105}$$

where d is the pathlength through the medium. This ray transfer matrix is for propagation within the medium and does not take into account the change in angle that occurs at the interfaces of the medium.

To see that this in fact a lenslike element, one can compare the limiting case of small a to ray transfer matrixes for propagation and a thin lens. In the limit as a approaches zero, the cosine terms approach 1 and the value of the sine of small angles approaches the angles. Thus

$$\lim_{a\to 0}\begin{bmatrix} \cos ad & (1/a) \sin ad \\ -a \sin ad & \cos ad \end{bmatrix} = \begin{bmatrix} 1 & d \\ -a^2 & 1 \end{bmatrix} \tag{106}$$

It is apparent that the upper right term of this limiting matrix is similar to that of the ray transfer matrix for propagation over a path d. Compared to the ray transfer matrix for a thin lens, the inverse focal length of this GRIN element is

$$\frac{1}{f} = a^2 d = -\frac{d}{n_0}\left(\frac{d^2 n}{dr^2}\right)_{r=0} \tag{107}$$

This is an important result in the context of photothermal lens spectroscopy. It relates the focal length of the lens directly to the spatial derivative of the refractive index.

The trigonometric functions in the ray transfer matrix are functions of the argument a, and this argument is proportional to -1 time the square root of d^2n/dr^2. When d^2n/dr^2 is negative (i.e., the refractive index decreases with position away from the propagation axis), the square root results in a real a, and the trigonometric ray transfer matrix results. In this case the lenslike element is a focusing element. The high refractive index on axis relative to the off-axis index results in longer optical pathlengths for on-axis rays. This is similar in form to a simple lens that is thicker at the middle, thus having a longer optical pathlength at the middle. When d^2n/dr^2 is positive, the refractive index increases with radial offset, and the square root results in an imaginary a. In this case the trigonometric functions in the ray transfer matrix are hyperbolic trigonometric functions of the absolute value of a. The resulting ray transfer matrix is

$$\mathbf{T} = \begin{bmatrix} \cosh ad & (1/a)\sinh ad \\ a\sinh ad & \cosh ad \end{bmatrix} \tag{108}$$

where the a term is

$$a = \sqrt{\frac{1}{n_0}\left(\frac{d^2 n}{d^2 r}\right)_{r=0}} \tag{109}$$

The optical pathlength is shorter along the axis of propagation and increases with radial distance. This is similar in function to a negative meniscus lens, where the material thickness increases with radial distance.

4.4. TRANSFORMATION OF GAUSSIAN MODES

The propagation characteristics of Gaussian modes being transformed by media with radial-dependent refractive indexes can also be described using the ray transformation matrixes. In this case there are no rays other than those that comprise the Gaussian beam. The beam is restricted to propagation along the axis of symmetry. The TEM_{00} mode can be written in terms of a complex Gaussian beam parameter, q:

$$E(r, z) = E_0 \frac{q(0)}{q(z)} \exp\left[-\frac{ikr^2}{2q(z)}\right] \times \exp\{-i[kz - \eta(z)]\} \tag{110}$$

Comparing the exponential term to the TEM_{00} mode equation, it is easy to see that the complex Gaussian beam parameter is

$$\frac{1}{q(z)} = \frac{1}{R(z)} - i\frac{\lambda}{n\pi w^2(z)} \tag{111}$$

Substituting definitions for $R(z)$ and z_0, $q(z)$ can also be written as

$$\frac{1}{q(z)} = \frac{z}{z^2 + z_0^2} - i\frac{z_0}{z^2 + z_0^2} = \frac{1}{z + iz_0} \tag{112}$$

Thus $q(z) = z + iz_0$. The field magnitude is found from the power calculation.

In the ray transfer matrix approach, only the ray offset and propagation angle are considered. This means that only radial-dependent terms of the Gaussian will be affected by the optical elements under consideration. It seems reasonable then that the transformation of a Gaussian beam can be described in terms of the complex beam parameter, since that is the only term that affects the radial-dependent exponent. In fact, the Gaussian beam is transformed by what is known as the *ABCD* law, according to

$$q_{\text{out}} = \frac{Aq_{\text{in}} + B}{Cq_{\text{in}} + D} \tag{113}$$

where q_{in} is the complex beam parameter going into the element, q_{out} is that exiting the element, and the A, B, C, and D are elements of the ray transfer matrixes that are of the general form

$$\mathbf{T} = \begin{bmatrix} A & B \\ C & D \end{bmatrix} \tag{114}$$

The *ABCD* law can also be written

$$\frac{1}{q_{\text{out}}} = \frac{C + D/q_{\text{in}}}{A + B/q_{\text{in}}} \tag{115}$$

This form may facilitate calculations for certain transformations. When used in the Gaussian beam propagation equation, the complex beam parameter transformed by these operations gives the electric field of the transformed Gaussian beam if the origin of the z axis is shifted similarly to compensate for the optical path difference. In other words, the $ABCD$ law does not account directly for the longitudinal phase shift. In using the $ABCD$ law, the z-axis origin is often shifted to a more convenient reference point. It is this reference point that should be used to calculate the longitudinal phase shift of the transformed beam. In most calculations the longitudinal phase shift is not important because the irradiance will be found from the resulting electric field.

4.4.1. Gaussian Mode Propagation in Isotropic Media

The simplest case of Gaussian beam propagation is that of translation in space. Consider a Gaussian beam with a beam waist of w_0 and confocal distance, z_0, that is initially at zero offset from the focus. In this case $z = 0$ and the beam has an initial complex beam parameter

$$q_{\text{in}} = iz_0 \tag{116}$$

where $z_0 = n\pi w_0^2/\lambda$. The beam then propagates over a distance z. The ray transfer matrix for propagation has $A = D = 1$, $B = z$, and $C = 0$. Using the $ABCD$ law, we obtain

$$q_{\text{out}} = \frac{Aq_{\text{in}} + B}{Cq_{\text{in}} + D} = \frac{1 \cdot q_{\text{in}} + z}{0 \cdot q_{\text{in}} + 1} = q_{\text{in}} + z \tag{117}$$

The inverse beam parameter is more facile to interpretation since the real part is the inverse radius of wavefront curvature and the imaginary part is related to the beam waist radius. The inverse beam parameter is found by completing the square in the denominator:

$$\frac{1}{q_{\text{out}}} = \frac{z}{z^2 + z_0^2} - i\,\frac{z_0}{z^2 + z_0^2} = \frac{1}{R(z)} - i\,\frac{\lambda}{n\pi w^2(z)} \tag{118}$$

which can be recognized as the complex beam parameter for a Gaussian beam at a distance z from the beam waist.

A more interesting case is the transformation that occurs at an interface between two media of different refractive indexes. In this case the $ABCD$ law reduces to

$$\frac{1}{q_{out}} = \frac{C + D/q_{in}}{A + B/q_{in}} = \frac{n_1}{n_2}\frac{1}{q_{in}} \tag{119}$$

or simply

$$q_{out} = \frac{n_2}{n_1}(z_{in} + iz_{0,in}) \tag{120}$$

It is clear that there is a direct relationship between the input and output z and z_0. Each term of the input complex beam parameter is simply multiplied by the refractive index ratio to give the corresponding term in the output parameter. Since the real and imaginary parts do not combine, the output has a confocal beam parameter of $z_{0,out} = n_2\pi w_0^2/\lambda$ and the distance to the minimum beam waist is effectively $z_{out} = (n_2/n_1)z_{in}$.

A change in refractive index often arises in focusing lasers into samples. Liquid samples have refractive indexes that are greater than air, so the beam propagation parameter will change at the interface to the sample. For example, a laser is focused in air so that the minimum spot size is w_0 and the focus is positioned in the center of a sample cuvette, 5 mm past the interface. The cuvette is then filled with an aqueous solution with $n_2 = 1.33$. Two questions to answer are: Where is the focus, and what is the confocal distance or focus beam waist of the transformed beam? These questions can be answered by propagating the beam transformed by the index change over a distance in the sample of z_{out}. Including the propagation of the beam over a distance z_{out}, the output beam parameter is

$$q_{out} = \frac{n_2}{n_1}(-z_{in} + iz_{0,in}) + z_{out} \tag{121}$$

z_{in} is negative in this case since the beam enters the sample prior to the focus, or equivalently, the focus is to the positive side of the interface. Again, the beam is analyzed in terms of the inverse beam parameter. The real part of the inverse beam parameter is the inverse radius of curvature,

$$\mathscr{R}\{q_{out}^{-1}\} = \frac{1}{R(z_{out})} = \frac{z_{out} - (n_2/n_1)z_{in}}{[z_{out} - (n_2/n_1)z_{in}]^2 + [(n_2/n_1)z_0]^2} \tag{122}$$

The radius of curvature is infinite at the focus. Solving for z_{out} at the focus yields

$$z_{out} = \frac{n_2}{n_1} z_{in} \tag{123}$$

The distance to the focus in air was 5 mm, and assuming that $n_1 = 1.00$ for air, the distance to the focus in aqueous solution is $z_{out} = 6.65$ mm. The focus beam waist radius in solution can be obtained from the transformed confocal distance,

$$z_{0,out} = \frac{n_2}{n_1} z_{0,in} = \frac{n_2 \pi w_0^2}{\lambda} \tag{124}$$

The spot size is the same in the aqueous solution as it is in air; however the confocal distance increases in the medium with the higher index.

4.4.2. Focusing Gaussian Beams

Focusing of Gaussian beams is somewhat more complicated than the simple lens equation used for rays. The geometry for focusing the Gaussian beam is illustrated in Figure 4-5. For any input complex beam parameter, the ray transfer matrix for a thin lens results in the relationship

$$\frac{1}{q_{out}} = \frac{C + D/q_{in}}{A + B/q_{in}} = \frac{1}{q_{in}} - \frac{1}{f} \tag{125}$$

The output beam is

$$\frac{1}{q_{out}} = \frac{1}{R_{in}} - \frac{1}{f} - i \frac{\lambda}{\pi n w^2} \tag{126}$$

The lens did not change the beam waist radius of the beam passing through it, but it did change the radius of curvature of the wavefront. The

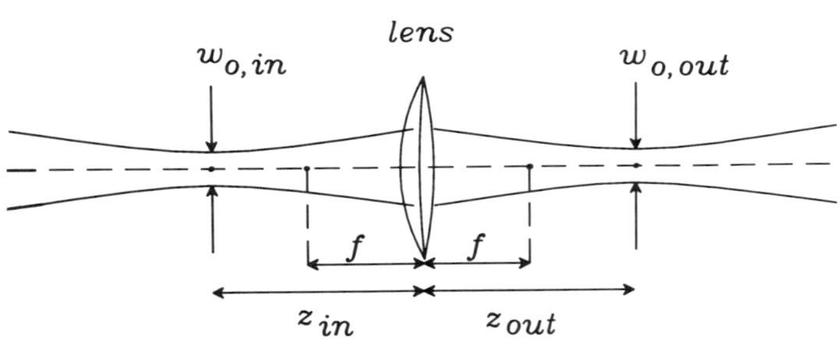

Figure 4.5. Geometry for focusing a Gaussian beam.

output curvature is

$$\frac{1}{R_{\text{out}}} = \frac{1}{R_{\text{in}}} - \frac{1}{f} \tag{127}$$

The change in wavefront curvature will affect both the position of the minimum beam waist and the minimum beam waist radius. To find the focus parameters, the complex beam parameter past the lens is allowed to propagate a distance, z. The total transform matrix for this case is

$$\mathbf{T} = \begin{bmatrix} 1 - z/f & z \\ -1/f & 1 \end{bmatrix} \tag{128}$$

The output beam parameter result, expressed in terms of the wavefront radius and beam waist radius, is

$$\begin{aligned} \frac{1}{R_{\text{out}}}(z) &= \frac{(1/R_{\text{in}} - 1/f)[1 + z(1/R_{\text{in}} - 1/f)] + z(\lambda/\pi n w_{\text{in}}^2)^2}{[1 + z(1/R_{\text{in}} - 1/f)]^2 + z^2(\lambda/\pi n w_{\text{in}}^2)^2} \\ \frac{\lambda}{\pi n w_{\text{out}}^2(z)} &= \frac{\lambda/\pi n w_{\text{in}}^{21}}{[1 + z(1/R_{\text{in}} - 1/f)]^2 + z^2(\lambda/n w_{\text{in}}^2)^2} \end{aligned} \tag{129}$$

The focus occurs at the position where $R_{\text{out}}(z) = \infty$. Solving for z_{out}, the focus position past the lens yields

$$z_{\text{out}} = -\frac{1/R_{\text{in}} - 1/f}{(1/R_{\text{in}} - 1/f)^2 + (\lambda/\pi n w_{\text{in}}^2)^2} \tag{130}$$

R_{in} and w_{in} are also functions of z. Substituting the definitions for R_{in} and w_{in} results in

$$z_{\text{out}} = f + \frac{f^2(z_{\text{in}} - f)}{(z_{\text{in}} - f)^2 + z_{0,\text{in}}^2} \tag{131}$$

where z_{in} is the distance to the minimum beam waist of the input beam. For a beam traveling from left to right, a positive z_{in} is to the left of the lens while a positive z_{out} is to the right of the lens. The equation is quadratic in terms of z_{in}. In general, there are two z_{in} for each z_{out}. For example, to obtain a focus at $z_{out} = f$, $z_{out} = f$ or ∞. This first result is quite different from that predicted from geometric optics. In geometric optics, $z_{in} = f$ only if $z_{out} = \infty$.

The difference between the two solutions lies in the focus beam waist radii. Substituting the focus position result into the equation for the beam waist radius gives the beam waist radius at the focus:

$$w_{0,out}^2 = \frac{f^2}{(z_{in} - f)^2 + z_{0,in}^2} w_{0,in}^2 \tag{132}$$

The beam waist radius when the input beam minimum radius is one focal length in front of the lens is

$$w_{0,out}(z_{in} = f) = \frac{f\lambda}{n\pi w_{0,in}} \tag{133}$$

On the other hand, the focus beam waist radius when $z_{in} = \infty$ is $w_{0,out} = 0$. This is not a physically realizable situation. A Gaussian beam starting at infinity would have an infinite beam waist at the lens. The inverse relationship between the input and focus beam waist radii is due to the diffractive nature of light.

Other interesting features of Gaussian beam focusing may be understood by examining the case where the input beam has a minimum beam waist radius at the lens. This corresponds to a plane wave and one might expect the focus to be at a distance f past the lens. Instead, the focus position occurs at

$$z_{out} = \frac{f}{1 + (f/z_{0,in})^2} \tag{134}$$

An infinite input beam confocal distance corresponds to a true plane wave. The focus spot position should be at the focal length of the lens. This is, in fact, predicted by equation 134. For finite input confocal distance, the focus position is farther from the lens than predicted for a plane wave. This is a somewhat curious effect. What this means is that the focus position of a lens is dependent on the extent of the source. Notice that for confocal distances that are much shorter than the focal length of the lens,

$$\lim_{f/z_{0,in} \gg 1} z_{out} = \frac{z_{0,in}^2}{f} \tag{135}$$

In this case the focus moves closer to the lens as the focal length of the lens is increased. For a focal length at infinity (e.g., a flat window), the focus is at the lens.

4.5. PARAXIAL DIFFRACTION THEORY

Many features of Gaussian beam transformation may be better understood through diffraction optics. In fact, diffraction optics theory predicts that lenses act as Fourier transform devices. In the Fresnel diffraction limit discussed below, the spatial pattern at a position one focal length beyond the lens is the Fourier transform of the spatial pattern one focal length in front of the lens. In this respect it makes sense that a large radius input would result in a small area at the focus, and vice versa. An infinite beam waist input could result in an infinitely small focus spot. However, there is no lens large enough to focus the input of infinite extent. The fact is that a Gaussian beam with a minimum beam waist located an infinite distance in front of the lens can also be accounted for in the Fourier transform. Any beam located at infinity would have an infinite beam waist at the lens. The Fourier transform of a "white" spectrum gives rise to a delta function and thus a focus beam waist radius of zero.

Diffraction theory can be used to determine the propagation of more complex beam shapes and the propagation of simple beams through more complex optical elements. Diffraction theory gives accurate predictions under certain conditions. The basis of this theory will be outlined in this section. A thorough discussion of the basis for this approach can be found in Klein (1970). Comprehensive treatments of transform properties can be found in Lee (1981). Collier et al. (1971) describe the system approach to diffraction optics and applications to periodic structures and holography.

4.5.1. Fresnel Diffraction

Monochromatic light passing through a complex aperture, Σ_1, will produce an electric field $\mathbf{E}(\xi, \eta)$ in the plane with transverse coordinates ξ and η. This is illustrated in Figure 4-6. The complex aperture may attenuate the radiation, shift the phase of the wave, or both. Propagation along the z axis will produce a distribution in the plane with transverse coordinates x and y. The electric field distribution will produce a pattern, Σ_2, in this plane. The electric field is given by the Huygens–Fresnel superposition integral,

$$\mathbf{E}_2(x, y) = \int_{-\infty}^{\infty} \int_{-\infty}^{\infty} h(x, y; \xi, \eta)\mathbf{E}_1(\xi, \eta)\, d\xi\, d\eta \tag{136}$$

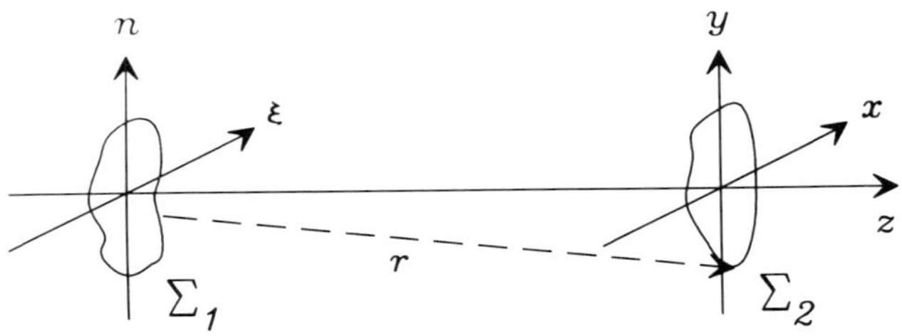

Figure 4.6. Diffraction from an aperture, Σ_1, producing an electric field distribution, Σ_2. The distance vector is measured from each position in the plane of the aperture to the plane of Σ_2. The field distribution is the superposition of all electric fields from Σ_1. (Adapted from Lee 1981.)

where $\mathbf{E}_1$ and $\mathbf{E}_2$ are the electric fields of Σ_1 and Σ_2, and $h(x, y; \xi, \eta)$ is the impulse response:

$$h(x, y; \xi, \eta) = \frac{ie^{-ikr}}{\lambda r} \cos(\mathbf{n}, \mathbf{r}) \tag{137}$$

where r is the distance between the points in the Σ_1 and Σ_2 planes. The angle is that between the wavefront normal at the Σ_1 plane, $\mathbf{n}$, and the radial vector, $\mathbf{r}$, to a point in the plane containing Σ_2. When the distance of propagation, z, is large compared to the dimensions of aperture Σ_1, $\cos(\mathbf{n}, \mathbf{r}) \approx 1$, and the r in the denominator of the impulse-response function is approximately equal to z. These approximations are good to within 5% as long as the angle does not exceed 18°. This condition is met in many situations and corresponds to the paraxial approximation used in ray tracing and the propagation of Gaussian beams. In this case the impulse-response function is

$$h(x, y; \xi, n) = \frac{ie^{-ikr}}{\lambda z} \tag{138}$$

but calculations using this function are still difficult, owing to the r dependence in the exponent.

One simplification is to use a binomial expansion of the square root associated with the calculation of r:

$$r = [z^2 + (x - \xi)^2 + (y - \eta)^2]^{1/2} \approx z\left[1 + \frac{(x - \xi)^2}{2z^2} + \frac{(y - \eta)^2}{2z^2}\right] \tag{139}$$

This simplication results in what is commonly known as Fresnel diffraction. In the case at hand, the integral is simplified somewhat since only paraxial rays are considered. Substitution of this expression into the impulse response results in the Fresnel diffraction integral:

$$\mathbf{E}_2(x, y) = \frac{ie^{-ikz}}{\lambda z} \int_{-\infty}^{\infty} \int_{-\infty}^{\infty} \mathbf{E}_1(\xi, \eta) e^{(-ik/2z)[(x - \xi)^2 + (y-\eta)^2]} \, d\xi \, d\eta \tag{140}$$

The form of this integral indicates that the field forming Σ_2 is related to the aperture Σ_1 by the convolution

$$\mathbf{E}_2(x, y) = h(x, y) \otimes \mathbf{E}_1(x, y) \tag{141}$$

where $\otimes$ represents the two-dimensional spatial convolution and the impulse response is

$$h(x, y) = \frac{ie^{-ik(z + x^2/2z + y^2/2z)}}{\lambda z} \tag{142}$$

The impulse response is often called the optical point-spread function. A point light source in the Σ_1 plane will spread to a distribution indicated by $h(x, y)$. The fact that the electric field in the Σ_2 plane can be represented by a convolution with a point-spread function shows that optical wave propagation is a linear space-invariant system.

A system is said to be linear if the law of superposition holds. The law of superposition states that an input of $c_1a_1(x, y) + c_2a_2(x, y)$ will result in an output of $c_1b_1(x, y) + c_2b_2(x, y)$ regardless of the values of the amplitude constants, c_1 and c_2, and spatial distribution of the functions $a_1(x, y)$ and $a_2(x, y)$. A system is space invariant if a shift in an input function, $a(x - x', y - y')$, results in an output, $b(x - x', y - y')$, that is shifted by an equal amount. Real optical systems typically have a magnification and may invert the input. This linear space variance can be accounted for by scaling the output by the positive or negative magnification. Moreover, optical systems are often not space invariant over their whole input and output planes. There are usually space invariant areas called isoplanic patches in which the system can be treated as linear space invariant.

The convolution representation may also be expressed in terms of Fourier transforms. Taking the Fourier transform of both sides of convolution results in a product of the electric field transform with the spatial frequency-dependent transfer function

$$\hat{\mathbf{E}}_2(\omega_x, \omega_y) = 2\pi \hat{h}(\omega_x, \omega_y) \hat{\mathbf{E}}_1(\omega_x, \omega_y) \tag{143}$$

where the spatial frequency point-spread function, often called the coherent transfer function (CTF), is

$$\hat{h}(\omega_x, \omega_y) = \frac{i}{2\pi} e^{-ikz} e^{+(iz/2k)(\omega_x^2 + \omega_y^2)} \tag{144}$$

The spatial frequency transfer function shows two effects of diffraction on electromagnetic wave propagation. The first effect, given in the first exponent, is just the longitudinal phase shift experienced by any electromagnetic wave propagating over a distance, z. The second exponent represents a phase dispersion with quadratic spatial frequency dependence.

An alternative representation of the diffraction integral is obtained by expanding the squared terms in the integration exponent:

$$\mathbf{E}_2(x, y) = \frac{ike^{-ik(z + x^2/2z + y^2/2z)}}{z} \frac{1}{2\pi} \int_{-\infty}^{\infty} \int_{-\infty}^{\infty} \mathbf{E}_1(\xi, \eta) e^{(-k/2z)(\xi^2 + \eta^2)} e^{(ik/z)(x\xi + yv)} \, d\xi \, d\eta \tag{145}$$

This integral is similar to the two-dimensional Fourier transform

$$\hat{\mathbf{E}}(\omega_x, \omega_y) = \mathscr{F}\{\mathbf{E}(x, y)\} = \frac{1}{2\pi} \int_{-\infty}^{\infty} \int_{-\infty}^{\infty} \mathbf{E}(x, y) e^{-i(x\omega_x y + \omega_y)} \, dx \, dy \tag{146}$$

The second exponent in the Fresnel diffraction integral is identical to the Fourier transform exponent with spatial frequencies, $\omega_x = -kx/z$ and $\omega_y = -ky/z$. The field in the Σ_2 plane is thus the product of the Fourier transform and the phase factor

$$\mathbf{E}_2(x, y) = \frac{ike^{-ik(z + x^2/2z + y^2/2z)}}{z} \mathscr{F}\{\mathbf{E}_1(x, y) e^{(-ik/2z)(x^2 + y^2)}\} \tag{147}$$

The Fourier transform variables, ω_x and ω_y, are replaced with the spatial terms prior to evaluation of the final electric field.

The important point illustrated by these somewhat complicated equations is that the electric field amplitude in a plane past that containing the aperture can be described by the Fourier transform of the field amplitude at the plane of the aperture. Practically speaking, the Fourier transform solutions resulting from the Fresnel approximations are much easier to obtain than is the exact solution obtained by integration of the Huygens–Fresnel integral. The mathematical relationship to the Fourier transform results in this branch of optics being called Fourier optics.

4.5.2. Fraunhofer Diffraction

The diffraction integral can be simplified further when the Σ_2 plane is in what is called the far field. This is the case of Fraunhofer diffraction. Fraunhofer diffraction can be used when the distance that the field propagates is very much greater than the size of the Σ_1 and Σ_2. In this case, $z \gg (k/2)(\xi^2 + \eta^2)$, and the first exponential term in the Fresnel diffraction integral is replaced by unity. The resulting Fraunhofer diffraction integral representation of the field is

$$\mathbf{E}_2(x, y) = \frac{ike^{-ik(z+x^2/2z+y^2/2z)}}{z} \frac{1}{2\pi} \int_{-\infty}^{\infty} \int_{-\infty}^{\infty} \mathbf{E}_1(\xi, \eta) e^{(ik/z)(x\xi + y\eta)} \, d\xi \, d\eta \quad (148)$$

Taking note of the exponent in the integral, the field in the Σ_2 plane is related to the Fourier transform of the field in the Σ_1 plane. The far-field electric field distribution can thus be expressed as

$$\mathbf{E}_2(x, y) = \frac{ike^{-ik(z+x^2/2z+y^2/2z)}}{z} \hat{\mathbf{E}}_1 \left(-\frac{kx}{z}, -\frac{ky}{z} \right) \quad (149)$$

Thus the far-field distribution is directly proportional to the Fourier transform. The field is multiplied by a phase factor that is eliminated when the irradiance is detected with a square-law detector. The far-field irradiance is proportional to

$$|\mathbf{E}_2(x, y)|^2 = \frac{k^2}{z^2} \left|\hat{\mathbf{E}}_1 \left(-\frac{kx}{z}, -\frac{ky}{z} \right)\right|^2 \quad (150)$$

4.5.3. Diffraction Elements

Diffraction theory is more flexible than ray tracing and Gaussian beam propagation in that the effect of virtually any optical element can be accounted for. The procedures used in diffraction are essentially the same as those used in ray tracing. An optical system is modeled as a series of elements and electric fields. The effect of the elements on the propagation of light is accounted for each in turn. Diffraction theory can be applied to spatially complicated electric fields and elements. In fact, results can be obtained for arbitrary input electric fields and then applied to specific cases.

The two main optical elements used in diffraction theory optics are the linear displacement, which can be accounted for with the convolution of the impulse-response function, and the transmission optical elements, which are used to represent lenses and transmitting apertures. The transmission elements can be real or complex. Components of optical elements that result in a phase shift are represented by an imaginary exponential term. Elements or parts of elements that attenuate the electric field are represented by real functions. Thus, in general, any optical element resulting in a change in propagation character of the input field can be represented by a transmission

$$t(x, y) = t_r(x, y)e^{-i\phi(x,y,z)} \tag{151}$$

where $t_r(x, y)$ is the real attenuation that occurs in the plane of the element and $\phi(x, y, z)$ is the phase shift. The longitudinal component of the phase shift transforms the electric field in a different fashion than does the perpendicular terms. The phase shift is subsequently divided into z-dependent and perpendicular terms for bookkeeping during the calculations.

4.5.4. Fourier Transform Properties of Apertures

The diffraction that occurs from apertures can be understood in terms of the Fourier transforms of their mathematical representations. The Fourier transforms used here are two-dimensional. Each of the two spatial coordinates has a transform defined by

$$\hat{t}(\omega_x) = \mathscr{F}\{t(x)\} = \frac{1}{\sqrt{2\pi}} \int_{-\infty}^{\infty} e^{-ix\omega_x} t(x)\, dx \tag{152}$$

with an associated inverse transform of

$$t(x) = \mathscr{F}^{-1}\{\hat{t}(\omega_x)\} = \frac{1}{\sqrt{2\pi}} \int_{-\infty}^{\infty} e^{ix\omega_x} \hat{t}(\omega_x)\, d\omega_x \tag{153}$$

Functions of x and y that are separable can be transformed using

$$\mathscr{F}\{t(x, y)\} = \mathscr{F}\{t(x)\}\mathscr{F}\{t(y)\} = \hat{t}(\omega_x)\hat{t}(\omega_y) \tag{154}$$

The diffraction integrals correspond only to forward transforms. However, the inverse transforms are used for function products produced by the convolution theorem.

Properties of Fourier transforms can give insight as the physical properties of optical systems. Some useful properties are:

1. The transform of the product of a constant times a function is the transform of the function multiplied by the same constant:

$$\mathscr{F}\{at(x)\} = a\hat{t}(\omega_x) \tag{155}$$

   The magnitude of the field or the transmission of the aperture does affect the form of the field in the transform plane.

2. Translation of an aperture pattern produces a linear phase shift in the spatial frequency domain:

$$\mathscr{F}\{t(x-a)\} = \hat{t}(\omega_x)e^{-i\omega_x a} \tag{156}$$

   The linear phase-shift term does not affect the irradiance. Thus the irradiance in the transform plane is invariant with respect to a linear displacement in the aperture or input beam.

3. Scaling the input spatial pattern produces an inversely scaled frequency-domain pattern:

$$\mathscr{F}\{t(ax)\} = \frac{1}{|a|}\,\hat{t}\left(\frac{\omega_x}{a}\right) \tag{157}$$

   Thus a smaller input pattern produces a larger pattern in the transform plane. This effect can be used to explain why a Gaussian with a larger beam waist radius produces a smaller spot at the focus of a lens.

4. Energy or power must be conserved in the optical system. This is accounted for by integration over the entire area of the plane. For electric fields related by Fourier transforms

$$\int_{-\infty}^{\infty}\int_{-\infty}^{\infty} |\mathbf{E}(x, y)|^2\,dx\,dy = \int_{-\infty}^{\infty}\int_{-\infty}^{\infty} |\hat{\mathbf{E}}(\omega_x, \omega_y)|^2\,d\omega_x\,d\omega_y \tag{158}$$

   Thus the total power or energy in the Fourier plane must be equal to that of the input plane. The transforms defined above satisfy this requirement as shown by Parseval's theorem.

5. The central value of the Fourier transform is

$$\hat{\mathbf{E}}(0, 0) = \frac{1}{2\pi} \int_{-\infty}^{\infty} \int_{-\infty}^{\infty} \mathbf{E}(x, y)\, dx\, dy$$
$$\mathbf{E}(0, 0) = \frac{1}{2\pi} \int_{-\infty}^{\infty} \int_{-\infty}^{\infty} \hat{\mathbf{E}}(\omega_x, \omega_y)\, d\omega_x\, d\omega_y \qquad (159)$$

   These equations show that the average of the variable to be transformed is the central value in the transformed variable space.

6. The transform of product of two functions is related by the convolution theorem

$$\mathscr{F}\left\{\int_{-\infty}^{\infty} t_1(\xi) t_2(x - \xi)\, d\xi\right\} = \mathscr{F}\{t_1(\xi) * t_2(x)\} = \sqrt{2\pi}\,\hat{t}_1(\omega_x)\hat{t}_2(\omega_x) \qquad (160)$$

   The factor of $(2\pi)^{1/2}$ is required for proper scaling of the result.

7. The transform of the autocorrelation is the power spectrum

$$\mathscr{F}\left\{\int_{-\infty}^{\infty} t(\xi) t(\xi - x)\, d\xi\right\} = \frac{1}{\sqrt{2\pi}} |\hat{t}(\omega_x)|^2 \qquad (161)$$

   Again, the factor of $(2\pi)^{1/2}$ is required for proper scaling. The power spectrum of the electric field is directly proportional to the irradiance in the transform plane.

4.5.5. One and Two-Dimensional Fourier Transforms

A few common one-dimensional Fourier transform pairs are given in Table 4.1. For two-dimensional transforms of functions with separable x and y terms, the product of two transforms will give the correct result. If the transmission or electric field varies in only one of the coordinates, the two-dimensional transform is the transform of that coordinate multiplied by a delta function in the other transformed variable. The special functions shown in this table include the rect function, which is a rectangle centered

Table 4.1. One-Dimensional Fourier Transform Pairs

$t(x)$	$\hat{t}(\omega_x)$
1	$(2\pi)^{-1/2}\delta(\omega_x)$
$\delta(x)$	$(2\pi)^{-1/2}$
$\delta(x-a)$	$(2\pi)^{-1/2}\exp(-ia\omega_x)$
$\cos(\omega_0 x)$	$(2\pi)^{-1/2}\frac{1}{2}[\delta(\omega_x-\omega_0)+\delta(\omega_x+\omega_0)]$
$\text{rect}(x/a)$	$(2/\pi)^{1/2}\sin(a\omega_z/2)/\omega_x$
$\exp(-x^2/a^2)$	$\lvert a\rvert 2^{-1/2}\exp(-a^2\omega_x^2/4)$
$\text{comb}_a(x)$	$(2\pi/a^2)^{1/2}\,\text{comb}_{2\pi/a}(\omega_x)$

at the origin defined by

$$\text{rect}\left(\frac{x}{a}\right)=\begin{cases}1 & \text{for } \left|\frac{x}{a}\right|\leqslant\frac{1}{2}\\ 0 & \text{for } \left|\frac{x}{a}\right|>\frac{1}{2}\end{cases} \tag{162}$$

This function is used to model the effect of slit-like apertures in an optical system. Also included is the comb function, a periodic array of delta functions defined by

$$\text{comb}_a(x)=\sum_{i=-\infty}^{\infty}\delta(x-na) \tag{163}$$

This function is useful in modeling periodic arrays, such as rhonchi rulings and venitian blind–style optical components. A special class of nonseparable functions often encountered in optics is that of cylindrically symmetric functions of the radius. The Fourier transform functions with radial symmetry in two-dimensional Cartesian coordinates is

$$\hat{t}(\omega_x,\omega_y)=\frac{1}{2\pi}\int_{-\infty}^{\infty}\int_{-\infty}^{\infty}e^{-i(\omega_x x+\omega_y y)}\,t(\sqrt{x^2+y^2})\,dx\,dy \tag{164}$$

As shown in Appendix 3B, transformation into polar cylindrical coordinates yields

$$\hat{t}(\omega_r)=\int_{-\infty}^{\infty}J_0(\omega_r r)t(r)\,dr \tag{165}$$

Table 4.2. Transform Pairs of Radial Symmetric Functions

$t(r)$	$\hat{t}(\omega_r)$
$\delta(r - a)$	$a\mathbf{J}_0(a\omega_r)$
$\text{circ}(r/a)$	$a\mathbf{J}_1(a\omega_r)/\omega_r$
$(a^2 + r^2)^{-1/2}$	$\omega_r^{-1} \exp(-a\omega_r)$
$(a^2 + r^2)^{-3/2}$	$a^{-1} \exp(-a\omega_r)$
$\exp(-r^2/a^2)$	$\lvert a \rvert 2^{-1/2} \exp(-a^2\omega_r^2/4)$

where $J_0(\omega_r r)$ is the Bessel's function of order zero. The transform given by this integral is a Hankel transform. Because of the relationship to the Cartesian coordinate Fourier transform, it has symmetrical transform properties useful for solving optical propagation problems. For example, the inverse transform is given by

$$t(r) = \int_0^\infty J_0(\omega_r r)\hat{t}(\omega_r)\omega_r \, d\omega_r \tag{166}$$

Like the Cartesian Fourier transforms, the radial transforms are symmetrical. The transform pairs shown in Table 4.2 can be used interchangeably simply by exchanging r with ω_r.

Two special functions shown in Table 4.2 are the delta function and the circ function. The delta function, $\delta(r - a)$, corresponds to an infinitely narrow transmitting ring at $r = a$. The circ function is similar to the rect function in cartesian coordinates except that it is for a disk of radius a. It is defined by

$$\text{circ}\left(\frac{r}{a}\right) = \begin{cases} 1 & \text{for } \left|\dfrac{r}{a}\right| \leqslant 1 \\ 0 & \text{for } \left|\dfrac{r}{a}\right| > 1 \end{cases} \tag{167}$$

The transform of this function is in terms of the Bessel's function, $\mathbf{J}_1(a\omega_r)$.

4.5.6. Diffraction Properties of the Ideal Thin Lens

It was shown in Section 4.4 that an ideal thin lens introduced a radial phase-shift equivalent to adding a term, $-1/f$, where f is the focal length, to the inverse radius of curvature, $1/R(z)$. This ray tracing operation is equivalent to multiplying the electric field by a complex exponential transmission

$$t_l(x, y) = e^{ik(x^2+y^2)/2f} \tag{168}$$

This same result is obtained in the generalized Abbe theory of lenses, which describes the effect of lens elements in terms of diffraction theory (Klein 1970). The Abbe theory is a generalized result that shows that this complex phase function is valid not only for Gaussian beams but for all electromagnetic waves that are propagating paraxial.

The effect of a lens in an optical system can be formulated by considering a plane, monochromatic electric field passing first through an aperture, Σ_1, with a transmission t_1, then immediately passing through a lens with focal length f. (Actually, it does not matter whether the aperture is in front of or behind the lens.) The electric field distribution past the lens, at a distance equal to the back focal length of the lens, is exactly the Fraunhofer diffraction result. First, the electric field distribution just beyond the lens is the product of aperture and lens transmissions with the incident electric field of the plane wave:

$$\mathbf{E}_1(\xi, \eta) = \mathbf{E}_0 t_1(\xi, \eta) t_l(\xi, \eta) = \mathbf{E}_0 t_1(\xi, \eta) e^{ik(\xi^2+\eta^2)/2f} \tag{169}$$

Using the Fresnel diffraction integral for with z equal to the focal length gives

$$\mathbf{E}_2(x, y) = \frac{ie^{-ik(f+x^2/2f+y^2/2f)}}{\lambda f} \int_{-\infty}^{\infty}\int_{-\infty}^{\infty} [\mathbf{E}_0 t_1(\xi, \eta) e^{(ik/2f)(\xi^2+\eta^2)}] e^{(-ik/2f)(\xi^2+\eta^2)} e^{(ik/f)(x\xi+y\eta)}\, d\xi\, d\eta \tag{170}$$

With some simplification

$$\mathbf{E}_2(x, y) = \mathbf{E}_0 \frac{ike^{-ik(f+x^2/2f+y^2/2f)}}{f} \frac{1}{2\pi} \int_{-\infty}^{\infty}\int_{-\infty}^{\infty} t_1(\xi, \eta) e^{(ik/f)(x\xi+y\eta)}\, d\xi\, d\eta$$

$$\mathbf{E}_2(x, y) = \mathbf{E}_0 \frac{ike^{-ik(f+x^2/2f+y^2/2f)}}{f} \hat{t}_1\left(-\frac{kx}{f}, -\frac{ky}{f}\right) \tag{171}$$

$$\mathbf{E}_2(x, y) = \mathbf{E}_0 \frac{ike^{-ik(f+x^2/2f+y^2/2f)}}{f} \hat{t}_1(\omega_x, \omega_y)$$

This result is equivalent to the Fraunhofer far-field result but with $\omega_x = -kx/f$ and $\omega_y = -ky/f$. The electric field at a distance one focal length beyond the lens is the Fourier transform of the field at the lens, multiplied by a phase factor. It should be noted that this is not the position of the minimum beam waist of the focused input. The radial-dependent phase-shift term indicates that the beam is diverging at this point.

4.5.7. Fourier Transform Properties of the Lens

When Σ_1 is in front of the lens, a Fourier transform forms at the Σ_2 plane at the back focal length. Refering to Figure 4-7, we need to find in the electric field produced in the Σ_2 plane an aperture in the Σ_1 plane at any distance in front of the lens. For a plane monochromatic wave illuminating an aperture in the Σ_1 plane, a distance of d in front of the lens, the transform of the electric field at the lens is

$$\hat{\mathbf{E}}_1(\omega_x, \omega_y) = \mathbf{E}_0 \hat{t}_1(\omega_x, \omega_y) i e^{-ikd + i\omega_x^2 d/2k + i\omega_y^2 d/2k} \tag{172}$$

The electric field at the lens can be represented by an effective transmission. The effective transmission aperture at the lens found by dividing out $\mathbf{E}_0$ is

$$\hat{t}_{\text{effective}}(\omega_x, \omega_y) = \hat{t}_1(\omega_x, \omega_y) i e^{-ikd + i\omega_x^2 d/2k + i\omega_y^2 d/2k} \tag{173}$$

Substituting the effective transmission into the equation for an aperture placed at the lens, followed by substitution for ω, results in the image formed one focal length beyond the lens:

$$\hat{\mathbf{E}}_2(x, y) = -\mathbf{E}_0 \frac{k}{f} e^{-ik(d+f)} e^{ik(1-d/f)(x^2/2f + y^2/2f)} \hat{t}\left(-\frac{kx}{f}, -\frac{ky}{f}\right) \tag{174}$$

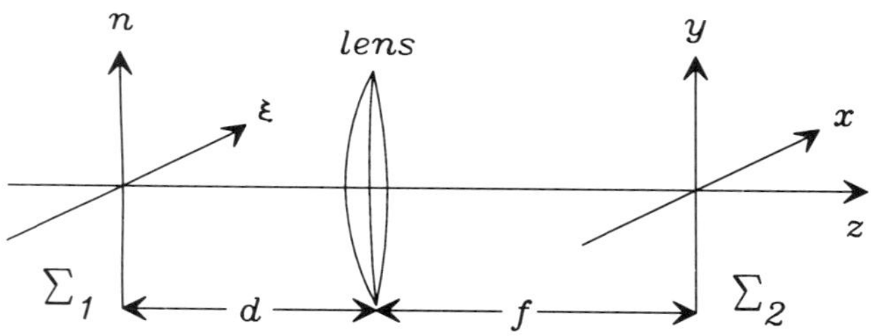

Figure 4.7. Aperture placed in front of a focusing lens will produce a spatial Fourier transform at the back focal plane of the lens.

The electric field at Σ_2, the back focal length of the lens, is composed of the product of a magnitude term, a linear phase shift, a quadratic radial phase shift, and the transform of the input aperture transmission (or electric field).

When Σ_1 is one focal length in front of the lens, the quadratic radial phase-shift term is zero:

$$\mathbf{E}_2(x, y) = -\frac{ke^{-i2kf}}{f}\mathbf{E}_0\hat{t}_1(-kx/f, -ky/f) = -\frac{ke^{-i2kf}}{f}\hat{\mathbf{E}}_1\left(-\frac{kx}{f}, -\frac{ky}{f}\right) \tag{175}$$

Except for the longitudinal phase shift and scaling terms, the electric field at a plane one focal length beyond the lens is equal to the Fourier transform of the field one focal length in front of the lens. Rays making up the electric field in the back focal plane are neither diverging nor converging since there is a radial-dependent phase-shift term. This indicates that a collimated beam is formed in the back focal plane for an input collimated at the front focal plane. This is similar to the result found for Gaussian beam propagation through a lens wherein the minimum beam waist was formed at the back focal plane of the lens when the input beam had a minimum beam waist at the front focal plane.

4.5.8. Imaging with Lenses

Although imaging is not common in photothermal spectroscopy, there are certain situations when the image of an aperture has to be formed at the sample (e.g., Fotiou and Morris 1987). An imaging system can be constructed using one or two lenses, as shown in Figure 4.8. In the two-lens system, the input aperture with transmission, $t_1(\xi, \eta)$, is illuminated with a plane wave. The electric field just past the aperture is $\mathbf{E}_1(\xi, \eta) = \mathbf{E}_0 t_1(\xi, \eta)$. The aperture Σ_1 plane is placed one focal length in front of the first lens, with a focal length of f_1. The transform of this aperture formed in the Σ_2 plane at the back focal length distance is

$$\mathbf{E}_2(x, y) = -\frac{ke^{-i2f_1}}{f_1}\mathbf{E}_0\hat{t}_1\left(-\frac{kx}{f_1}, -\frac{ky}{f_1}\right) \tag{176}$$

The second lens has a focal length of f_2. It is positioned a distance of f_2 beyond the back focal length of the first lens. Thus the image formed at the back focal length of the second lens, Σ_3, will be the Fourier transform of

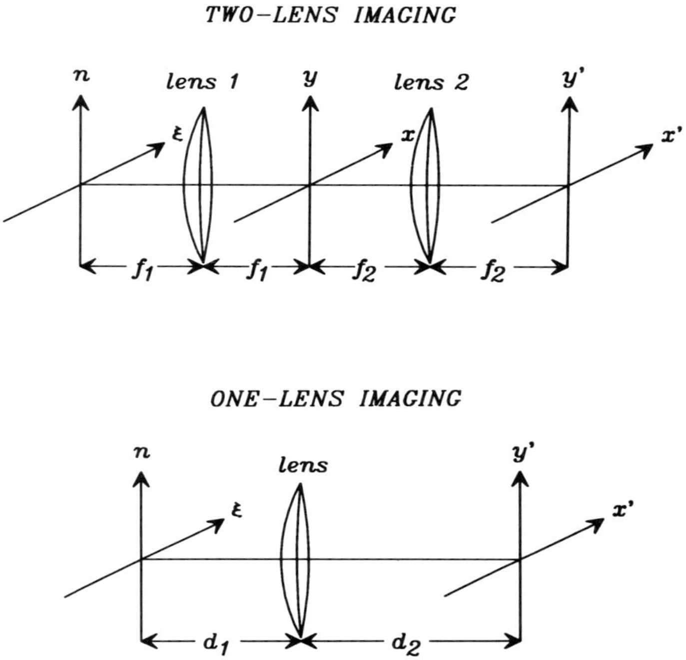

Figure 4.8. Two-lens and one-lens imaging system.

$\mathbf{E}_2(x, y)$ at Σ_2. The electric field at Σ_3 is

$$\mathbf{E}_3(x', y') = \frac{f_1}{f_2} e^{-i2k(f_1+f_2)} \mathbf{E}_0 t_1 \left(-\frac{x' f_1}{f_2}, -\frac{y' f_1}{f_2}\right) \tag{177}$$

The negative sign is retained because only forward transforms are performed with diffraction and $\mathscr{F}^{-1}\{t(x, y)\} = \mathscr{F}\{t(-x, -y)\}$. The magnification of the two-lens system is $M = -f_2/f_1$. The image is inverted. Using this definition gives

$$\mathbf{E}_3(x', y') = -\frac{1}{M} e^{-i2k(f_1+f_2)} \mathbf{E}_0 t_1 \left(\frac{x'}{M}, \frac{y'}{M}\right) \tag{178}$$

The impulse response of an optical system is a function that specifies the mapping of the input plane coordinated with the output plane. The input and image plane electric fields for complex apertures are related with the impulse response through the superposition integral,

$$\mathbf{E}_{\text{out}}(x',\ y') = \int_{-\infty}^{\infty}\int_{-\infty}^{\infty} h(x',\ y';\ \xi,\ \eta)\mathbf{E}_1(\xi,\ \eta)\ d\xi\ d\eta \tag{179}$$

The impulse response for the two-lens imaging system is

$$h(x',\ y';\ \xi,\ \eta) = -\frac{1}{M}\ e^{-i2k(f_1+f_2)}\ \delta(x' + M\xi,\ y' + M\eta) \tag{180}$$

This impulse response has three terms. The first is the magnitude term, $1/M$, which accounts for the decrease (or increase) in the electric field that results from a change in the image size. The second term is the phase shift that occurs due to propagation over a distance of $z = 2(f_1 + f_2)$. This distance is for ideal thin lenses. Technically, the distance is greater than this due to the larger refractive index of the lens material. The last term is the delta function that maps each point in the input plane to a point in the image plane. The mapping is linearly proportional to the magnification and the image is inverted.

Imaging with a single lens produces a similar impulse response (Lee 1981). Through a straightforward, although lengthy calculation, it can be shown that the impulse response for a single lens is

$$\begin{aligned} h(x',y';\xi,\eta) &\\ = -\frac{k^2}{d_1 d_2}&\exp(-ikd_2)\exp\left[-i\left(\frac{k}{2d_1}\right)(\xi^2+\eta^2)\right]\exp\left[-i\left(\frac{k}{2d_2}\right)(x'+y'^2)\right] \\ \times &\int_{-\infty}^{\infty}\int_{-\infty}^{\infty}\exp\left[-i\left(\frac{k}{2}\right)\left(\frac{1}{d_1}+\frac{1}{d_2}-\frac{1}{f}\right)(x^2+y^2)\right] \\ &\exp\left\{ik\left[\left(\frac{\xi}{d_1}+\frac{x'}{d_2}\right)x+\left(\frac{\eta}{d_1}+\frac{y'}{d_2}\right)y\right]\right\}dx\,dy \end{aligned} \tag{181}$$

where d_1 is the distance between Σ_1 and the lens and d_2 is the distance to the Σ_2 image plane. When these distances to the lens satisfy the geometric optics image formula,

$$\frac{1}{f} = \frac{1}{d_1} + \frac{1}{d_2} \tag{182}$$

the impulse-response simplifies to

$$h(x', y'; \xi, \eta) = -\frac{1}{M} e^{-ik(d_1+d_2)} e^{-i(k/2d_1)(\xi^2+\eta^2)} e^{-i(k/2d_2)(x'^2+y'^2)} \delta(x' + M\xi, y' + M\eta) \tag{183}$$

The magnification is defined by $M = -d_2/d_1$ in this case. The main difference between the double-lens and single-lens imaging systems is the quadratic radial phase factor present with the single-lens system.

4.5.9. Diffraction Transforms with Coherent and Incoherent Light

Up to this point the response of the systems has been explored by staring with a monochromatic plane wave. This light is both spatially and temporally coherent. Light generated by real sources, including most lasers, is not monochromatic and is not a plane wave. Nonmonochromatic sources of finite spatial extent have field amplitude and phase that varies randomly with time and in space. The random field amplitude is often characterized by a mutual coherence function. The mutual coherence function is defined as the time-averaged cross-correlation of the complex electric field amplitudes,

$$\Gamma(x_1, x_2; \tau) = \langle \mathbf{E}(x_1; t + \tau)\mathbf{E}^*(x_2; t) \rangle \tag{184}$$

where x_1 and x_2 are the positions at which the electric field is measured, τ is the time difference between the two sets of measurements, and the angle brackets indicate the time average. This function is written in terms of only the x dimension. A general mutual coherence function will be dependent on both x and y dimensions. The mutual coherence function has two components. Spatial coherence describes the field phase at two different locations but at the same time. Spatial coherence is determined from the fringe contrast using a double-slit interferometer. The temporal coherence component is measured at a single position but at different times. Temporal coherence is often measured by the fringe contrast in a Michelson interferometer.

The mutual coherence function satisfies a wave equation and so should also satisfy the conditions for application of the diffraction equations (Lee 1981). It can be shown that the temporal coherence of the a wave is maintained through a linear system. Of interest here is the spatial coherence of the field passing through the linear space-invariant optical system. Application of the Fresnel diffraction equations results in the mutual coherence function after passing through the system,

$$\Gamma(x_1, x_2; 0) = \int_{-\infty}^{\infty}\int_{-\infty}^{\infty} h(x_1 - \xi_1)h^*(x_2 - \xi_2)\Gamma(\xi_1, \xi_2; 0)\, d\xi_1\, d\xi_2 \qquad (185)$$

where the h are the impulse-response or point-spread functions of the system. The irradiances at the input and output planes of the optical system are proportional to the squared complex electric fields. Since the power is conserved, irradiance is proportional to $\Gamma(x_1, x_1; 0)$ in the output plane. The output plane irradiance can be found from

$$\mathbf{E}(x_1) \propto \Gamma(x_1, x_1; 0) = \int_{-\infty}^{\infty}\int_{-\infty}^{\infty} h(x_1 - \xi_1)h^*(x_1 - \xi_2)\Gamma(\xi_1, \xi_2; 0)\, d\xi_1\, d\xi_2 \qquad 186)$$

where the proportionality constant is $\varepsilon_0 c/2$.

The mutual coherence function system response can be used to show differences between the system response of spatially coherent and incoherent radiation. For coherent radiation, the mutual coherence function is simply the product of the two electric fields. In the input plane

$$\Gamma(\xi_1, \xi_2; 0) = \langle \mathbf{E}(\xi_1; t)\mathbf{E}^*(\xi_2; t)\rangle = \mathbf{E}(\xi_1)\mathbf{E}^*(\xi_2) \qquad (187)$$

The irradiance at a point in the output plane is proportional to

$$\begin{aligned}\Gamma(x_1, x_1; 0) &= \mathbf{E}(x_1)\mathbf{E}^*(x_1)\\ &= \int_{-\infty}^{\infty}\int_{-\infty}^{\infty} h(x_1 - \xi_1)h^*(x_1 - \xi_2)\mathbf{E}(\xi_1)\mathbf{E}^*(\xi_2)\, d\xi_1\, d\xi_2 \qquad (188)\end{aligned}$$

The two integrals are separable and

$$\mathbf{E}(x_1) = \int_{-\infty}^{\infty} h(x_1 - \xi_1)\mathbf{E}(\xi_1)\, d\xi_1 \qquad (189)$$

This shows that an optical system response to coherent light is linear in electric field amplitudes.

The mutual coherence function for spatially incoherent light is

$$\Gamma(\xi_1, \xi_2; 0) = \langle \mathbf{E}(\xi_1; t)\mathbf{E}^*(\xi_2; t)\rangle = |\mathbf{E}(\xi_1)|^2\delta(\xi_1 - \xi_2) \qquad (190)$$

In this case the irradiance in the output plane is

$$\begin{aligned}\Gamma(x_1, x_1; 0) &= |\mathbf{E}(x_1)|^2 \\ &= \int_{-\infty}^{\infty}\int_{-\infty}^{\infty} h(x_1 - \xi_1)h^*(x_1 - \xi_2)|\mathbf{E}(\xi_1)|^2\delta(\xi_1 - \xi_2)\,d\xi_1\,d\xi_2 \\ &= \int_{-\infty}^{\infty} |h(x_1 - \xi_1)|^2|\mathbf{E}(\xi_1)|^2\,d\xi_1 \end{aligned} \tag{191}$$

The incoherent illuminated system is thus linear in irradiance. The irradiance in the output plane is found from the convolution of the input plane irradiance profile with the square of the coherent system impulse response. For the two-dimensional case, this can be written in terms of irradiance:

$$\mathbf{E}(x, y) = \int_{-\infty}^{\infty}\int_{-\infty}^{\infty} |h(x - \xi, y - \eta)|^2\mathbf{E}(\xi, \eta)\,d\xi\,d\eta \tag{192}$$

The incoherent impulse response is $|h|^2$.

Referring back to the single-lens imaging system, the radial phase-shift factor will be present when imaging coherent light. This indicates that the wavefront has a radius of curvature and thus the focus is not at the positions indicated by geometric optics. On the other hand, incoherent illumination results in a system linear in irradiance and with a square impulse-response dependence. The radial phase-shift factor is reduced to unity in the squared impulse response and the single-lens system will produce an image in the plane that satisfies the geometric optics condition. As a point of comparison, the double-lens imaging system produces an image at the back focal length of the second lens for both coherent and incoherent radiation.

4.5.10. Finite-Diameter Lens Aperture Effects

Focusing with a finite-diameter lens will produce an electric field or irradiance of finite resolution. Resolution is defined by a separation distance of images or features in the focus plane of an optical system. To be resolved, the two features must be discernible. The criterion for this is that the features be separated by the diameter of the impulse-response function of the system. The effect of finite lens diameter on resolution at the focus is most easily calculated for coherent plane wave illumination of a lens. In the case where there is no aperture, corresponding to a lens of infinite diameter, the input electric field is a constant, $\mathbf{E}_0$. The electric field at the plane positioned at the back focal length is proportional to the Fourier transform of the input field. Since the transform of a constant is a delta function,

$$\mathbf{E}_2(x, y) = \mathbf{E}_0 \frac{ike^{-ik(f+x^2/2f+y^2/2f)}}{f} \delta\left(-\frac{kx}{f}\right) \delta\left(-\frac{ky}{f}\right) = \mathbf{E}_0 ie^{-ikf} \delta(x)\delta(y) \tag{193}$$

A superposition of two plane waves of infinite extent, each traveling at a different direction angle, δ, would result in two points in the back focal length plane of this lens, separated by $f \sin \delta$. Because the size of each spot is infinitely small in this plane, the resolution of the image is infinite. Subsequently, one would be able to determine infinitely small differences in the direction angles of the incident plane waves.

On the other hand, a coherent plane wave passing through a lens of finite radius a is equivalent to an aperture placed at the lens. For the circular aperture, the transmission, t_1, is a circ function. From the table of radial symmetric transforms, the finite circular aperture produces the radial symmetric function

$$\mathbf{E}_2(r) = \mathbf{E}_0 ie^{-ik(f+r^2/2f)} \frac{a}{r} J_1\left(\frac{akr}{f}\right) \tag{194}$$

A series of plane waves with slightly different direction angles would produce a series of "spots" at the back focal plane. The electric field of each spot is related to $2J_1(X)/X$. This function is illustrated in Figure 4-9 for reference. It is similar to the rectangular coordinate sinc function, $\sin(x)/x$, that is produced when a plane wave passes through a rectangular aperture. These functions are oscillatory with magnitudes that decrease inverse to the distance from the maximum. The first zero of the Bessel's J_1 function occurs at $X \approx 3.83$. Thus a plane wave at the lens will result in a spot with a radial distance to the first zero of $3.83f/ka$. Clearly, the smaller the radius of the lens, the larger will be the spot radius. Moreover, since the f-number is f/a, the spot size is directly proportional to the f-number of the lens.

One criterion for image resolution is that two neighboring points in the image plane be separated by a distance equal to the radius to the first zero of the Bessel's function. Since the plane waves are imaged to points separated by $f \sin \delta \approx f\delta$, *for small angles the angle resolution of the lens is* $\delta \approx 3.83/ka$. The angular resolution is independent of focal length but scales inversely proportional to a, and directly proportional to λ, since $k = 2\pi/\lambda$.

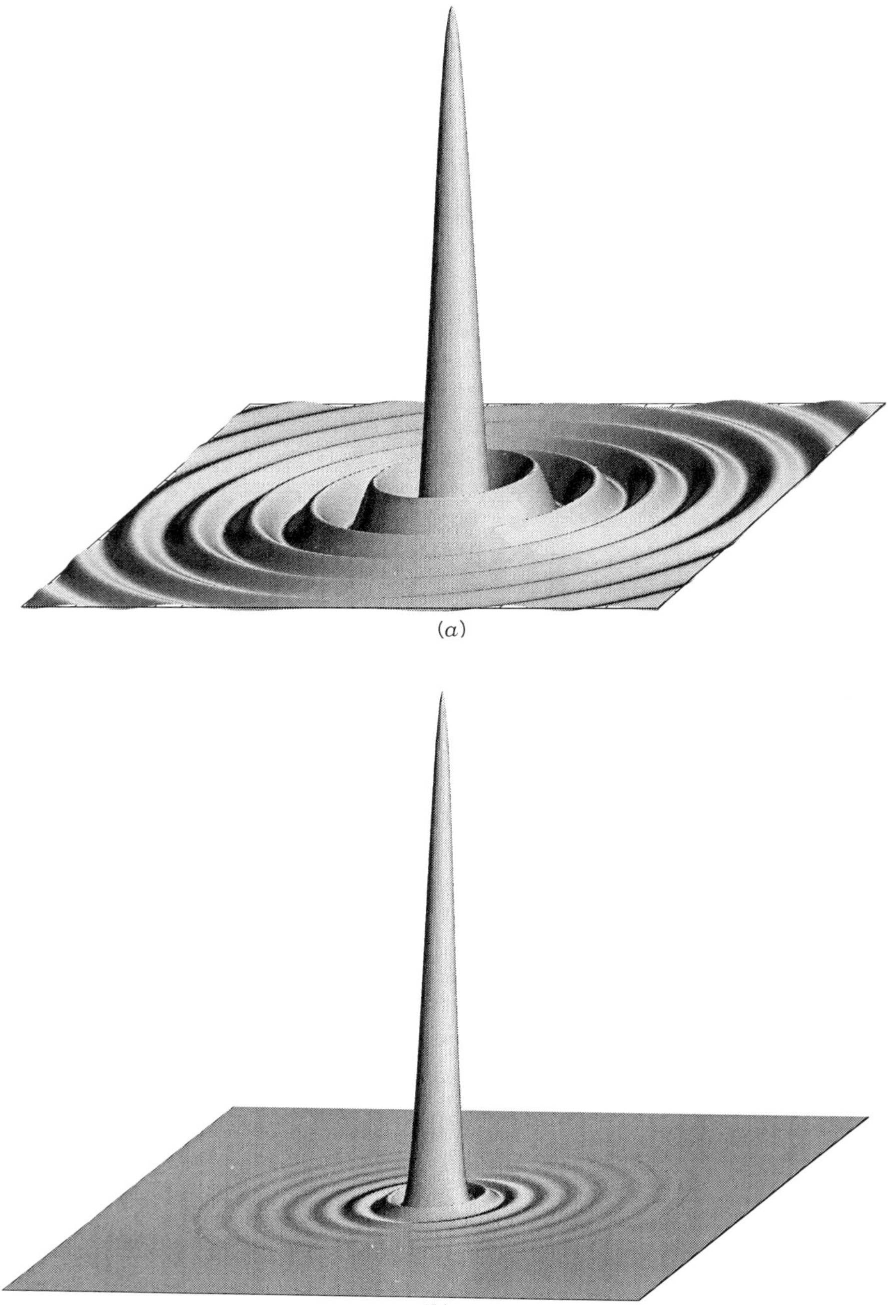

(*a*)

(*b*)

Figure 4.9. (*a*) Electric field distribution produced by a circular aperture; (*b*) irradiance that results by squaring the electric field. The circular aperture of a lens limits the resolution of the image.

4.5.11. Lens Pupil and Optical Transfer Functions

The finite-diameter-lens aperture effect can be generalized to include other aberrations that the optical system may introduce. The lens pupil function is expressed in terms of a real transmission and a phase function of a variable, W:

$$P(x, y) = \operatorname{circ}\left(\frac{\sqrt{x^2 + y^2}}{a}\right) e^{-ikW(x,y)} \tag{195}$$

For random surface roughness, W has the statistical properties of $\langle W \rangle = 0$ and $\langle W^2 \rangle = \sigma_W$. W can also be finite for defocusing:

$$W_{\text{defocus}} = \left(\frac{1}{d_1} + \frac{1}{d_2} - \frac{1}{f}\right)\frac{x^2 + y^2}{2} \tag{196}$$

and spherical aberrations:

$$W_{\text{spherical}} = W_{40}(x^2 + y^2)^2 + W_{60}(x^2 + y^2)^3 \tag{197}$$

where W_{40} and W_{60} are proportionality constants for the spherical aberrations. In general, all of these aberrations will affect the resolution of the optical system. The pupil function, $P(x, y)$, will be equivalent to an aperture placed at a lens and the Fraunhofer diffraction result will apply.

The effect of the lens pupil function on the Fourier transform properties of a lens can be formulated. Recall that the Fourier relationship was derived by projecting an effective aperture to the position of the lens, then using the Fraunhofer relationship for the electric field at the back focal plane of the lens. When there is also an aperture due to the lens pupil function, the aperture is the product of the projected aperture and the lens pupil function. Hence

$$t_1(\xi, \eta) = t_{\text{effective}}(\xi, \eta)P(\xi, \eta) \tag{198}$$

For an aperture placed one focal length in front of the lens, the electric field at the back focal length is

$$\mathbf{E}_2(x, y) = -\mathbf{E}_0 \frac{ke^{-ik2f}e^{-(ik/2f)(x^2+y^2)}}{f} \hat{P}(\omega_x, \omega_y) \otimes [\hat{t}_1(\omega_x, \omega_y)e^{(if/2f)(\omega_x^2+\omega_y^2)}] \tag{199}$$

The convolution is over the frequency variables. Assuming that the convolution does not affect the phase-shift term,

$$\mathbf{E}_2(x, y) = -\mathbf{E}_0 \frac{ke^{-ik2f}}{f} \hat{P}(\omega_x, \omega_y) \otimes \hat{t}_1(\omega_x, \omega_y) \tag{200}$$

Thus the transform in the conjugate plane is a convolution of the ideal transform with the diffraction pattern formed by the pupil function for plane wave illumination in the Σ_1 plane. For an aberration-free lens, the pupil function is a delta function.

The effect of the lens pupil function on an imaging system can be formulated in a similar fashion. For a single-lens imaging system, the pupil function will result in the impulse response

$$h(x', y'; \xi, \eta) = -\frac{1}{M} e^{-ik(d_1+d_2)} e^{-i(k/2d_1)(\xi^2+\eta^2)} e^{-i(k/2d_2)(x'^2+y'^2)} \times \int_{-\infty}^{\infty} \int_{-\infty}^{\infty} P(x, y) e^{(ik/d_2)[(x'-M\xi)x+(y'-M\eta)y]} \, dx\, dy \tag{201}$$

The integral is a Fourier transform and the result can be written

$$h(x', y'; \xi, \eta) = -\frac{2\pi}{M} e^{-ik(d_1+d_2)} e^{-i(k/2d_1)(\xi^2+\eta^2)} e^{-i(k/2d_2)(x'^2+y'^2)} \times \hat{P}\left[\left(\frac{k}{d_2}\right)(M\xi - x'), \left(\frac{k}{d_2}\right)(M\eta - y')\right] \tag{202}$$

Similar to the effect on the Fourier transform properties of the lens, the impulse response of the imaging system is no longer a delta function but rather, is the diffraction pattern of the point-source illuminated pupil function.

The impulse response of the aberration-free imaging system is a delta function that accounts for magnification. The impulse response of the aberrant imaging system is the convolution of the pupil function with this delta function:

$$h(x, y; \xi, \eta) = h_{\text{ideal}}(x, y; \xi, \eta) \otimes \left[\frac{ike^{-ik(f+x^2/2f+y^2/2f)}}{f} \hat{P}_1\left(-\frac{kx}{f}, -\frac{ky}{f}\right)\right] \tag{203}$$

where h_{ideal} is the impulse response of the aberration-free imaging system.

Equivalently, the electric field in the image plane is the electric field of the nonaberrant imaging system convoluted with the transform of the pupil function. The phase factor in the convolution is traceable to the lens pupil being at the lens. The additional phase factor can cause errors in the estimated focus position of lenses. In particular, when focusing Gaussian beams, if the lens aperture is on the order of or less than the radius of the Gaussian beam, the focus will not be at the position predicted from the *ABCD* law. This effect is known as the *focus shift* for Gaussian beams (Li 1993).

The effects of the lens pupil function may also be expressed in terms of the spatial frequency response of the system. The spatial frequency response is often much easier to understand and work with than is the impulse response of the system. Spatial frequency responses are called transfer functions. The spatial frequency content of an electric field in some plane of the optical system is the product of the spatial frequency content of the input signal with the frequency response of the system. From the convolution theorem, the spatial Fourier transform of the imaging system is

$$\hat{\mathbf{E}}_2(\omega_x, \omega_y) = 2\pi \hat{\mathbf{E}}_1(\omega_x, \omega_y)\hat{h}(\omega_x, \omega_y) \tag{204}$$

where ω are the spatial frequencies of the object at the input plane. The Fourier-transformed impulse response is called the coherent transfer function (CTF). This function gives the spatial frequency response of the imaging system. Since the transform of the lens aperture impulse response will yield a circ function, the spatial frequency response of an imaging system with finite-diameter lenses will be constant up to a cutoff value ω_c. Beyond ω_c, the system cannot resolve the input details. In the image plane the cutoff frequency is $\omega_c = ak/f$. The cutoff frequency in the input plane can be related to this result by the magnification (Lee 1981). The input plane spatial frequency cutoff is ak/f.

In the case of incoherent illumination, the imaging system response is the square of the impulse-response function. The Fourier transform of the squared impulse-response function, scaled to unit dc frequency response, is called the optical transfer function (OTF). Using the OTF, the transformed image is

$$|\hat{\mathbf{E}}_2(\omega_x, \omega_y)|^2 = H(\omega_x, \omega_y)|\hat{\mathbf{E}}_1(x, y)|^2 \tag{205}$$

where

$$H(\omega_x, \omega_y) = \frac{\mathscr{F}\{|h(x, y)|^2\}}{\iint |h(x, y)|^2 \, dx \, dy} \tag{206}$$

The modulus of the OTF is the modulation transfer function (MTF). The OTF is related to the CTF through

$$H(\omega_x, \omega_y) = \frac{\hat{h}(\omega_x, \omega_y) \otimes \hat{h}^*(\omega_x, \omega_y)}{|\hat{h}(\omega_x, \omega_y) \otimes \hat{h}^*(\omega_x, \omega_y)|_{\omega_x=0,\omega_y=0}} \tag{207}$$

The differences between the OTF and CTF can be understood in terms of the convolution needed to find the OTF. For example, an incoherent illuminated spatial frequency response of the imaging system with finite-diameter lenses will be the convolution of two equivalent circ functions. Just as the convolution of two rectangle functions yields a triangular function, the convolution of two circ functions results in the cone-shaped function

$$H(\omega_x, \omega_y) = \begin{cases} \dfrac{2}{\pi} \cos^{-1} \dfrac{\omega_r}{\omega_c} - \dfrac{2\omega_r}{\omega_c} \sqrt{1 - \left(\dfrac{\omega_r}{\omega_c}\right)^2} & \omega_r \leqslant \omega_c \\ 0 & \omega_r > \omega_c \end{cases} \tag{208}$$

The frequency response under incoherent illumination is in general worse than that using coherent illumination.

These transfer functions are the Fourier transform of the system impulse-response functions. As such, they are easily related to the images that are formed at the image plane. Image degradation or blur occurs due to a loss of higher spatial frequency response. In general, any aberrations in the lenses used in an optical system will result in losses in the higher spatial frequencies. The product of the transfer functions will indicate the spatial frequency response of the optical system and thus serves as a convenient method for determining the resolution of images formed. Since the frequency response of the OTF will, in general, decay more rapidly with spatial frequency than the CTF because of the convolution, the minimum irradiance spot size will, in general, be smaller for coherent illumination.

4.6. DIFFRACTION OF GAUSSIAN BEAMS

Gaussian beams lend themselves to calculation using diffraction optics since the Fourier transform of a Gaussian is itself a Gaussian. Subsequently, the Gaussian may also be considered as a fundamental mode in diffraction theory since the shape of the beam profile does not change with propagation through space. To illustrate the application of Fresnel diffraction, consider a plane wave passing through an aperture with a radial Gaussian transmission profile. The Gaussian transmission profile will result in a beam with electric field that decreases with radius in a Gaussian fashion. The Gaussian beam is formed from the passage of a plane wave through the aperture will have a wavefront radius of curvature of infinity. This condition corresponds to the minimum beam waist. Beyond the aperture, the beam should be described by the equation for TEM_{00} beam propagation. Since the results are already known, this situation is used to explore the procedures for using, and the accuracy of the diffraction equations.

4.6.1. Free-Space Propagation of Gaussian Beams

In diffraction optics, a plane wave is transformed by both transmission through apertures and propagation through space. To describe Gaussian beam propagation, there is only one optical element, that of the real transmitting aperture. As with the case of ray tracing, the direction of propagation is chosen to be along the z axis. Prior to the aperture, the plane wave has an electric field of $\mathbf{E}_0 \exp(-ikz)$. The origin of the z axis is chosen to be at the aperture. The electric field transmission, t_1, for the Gaussian profile is

$$t_1(\xi, \eta) = e^{-(\xi^2+\eta^2)/w_0^2} \tag{209}$$

Here the maximum electric field transmission amplitude is 1 and there is no phase change. The electric field just past the aperture is

$$\mathbf{E}_1(\xi, \eta) = t_1(\xi, \eta)\mathbf{E}_0 = \mathbf{E}_0 e^{-(\xi^2+\eta^2)/w_0^2} \tag{210}$$

The Fresnel diffraction equation is used to obtain the electric field at a plane along the z axis. First, the Fourier transform is found:

$$\mathscr{F}\{\mathbf{E}_0 t_1(x, y) e^{-(ik/2z)(x^2+y^2)}\} = \mathbf{E}_0 \frac{w_0^2 z}{2z + iw_0^2 k} \exp\left[-\frac{w_0^2 z}{4z + 2iw_0^2 k}(\omega_x^2 + \omega_y^2)\right] \tag{211}$$

Substituting the ω and multiplying by the phase factor results in

$$\mathbf{E}_2(x, y) = \mathbf{E}_0 \frac{i\pi w_0^2}{z\lambda + i\pi w_0^2} \exp\left(-\frac{i\pi r^2}{z\lambda + i\pi w_0^2} - \frac{i2\pi z}{\lambda}\right) \tag{212}$$

This result is simplified by substitution of the definition for z_0:

$$\mathbf{E}_2(x, y) = \mathbf{E}_0 \frac{z_0}{z_0 - iz} \exp\left(-\frac{iz_0 r^2}{w_0^2 z + iw_0^2 z_0} - \frac{i2\pi z}{\lambda}\right) \tag{213}$$

To convert this into the form of the TEM_{00} beam equation, the exponent can be recast in a form that uses the definitions for $w(z)$ and $R(z)$. Using the definitions for these parameters given earlier yields

$$\mathbf{E}_2(x, y) = \mathbf{E}_0 \frac{z_0}{z_0 - iz} e^{-r^2/w^2(z)} e^{-ikz} e^{-ikr^2/2R(z)} \tag{214}$$

Since the electric field magnitude is real, the complex preexponential part, $z_0/(z_0 - iz)$, is interpreted in terms of the real part. This is done by taking the square root of the complex square,

$$\sqrt{\mathbf{E}_0^2 \frac{z_0}{z_0 - iz} \frac{z_0}{z_0 + iz}} = \mathbf{E}_0 \sqrt{\frac{1}{1 + z^2/z_0^2}} = \mathbf{E}_0 \frac{w_0}{w(z)} \tag{215}$$

using the definition for $w(z)$. This results in an electric field of

$$\mathbf{E}_2(x, y) = \mathbf{E}_0 \frac{w_0}{w(z)} e^{-r^2/w^2(z)} e^{-ikz} e^{-ikr^2/2R(z)} \tag{216}$$

Comparing this to the equation for TEM_{00} mode propagation, one may see that Fresnel diffraction through a Gaussian transmission aperture results in a wave that is nearly identical to the TEM_{00} mode. The subtle distinction is that Fresnel diffraction does not have a tangent phase-shift term. It should be noted that this term is not accounted for in paraxial ray tracing either. The tangent phase-shift term does not alter the irradiance of the field, and neglecting its contribution should not affect the imaging characteristics of lenses.

4.6.2. Focusing Gaussian Beams

The focus position for a Gaussian input beam can be determined by finding the electric field as a function of distance from the lens. The optical system is modeled by starting with a plane monochromatic wave, then pass through a Gaussian aperture placed directly adjacent to a lens. This model is equivalent to an electric field with a Gaussian profile and a minimum beam waist radius at the lens. For a transmission of the Gaussian aperture of t_1, and of the lens, t_l, the electric field immediately past the lens is

$$\mathbf{E}_1(\xi, \eta) = \mathbf{E}_0 t_1(\xi, \eta) t_l(\xi, \eta) = \mathbf{E}_0 e^{-(\xi^2+\eta^2)/w_0^2} e^{ik(x^2+y^2)/2f} \tag{217}$$

One way to solve for the field at some distance past the lens is to use the convolution property of Fresnel diffraction. The Fourier transform is readily found by assuming that a perfect lens is used:

$$\hat{\mathbf{E}}_1(\omega_x, \omega_y) = \mathbf{E}_0 \frac{w_0^2 f}{2f - i w_0^2 k} \exp\left[-\frac{w_0^2 f(\omega_x^2 + \omega_y^2)}{4f - 2iw_0^2 k} \right] \tag{218}$$

Multiplying by the point-spread function results in the electric field frequency distribution:

$$\hat{\mathbf{E}}_1(\omega_x, \omega_y) = \mathbf{E}_0 \frac{i w_0^2 f}{2f - i w_0^2 k} \exp\left[-(\omega_x^2 + \omega_y^2) \left(\frac{w_0^2 f}{4f - 2iw_0^2 k} - \frac{iz}{2k} \right) - ikz \right] \tag{219}$$

The transform of this frequency distribution is found and the substitution $z_0 = \pi w_0^2/\lambda$ is made. The electric field immediately past the lens is

$$\mathbf{E}_2(x, y) = \mathbf{E}_0 \frac{i z_0 f}{ifz + zz_0 - fz_0} \exp\left[-(x^2+y^2) \frac{z_0(f - iz_0)}{w_0^2(ifz + zz_0 - fz_0)} - ikz \right] \tag{220}$$

This is analyzed in terms of the form of the Gaussian beam equation. The preexponential term is related to the beam waist radius. Taking the square root of the complex square, the beam waist radius is found to be

$$\frac{w^2(z)}{w_0^2} = \frac{f^2 z^2 + (zz_0 - fz_0)^2}{f^2 z_0^2} \tag{221}$$

This is the same result as was found using the *ABCD* method of Gaussian beam propagation. An alternative to determining the beam waist radius is to examine the real part of the quadratic radially dependent exponent. The result of this alternative is equivalent to examining the preexponential term.

The imaginary part of the exponent with quadratic radial dependence is related to the radius of curvature. The radially dependent imaginary part of the exponent can be solved for the point at which there is no radial phase shift. This point corresponds to the focus of the Gaussian beam. Solving for the focus position in this manner yields

$$z_{\text{focus}} = \frac{f}{1 + f^2/z_0^2} \tag{222}$$

This again is the same result as that obtained using the *ABCD* law for Gaussian beam propagation.

4.6.3. Lens Transform Properties with Gaussian Beams

When the minimum beam waist of a Gaussian beam is at the front focal length of the lens, the electric field profile at a distance equal to the back focal length of the lens is simply the Fourier transform of the input beam. For an input electric field beam waist of w_{01}, the field at the back focal length distance is

$$\begin{aligned} \mathbf{E}_2(x, y) &= -\frac{ke^{-ik2f}}{f} \mathscr{F}\{\mathbf{E}_0 e^{-(x^2+y^2)/w_{01}^2}\}\big|_{\omega_x = -kx/f, \omega_y = -ky/f} \\ &= -e^{-ik2f} \frac{kw_{01}^2 \mathbf{E}_0}{2f} e^{-(k^2 w_{01}^2/4f)(x^2+y^2)} \end{aligned} \tag{223}$$

Interpretation of the preexponential term or the exponential term in terms of the focus beam waist radius clearly shows that the beam waist radius at the focus is $w_{02} = f\lambda/\pi w_{01}$.

4.6.4. Gaussian Beam Attenuation

An efficient means to obtain variable attenuation of a Gaussian beam using a venetian blind–style infrared beam attenuator has been described by Bialkowski (1987). This attenuator does not result in beam walk-off or pointing errors that could destroy the critical alignments necessary for accurate photothermal spectroscopy. The operational principles of this attenuator are simple in light of the results of diffraction optics theory. A schematic of the attenuator is shown in Figure 4.10. The output of a laser, operating in TEM_{00} mode, is collimated to a relatively large electric field beam waist radius at the front focal length of a lens. The venetian blind attenuator is placed in the beam at this position. The electric field directly past the attenuator is the product of the attenuator transmission, t_a, with the Gaussian beam with a minimum beam waist radius of w_0:

$$\mathbf{E}_1(x, y) = t_a(x, y)\mathbf{E}_o e^{-(x^2+y^2)/w_0^2} \tag{224}$$

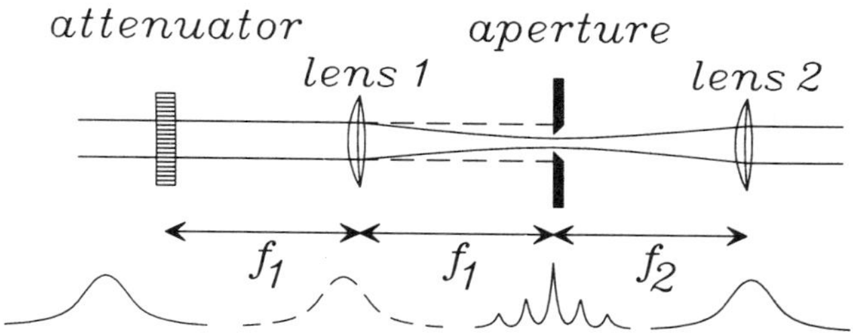

Figure 4.10. Method for attenuation of high-powered laser beams without causing beam walk-off or wavefront distortion.

The attenuator has a periodic transmission along one of the two orthogonal axes. The transmission can be modeled as the convolution of a comb function with a rectangle function:

$$t_a(x, y) = \text{comb}_a(x) \otimes 2b \text{ rect}\left(\frac{x}{b}\right) \tag{225}$$

where a is the spacing between and $2b$ is the width of the slats of the attenuator. The factor of $2b$ accounts for the total transmission change that occurs by changing b. Also, for $b < a$, the transmission will be zero. The electric field in the transform plane of the lens is

$$\begin{aligned}\mathbf{E}_2(x, y) &= -\frac{ke^{-ik2f}}{f}\mathscr{F}\{t_a(x, y)\mathbf{E}_0 e^{-(x^2+y^2)/w_{01}^2}\}\\ &= -e^{-ik2f}\left[\mathbf{E}_0\frac{w_0^2}{2}e^{-w_0^2(\omega_x^2+\omega_y^2)/4}\otimes \hat{t}_a(\omega_x, \omega_y)\right]\end{aligned} \tag{226}$$

The transformed transmission function is the product of the transformed rect and comb functions:

$$\hat{t}_a(\omega_x, \omega_y) = \frac{8\pi b}{a}\,\delta(\omega_y)\,\frac{\sin(\omega_x b/2)}{\omega_x}\,\text{comb}_{2\pi/a}(\omega_x) \tag{227}$$

The convolution of this function with the Gaussian produces a series of Gaussians spread out in the x dimension. Since the distances between the venetian blind slats are less than the width of the collimated Gaussian, the distances between Gaussians in the transform plane are greater than their widths. The series of Gaussians correspond to the diffraction orders produced by the periodic transmission aperture.

A rectangular slit is placed in the transform plane and positioned to allow the zero-order Gaussian to be transmitted unaltered while blocking the higher-order Gaussians. Since the width of venetian blind slats is much less than the width of the input Gaussian, the sin function has little effect on the zero-order Gaussian. Past this slit the electric field is

$$\mathbf{E}_2(x, y) = -e^{-ik2f}\frac{8\pi b}{a}\mathbf{E}_0 w_0^2 e^{-w_0^2(\omega_x^2+\omega_y^2)/4} \tag{228}$$

A second lens is used to perform another Fourier transform. For equal-focal-length lenses, the beam in the back focal plane of the second lens will be a Gaussian given by

$$\mathbf{E}_e(x, y) = \tau\mathbf{E}_0 e^{-(x^2+y^2)/w_0^2} \tag{229}$$

where $\tau = 8\pi b/a$ is the transmission.

If the output of the laser is not a Gaussian to begin with, a circular aperture may be used at the transform plane. The diameter of this aperture is chosen not only to block the high-order diffraction components but also to block the spatial noise components in the less-than-ideal laser output. This technique of transforming the spatially complex laser output into a Gaussian beam by placing a circular aperture at the focus of a lens is called spatial filtering. By using a circular aperture in the attenuator, not only will the laser be attenuated but the beam spatial quality may also be improved.

4.6.5. Effects of Spatial Noise

Even the best lasers produce less than pure monochromatic TEM_{00} mode outputs. In addition, the beam quality can be corrupted after reflecting off mirrors, passing through lenses and windows with dust or aberrations, and scattering by dust in the air. In addition to these determinant noise sources, there will be quantum noise fluctuation in the photon count at any given point in the beam profile. These spatial noise-producing factors will all result in a beam profile that is less than that of an ideal Gaussian at the final lens of a system.

In much laser-based spectroscopy apparatus, including photothermal spectroscopy, the signal is proportional to the irradiance of the excitation laser at the focus of a lens. This irradiance is often measured as total power prior to the lens. However, the measured power is not strictly related to the irradiance at the focus if there is spatial noise. The spatial noise in the beam profile due to scattering and absorption loss, and quantum fluctuation is proportional to the electric field of the beam. Spatial noise can produce a beam with a profile at the position of the minimum beam waist represented by

$$\mathbf{E}_1(x, y) = \mathbf{E}_0[1 + v(x, y)]e^{-(x^2+y^2)/w_0^2} \tag{230}$$

where $v(x, y)$ is a random function of the spatial coordinates with zero mean, $\langle v(x, y)\rangle = 0$, but finite variance. The power is obtained by integrating the irradiance over the spatial coordinates:

$$\phi_1 = c\varepsilon_0 \int_{-\infty}^{\infty} \int_{-\infty}^{\infty} |\mathbf{E}_1(x, y)|^2 \, dx \, dy = \phi + \sigma^2 \tag{231}$$

The noise term is defined by

$$c\varepsilon_0 \int_{-\infty}^{\infty} \int_{-\infty}^{\infty} |e^{-(x^2+y^2)/w_0^2} v(x, y)|^2 \, dx \, dy = \sigma^2 \tag{232}$$

and the cross-product terms in the $\mathbf{E}_2$ are zero because of the zero-mean character of $v(x, y)$. Measurement of the beam power with a wide-area detector will produce an estimate of $\phi + \sigma^2$.

In photothermal and other spectroscopies, the excitation laser beam is often focused into the sample cell. The spectroscopic signal is proportional to the irradiance at the center of the focused beam. A theoretical optical setup is illustrated in Figure 4.11. For simplicity, we consider the effects of focusing using the Fourier transform properties of diffraction optics. For a minimum beam waist radius one focal length in front of the lens, the

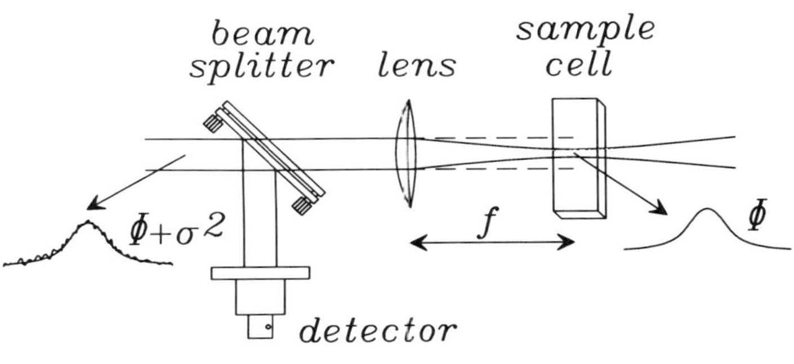

Figure 4.11. Effect of spatial noise on the quantitative estimation of the power at the focus of a laser beam. The detector responds to the total power, while the power at the focus center is proportional only to the noiseless power.

minimum beam waist will occur one focal length beyond the lens. Moreover, the electric field distribution at the focus is the Fourier transform of the input beam. According to the central ordinate property of Fourier transforms, the irradiance at the center of the focused beam is

$$\mathbf{E}_2(0, 0) = \frac{c\varepsilon_0}{2\pi} \left| \int_{-\infty}^{\infty} \int_{-\infty}^{\infty} \mathbf{E}_1(x, y)\, dx\, dy \right|^2 \tag{233}$$

For random $v(x, y)$, the irradiance at the center of the focused beam is on the average $\mathbf{E}_2(0, 0) \approx w_0^2 \phi$. Since the beam power measured prior to the lens is $\phi + \sigma^2$, and the irradiance at the center of the focused beam is independent of σ^2, estimates of the irradiance at the focus are in error by an amount equal to σ^2. If σ^2 is a random variable in time, say due to scattering from dynamic particles and quantum noise, the focus center irradiance estimates will vary randomly with time. However, the error in the focus irradiance estimate is not normally distributed but is biased to high values. The true focus irradiance will always be less than or equal to the estimate when spatial noise is present. The estimate error is also proportional to the beam power (in this model). This, in conjunction with the nonnormal error distribution, makes signal processing difficult, if it is valid at all.

There are several ways to avoid this problem. One is to measure the irradiance at the center of the focus using a beamsplitter placed after the lens and a pinhole placed at the center of this reference beam. This is often difficult, owing to the precision at which the pinhole can be placed. Moreover, there may be pointing noise in the beam. Pointing noise arises due to thermal drifts and acoustic disturbances in the laser source and the laboratory environment. With pointing noise the pinhole may not sample the central point of the focus at every moment. A better way to avoid the measurement bias is to use spatial filtering to reduce the spatial noise.

The basic components of the spatial filter are shown in Figure 4.12. It is difficult to analyze the effects of the general spatial filter for any input noise

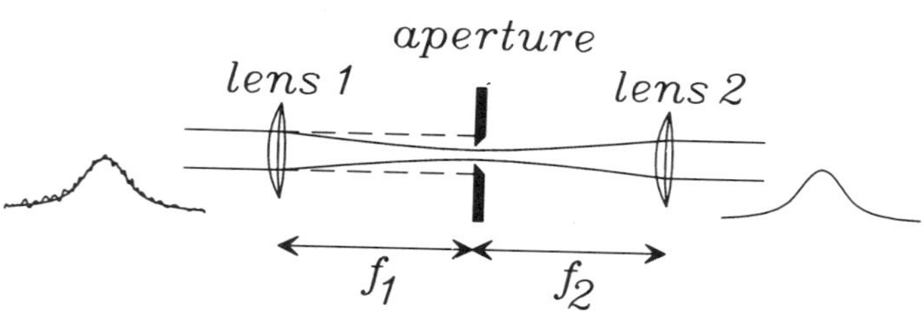

Figure 4.12. Spatial filter used to reduce spatial noise in a laser beam. The second lens is needed to collimate the beam.

structure. Nonetheless, the general operating principles can be understood in terms of the results from diffraction theory. Spatial filtering operates on the principle that the Fourier transform of "white" spatial noise will be distributed across the entire Fourier space plane. By placing a pinhole aperture at the focus of a lens, only the lower spatial frequencies are allowed to pass. The pinhole diameter is large enough to pass the Gaussian beam unaltered. In this fashion the spatial noise in the beam can be reduced substantially. Once the spatial noise is reduced, measurement of the beam power will give a more accurate estimate of the focused beam irradiance.

In optical signal processing, spatial filters more complex than the simple pinhole are placed in the first Fourier plane. This results in a correlation between the input electric field and the filter transmission being produced at the second, or inverse, transform plane. The basic spatial filter illustrated in Figure 4.12 is identical to the 4F optical correlator. This basic optical arrangement is extremely versatile and can be used to describe many optical effects and devices.

4.7. DIFFRACTION GRATINGS

Diffraction gratings are periodic structures with a reflection, transmission, or index of refraction that varies in space. Referring to the grating shown in Figure 4.13, the angle at which light incident at an angle θ_i relative to the normal of a periodic structure is diffracted is given by the grating equation:

$$n\Lambda(\sin\theta_i + \sin\theta_d) = m\lambda \qquad m = 0, \pm 1, \pm 2, \ldots \tag{234}$$

where n is the refractive index, Λ the distance between the lines or planes in the periodic structure of the grating, λ the vacuum wavelength of light, and

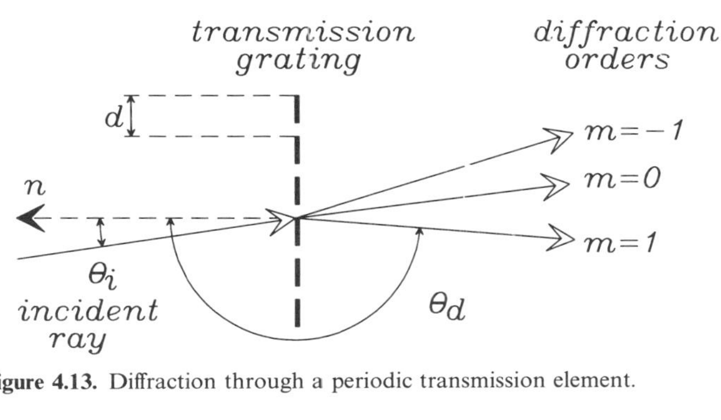

Figure 4.13. Diffraction through a periodic transmission element.

θ_d the diffraction angle, again measured relative to the normal. The integer, m, is the grating order and can have values 0, ± 1, $\pm 2, \ldots$ corresponding to the "zeroth" order, plus/minus first order, plus/minus second order, and so on. The grating equation is useful for determining the angular dependence between input and output rays. It does not, however, indicate how much of the light will be diffracted into angles specified by a given order number.

When the incident and diffracted angles are equal, first-order diffraction results in the Bragg's law:

$$2n\Lambda \sin \theta = \lambda \tag{235}$$

Bragg's law is used for geometric interpretation of grating structures in a number of different cases. It gives incident and diffracted angles for first-order diffraction and also gives the geometric relationship between the electric field and the two intersecting plane waves that form the field. In this case, two plane waves, intersecting at a half-angle, θ, will produce a periodic electric field with a spacing of Λ. The resulting periodic electric field wave vector will be in the plane containing the wave vectors of the two intersecting plane waves and will point in a direction perpendicular to the normal relative to which the half-angles are measured.

Several different types of grating are encountered in optics. Gratings are classified by the type of modulation, where the modulation occurs, and whether the diffracted light is transmitted or reflected through the gratings. Surface reflection gratings are most often used in monochromators and spectrographs. These gratings are made by scribing or embossing lines into a reflective substrate or by coating a periodic polymer or glass structure with a reflective material. Less familiar to the spectroscopist are volume transmission and volume reflection gratings. In volume gratings the spatially periodic structure occurs within the transparent element, not at the surface. There are two types of volume gratings, those with a periodic, spatially dependent transmission, and those with a periodic refractive index or optical pathlength modulation. Gratings with spatially periodic transmission are called amplitude gratings, and refractive index gratings are called phase gratings (index changes alter the the phase of light passing through the element). Gratings formed by the photothermal effect are usually volume phase gratings.

Volume grating can be of either transmission or reflection type, depending on the direction in which the diffracted light exits the element. In transmission gratings, the diffracted light exits the element on the opposite side, whereas in reflection gratings, the diffracted light exits the element on the same side as the incident radiation. For transmission gratings, the direction of the periodic index or transmission modulation, often given by a vector, **κ**, is perpendicular to the incident light. Reflection gratings have index or transmission modulation **κ** in a direction parallel to the incident light. Multiple-layer dielectric filters are one form of volume-phase reflection gratings. The grating **κ** can be slanted with respect to the surface of the physical grating element. The resulting slant gratings can act as either transmission or reflection gratings, depending on the direction of the incident light. When the incoming light rays form obtuse angles with the normal of the planes of refractive index modulation, the **κ** vector, transmission gratings are formed. When these angles are acute, reflection gratings are formed.

4.7.1. Thin-Amplitude-Transmission Gratings

The propagation of light through periodic structures is normally treated using a geometric approach–based diffraction optics. The geometric approach is easily understood by comparing the propagation of a plane wave moving at an arbitrary angle relative to the Cartesian coordinates axes to a plane wave passing through a periodic transmission structure. This situation is shown in Figure 4.14. A plane monochromatic wave traveling at an angle

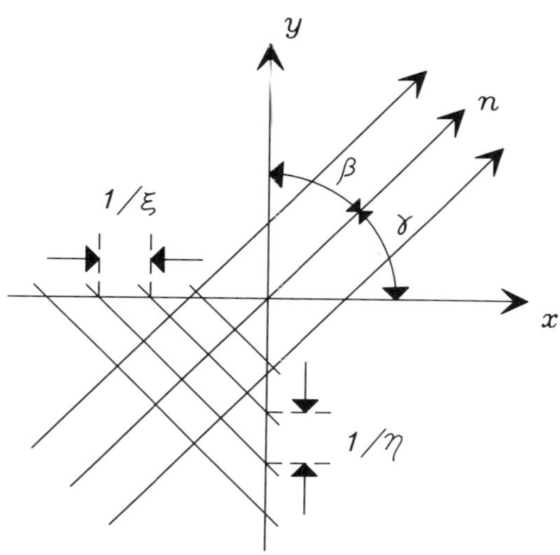

Figure 4.14. Electromagnetic wave propagation in the x–y plane with a propagation vector n. Wave maxima are shown as diagonal lines. The wave produces periodic maxima along the x and y axes given with spatial frequencies ξ and η, respectively.

α relative to the x axis, β relative to the y axis, and γ relative to the z axis can be represented by the equation

$$\mathbf{E}(x, y, z) = \mathbf{E}_0 \exp[-ik(x \cos \alpha + y \cos \beta + z \cos \gamma)]) \tag{236}$$

The three angles are not independent and, in fact,

$$\cos^2 \alpha + \cos^2 \beta + \cos^2 \gamma = 1 \tag{237}$$

Substitution into the plane wave equation followed by simplification results in

$$\begin{gathered} \mathbf{E}(x, y, z) = \mathbf{E}(x, y, 0) \exp[-ikz(1 - \lambda^2\xi^2 - \lambda^2\eta^2)^{1/2}] \\ \xi = \frac{\cos \beta}{\lambda} \qquad \eta = \frac{\cos \gamma}{\lambda} \end{gathered} \tag{238}$$

where $\mathbf{E}(x, y, 0)$ is the electric field amplitude at $z = 0$. This equation is, of course, equivalent to the Huygens–Fresnel result. It shows that a plane wave propagating in any direction can be described by the field amplitude at $z = 0$ multiplied by a z-dependent complex exponential term. The complex exponential is given in terms of the parameters ξ and η. These are the spatial frequencies that the electromagnetic wave maxima make along the x and y axes, respectively. The spacing between wave maxima are $1/\xi$ along the x axis and $1/\eta$ along the y axis.

Consider next the effect of a grating placed at the z axis origin with a periodic transmission dependent on the distance cosine along the y axis. The transmission of the grating is

$$t_g = t_0 + t_1 \cos 2\pi\eta y \tag{239}$$

When illuminated by a plane wave propagating in the z direction, the electric field immediately past the grating will be

$$\begin{aligned} \mathbf{E}_1(x, y, 0) &= \mathbf{E}_0 t_g = \mathbf{E}_0 t_0 + \mathbf{E}_0 t_1 \cos 2\pi\eta y \\ \mathbf{E}_1(x, y, 0) &= \mathbf{E}_0 t_0 + \tfrac{1}{2}\mathbf{E}_0 t_1 e^{i2\pi\eta y} + \tfrac{1}{2}\mathbf{E}_0 t_1 e^{-i2\pi\eta y} \end{aligned} \tag{240}$$

Comparing this to the equation for plane wave propagation, it may be seen that the exponential terms are equivalent to two plane waves propagating at two angles relative to the z axis. The waves are propagating in the y–z plane and the absolute values of these angles are the same, $\sin|\theta| = \lambda\eta$. The positive exponent corresponds to a wave moving in the negative y direction and the second exponent to one with propagation in the positive y direction. The amplitudes of these two waves are both $\frac{1}{2}t_1\mathbf{E}_0$. These two plane waves are the ± 1 orders of the diffracted wave. The first term in the electric field equation is a plane wave moving in the same direction as the illuminating plane wave (e.g., along the z axis). This is the undiffracted component of the wave, or the zero-order diffracted wave.

If the grating does not have a spatial cosine-dependent transmission but is still periodic, the electric field just past the grating for normal plane wave illumination is

$$\mathbf{E}_1 = \mathbf{E}_0 t_0 + \mathbf{E}_0 \sum_m A_m e^{-ik_m y} \tag{241}$$

where $k_m = 2\pi m\eta$ is the x-direction component of the wave vector for the mth component of the diffracted wave and the amplitude of this component is A_m. The direction of mth component is $\sin\theta_m = k_m/k = m\lambda\eta$. When the plane wave is not at normal incidence to the grating direction, the grating equation is used to determine the relationship between incident and diffracted waves. The amplitude is found from the Fourier component of the transmission,

$$A_M = \frac{\mathbf{E}_0}{\Lambda} \int_0^{\Lambda} t(y) e^{-ik_m y}\, dy \tag{242}$$

where $\Lambda = 1/\eta$ is the spatial extent of the periodic structure. The Fourier transform is the decomposition of the transmission into a series of cosines (e.g., Fourier cosine series decomposition). The m are the diffraction orders of the grating equation. This Fourier series analysis of the periodic structure is an alternative method to transforming the transmission, composed of the convolution of a periodic comb function and the unit reproduced in the grating structure.

Diffraction efficiency is defined as the irradiance diffracted in a given grating order relative to that of the plane wave incident on the grating. The maximum diffraction efficiency for a cosine amplitude transmission grating is $(\mathbf{E}_{\pm 1}/\mathbf{E}_0)^2 = 6.25\%$ for maximum amplitude modulation. Maximum modulation occurs for t_0 and t_1 each equal to $\frac{1}{2}$. The irradiance in the ± 1-order diffracted waves should be compared to the transmission of the zero-order component, which is only 25%. The total irradiance transmission of the amplitude grating, including all orders, is 37.5%, not accounting for any reflection at the surface of the element.

4.7.2. Thin-Phase-Transmission Gratings

Phase gratings are formed when the refractive index varies in a periodic fashion. Diffraction from a phase grating is shown in Figure 4.15. Refractive index variations are represented by complex exponent transmissions. A periodically modulated refractive index of $n = n_0 + \delta n \cos 2\pi\eta y$ will result in the complex transmission

$$t(y) = e^{i\delta\phi \cos 2\pi\eta y} \tag{243}$$

The relative phase shift angle is defined by

$$\delta\phi = \frac{2\pi l \delta n}{\lambda} \tag{244}$$

where l is the pathlength. The amplitudes of the diffracted waves, found from the Fourier transform of the transmission, are (Eichler et al. 1986)

$$A_M = \mathbf{E}_0 i^m J_{\mathrm{m}}(\phi) \tag{245}$$

where $J_m(\phi)$ is the mth-order Bessel's function.

The peak diffraction efficiency in the ± 1 order, found by evaluating $|J_1(\phi)|^2$, is found to be 34% at a phase shift of $\phi \approx 1.8$. The two apparent differences between the cosine amplitude and phase gratings are (1) that the

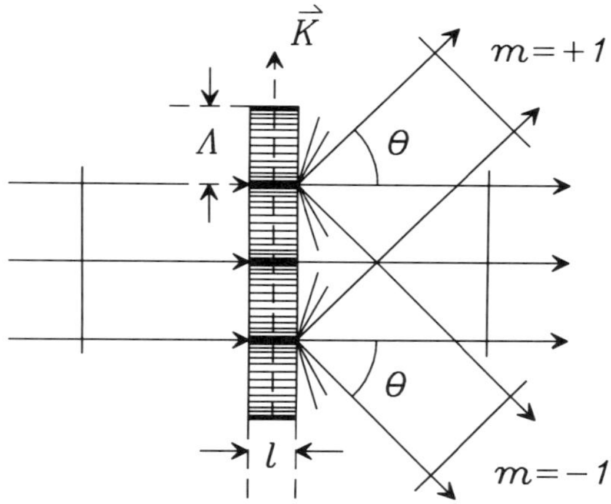

Figure 4.15. A plane wave passing through a thin phase grating produces waves that diffract through several diffraction orders. Only the $\pm$ 1 and 0 orders are shown completely.

efficiency of the phase grating is significantly higher than that of the amplitude grating, and (2) that the phase grating will produce diffraction at several orders, not just the ± 1 order produced for the amplitude grating.

This complex exponent treatment can also be used to evaluate the maximum diffraction efficiency for amplitude gratings. The imaginary part of the refractive index change is related to the loss through optical absorption. Using an imaginary δn then represents an amplitude grating with absorption loss. The difference between this grating and the cosine transmission grating is that absorption loss is exponentially dependent on grating thickness. The maximum first-order diffraction efficiency in this case is 7% and occurs for imaginary phase shifts of $i\phi \approx 0.8$.

4.7.3. Thick Gratings

The difference between thick and thin gratings can be described using a phase-shift difference between waves diffracted at the front of the grating versus those diffracted at the back of the grating. The phase-shift difference is given by a parameter Q:

$$Q = \frac{2\pi l \lambda}{\Lambda^2 n} \tag{246}$$

If the phase-shift difference is small ($Q \ll 1$), the diffracted waves will constructively interfere. In this case several constructive interference cases can be found for any input angle through the grating equation. The thin grating is subsequently not sensitive to angle in the respect that diffraction will occur at any incident angle. It also has low efficiency in that light is diffracted through several orders. If the phase-shift difference is large ($Q \gg 1$), the light diffracted from the front of the grating can interfere destructively with light diffracting from the back of the grating. In this case there is a narrow range of incident angles that will result in diffraction with high efficiency. Moreover, only one diffraction order is observed to occur with high efficiency. The condition for this maximum diffraction efficiency is given by Bragg's equation with the angles measured relative to κ.

Diffraction efficiencies from thick gratings, in particular for $Q \geqslant 0$, are given by coupled wave analysis. In this type of analysis, Maxwell's equations for electromagnetic wave propagation in the periodic dielectric medium are solved. Analytic results for coupled wave analysis were obtained by Kogelnik (1969) and are described in the texts by Collier et al. (1971) and Eichler et al. (1986). The equations serve as the basis for understanding optical holography, wave mixing, and of course, transmission gratings. Kogelnik's coupled wave analysis included phase, amplitude, and mixed gratings in lossless and lossy dielectric media.

There are two main types of thick-volume phase gratings: transmission grating and reflection grating, illustrated in Figures 4.16 and 4.17. In transmission grating, light is transmitted through the grating structure. Reflection grating is very similar to a multiple-layer dielectric filter in that the periodic index modulation reflects certain components of the spectrum. The main distinction between transmission and reflection gratings is in the distance between the index planes, which in turn is related to the incident and diffracted angles through Bragg's law. When the spacing between index planes is larger than the wavelength of light, the resulting grating is a transmission grating. On the other hand, when the spacing between the planes is on the order of $\lambda/2$, the grating is a reflection grating. Although this distinction may not seem that important, Kogelnik's coupled wave analysis yields different solutions for reflection and transmission gratings.

The geometry for diffraction through a thick-volume transmission grating is shown in Figure 4.16. Eichler et al. (1986) show that the diffracted electric field for the thick-volume transmission grating is

$$\mathbf{E}_1 = i\mathbf{E}_0 \exp\left(-\frac{\alpha l}{2 \cos \theta_b}\right) \sin \frac{\pi l \delta \varepsilon}{2n\lambda \cos \theta_b} \tag{247}$$

where α is the bulk absorption coefficient of the grating material, l the grating thickness, θ_b the Bragg angle inside the grating material (the internal

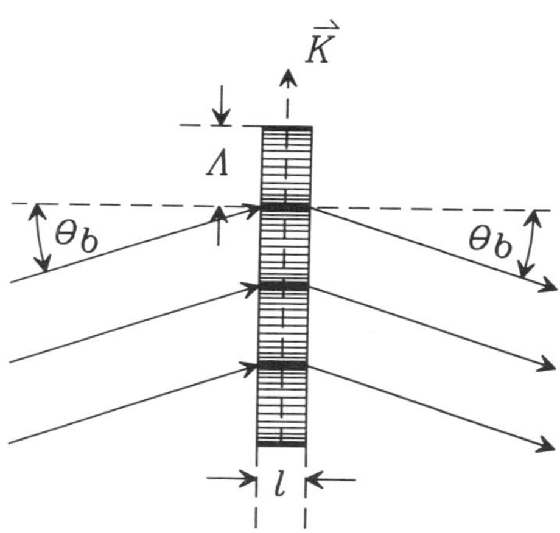

Figure 4.16. A plane wave passing through a thick phase grating produces a single diffracted wave in addition to the zero-order component. The Bragg condition for diffraction must be met.

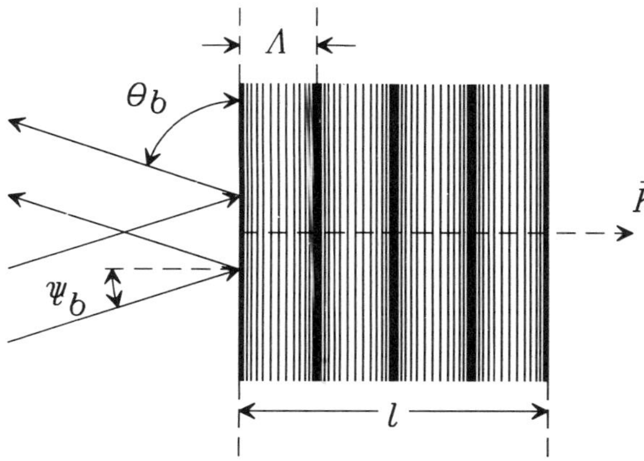

Figure 4.17. Reflective diffraction of the plane wave with a thick phase grating produces a single diffracted wave in addition to the zero-order component. Again, the Bragg condition must be met. Refection grating occurs when $\Lambda \approx \lambda/2$.

angle), and $\delta\varepsilon$ the periodically modulated dielectric coefficient. The internal angle can be related to the external angle through Snell's law and the bulk indexes of the internal and external media. The dielectric coefficient is the square of the refractive index, and the refractive index is, in general, composed of a real part and an imaginary part that is proportional to the absorption coefficient:

$$\begin{aligned} \sqrt{\varepsilon} &= \tilde{n} = n_r + \frac{i\alpha\lambda}{4\pi} \\ \delta\varepsilon &\approx 2n\left(\delta n + \frac{i\lambda\delta\alpha}{4\pi}\right) \end{aligned} \tag{248}$$

With absorption and refractive index modulation, the resulting grating is a mixed grating. The diffraction efficiency for this grating is

$$\frac{|\mathbf{E}_1|^2}{|\mathbf{E}_0|^2} = \exp\left(-\frac{\alpha l}{\cos\,\theta_b}\right)\left(\sin^2\frac{\pi l\delta n}{\lambda\,\cos\,\theta_b} + \sinh^2\frac{l\delta\alpha}{4\,\cos\,\theta_b}\right) \tag{249}$$

For pure phase gratings, $\delta\alpha$ is zero and the sinh term is zero. Amplitude gratings have zero δn. In this case the sin term is zero.

For a lossless medium, $\alpha = 0$, and a pure phase grating can have a theoretical diffraction efficiency of 100% when the sin term is 1. This occurs when

$$\frac{\pi l \delta n}{\lambda \cos \theta_b} = \frac{\pi}{2} \tag{250}$$

Subsequently, 100% diffraction efficiencies can be obtained for any refractive index modulation, δn, provided that the grating is thick enough.

Thick-volume phase reflection gratings are similar to transmission gratings. The main difference is that the light intersects the grating at angles that are acute with respect to the normal of the planes of phase modulation. A volume phase reflection grating is shown in Figure 4.17. Collier et al. (1971) show that the diffracted electric field for a lossless volume phase grating can be expressed by

$$\mathbf{E}_1 = -i\mathbf{E}_0 \frac{\exp(-i\xi)\ \sin(\xi^2 + v^2)^{1/2}}{(1 + \xi^2/v^2)^{1/2}} \tag{251}$$

The parameters used in this equation are defined by

$$\xi = \delta\theta k l \sin \theta_b \qquad v = \frac{\pi l \delta n}{\lambda \cos \theta_b} \tag{252}$$

where k is the wave vector in the medium, $2\pi n/\lambda$. This result can be used to determine the diffraction efficiency at angles that deviate from the Bragg angle. For the volume phase reflection grating, the parameters are

$$\xi_r = \delta\theta k l \cos \theta_b \qquad v_r = \frac{\pi l \delta n}{\lambda \cos \psi_b} \tag{253}$$

where ψ_b is the internal angle measured relative to the $\boldsymbol{\kappa}$ vector, or the normal to the refractive index planes of the grating. The amplitude of the reflected wave for a lossless medium is

$$\mathbf{E}_1 = -\mathbf{E}_0 \frac{i}{i\xi_r^2/v_r^2 + [1 - (\xi_r/v_r)^2]^{1/2} \coth(v_r^2 - \xi_r^2)^{1/2}} \tag{254}$$

For Bragg incidence, $\xi_r = 0$, and the reflected wave has the amplitude

$$\mathbf{E}_1 = -i\mathbf{E}_0 \tanh v_r \tag{255}$$

The volume phase reflection grating can have a 100% diffraction efficiency only in the asymptotic limit as v_r approaches infinity.

REFERENCES

Abramowitz, M. and Stegun, I. A., eds. *Handbook of Mathematical Functions*, Dover, New York (1965).

Bialkowski, S. E. *Rev. Sci. Instrum.* **58** 2338 (1987).

Collier, R. J.; Burckhardt, C. B.; and Lin, L. H. *Optical Holography*, Academic Press, New York (1971).

Eichler, H. J.; Günter, P.; and Pohl, D. W. *Laser-Induced Dynamic Gratings*, Springer-Verlag, New York (1986).

Flygare, W. H. *Molecular Structure and Dynamics*, Prentice Hall, Englewood Cliffs, NJ (1978).

Flytanis, C. In *Quantum Electronics*, Vol. 1, Part A, Rabin, H. and Tang, C. L., eds., Academic Press, New York (1975).

Fotiou, F. K. and Morris, M. D. *Anal. Chem.* **59** 185 (1987).

Fröhlich, H. *Theory of Dielectrics*, Clarendon Press, Oxford (1949).

Guillory, W. A. *Introduction to Molecular Structure and Spectroscopy*, Allyn and Bacon, Boston (1977).

Handbook of Chemistry and Physics, 61st ed., Weast, R. C., ed., CRC Press, Boca Raton, FL (1980).

Klein, M. V. *Optics*, Wiley, New York (1970).

Kogelnik, H. *Bell Syst. Tech. J.* **48** 2909 (1969).

Kogelnik, H. and Li, T. *Appl. Opt.* **5** 1550 (1966).

Lee, S. H., Ed. *Optical Signal Processing Fundamentals*, Springer-Verlag, New York (1981).

Li, Y. *Opt. Eng.* **32** 774 (1993).

Verdeyen, J. T. *Laser Electronics*, Prentice Hall, Englewood Cliffs, NJ (1981).

Yariv, A. *Optical Electronics*, 3rd ed., Holt, Rinehart and Winston, New York (1985).

CHAPTER

5

TEMPERATURE CHANGE AND OPTICAL ELEMENTS IN HOMOGENEOUS SAMPLES

Photothermal spectroscopy can be used to detect optical absorption in a variety of samples. The samples may be homogeneous or heterogeneous in terms of both optical and thermal properties. A homogeneous sample is one that transmits light and has a composition that is consistent throughout. Homogeneous samples can be solid, liquid, or gas. Some common examples are liquids, liquid solutions, and gas-phase mixtures. Although solids are generally heterogeneous in nature, certain solid glass solutions and symmetric crystalline substances can be considered to be homogeneous. For homogeneous samples, the thermal change is normally monitored within the sample itself. In contrast, heterogeneous samples may not transmit light or may not be of constant composition. In these samples the thermal change is monitored outside the sample or that portion of the sample that is being excited. The models for signal generation and the experimental provisions for monitoring the thermal changes that occur in these two classes of samples are different.

Optical elements (e.g., prisms, lenses, and gratings) all rely on a spatially dependent refractive index change. The strengths of the optical elements are proportional to the magnitude of the refractive index change and inversely proportional the spatial scale of the change. In homogeneous samples, the optical elements must be produced using spatially anisotropic excitation sources. Excitation anisotropy results in spatially dependent absorbed energy or power and subsequently, a spatially dependent refractive index change. The refractive index change is normally thought to be due entirely to the density or temperature change. In certain cases, additional considerations of the refractive index change are necessary. In particular, optical property changes arising as a consequence of the changing excited-state populations may affect the refractive index. Again, it is necessary to account for these changes only in homogeneous samples where the probe light passes through the volume excited.

In this chapter we attempt to assemble elements required for a theoretical description of the signals encountered in the photothermal spectroscopy of homogeneous samples. In the first part we discuss the temperature changes that can occur for Gaussian laser excitation source. Gaussian laser beams are important in photothermal spectroscopy since they give the smallest possible focus spot size. The temperature change is calculated for a number of irradiation conditions. Next, optical elements used to generate the photothermal signals are formulated in terms of the temperature change. The connection between the temperature and refractive index change is the topic of the last part of the chapter.

5.1. TEMPERATURE CHANGES FROM GAUSSIAN EXCITATION SOURCES

A common thermal diffusion problem encountered in photothermal spectroscopy is that arising from sample excitation with a Gaussian spatial profile laser. Solutions to this problem have been shown by Gordon et al. (1965), Twarowski and Kliger (1977), and Bailey et al. (1980). These are succinctly summarized by Fang and Swofford (1983), Dovichi (1987), and Gupta (1989). These derivations assume that the laser beam is cylindrically symmetric (i.e., nondivergent) and that attenuation along the path of the laser does not result in a significant difference in the temperature change. It is also assumed that the sample cell dimensions are much greater than the diffusion length. Subsequently, the problem is one of two-dimensional diffusion in thermal conduction media of infinite extent.

The cylindrically symmetric nondivergent excitation laser beam approximation is maintained as long as the laser is focused near the center of the sample cell and has a confocal distance, z_0, greater than the sample pathlength. The infinite sample extent approximation is valid when relatively large-volume samples are excite with focused laser sources. The sample will behave as if it were of infinite extent as long as the maximum observation time is less than the time required for heat to diffuse to the cell walls. This radial diffusion approximation is valid for times $t \leqslant r_{\min}^2/4D_T$, where $r_{\min}$ is the minimum distance between the laser focus location and the cell wall. Thermal diffusion at the sample cell windows is also neglected. Heat diffusion through the windows can account for a major heat loss mechanism for short-pathlength sample cells. Rarely accounted for, the relative contribution of this loss mechanism to the total thermal diffusion can be minimized using sample cells with relatively long pathlengths.

5.1.1. Thermal Diffusion Approximation

In Chapter 3 density was used to examine a sample's response to optical excitation because of the relative simplicity of the resulting expressions. Temperature, on the other hand, is a more complex function of time and space. The impulse excited temperature change response is complicated, owing to both the kinetic rate of heat production following optical absorption and the sample expansion due to the propagating mode. With a few notable exceptions (e.g., Longaker and Litvak 1969, Siebert et al. 1974, Barker and Rothem, 1982), temperature change has been used as the basic parameter in most theoretical descriptions of photothermal spectroscopy. Persons working with gas-phase samples tend to use a density calculation approach, while those working with condensed-phase samples use temperature to calculate the sample response. This may be more a matter of convenience and historical precedent than one of accuracy. Historically, the thermal diffusion equation that describes temperature change was an empirically derived relationship used to model heat flow. Hydrodynamic model solutions are less tractable and somewhat difficult to interpret.

Density change solutions of the hydrodynamic model can be separated into two modes of relaxation: the thermal diffusion mode and the acoustic-wave propagating mode. There are several situations where the propagating mode does not have to be considered. When the sample is at local equilibrium (e.g., each volume element is at constant temperature) and the pressure is equilibrated throughout the sample, the hydrodynamic equations essentially reduce to the thermal diffusion equation, $\delta\rho = (\partial\rho/\partial T)_p \, \delta T$. Under these conditions there is no real difference between models based on temperature change and those based on density change. In this case the density is related directly to temperature through the volume expansion coefficient. Moreover, the thermal diffusion equation, which describes the temperature change that occurs in homogeneous fluid, will adequately describe the magnitude of the signal under many conditions. The magnitude of the thermal diffusion density perturbation is independent of that due to the propagating mode. Density, and therefore refractive index changes, due to both modes is proportional to the rate of heat production in the sample. If the time-dependent parameters of the experiment (e.g., relaxation kinetics and excitation profile) remain constant from sample to sample and the hydrodynamic properties of all samples are the same, the thermal diffusion equation should adequately describe signal magnitudes and the long-term signal behavior.

On one hand, we know that the density change is primarily responsible for refractive changes that result in the photothermal spectroscopy signal and that the density change more naturally follows from the hydrodynamic equations. On the other hand, there is historical precedent for using the temperature-change solutions, and most of the analytical photothermal spectroscopy literature follows this convention. The greatest motivation for using the temperature-change convention is that most theoretical descriptions of the photothermal signal are based on it. This convention is used in this chapter with the understanding that it is less accurate than the density-change solutions.

5.1.2. Gaussian Laser Excitation of Optically Thin Samples

An optically thin sample is one where little attenuation of the excitation source occurs upon passing through the sample. The energy or power absorbed by the sample per unit length is essentially independent of pathlength. This approximation is commonly used in modeling photothermal spectroscopy of low-absorbance samples. Quantitatively, it results in an error of less than 1% for samples with transmissions greater than about 90% *T*.

Consider a weakly absorbing sample being irradiated with a Gaussian or TEM_{00} laser beam. The laser is focused into the sample and the sample pathlength is much shorter than the confocal distance. The laser light is propagating along the $+z$-axis direction. For simplicity, the position where the laser first passes into the sample volume is the origin of the z axis. The irradiance at any point through the homogeneous sample can be found from the differential expression for absorption long the sample path

$$-\frac{dE(x, y, z, t)}{dz} = \alpha E(x, y, z, t) \tag{1}$$

$E(x, y, z, t)$ ($W\,m^{-2}$) is the irradiance in the sample at a distance z past the origin, and α (m^{-1}) is the absorption coefficient. For an optically thin sample (i.e., small absorption coefficient) the irradiance does not change significantly on passing through the sample. The irradiance on the right-hand side of this equation is thus approximately constant and may be substituted for by $E_0(x, y, t)$, the irradiance incident on the sample. The equation can be integrated without accounting for pathlength-dependent absorption loss. The power delivered to the sample is simply

$$E_l(x, y, t) - E_0(x, y, t) \approx E_0(x, y, t)\alpha l \tag{2}$$

where l (m) is the sample pathlength and $E_l(x, y, t)$ is the irradance leaving the sample. The power density transferred to the sample by absorption is the irradiance loss per unit length, or

$$q_H(x, y, t) = Y_H \frac{E_l(x, y, t) - E_0(x, y, t)}{l} = \alpha Y_H E_0(x, y, t) \tag{3}$$

where Y_H is the heat yield. It is included to account for energy loss through reemission of radiation, or through the production of stable photoproducts and long-lived metastable states, as discussed in Chapter 2.

The irradiance of the Gaussian laser used to excite the sample can be expressed as

$$E_0(x, y, t) = \frac{2\Phi(t)}{\pi w^2} e^{-2(x^2+y^2)/w^2} \tag{4}$$

$\Phi(t)$ (W) is the time-dependent power and w is the Gaussian beam waist radius. The z dependence of the beam waist radiance is not incorporated in this expression since it is assumed that the excitation laser is focused into the sample cell and the sample pathlength is less than the confocal distance. The factor $2/\pi w^2$ is a normalization constant. The Gaussian irradiance is normalized such that integration of $E_0(x, y, t)$ over x and y space is equal to $\Phi(t)$, the total power of the beam.

The time-dependent temperature change is found by solving the thermal diffusion equation

$$\frac{\partial}{\partial t}\, \delta T(x, y, t) - D_T \nabla^2\, \delta T(x, y, t) = \frac{q_H(x, y, t)}{\rho C_P} = \frac{\alpha Y_Y E_0(x, y, t)}{\rho C_P} \tag{5}$$

As was shown in Chapter 3, for media of infinite extent along the x and y coordinates, this differential equation has a solution that can be expressed as the convolution of the point-source solution with the heat source. Designating the spatially dependent Gaussian term as $S(x, y)$, the Gaussian laser beam can be written as

$$E_0(x, y, t) = \Phi(t) S(x, y) \qquad S(x, y) = \frac{2}{\pi w^2} e^{-2(x^2+y^2)/w^2} \tag{6}$$

Assuming that the sample is of infinite extent, the solution to the thermal diffusion equation is the convolution of the point-source solution,

$$\delta T(x, y, t) = \alpha Y_H \Phi(t) * [S(x, y) \otimes \delta T_{\text{point}}(x, y, t)] \tag{7}$$

where the two-dimensional point-source temperature change is

$$\delta T_{\text{point}}(x, y, t) = \frac{1}{\rho C_P} \frac{1}{4\pi D_T t} e^{-(x^2+y^2/4D_T t} \tag{8}$$

The spatial convolution, $\otimes$, results in the time- and space-dependent solution

$$\delta T(x, y, t) = \Phi(t) * \left[\frac{2 Y_H \alpha}{\pi \rho C_P} \frac{1}{w^2(1 + 2t/t_c)} e^{-2(x^2+y^2)/w^2(1+2t/t_c)} \right] \tag{9}$$

where the characteristic thermal diffusion time is defined as $t_c = w^2/4D_T$, mostly for simplicity, and D_T is the thermal diffusion coefficient. The t_c definition is consistent with other measures of diffusion length and time. One can see, for example, that the on-axis temperature produced by impulse excitation will decay to one-third of its original value when $t = t_c$. The term in brackets is the impulse response for TEM_{00} mode laser excitation of a sample. The temperature impulse response is convoluted with that for kinetic relaxation and with the time-dependent excitation power. Thus by defining

$$\delta T_{\text{impulse}}(x, y, t) = \frac{2}{\pi \rho C_P} \frac{1}{w^2(1 + 2t/t_c)} e^{-2(x^2+y^2)/w^2(1+2t/t_c)} \tag{10}$$

the time-dependent temperature is the convolution over the time-dependent power:

$$\delta T(x, y, t) = \alpha Y_H \Phi(t) * \delta T_{\text{impulse}}(x, y, t) \tag{11}$$

5.1.3. Short-Pulse Laser Excitation

The temperature change occurring with short-pulse sample excitation based on solutions to the thermal diffusion equation was shown originally by Twarowski and Kliger (1977). Twarowski and Kliger derived the temperature change for multiphoton absorbing species. Their results are general and include the single-photon absorption case considered here. They assumed that the sample was optically thin in absorbance and that the beam waist does not change over the length of the sample. Further, it was assumed that the excited state relaxes instantaneously. The temporal and spatial behavior of temperature change, and thus the sample enthalpy change due to thermal diffusion for pulsed TEM_{00} mode laser excitation for laser pulse durations much less than t_c and at times greater than that required for the acoustic perturbation to dissipate, is simply the impulse response. In this case, time integration of $\Phi(t)$ is the pulse energy, Q (J), and the temperature change at times $t \geqslant 0$ is

$$\delta T_{\text{pulsed}}(x, y, t) = \frac{2\alpha Y_H Q}{\pi \rho C_P} \frac{1}{w^2(1 + 2t/t_c)} e^{-2(x^2+y^2)/w^2(1+2t/t_c)} \tag{12}$$

It is clear that short-pulse excitation produces an initial temperature perturbation that is equivalent in spatial form to the Gaussian-shaped beam used to excite the sample. Initially, the radius of the Gaussian distribution is equal to that of the excitation source, w. Thermal diffusion broadens this spatial distribution with time. The temperature change distribution remains Gaussian in form for times greater than zero. In fact, the squared radius increases linearly with time. The time-dependent square radius is $w^2(1 + 2t/t_c)$. The temperature change that occurs for multiphoton absorption is found in a similar fashion. For multiphoton absorption, the power density transferred to the sample is

$$q_H(x, y, t) = \alpha^{(n)} Y_H E_0^n(x, y, t) \tag{13}$$

where $\alpha^{(n)}$ is the absorption coefficient for the n-photon absorption transition. Using this heat source, the temperature change is

$$\delta T_{\text{pulsed}}^{(n)}(x, y, t) = \frac{\alpha^{(n)} Y_H Q}{\rho C_P} \left(\frac{2}{\pi w^2}\right)^n \frac{1}{1 + 2nt/t_c}\, e^{-2n(x^2+y^2)/w^2(1+2nt/t_c)} \tag{14}$$

The initial n-photon temperature change profile is narrower than that for the single-photon absorption. This results in the thermal decay being faster than that for single-photon absorption. The t_c value effectively decreased since everywhere that t_c occurs, the time-dependent decay factor is t_c/n.

Bailey et al. (1980) undertook a derivation similar to that of Twarowski and Kliger but included first-order excited-state relaxation. Their results explicitly show that the temperature rise time is the time convolution of the instantaneous temperature response with the first-order relaxation rate. The temperature change was found from the thermal diffusion equation. Thermal diffusion is linear and additive. The effect of thermal diffusion on sample temperature constitutes a linear system. The input or impulse to the system is the laser energy, and the response is the resulting time- and space-dependent temperature change. The most general way to express the thermal diffusion–controlled temperature change is in the form of a convolution of the time- and space-dependent terms

$$\delta T(x, y, t) = q_H(x, y, t) \otimes \delta T_{\text{point}}(x, y, t) \tag{15}$$

In many cases the excited-state relaxation is a sum of first-order time-dependent processes, and the spatial dependence of heat production is just that of the laser beam used to excite the sample. In the absence of nonlinear kinetic effects such as optical bleaching or absorption saturation, the time-dependent power is the convolution of the irradiance with the kinetic relaxation impulse-response function

$$q_H(x, y, t) = \alpha\phi(t) * R'(t) * E_0(x, y, t) \tag{16}$$

where $R'(t)$ (s^{-1}) is the generalized relaxation or heat production rate discussed in Chapter 2 and incorporates the heat yield. The temperature change is obtained in a manner equivalent to that for impulse excitation:

$$\delta T(x, y, t) = \alpha[\Phi(t) * R'(t)] * [S(x, y) \otimes \delta T_{\text{point}}(x, y, t)] \tag{17}$$

As with impulse excitation, the time- and space-dependent temperature change may be written in terms of the impulse response to the Gaussian excitation beam as the triple time convolution

$$\delta T(x, y, t) = \alpha \Phi(t) * R'(t) * \delta T_{\text{impulse}}(x, y, t) \tag{18}$$

It should be remembered that this result is valid only for times greater than the acoustic propagation time, $t > t_a$, where $t_a = w/c$ and c is the sound velocity. For times less than the acoustic-limited rise time, the effect of the propagating mode must be taken into account. The effect of the propagating mode on temperature and density change is discussed in Chapter 3.

The triple convolution explicitly takes into account the time dependence of both the excitation pulse and the excited-state relaxation kinetics. For experiments designed to measure the rates of excited-state relaxation, the laser pulse duration should be much less than the rise time for kinetic decay. When this is the case, the laser power may be integrated independent of the kinetic and thermal diffusion rates. The temperature change may then be approximated by

$$\delta T(x, y, t) = \alpha Q R'(t) * \delta T_{\text{impulse}}(x, y, t) \tag{19}$$

5.1.4. Continuous Laser Excitation

5.1.4.1. Laser Heating

For continuous excitation, $\Phi(t)$ is a constant, Φ_0, and convolution of the time-dependent laser power with the kinetic response will be equal to the laser power for times much longer than the longest kinetic time constant. For irradiation times longer than the longest relaxation time constant, the temperature change is given by the general result

$$\begin{aligned}
\delta T(x, y, t) &= Y_H \Phi(t) * \delta T_{\text{impulse}}(x, y, t) \\
\delta T(x, y, t) &= Y_H \int_0^t \Phi(t - t')\, \delta T_{\text{impulse}}(x, y, t')\, dt' \qquad (20) \\
\delta T_{\text{cw}}(x, y, t) &= \frac{2\Phi_0 \alpha Y_H}{\pi \rho C_P} \int_0^t \frac{1}{w^2(1 + 2t'/t_c)}\, e^{-2(x^2+y^2)/w^2(1+2t'/t_c)}\, dt'
\end{aligned}$$

The time integration result can be cast in terms of the exponential integral (Whinnery 1974)

$$\delta T_{\mathrm{cw}}(x, y, t) = \frac{\Phi_0 \alpha Y_H t_c}{\pi w^2 \rho C_P} \left\{ E_1 \left[\frac{2(x^2 + y^2)}{w^2(1 + 2t/t_c)} \right] - E_1 \left[\frac{2(x^2 + y^2)}{w^2} \right] \right\} \tag{21}$$

This result is valid only for time up to t. E_1 is the exponential integral defined by

$$E_1(z) = \int_z^{\infty} \frac{1}{z'} e^{-z'} dz' \tag{22}$$

and is equal to the incomplete gamma function, $\Gamma(0, z)$, often used in symbolic language processors. The exponential integral is just a symbolic representation of the time integral, although certain analytic properties have been described (Gautschi and Cahill 1972). Two important properties of $E_1(z)$ are that it decreases with z in a hyperbolic fashion and that $E_1(0)$ is infinite. The latter property makes analysis of the temperature change difficult to decipher since both exponential integral terms diverge for the on-axis (i.e., $x = 0$, $y = 0$) condition.

5.1.4.2. On-Axis Temperature Change

The time-dependent on-axis temperature change may be found by using series representations of this integral. A useful series expansion relationship is

$$E_1(z) = -\gamma - \ln(z) - \sum_{n=1}^{\infty} \frac{(-1)^n z^n}{n \cdot n!} \tag{23}$$

where γ is Euler's number. Since the step function temperature-change response is a difference of two exponential integrals, the on-axis temperature change may be obtained from the ratio resulting from the difference in the natural logarithm terms

$$\delta T_{\mathrm{cw}}(0, 0, t) = \frac{\Phi_0 \alpha Y_H t_c}{\pi w^2 \rho C_P} \ln \frac{t_c + 2t}{t_c} \tag{24}$$

The on-axis temperature change increases with time in a logarithmic fashion. According to this expression, the temperature will reach infinity as time approaches infinity for any excitation source power and any absorption coefficient. Of course, infinite temperature cannot be obtained, but this result would indicate that any sample may boil or decompose. Needless to say, this could cause problems in photothermal measurement. The time-dependent temperature change is tempered by the logarithmic response. The time rate of change decreases with increasing time. So the problem of sample overheating is not as bad as might be indicated from only the limiting behavior analysis. The thermal diffusion equations were solved with the assumption that the sample was of infinite extent. However, sample cells are of a finite extent and heat transer can occur at the surfaces of the cell. The thermal conductivity of air is high and the external sample cell walls may efficiently couple heat to the surrounding air. In addition, nonlinear heat transfer mechanisms such as radiative transfer and convection will be operative if the temperature in the excitation volume gets too high.

As given above, the temperature change is proportional to the characteristic thermal time constant, t_c. However, since $t_c = w^2/4D_T$, the temperature-change-magnitude term can also be expressed by

$$\delta T_{\mathrm{cw}} \propto \frac{\Phi_0 \alpha Y_H t_c}{\pi w^2 \rho C_P} = \frac{\Phi_0 \alpha Y_H}{4\pi\kappa} \tag{25}$$

where κ ($\mathrm{J\,m^{-1}\,s^{-1}\,K^{-1}}$) is the thermal conductivity of the medium and is related to the thermal diffusion coefficient by $D_T = \kappa/\rho C_P$. For continuous excitation, the thermal conductivity is often used to compare temperature-change magnitudes of different sample media since it is independent of density, heat capacity, and excitation volume. The temperature-change magnitude for continuous excitation is clearly different from that for pulsed laser excitation. In addition to the time at which the temperature maximum is reached, continuous excitation produces a temperature change that is inversely proportional to the thermal conductivity and independent of excitation volume, while pulsed excitation produces a change that is inverse proportional to the heat capacity and volume.

5.1.4.3. Postexcitation Cooling

Equation (21) gives the time- and space-dependent temperature change during the excitation process. Thermal diffusion will cause the heat to dissipate throughout the sample when the excitation laser source is turned off. The spatially dependent temperature change that persists after the excitation source is turned off or blocked can be found by convolution of the final temperature change with the point-source solution thermal diffusion equation solution. From the last part of (20),

$$\delta T_{\rm cw}(x, y, t \geqslant t_1) = \frac{2\Phi_0 \alpha Y_H}{\pi \rho C_P} \left[\int_0^{t_1} \frac{1}{w^2(1 + 2t'/t_c)} e^{-2(x^2+y^2)/w^2(1+2t'/t_c)} dt' \right] \otimes \delta T_{\rm point}(x, y, t - t_1) \tag{26}$$

where t_1 is the time that the excitation source is on. After performing the spatial convolution, the above results in the temperature change

$$\delta T_{\rm cw}(x, y, t) = \frac{2\Phi_0 \alpha Y_H}{\pi \rho C_P} \int_0^{t_1} \frac{1}{w^2(1+2t'/t_c+2\delta t/t_c)} e^{-2(x^2+y^2)/w^2(1+2t'/t_c+2\delta t/t_c)} dt'$$

$$\delta t = \begin{cases} t - t_1 & \text{for } t \geqslant t_1 \\ 0 & \text{for } t < t \end{cases} \tag{27}$$

The time integration, cast in terms of the exponential integral, is

$$\delta T_{\rm cw}(x, y, t) = \frac{\Phi_0 \alpha Y_H}{4\pi\kappa} \left\{ E_1 \left[\frac{2(x^2 + y^2)}{w^2(1 + 2t_1/t_c + 2\delta t/t_c)} \right] - E_1 \left[\frac{2(x^2 + y^2)}{w^2(1 + 2\delta t/t_c)} \right] \right\} \tag{28}$$

For $t \leqslant t_2$, δt is zero, and (28) is equal to the continuous heating equation given in (21). For times greater than t_1, the on-axis cooling is described by

$$\delta T_{\rm cw}(x, y, t \geqslant t_1) = \frac{\Phi_0 \alpha Y_H}{4\pi\kappa} \left\{ E_1 \left[\frac{2(x^2+y^2)}{w^2(1+2t/t_c)} \right] - E_1 \left[\frac{2(x^2+y^2)}{w^2(1+2t/t_c-2t_1/t_c)} \right] \right\} \tag{29}$$

Figure 5.1 illustrates the relative temperature-change profiles during the heating part of the cycle. The computed temperature changes for $t/t_c = 5$ and $t/t_c = 25$ are scaled to a maximum value of 1. The y-axis coordinate was set equal to zero for these data. This figure illustrates the broadening of the temperature-change profile due to thermal diffusion during continuous excitation. The profile of the Gaussian heating source is shown for reference. Figure 5.2 shows the relative temperature-change profile during the cooling half of the cycle. The Gaussian excitation source profile and that for the relative temperature change corresponding to a heat for $t = 25t_c$ are illustrated along with the relative temperature changes for cooling at times equal to $5t_c$, $10t_c$, and $25t_c$. The temperature-change profile broadens very rapidly during cooling.

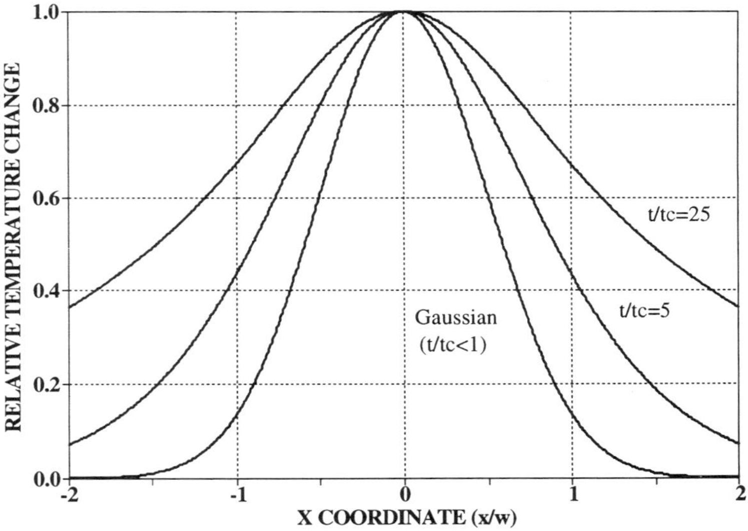

Figure 5.1. Temperature-change profiles produced during continuous Gaussian laser sample excitation at times relative to t_c. The profiles are scaled to the same maximum.

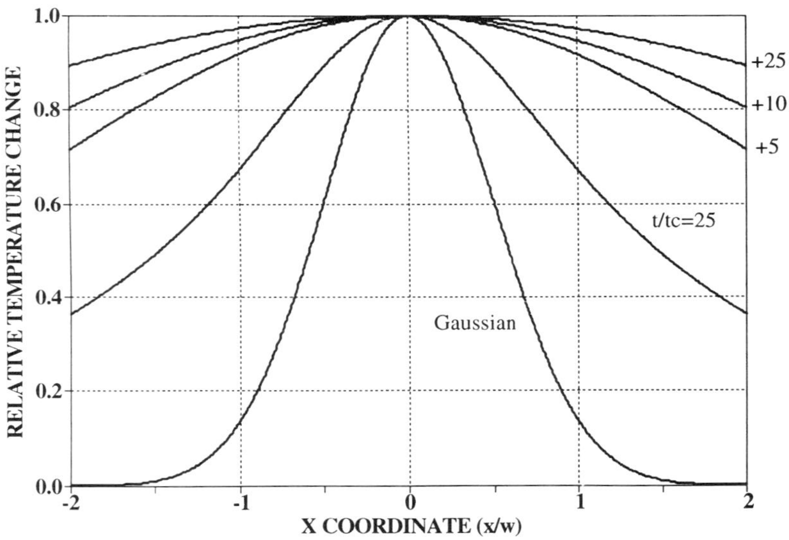

Figure 5.2. Temperature-change profiles at various times after continuous Gaussian laser sample excitation for $t = 25t_c$. Cooling times are 5, 10, and $25t_c$. The profiles are scaled to the same maximum.

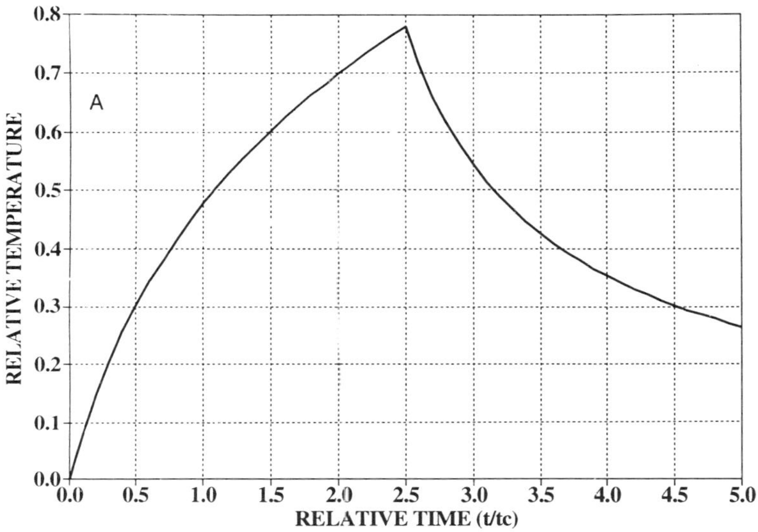

(*a*)

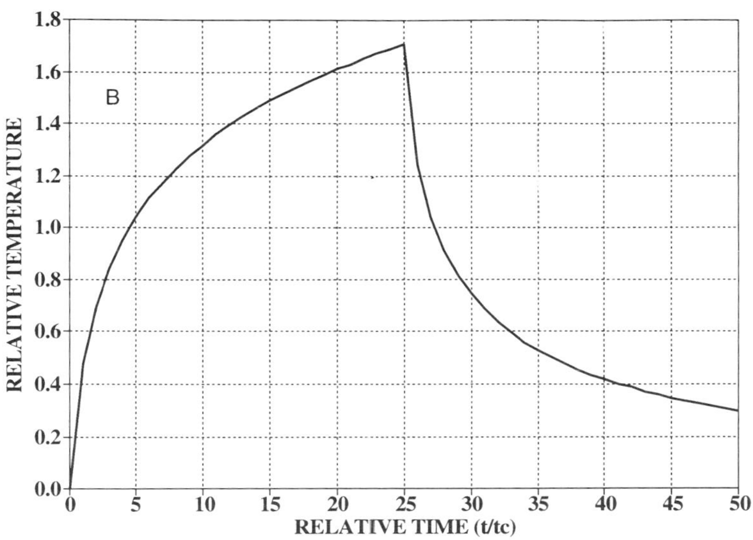

(*b*)

Figure 5.3. On-axis heating and cooling for laser sample excitation: (*A*) heating and cooling for a irradiation time of $2.5t_c$; (*B*) for an irradiation time of $25t_c$.

The on-axis temperature change during cooling is found using the series approximation to the exponential integral:

$$\delta T_{\mathrm{cw}}(0, 0, t \geqslant t_1) = \frac{\Phi_0 \alpha Y_H}{4\pi\kappa} \ln \frac{t_c + 2t}{t_c + 2(t - t_1)} \tag{30}$$

Temperature changes for on-axis heating and cooling are plotted in Figure 5.3. Two different relative irradiation times, t_1, are shown. In part (*A*), the heating period is $t_1 = 2.5t_c$ and in part (*B*), $t = 25t_c$. Longer heating produces a greater relative temperature change. This is easily seen by comparing the relative temperature change magnitudes of parts (*A*) and (*B*). On the other hand, the relative temperature after the cooling period, $t_2 = t_1$, is higher for the shorter heating and cooling period.

5.1.5. Chopped Laser Excitation

Using chopped continuous laser excitation, the temperature will increase and decrease periodically with the phase of the chopper cycle. Fang and Swofford (1983) used the Green's function for sample heating to determine the photothermal lens signal for a chopped Gaussian excitation laser. A similar calculation can be performed to determine the temperature change. The temperature change induced by a chopped Gaussian laser excitation source can be described by convoluting the laser-induced heating and cooling cycle with a repetition function. For a chopper with a duty cycle such that t_1 is the time that the chopper allows the beam to pass and t_2 is the time that the chopper blocks the beam, a repetition function for a periodic source can be cast in terms of delta functions

$$C(t) = \sum_{m=1}^{M} \delta(t - m\tau) \tag{31}$$

where τ is the cycle period and a total of M cycles have occurred since the start of the experiment. In this case the cycle period is $\tau = t_1 + t_2$. The temperature change at the end of the Mth cycle is

$$\delta T_{\mathrm{chopped}}(x, y, M\tau) = \delta T_{\mathrm{cw}}(x, y, t) * C(t) \tag{32}$$

where $\delta T_{\mathrm{cw}}(x, y, t)$ is that found in Section 5.1.4 for a single laser-induced heating and cooling cycle.

Convolution of the heating/cooling cycle with the delta function results in the temperature change

$$\delta T_{\text{chopped}}(x, y, M\tau) = \frac{\Phi_0 \alpha Y_H}{4\pi\kappa} \sum_{m=1}^{M} \left\{ E_1 \left[\frac{2(x^2 + y^2)}{w^2(1 + 2m\tau/t_c)} \right] - E_1 \left[\frac{2(x^2 + y^2)}{w^2(1 + 2m\tau/t_c - 2t_1/t_c)} \right] \right\} \tag{33}$$

The function shown is the temperature change at the end of M complete cycles. The temperature is an additive linear combination of the temperature change due to all previous cycles. Each cycle heats the sample for a fixed time of t_1 and thus increases the temperature by a fixed amount. The contribution that each complete cycle makes to the temperature at the end of the Mth cycle is the temperature change of that cycle at a time $m\tau$ after heating. Thus the final temperature is the sum over the previous cycles, each preceding the final time by a factor of $\delta t = m\tau$.

The on-axis temperature change at the end of the Mth cycle of the chopper is

$$\delta T_{\text{chopped}}(0, 0, M\tau) = \frac{\Phi_0 \alpha Y_H}{4\pi\kappa} \sum_{m=1}^{M} \ln \frac{t_c + 2m\tau}{t_c + 2(m\tau - t_1)} \tag{34}$$

The on-axis temperature change apparently has a finite limiting value since

$$\lim_{m\to\infty} \ln \frac{t_c + 2m\tau}{t_c + 2(m\tau - t_1)} = \ln \frac{t_c + 2m\tau}{t_c + 2m\tau} = 0 \tag{35}$$

However, model calculations show that a steady-state on-axis temperature change is not reached even with as many as 10^8 cycles, even for favorable values of t_1 and t_c. With chopper frequencies on the order of kHz, the time required to attain an equilibrium temperature would be several hours. The infinite medium approximation breaks down with these long times since the diffusion distance would be greater than the sample cell dimensions. In this case the maximum temperature reached will be that where the rate of heat transfer from the sample cell to the surroundings equals the rate at which heat is added to the sample.

The differential temperature change resulting from each individual cycle is independent of the initial temperature. Subsequently, the temperature change during the last heating cycle can be obtained by adding the temperature due to previous cycles to the temperature rise of the current cycle. To obtain the temperature increase that occurs during $(M + 1)$th heating cycle, the temperature change due to the present cycle is added to that from the previous, completed cycles

$$\begin{aligned}\delta T_{\text{chopped}}(x, y, M\tau + \delta t) &= \frac{\Phi_0 \alpha Y_H}{4\pi\kappa} \sum_{m=1}^{M} \left\{ E_1 \left[\frac{2(x^2 + y^2)}{w^2(1 + 2m\tau/t_c + 2\delta t/t_c)} \right] \right. \\ &\quad \left. - E_1 \left[\frac{2(x^2 + y^2)}{w^2(1 + 2m\tau/t_c + 2\delta t/t_c - 2t_1/t_c)} \right] \right\} \\ &\quad + \frac{\Phi_0 \alpha Y_H}{4\pi\kappa} \left\{ E_1 \left[\frac{2(x^2 + y^2)}{w^2(1 + 2\delta t/t_c)} \right] - E_1 \left[\frac{2(x^2 + y^2)}{w^2} \right] \right\}\end{aligned} \tag{36}$$

The last term is the same as the continuous heating given in (28) but with $\delta t = t - M\tau \leqslant t_1$. The cooling cycle is described by a similar equation:

$$\begin{aligned}\delta T_{\text{chopped}}(x, y, M\tau + \delta t) &= \frac{\Phi_0 \alpha Y_H}{4\pi\kappa} \sum_{m=1}^{M} \left\{ E_1 \left[\frac{2(x^2 + y^2)}{w^2(1 + 2m\tau/t_c + 2\delta t/t_c)} \right] \right. \\ &\quad \left. - E_1 \left[\frac{2(x^2 + y^2)}{w^2(1 + 2m\tau/t_c + 2\delta t/t_c - 2t_1/t_c)} \right] \right\} \\ &\quad + \frac{\Phi_0 \alpha Y_H}{4\pi\kappa} \left\{ E_1 \left[\frac{2(x^2 + y^2)}{w^2(1 + 2\delta t/t_c)} \right] \right. \\ &\quad \left. - E_1 \left[\frac{2(x^2 + y^2)}{w^2(1 + 2\delta t/t_c - 2t_1/t_c} \right] \right\}\end{aligned} \tag{37}$$

In this case $\delta t \geqslant t_1$. The on-axis temperature change that occurs during the $(M + 1)$th cycle is

$$\delta T_{\text{chopped}}(0, 0, M\tau + \delta t) = \frac{\Phi_0 \alpha Y_H}{4\pi\kappa} \left[\ln \frac{t_c + 2\delta t}{t_c} + \sum_{m=1}^{M} \ln \frac{t_c + 2(m\tau + \delta t)}{t_c + 2(m\tau + \delta t - t_1)} \right] \tag{38}$$

for heating and

$$\delta T_{\text{chopped}}(0, 0, M\tau + \delta t) = \frac{\Phi_0 \alpha Y_H}{4\pi\kappa} \left[\ln \frac{t_c + 2\delta t}{t_c + 2(\delta t - t_1)} + \sum_{m=1}^{M} \ln \frac{t_c + 2(m\tau + \delta t)}{t_c + 2(m\tau + \delta t - t_1)} \right] \tag{39}$$

for the cooling part of the cycle. The on-axis temperature change that occurs for the first several chopper cycles is illustrated in Figure 5.4. Here the

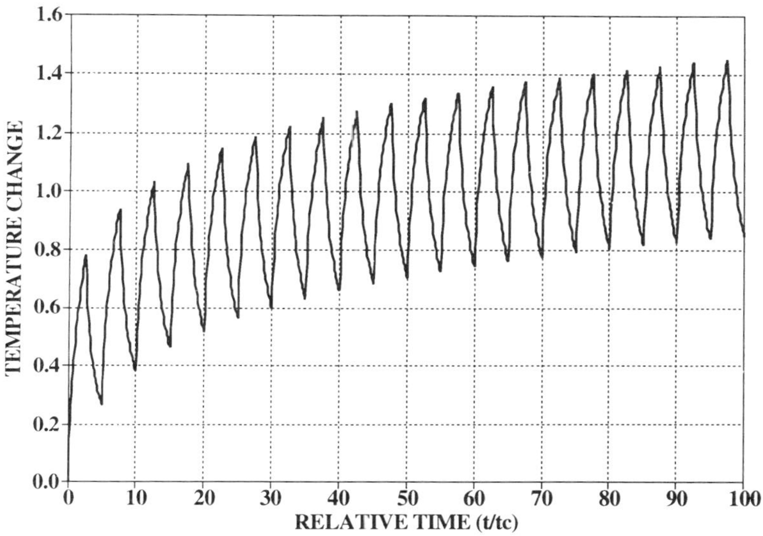

Figure 5.4. On-axis heating for chopped excitation. The duty cycle is 50% with an irradiation time of $2.5t_c$. The average temperature continues to increase, although the temperature change produced for a single chopper cycle is about the same after the first few excitation/cooling cycles.

chopper duty cycle was 50% and the excitation and cooling durations were each $2.5t_c$. As with the temperature change at the end of the cycle, these temperatures do not reach limiting values in practical time scales. This results in a problem when attempting to determine the magnitude of a photothermal signal that is proportional to the temperature. The temperature change can be thought of a sum of two components: an average temperature increase, proportional to the average excitation power, and an oscillating component. The average temperature increase will be logarithmic and the magnitude of the oscillating component is approximately constant. Problems resulting from the long time constant required to reach an equilibrium temperature change can be overcome by designing the photothermal apparatus to respond to the change in the temperature change, not the absolute temperature.

5.1.6. On-Axis Temperature Change for Periodic Excitation

In certain experiments the excitation source can be modulated periodically at a single angular frequency, $\omega = 2\pi f$. One can determine the amplitude of the periodic temperature change that occurs at this frequency by convolution of the impulse response with the periodic excitation source. The response to an oscillating light source can be obtained using

$$\delta T_{\mathrm{osc}}(t, r) = \Phi_{\omega} \int_{0}^{\infty} \delta T_{\mathrm{impulse}}(r, t') e^{i\omega(t-t')} \, dt' \tag{40}$$

where Φ_{ω} ($\mathrm{W\,m^{-2}}$) is the peak power of the oscillating source. This integral is similar to the Laplace transform,

$$\mathscr{L}\{\delta T_{\mathrm{impulse}}(r, t)\} = \int_{0}^{\infty} \delta T_{\mathrm{impulse}}(r, t) e^{-\zeta t} \, d\xi \tag{41}$$

Subsequently, the relationship

$$\delta T_{\mathrm{osc}}(r, t) = \Phi_{\omega} e^{i\omega t} \mathscr{L}\{\delta T_{\mathrm{impulse}}(r, t)\}|_{\zeta = i\omega} \tag{42}$$

may be used to find the response for an excitation source modulated at frequency ω. Laplace transform pairs are tabulated in many books, thus allowing the periodic response to be determined relatively easily for a wide variety of functions

The time-dependent change is found using the procedure described above is usually not as important as the amplitude of this change. The amplitude of the change is found from the complex absolute value of the time-dependent periodic response. This amplitude is what would be measured if the change were monitored and the resulting electronic signal processed with a narrow-bandpass filter or a lock-in amplifier. Similarly, if a complex periodic excitation source can be decomposed into several frequency components (e.g., a square wave can be decomposed into several harmonic frequencies), the temperature-change amplitude can be determined for each angular frequency comprising the excitation source. Any real optical periodic excitation source cannot be purely periodic. A periodic source, at best, contains a constant term and a periodic term. In this case the excitation power can be defined by

$$\Phi(t) = \Phi_0(1 - e^{i\omega t}) \tag{43}$$

where Φ_0 is the average power. The temperature change that would result using this excitation source is simply the sum of the continuous and the oscillating components,

$$\delta T(r, t) = \delta T_{\mathrm{cw}}(r, t) + \delta T_{\mathrm{osc}}(r, t) \tag{44}$$

The temperature change is thus the sum of a component that continues to increase and an oscillating component that reaches a steady-state value.

The temperature-change impulse-response function does not lend itself to Laplace transform. This is because the exponential term has an inverse time dependence. However, the on-axis temperature periodic temperature-change amplitude can be determined. In this case

$$\begin{aligned}\delta T_{\mathrm{osc}}(0, t) &= \frac{2\alpha Y_H \Phi_0}{\pi w^2 \rho C_P}\, e^{i\omega t} \mathscr{L}\left\{\left(1+\frac{2t}{t_c}\right)^{-1}\right\}\bigg|_{\zeta = i\omega} \\ &= \frac{\alpha Y_H \Phi_0}{4\pi\kappa}\, e^{i\omega t} E_1\left(\frac{i\omega t_c}{2}\right)\end{aligned} \tag{45}$$

This result is the steady-state periodic inverse focal length since the integration limits on the convolution extend to infinite time (Fournier and Boccara 1989). The amplitude of the periodic on-axis temperature change is found using the relationships $E_1(ix) = i\,\mathrm{si}(x) - \mathrm{ci}(x)$, where the $\mathrm{ci}(x)$ and $\mathrm{si}(x)$ functions are integrals of $-(\cos t)/t$ and $-(\sin t)/t$ from x to ∞ (Gradshteyn and Ryzhik 1980). The amplitude of the oscillating temperature change is thus

$$|\delta T_{\mathrm{osc}}(\omega)| = \frac{\Phi_0 \alpha Y_H}{4\pi\kappa}\sqrt{\mathrm{si}^2\left(\frac{\omega t_c}{2}\right) + \mathrm{ci}^2\left(\frac{\omega t_c}{2}\right)} \tag{46}$$

where the vertical bars indicate the amplitude.

5.1.7. Thermal Gratings

The intersection of two mutually coherent beams produces a periodic irradiance or energy profile due to interference of the beam's electric field. Optical absorption of the periodic irradiance or energy produces a photothermal grating. The geometry for defining the experimental parameters is illustrated in Figure 5.5. In the experimental arrangement used to produce these gratings, the excitation laser is split into two components, A and B, which cross within the sample cell at an angle of 2θ internal to the sample. The internal angle can be related to that outside the sample through Snell's law. In keeping with the geometry used to describe the Gaussian beams, these two beams are propagating in a direction making an angle $\pm\theta$ with respect to the z axis. Bragg's law predicts that these beams will produce a periodic electric field with a spatial period $\Lambda = \frac{1}{2}\lambda/n \sin\theta$. The spatial periods range from infinity when the two beams are coincident to a minimum of $\Lambda = \frac{1}{2}\lambda/n$ when $\theta = 90°$. For small angles, $\Lambda \approx \frac{1}{2}n\lambda/\theta$. The spatial angular frequency due to this period is $q = 2\pi/\Lambda$. The grating vector $\mathbf{q}$ is

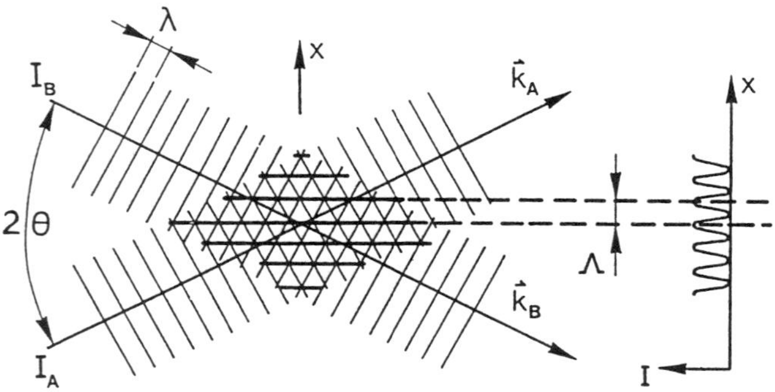

Figure 5.5. Interference pattern generated by two coherent waves. The intersection angle, θ, is defined relative to the grating tangent. (Adapted from Eichler et al. 1986, Fig. 2.2.)

related to the wave vectors of the two intersecting beams by $\mathbf{q} = \pm(\mathbf{k}_A - \mathbf{k}_B)$. The amplitude and polarization of the periodic electric field depend on those of the two beams. The simplest case and the one most pertinent to photothermal spectroscopy is when the amplitudes of the beams are equal and the polarizations are both perpendicular to the plane containing the two wave vectors.

For Gaussian laser beams collimated to a wide diameter and with a minimum beam waist at the intersection, the beams can be approximated as two plane waves. The irradiance at the crossing is generlly found from the square of the complex electric fields of the two waves. In general, this results in a tensor quantity with nine elements, each specifying an electric field polarization component (Eichler et al. 1986). For two plane waves with irradiances E_A and E_B and with equivalent polarizations perpendicular to the plane containing the intersection angle, the irradiance is

$$E(x) = E_A + E_B + 2\sqrt{E_A E_B}\cos qx \tag{47}$$

The grating vector is in the plane made by the two intersecting beams, given as the x direction in equation 47. For a fixed amount of power distributed between the two beams, a maximum periodic irradiance occurs when the two beams are of equal magnitude. In this case the irradiance at the intersection is $E(x) = \frac{1}{2}E_0[1 + \cos(2\pi x/\Lambda)]$.

If the beams may have sufficiently large beam waists to consider the periodic irradiance as being of infinite extent along the x axis, the time-dependent temperature change can be found by convoluting the one-dimensional thermal diffusion point-source solution with the periodic power source. We will consider the impulse temperature response that occurs for pulsed irradiation of the sample. For pulsed excitation, the pulse duration has to be sufficiently long that the coherence is not affected. For an interference pattern of infinite extent in the x dimension, the time-dependent temperature change for pulsed excitation is

$$\delta T_{\text{grating}}(x, t) = \frac{H\alpha Y_H}{\rho C_P (4\pi D_T t)^{1/2}} \int_{-\infty}^{\infty} \{1 + \cos[q(x - x_0)]\} \exp\left(-\frac{x_0^2}{4D_T t}\right) dx_0$$
$$= \frac{H\alpha Y_H}{\rho C_P}[1 + \cos qx \exp(-q^2 D_T t)] \tag{48}$$

where H (J m^{-2}) is the integrated irradiance of the pulsed laser. The initial temperature-change profile is periodic with a spatial temperature-change modulation between zero and the maximum possible, $2H\alpha Y_H/\rho C_P$. The spatially periodic component decays with a time constant that depends inversely on the spatial angular frequency of the interference pattern. The decay is exponential with a $1/e$ grating time constant of $\tau_G = q^{-2}D_T^{-1} = \Lambda^2/(4\pi^2 D_T)$. Thus the higher the spatial period, the longer the lifetime of the thermal grating.

The grating produced by two continuous plane waves with equivalent linear electric field polarizations pointing out of the plane of intersection is found by integrating the impulse response over the duration of irradiation:

$$\delta T_{\text{grating}}(x, t) = \frac{E\alpha Y_H}{\rho C_P}\int_0^t [1 + \cos qx \exp(-q^2 D_T t')]\, dt'$$
$$= \frac{H\alpha Y_H}{\rho C_P} + \frac{E\alpha Y_H}{\rho C_P}\frac{\cos qx}{q^2 D_T}[1 - \exp(-q^2 D_T t)] \tag{49}$$

The temperature change consists of two components: a nonperiodic component with a magnitude that increases linearly with time or integrated irradiance, and a periodic component with an exponential time dependence. The long-term limiting value of the periodic component is

$$\lim_{t\to\infty} \delta T_{\text{grating}}(x, t) = \frac{E\alpha Y_H}{q^2 \kappa}\cos qx \tag{50}$$

High-spatial-frequency interference patterns produce steady-state temperature changes that are smaller than those produced with low spatial frequencies.

The intersection of plane waves is an ideal case that may be approximated when the Gaussian laser beams have large waist radii. This approximation may not hold when the beams are focused into the sample cell. The integrated irradiance pattern that occurs at the focus of Gaussian beams is the product of a Gaussian with the periodic diffraction pattern formed from plane waves. Siegman (1977) has addressed in detail the diffraction pattern produced by Gaussian beams. A more simplified treatment is given here. The irradiance profile for intersecting Gaussian beams is given by Zhu and Harris (1989, 1990) in their evaluation of the effects of nonlinear optical absorption on thermal gratings. Accounting for the beam waist distortion that occurs due to the crossing angle, the interference of two pulsed laser beams with equal energy and polarizations will produce an integrated irradiance:

$$H_{\text{grating}}(x, y) = \frac{2Q}{\pi w_x w_y} \frac{\exp(-2x^2/w_x^2 - 2y^2/w_y^2)}{1 + \exp(-q^2 w_x^2/8)} (1 + \cos qx) \tag{51}$$

The exponential term in the denominator is included to normalize the modified Gaussian. The beam waists along the x and y dimensions are different since the beams intersect a plane at $z = 0$ at an angle θ with respect to the plane normal. The beam waists are $w_x = w_0/\cos\theta$ and $w_y = w_0$. The temperature change that occurs with absorption, found by convolution with the two-dimensional thermal diffusion point-source solution and including the absorption coefficient-heat yield term is

$$\begin{aligned} \delta T_{\text{grating}}(x, y, t) = {} & \frac{2Q\alpha Y_H}{\pi w_x w_y \rho C_P (1 + 2t/t_x)(1 + 2t/t_y)} \\ & \times \exp\left[-\frac{2x^2}{w_x^2(1 + 2t/t_x)} - \frac{2y^2}{w_y^2(1 + 2t/t_y)} \right] \\ & \times \frac{1 + \exp[-q^2 D_T t/(1 + 2t/t_x)] \cos[qx/(1 + 2t/t_x)]}{1 + \exp(-q^2 w_x^2/8)} \end{aligned} \tag{52}$$

Thermal diffusion in the z dimension is not accounted for, so this result is valid only for small angles. However, diffusion in the z dimension should not affect the grating period, only its time-dependent magnitude. t_x and t_y are the characteristic thermal decay-time constants for the x and y dimensions, defined by $w_x^2/4D_T$ and $w_x^2/4D_T$, respectively. The first part of this result is what is expected from pulsed sample excitation using an elliptical beam. The second part is the periodic component resulting from interference. The latter differs from the plane wave solution in two respcts. First, the lifetime of the periodic temperature distribution is no longer exponential but decays in a complex fashion. Second, the spatial frequency changes with time. The factor qx in the cosine is divided by $1 + 2t/t_x$, indicating that the time-dependent spatial frequency is $q/(1 + 2t/t_x)$.

In photothermal diffraction, the periodic temperature change is monitored with a probe laser set at the Bragg angle for diffraction. If the period of the temperature change is dynamic, the Bragg condition cannot be met at all times. The time dependence of the spatial frequency change is less important as the characteristic time constant, t_x, gets larger. This, in turn, increases with w_x^2. To avoid problems that may arise due to the dynamic spatial frequency, a large Gaussian beam waist should be used and the beams should intersect at large angles. The large intersection angle will not only increase w_x but will also increase the periodic temperature-change frequency. The latter, in turn, decreases the time required for the periodic structure to decay.

5.2. THERMODYNAMIC PARAMETERS

5.2.1. Thermodynamic Parameters Affecting Temperature

Depending on whether continuous or pulsed excitation is used, the temperature-change magnitude is inversely proportional to either the thermal conductivity or heat capacity of the sample. Thermodynamic parameters for several gases, liquids used as solvents, and solids are given in Tables 5.1, 5.2, and 5.3. These data are primarily from Dovichi (1987). Parameters for water have been changed to be consistent with *Handbook of Chemistry and Physics*

Table 5.1. Thermooptical Constants for Gases

Substance	ρ (kg m^{-3})	$C_P \times 10^{-3}$ (J kg^{-1} K^{-1})	$\kappa \times 10^2$ (W m^{-1} K^{-1})	$D_T \times 10^5$ (m^2 s^{-1})	$dn/dT \times 10^6$ (K^{-1})
Air	1.17	1.01	2.61	2.21	−0.88
N_2	1.14	1.04	2.60	2.19	−0.90
O_2	1.30	0.92	2.67	2.23	−0.82
Ar	1.62	0.52	1.77	2.10	−0.86
He	0.16	5.32	15.00	17.62	−0.11
H_2	0.08	14.4	18.15	15.75	−0.42
CO_2	1.79	0.84	1.66	1.10	−1.38
CH_4	0.65	2.24	3.43	2.36	−1.35
C_2H_6	1.22	1.76	2.18	1.02	−2.35

Source: Data from Dovichi (1987).

Table 5.2. Thermooptical Constants for Common Solvents

Substance	$\rho \times 10^{-3}$ ($kg\,m^{-3}$)	$C_P \times 10^{-3}$ ($J\,kg^{-1}\,K^{-1}$)	$\kappa \times 10$ ($W\,m^{-1}\,K^{-1}$)	$D_T \times 10^7$ ($m^2\,s^{-1}$)	n	$dn/dT \times 10^4$ (K^{-1})
n-Pentane	0.63	2.29	1.18	0.82	1.360	−5.5
n-Hexane	0.66	2.23	1.23	0.84	1.377	−5.2
n-Heptane	0.67	2.21	1.27	0.86	1.390	−4.9
n-Octane	0.71	2.19	1.32	0.85	1.400	−4.7
n-Nonane	0.72	2.18	1.36	0.87	1.407	−4.5
n-Decane	0.73	2.21	1.40	0.87	1.414	−4.2
Cyclohexane	0.78	1.81	1.23	0.87	1.43	−5.56
CS_2	1.26	1.02	1.50	1.17	1.63	−8.09
CCl_4	1.60	0.85	1.03	0.75	1.46	−6.12
Chloroform	1.52	0.95	1.17	0.81	1.446	−6.03
CH_2Cl_2	1.33	1.08	1.22	0.85	1.425	−5.5
Benzene	0.88	1.71	1.37	0.90	1.504	−6.52
Toluene	0.87	1.69	1.35	0.92	1.500	−5.68
ortho-Xylene	0.88	1.77	1.43	0.92	1.511	−5.0
Ethyl ether	0.72	2.31	1.37	0.82	1.35	−6.06
Acetone	0.80	2.18	1.90	1.09	1.362	−5.42
Ethyl acetate	0.91	1.96	1.49	0.84	1.372	−4.9
Methanol	0.79	2.46	2.02	1.04	1.311	−3.94
Ethanol	0.80	2.36	1.67	0.89	1.361	−4.0
n-Propanol	0.80	2.22	1.56	0.88	1.385	−3.7
Acetic acid	1.04	2.05	1.72	0.81	1.372	−3.9
Water (0°C)	1.00	4.21	5.64	1.34	1.334	0
Water (20°C)	1.00	4.18	5.98	1.43	1.333	−0.091
Water (80°C)	0.97	4.20	6.70	1.64	1.332	−2.31

Source: Data from Dovichi (1987).

(1980) and to reflect the more recent data of Franko and Tran (1989). Mori et al. (1982, 1983) also give extensive listings of gases and liquid thermal and thermooptical constants. Another source of data is the *Handbook of Chemistry and Physics*, which lists refractive indexes and expansion coefficients for several gases, liquids, and solid materials.

For continuous excitation, the thermal conductivity regulates the temperature change. Thermal conductivities often vary with temperature. In the gas phase, the temperature dependence of the thermal conductivity can be estimated from the hard-sphere formula, $\kappa \approx \frac{2}{3}(mC_P/\gamma)(k_BT/\pi m)^{1/2}/\pi d^2$, where m is the mass and d is the collision diameter of the gas molecules. The factor mC_P/γ is the constant-volume heat capacity per molecules, $\sim\frac{1}{2}k_B \cdot \mathrm{df}$, where df is the number of degrees of freedom, and so does not depend on molecular weight. For small molecules the number of degrees of freedom is

Table 5.3. Thermooptical Constants for Common Solids

Substance	$\rho C_P \times 10^{-4}$ ($J\,m^{-3}\,K^{-1}$)	$\kappa \times 10$ ($W\,m^{-1}\,K^{-1}$)	$D_T \times 10^7$ ($m^2\,s^{-1}$)	n	$dn/dT \times 10^5$ (K^{-1})
LiF	4.12	113	27.4	1.39	−1.27
NaF	3.04	100	32.9	1.32	−1.6
NaCl	1.845	65	35.2	1.54	−3.65
KCl	1.347	65	48.3	1.49	−3.4
KBr	1.197	48	40.1	1.52	−4.0
CsBr	1.171	09.6	08.2	1.66	−6.3
CsI	0.9094	11.3	12.4	1.79	−10.0
AgCl	1.984	11.5	05.8	2.06	−6.1
CaF_2	2.715	97	37.5	1.43	−0.77
MgO	3.129	250	79.9	1.66	1.89
Al_2O_3	2.999	250	83.4	1.78	1.41
Quartz	2.084	85	40.8	1.54	−0.539
ZeSe	2.646	130	49.1	2.43	4.8
CdTe	1.224	70	57.2	2.67	11.75
GeAs	1.420	370	261	3.30	18.7
Diamond	1.561	1510	967	2.42	0.98

Source: Data from Dovichi (1987).

Table 5.4. Convection Limits for Common Solvents and Gases at 25°C

Substance	$10^4 \times \beta^a$ (K^{-1})	$10^7 \times v$ ($m^2\,s^{-1}$)	$10^8 \times D_T$ ($m^2\,s^{-1}$)	ρ ($kg\,m^3$)	C_P ($J\,kg\,K^{-1}$)	$\alpha \Phi Y_H{}^b$ ($mW\,m^{-1}$)
Air	−33.5	156	2100	1.18	1010	8
Ar	−33.5	124	2200	1.78	520	3
N_2	−33.5	142	2200	1.25	1140	11
$(C_2H_5)_2O$	−1.6	3.11	8.2	713	2310	0.02
CH_3OH	−1.2	6.92	10	790	2460	0.09
C_2H_5OH	−1.1	13.8	8.9	800	2360	0.1
$(CH_3)_2CO$	−1.5	3.95	11	800	2180	0.04
H_2O	−0.26	8.93	14	997	4180	4
$CHCl_3$	−1.4	3.57	8.1	1520	950	0.007
CCl_4	−1.2	5.66	7.5	1600	850	0.01

[a]Assuming ideal behavior for gases, $\beta = -T^{-1}$.
[b]For excitation laser beam waist radius of $w = 1$ mm.

essentially the sum of translation and rotation degrees unless the temperature is sufficiently high that the vibrational modes are populated. For an ideal hard-sphere gas, the thermal conductivity increases with square-root temperature and decreases with that of the mass because of the increase in the mean velocity of the gas atoms or molecules that must conduct the heat. For a given amount of energy, the temperature change increases with the mass of the gas and decreases with the average temperature. Subsequently, Ar is a better buffer gas than He if the temperature is to be maximized. The temperature change is almost an order of magnitude higher with Ar versus that using He. Being a constituent of air, Ar is a renewable resource and less expensive than He, one of the rare cases where the less expensive substance is desirable. Most gas species follow the general trend predicted from hard-sphere collision theory at temperatures and pressures around standard conditions.

In condensed phases, thermal conduction takes the form of transmission of energy through intermolecular energy transfer. Simple kinetic arguments give the thermal conductivity of a liquid as $\kappa \approx 2.8 k_B N^{2/3} \gamma^{1/2} c$, where N (m^{-3}) is the number of molecules per unit volume and c ($m\ s^{-1}$) is the speed of sound (Herschfelder et al. 1954). This formula predicts the thermal conductivities of a large number of liquids to within 10%. It can be simplified to illustrate the trend with density. Sound speed is related to the density through $c^2 = \gamma/\rho K_T$, and overall, the thermal conductivity is pro-

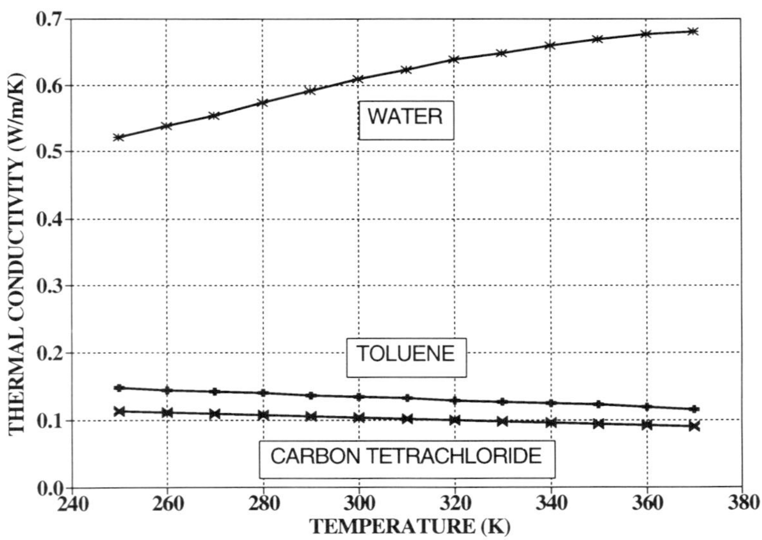

Figure 5.6. Temperature-dependent thermal conductivities for water, toluene, and carbon tetrachloride. Water has anomalously high conductivity, which increases with temperature.

portional to $\kappa \propto \rho^{1/6} K_T^{-1/2}$ For most liquids, ρ decreases while K_T increases with increasing temperature. κ will thus decrease with temperature for most liquids, water, of course, being anomaly. Measured temperature-dependent thermal conductivities for liquid water, carbon tetrachloride, and toluene are plotted in Figure 5.6. (*Handbook of Chemistry and Physics* 1980). The conductivities for carbon tetrachloride and toluene are reasonably constant around room temperature, decreasing slightly presumably due to the volume expansion with increasing temperature.

Water exhibits anomalous behavior, both in its magnitude relative to the nonpolar solvents and increasing with temperature in a nonlinear fashion. The temperature-dependent thermal conductivity is approximately $\kappa(T°\mathrm{C}) \approx 0.5641 + 0.00182T - 6.199 \times 10^{-6}T^2$ ($\mathrm{W\,m^{-1}\,K^{-1}}$) with T in °C. The increase in water's thermal conductivity with temperature is more like the trend for gas thermal conductivity and does not follow the trend predicted by the density change. The density of liquid water changes with temperature in a nonlinear fashion from -30 to 80°C (Angell 1982). The density is a maximum at 4°C. At temperatures below 4°C, density increases with temperature, and above this temperature, density decreases with temperature. The density of water can be predicted from $\rho(T°\mathrm{C}) = (999.83852 + 16.945176T - 7.9870401 \times 10^{-3}T^2 - 4.6170461 \times 10^{-5}T^3 + 1.0556302 \times 10^{-7}T^4 - 2.8054253 \times 10^{-10}T^5)/(1 + 1.6879850 \times 10^{-2}T)$, where T is in °C (*Handbook of Chemistry and Physics* 1980). This density is plotted in Figure 5.7. The density of water reaches a maximum at $\sim$4°C (277 K)

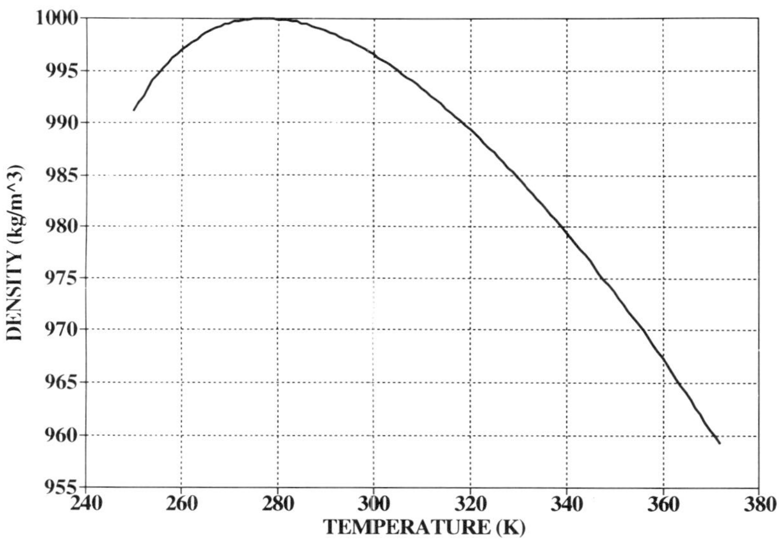

Figure 5.7. Temperature-dependent water density.

while the thermal conductivity shows no sign of decreasing even above this temperature. On the other hand, the isothermal compressibility of water does decrease with temperature above 0°C. However, this alone cannot account for the anomalous κ. The increased magnitude relative to the nonpolar liquids is presumably due to enhanced energy transfer through the hydrogen-bonded structure of the liquid. Thermal conductivities for methanol and ethanol are also greater than that of the nonpolar solvents, adding credence to this postulate.

For pulsed excitation, the temperature change is inversely proportional to the volume heat capacity, ρC_P. A good source of temperature-dependent heat capacity data is the *JANAF Thermochemical Tables* (1971). C_P is fairly constant over a wide range of temperatures for most gases, liquids, and solids, changing mostly due to thermal population of vibrational degrees of freedom. Of course, at constant pressure, the density is inversely proportional to temperature. The specific heat of water has been measured precisely at temperatures between 0 and 100°C. A plot of part of this data is shown in Figure 5.8. Although the data do show anomalous behavior, with a minimum at 34.5°C, C_P changes by less than 1% within this range.

In summary, to increase the temperature change for a given amount of absorbed power or energy, one should use heavier gases and nonpolar solvents. For gases, increasing the average temperature of the sample will increase κ, thereby lowering the temperature change induced by continuous

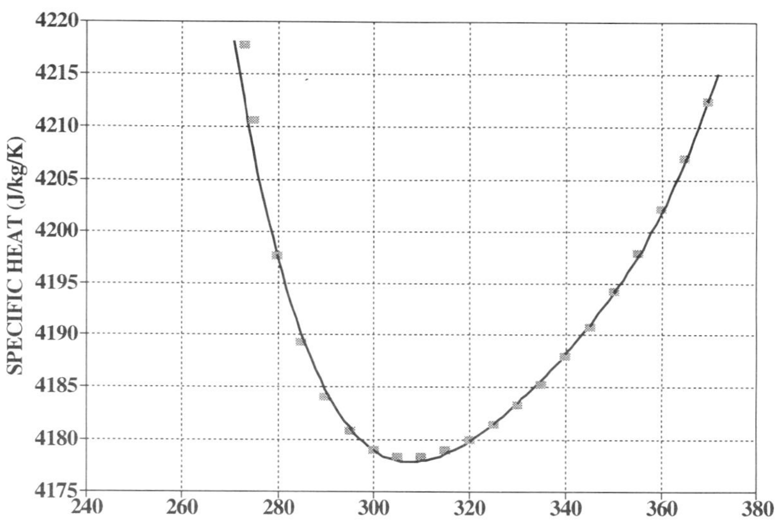

Figure 5.8. Temperature-dependent specific heat for water.

irradiation. On the other hand, increasing the average temperature at constant pressure will decrease the density. The density decrease will lower the heat capacity, thereby increasing the temperature change for pulsed excitation. This effect can be counteracted by the heat capacity per molecule, which tends to increase with temperature if the molecule contains low-energy vibrations. Increasing the sample temperature of most liquid solvents decreases both thermal conductivity and density. The thermal conductivity decrease with increasing sample temperature will enhance the photothermal--induced temperature change. Decreases in the density with increasing temperature will also enhance the pulsed temperature change. Again, the increasing heat capacity with increased temperature can counter this effect. Water is the anomaly. The thermodynamic parameters for water predict that the photothermal-induced temperature change is maximized at low temperatures, using continuous excitations, while high sample temperatures are favored with pulsed excitation.

5.2.2. Convection Heat Transfer

When the rate of energy absorption by the sample is too high, approximations used to simplify the hydrodynamic equations break down. What usually happens in this case is that the density change induced by the large temperature change becomes significant enough that gravitational effects come into play. The effect of gravity on samples that have large inhomogeneous density perturbations is convection. Convection is a mass transport phenomenon whereas the less dense, heated portion of the sample rises while the cooler, more dense sample sinks. This can occur in a coherent, cooperative fashion. In extreme cases, the convective heat transfer can result in turbulence.

The convection limit was discovered early in the development of continuous laser-excited photothermal spectroscopy (Whinnery et al. 1967). Cremmers and Keller (1982) give a limit to the onset of convection. They find that convection will not distort the temperature-change profile as long as $\alpha\Phi_0 Y_H \ll -32\eta k^2/\beta g w^3 \rho^2 C_P$, where η is the viscosity, k the thermal conductivity, β the expansion coefficient, and g the acceleration due to gravity (9.81 m s^{-2}). In terms of the thermal diffusion coefficient and kinematic viscosity, $\alpha\Phi_0 Y_H \ll -32\rho C_P v' D_T^2/\beta g w^3$. Davis and Petuchowski (1981) obtain a similar result based on the natural convention velocity for a gas, $v_c \approx (2\alpha\Phi Y_H g/\rho C_P T)^{1/3}$. For $w/v_c \ll t_c$, $\alpha\Phi_0 Y_H \ll 32\rho C_P v' T D_T^3/g w^3$. Using the data in Tables 3.4 and 3.7, and for an excitation beam waist radius of 1 mm, the power required for convection heat transfer is seen to vary widely for the different solvents and gases commonly used in photothermal spectroscopy, as shown in Table 5.4. It is clear that gases are less prone to convection than are the solvents, and solvents with high kinematic viscosity and/or thermal diffusion coefficient are less likely to convect.

Buffett and Morris (1983) examined some effects of convection in photothermal lens spectroscopy. They found that convection increases the low frequency, or flicker noise in the measurement. Because convention occurs due to the absorbed power, they speculated that continuously excited photothermal spectroscopy could not be used to measure small absorption in the presence of a large solvent or background absorption. Methods which use differential sample cell arrangements to subtract solvent and interferant absorbance will be limited by the onset of convection heat transfer. However, Simo Alfonso et al. (1990) have described an experimental method that can be used to decrease flicker noise in photothermal lens spectroscopy with convection heat transfer. They use a slit instead of the usual pinhole aperture to spatially filter the diverging probe beam. The $1/f$ character of flicker noise implies low noise for short measurement times. Thus pulsed laser–excited photothermal signals should be rather immune to the effects of convection heat transfer at during the early parts of the signal. Convective heat transfer should ultimately limit the long-time pulsed laser–excited photothermal spectroscopy signals. It has not yet been reported, probably owing to the lower energies required for substantial signals in the pulsed laser experiments.

5.3. OPTICAL ELEMENTS

Photothermal signals are developed from refractive index changes that occur in a sample due to light absorption. This refractive index change is normally made spatially dependent. In homogeneous samples, the spatial dependence arises from sample illumination geometry and thermal diffusion. When the refractive index change varies in space, the path that a ray of light makes through the sample will be altered over the normal ray path. The ray path change will be a function of both the degree to which the refractive index has changed and the spatial distribution of the refractive index change. It is convenient to model the refractive index change spatial distribution in terms of familiar optical elements. The action that the refractive index change has on the propagation of light may then be understood in terms of the corresponding optical elements. Optical elements commonly used to model the spatially dependent refractive index perturbation in photothermal spectroscopy are the lens, giving rise to photothermal lens spectroscopy; the prism, resulting in photothermal deflection; and the volume grating, which results in photothermal diffraction spectroscopy. In addition, bulk changes in the sample's refractive index that do not change the path of the light used to monitor the index change can be detected with several interferometric techniques.

5.3.1. Phase Shift and Optical Pathlength Difference

Photothermal interferometry monitors a change in refractive index by measuring the phase shift of light passing through the sample cell. A ray of light entering the sample with an initial phase angle of zero will have a phase angle, ϕ (rad), after passing through the sample with a spatially dependent refractive index distribution $n(x, y, z)$ of

$$\phi = \frac{2\pi}{\lambda} \int_{\text{path}} n(x, y, z)\, ds \tag{53}$$

ds is the optical path element and integration is over the path that a ray makes through the sample. For Gaussian laser beam excitation, the temperature change is assumed to be a function of the x and y coordinates only and is symmetric about the z axis.

Since refractive index change is directly proportional to the temperture change:

$$n(x, y, t) = n_0 + \delta n(x, y, t) = n_0 + \left(\frac{dn}{dT}\right) \delta T(x, y, t) \tag{54}$$

n_0 is not dependent on the spatial coordinates for the homogeneous sample. The change in phase angle between the perturbed and unperturbed samples is thus

$$\delta\phi = \frac{2\pi}{\lambda}\left(\frac{dn}{dT}\right) \int_{\text{path}} \delta T(x, y, t)\, ds \tag{55}$$

The differential phase shift, $\delta\phi$, is a function of the geometry of the probe ray relative to that of the excitation source. The simplest geometry to describe is that where probe ray propagates along the z axis, collinear with the excitation beam. Assuming that the refractive index change is small, a ray entering the sample along the z axis will not experience a significant path change over the length of the sample. Thus the integral over the path is reduced to the integral over the pathlength of the cell:

$$\delta\phi \approx \frac{2\pi}{\lambda}\left(\frac{dn}{dT}\right) \delta T(x, y, t) \int_0^l dz \tag{56}$$

and

$$\delta\phi \approx \frac{2\pi l}{\lambda}\left(\frac{dn}{dT}\right)\delta T(x, y, t) \tag{57}$$

where l is the pathlength through the sample cell.

Because the optical phase shift is not detected directly but rather, is observed by an interferometer fringe shift or coherent modulation, photothermal interferometry is perhaps easier to perform using optical apparatuses that measure optical pathlength differences. This avoids the periodic signals that occur when the phase shift is greater thn 2π. Defining the optical pathlength, d (m), as

$$d = \int_{\text{path}} n(x, y, z)\, ds \tag{58}$$

the optical pathlength difference, either between the sample with the excitation beam on and off, or through two separate sample chambers, is

$$\delta d \approx l\left(\frac{dn}{dT}\right)\delta T(x, y, t) \tag{59}$$

for small temperature changes.

In photothermal interferometry, it makes sense to measure the phase shift or optical pathlength difference occurring on the axis of the excitation laser. For example, the optical pathlength change that occurs for on-axis monitoring of the refractive index change for a pulsed TEM_{00} excitation laser is

$$d_{\text{pulsed}} = \left(\frac{dn}{dT}\right)\frac{2\alpha l Q Y_H}{\pi w^2 \rho C_P(1 + 2t/t_c)} \tag{60}$$

while that for chopped continuous excitation would be, from equations 38 and 63,

$$d_{\text{chopped}} = \left(\frac{dn}{dT}\right)\frac{\Phi_0 \alpha l Y_H}{4\pi\kappa}\left[\ln\frac{t_c + 2\delta t}{t_c} + \sum_{m=1}^{M}\ln\frac{t_c + 2(m\tau + \delta t)}{t_c + 2(m\tau + \delta t - t_1)}\right] \tag{61}$$

for the heating part of the cycle.

5.3.2. Deflection Angle

It was shown in Chapter 4 that the basic equation for ray propagation in a medium with radially symmetric refractive index gradient resulted in a propagation angle

$$\theta \approx \frac{1}{n_0} \int_{\text{path}} \left[\frac{dn(r)}{dr} \right]_{r_0} ds \tag{62}$$

where θ (rad) is the angle that the ray makes relative to the initial axis of propagation, $n(r)$ the refractive index at the radial offset r from the propagation axis, and r is perpendicular to the propagation axis. The gradient of the refractive index is evaluated at r_0, the offset of the ray at any distance through the medium.

This angle is internal to the sample. If the medium with the refractive index perturbation is not air, the angle at which the ray emerges from the medium will be altered at the interface. Snell's law can be used to relate the internal angle to the angle of ray propagation in air. However, the propagation angles are assumed to be small, so the approximation, $\sin\theta \approx \theta$, can be made if the ray excites the sample cell approximately parallel to the cell window normal. Thus assuming that $n_{\text{air}} = 1$, the ray angle that would be outside the sample medium is

$$\theta \approx \int_{\text{path}} \left[\frac{dn(r)}{dr} \right]_{r_0} ds \tag{63}$$

This equation gives the propagation or deflection angle relative to the initial propagation axis but cannot account for the second angle, that relative to some arbitrary axis perpendicular to the propagation axis. For a ray initially propagating along the z axis, the propagation angle equation can be generalized to Cartesian coordinates, resulting in angles in the x–z and y–z planes:

$$\begin{aligned} \theta_x &\approx \int_{\text{path}} \left[\frac{dn(x, y, z)}{dx} \right]_{x_0, y_0} ds \\ \theta_y &\approx \int_{\text{path}} \left[\frac{dn(x, y, z)}{dy} \right]_{x_0, y_0} ds \end{aligned} \tag{64}$$

in which case there is no ambiguity of the deflection-angle direction. Substituting the temperature dependence of the refractive index change for the Gaussian profile excitation laser source results in a ray deflection angle, for example, along the x axis, of

$$\theta_x \approx \left(\frac{dn}{dT}\right) \int_{\text{path}} \left[\frac{dT(x, y)}{dx}\right]_{x_0, y_0} ds \tag{65}$$

For index perturbations the deflection angle is also small. In this case optical path integration results in

$$\theta_x \approx l \left(\frac{dn}{dT}\right) \left[\frac{dT(x, y)}{dx}\right]_{x_0, y_0} \tag{66}$$

where l is the length through the sample.

5.3.3. Thermal Lens Focal Length

The focal length of the photothermal lens element can be found from the spatially dependent temperature change, and thus the refractive index change, using a variety of methods. The result for the circularly symmetric lens will first be derived using the results from Chapter 3. It was shown in Section 4.3.3 that ray propagation through a medium with a radially symmetric refractive index described by

$$n(r) = n_0 + \frac{r^2}{2} \left(\frac{d^2 n}{dn^2}\right)_{r=0} \tag{67}$$

resulted in the ray transfer matrix

$$\mathbf{T} = \begin{bmatrix} \cosh al & \dfrac{1}{a} \sinh al \\ a \sinh al & \cosh al \end{bmatrix} \tag{68}$$

where

$$a = \sqrt{\frac{1}{n_0} \left(\frac{d^2 n}{d^2 r}\right)_{r=0}} \tag{69}$$

and l is the pathlength. The hyperbolic form is used because the d^2n/dr^2 is usually positive in photothermal spectroscopy experiments due to the negative thermooptical coefficient of most samples. The index change is related to that for temperature using the thermooptical coefficient and thus

$$a = \sqrt{\frac{1}{n_0}\left(\frac{dn}{dT}\right)\left(\frac{d^2\,\pi\delta T}{d^2 r}\right)_{r=0}} \tag{70}$$

A sample with a refractive index, n_0, that is being excited by a Gaussian laser is considered. Probe rays enter the sample from air, propagate through the sample medium, and exit the sample again, making a transition to air. The total ray transfer matrix must take into account the refractive index change between the sample medium and the air on entering and exiting the cell. Assuming that the cell windows are negligibly thin, the total ray transfer matrix for a ray passing through the sample is

$$\mathbf{T} = \begin{bmatrix} \cosh al & \dfrac{n_A}{n_0 a}\sinh al \\ \dfrac{n_0 a}{n_A}\sinh al & \cosh al \end{bmatrix} \tag{68}$$

where n_A is the refractive index of air. For small temperature or index changes,

$$\lim_{a\to 0}\mathbf{T} = \begin{bmatrix} 1 & l\,\dfrac{n_A}{n_0} \\ la^2\,\dfrac{n_0}{n_A} & 1 \end{bmatrix} \tag{72}$$

Each element of this approximate matrix differs from the hyperbolic terms by less than 1% for $\alpha l \leqslant 0.3$. This matrix is identical to that for ray propagation through a thick lens made of a material with a refractive index n_0. The upper right-hand term accounts for propagation through the sample or lens. The lower left term is equal to the inverse focal length of the lens. Substituting for a^2, the inverse focal length of the photothermal lens is

$$\frac{1}{f} = -la^2\,\frac{n_0}{n_A} = -\frac{l}{n_A}\left(\frac{dn}{dT}\right)\left(\frac{d^2\,\delta T}{d^2 r}\right)_{r=0} \tag{73}$$

The refractive index of air is normally assumed to be 1; thus

$$\frac{1}{f} = -l\left(\frac{dn}{dT}\right)\left(\frac{d^2\,\delta T}{dr^2}\right)_{r=0} \tag{74}$$

An equivalent inverse focal length has been found using geometric optics (Gordon et al. 1965, Fang and Swofford 1983). Both of these used a geometric model to calculate the divergence angle of a ray passing through a sample with a radially symmetric refractive index perturbation.

This comes about because terms beyond second order were truncated in the power series expansion of the refractive index change (see Section 4.3.3). This approximation is called a parabolic approximation because the refractive index change profile is assumed to be described by a parabolic (e.g., second-order) equation. A Gaussian and its parabolic expansion are illustrated in Figure 5.9. At an offset of w, the Gaussian already deviates from the parabolic expansion by about 20%. Thus probe rays used to monitor the photothermal lens element should not deviate more than $\pm\frac{1}{2}w$ from the center of the temperature perturbation for accurate predictions based on the foregoing inverse focal length. Because the probe beam must be of a diameter less than that of the pump, a laser probe beam source is usually necessary. Using a laser source, a small probe beam diameter can be obtained.

Another shortcoming of this inverse photothermal lens focal length approximation is that the result is valid only for ray propagating on or near

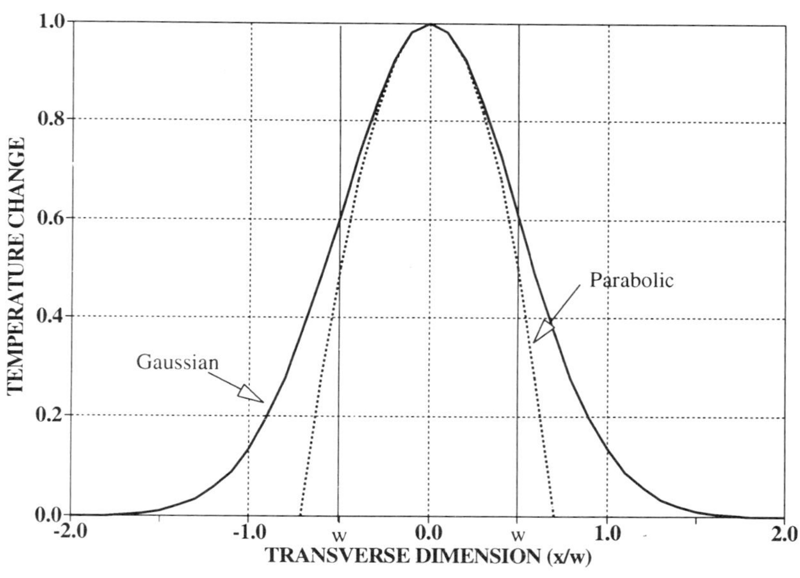

Figure 5.9. Gaussian temperature change produced by a pulsed laser. The parabolic approximation used to calculate the inverse thermal lens focal length is shown as a dashed line. The parabolic approximation fails for transverse offsets greater than about $w/2$.

the axis. This shortcoming can be removed by calculating the inverse focal length at any position relative to the excitation beam center. Jackson et al. (1981) integrate the basic equation for beam propagation discussed in Chapter 4 to obtain inverse focal lengths of

$$\frac{1}{f_i} = -\int_{\text{path}} \frac{d^2 n(x, y, z)}{ds_{\perp,i}^2}\, ds \tag{75}$$

f_i is the focal length outside the sample in one of two orthogonal directions perpendicular to propagation, $s_{\perp,i}$, and integration is over the ray path in the sample. (The expression obtained by Jackson et al. has an additional factor of $1/n_0$ that cancels for detection in air.) The inverse focal lengths are internal to the sample. Relating the temperature and refractive index changes with the thermooptic coefficient, we obtain

$$\frac{1}{f_i} = -\left(\frac{dn}{dT}\right) \int_{\text{path}} \frac{d^2\, \delta T(x, y, z)}{ds_{\perp,i}^2}\, ds \tag{76}$$

Compared to the radially symmetric results, this equation is much more flexible in that it allows for evaluation of the inverse focal length for any temperature distribution. For small ray deviations and no temperature change along the z coordinate, a bundle of rays centered at (x_0, y_0) will produce lenses with inverse focal lengths of

$$\frac{1}{f_i} = -l\left(\frac{dn}{dT}\right)\left[\frac{d^2\, \delta T(x, y)}{ds_{\perp,i}^2}\right]_{x_0, y_0} \tag{77}$$

The $s_{\perp,i}$ corresponds to orthogonal directions perpendicular to that for ray propagation. For rays that are collinear with the excitation beam, the $s_{\perp,i}$ are equal. In this case $s_{\perp,i} = r$, the radial direction, and the above are equivalent to that obtained using collinear geometry. When evaluated for on-axis rays, the result above is equivalent to that obtained earlier using the ray transfer matrix for a GRIN lens.

5.3.4. Grating Strength

The strength of the photothermal diffraction grating is related in a complex fashion to the index modulation of the grating. The index modulation is, in turn, proportional to the temperature change. Kogelnik's coupled wave diffraction efficiency equations were discussed in Chapter 4. These equations can be used to relate both refractive index and absorption coefficient modulation to the diffraction efficiency for a probe beam that satisfies the Bragg condition. The diffraction efficiency, η, of a thick-volume phase transmission grating for a probe laser at wavelength λ_p is

$$\eta = \frac{\Phi_{\text{diffracted}}}{\Phi_0} = \sin^2 \frac{\pi l \delta n}{\lambda_p \cos \theta_b} \tag{78}$$

where $\Phi_{\text{diffracted}}$ and Φ_0 are the diffracted and incident powers of the probe. The probe is incident on the phase grating at the internal Bragg angle, $2n\Lambda \sin \theta_b = \lambda_p$. The Bragg angle is related to the angle of the excitation laser by $\sin \theta_b = (\lambda_p/\lambda) \sin \theta$, where the angle and wavelength without the subscript are those of the excitation source. The index modulation, δn, is the maximum index change for the periodic phase structure of length l. The refractive index change can be related to the time-dependent temperature change derived in Section 5.1.6 through the thermooptical coefficient, $\delta n(x, y, t) = (dn/dT)\, \delta T_{\text{grating}}(x, y, t)$. The maximum index change for pulsed excitation occurs at zero time and for $x = y = 0$, for which

$$\delta n = \left(\frac{dn}{dT}\right) \frac{2 \cos \theta\, Q \alpha Y_H}{\pi w^2 \rho C_P} \tag{79}$$

where θ is the half-angle of the excitation laser beam intersection. The length of the grating, l, also depends on the beam intersection geometry. The length is approximated by $2w/\sin \theta$ (Zhu et al. 1992). Making these substitutions, the diffraction efficiency for the thick-volume phase grating produced by the photothermal effect is

$$\eta \approx \sin^2 \left[\left(\frac{dn}{dT}\right) \frac{4 \cot \theta\, Q \alpha Y_H}{w \rho C_P \lambda_p \cos \theta_b}\right] \tag{80}$$

The approximations made in deriving the diffraction efficiency above may not be accurate. However, the trends predicted by this result should be at least qualitatively correct. For small temperature changes, $\sin^2 x \approx x^2$, and the diffraction efficiency will increase as the square of $\alpha Q Y_H$ and inversely proportional to w^2, favoring focused excitation beams. The efficiency is also predicted to increase as $\cot^2\theta$, indicating that smaller excitation laser intersection angles will increase the diffraction efficiency.

Applying Kogelnik's coupled wave diffraction efficiency result to the index perturbation produced by two intersecting Gaussian laser beams is sort of like trying to force a round peg into a square hole. Siegman (1977) has examined the effect of using Gaussian excitation and probe beams in the diffraction experiment. The equation that describes the diffraction efficiency for weak phase gratings probed at the Bragg angle is

$$\eta = \left(\frac{\pi \delta n w}{\theta \lambda_p}\right)^2 S\left(\frac{w_p}{w}, \frac{\lambda_p}{\lambda}\right) \tag{81}$$

where $S(w_p/w, \lambda_p/\lambda)$ is a geometrical factor depending on the relative radii and wavelengths of the Gaussian excitation and probe lasers. The geometrical factor is not important if calibration samples are used to determine the relative diffraction efficiency. The relative efficiencies should still follow the same trends as that predicted from the ideal diffraction grating produced by plane waves. This approach has been used quite successfully by McGraw et al. (1987), Zhu and Harris (1989) and Zhu, et al. (1992) to describe nonlinear optical absorption effects in photothermal diffraction spectroscopy. McGraw et al. show that the diffraction efficiency for the photothermal experiment is

$$\eta = \left[\left(\frac{dn}{dT}\right)\frac{\alpha Q Y_H}{\theta w \lambda_p \rho C_P}\right]^2 S\left(\frac{w_p}{w}, \frac{\lambda_p}{\lambda}\right) \exp\left(-\frac{t}{\tau_G}\right) \tag{82}$$

for a sample with linear optical absorption. The time constant for thermal decay of the grating is the same as that given in Section 5.1.6, $\tau_G = q^{-2} D_T^{-1}$. McGraw et al. also give the efficiency for probe angles that do not satisfy the Bragg condition. They find that the decreased angular sensitivity due to the finite spatial frequency bandwidth of the probe and excitation beams results in a probe-angle-dependent diffraction efficiency of

$$\eta \propto \exp\left[\frac{-w^2 k_p^2}{2(1 + w^2/w_p^2)} (\theta_p - \theta_b)^2\right] \tag{83}$$

In this equation θ_p is the experimental probe angle and k_p is the probe wave vector. This equation predicts that the angular sensitivity of the diffraction efficiency is Gaussian centered about the Bragg angle. Their experimental observations, as well as those in subsequent studies (Zhu et al. 1992), show that this expression is at least qualitatively correct.

5.4. TEMPERATURE-DEPENDENT REFRACTIVE INDEX CHANGE

All of the dynamic optical elements, which are the basis of photothermal spectroscopy, are linearly proportional to the temperature-dependent refractive index change. The refractive index change is related to the temperature change by the differential thermooptic constant

$$\delta n = \left(\frac{dn}{dT}\right)\delta T \tag{84}$$

where δn is the refractive index change, δT the temperature change due to the photothermal effect, and dn/dT the thermooptic coefficient. The thermooptic coefficient is linear over the narrow temperature range of the photothermal temperature changes for most samples. Moreover, it usually has negative values. The refractive index change is primarily dependent on density and only secondarily dependent on temperature. Because density usually decreases with temperature, the effect of sample heating is to lower the refractive index. For small refractive-index perturbations, the total refractive index of the sample is given by

$$n = n_0 + \delta n = n_0 + \left(\frac{dn}{dT}\right)\delta T \tag{85}$$

where n_0 is the refractive index of the unperturbed sample. Using the temperature-change solutions found for Gaussian laser excitation of the sample, the spatially dependent refractive index can be found.

5.4.1. Density and Temperature Dependence of Refractive Index

The refractive index is usually thought to be a function of the temperature and density of the media, although technically, it is more directly related to the number density of species in their ground and excited states. The index change is related to temperature and density by considering two individual components. For small changes in temperature and density, the refractive index change is

$$\delta n(T, \rho) = \left(\frac{\partial n}{\partial T}\right)_\rho \delta T + \left(\frac{\partial n}{\partial \rho}\right)_T \delta \rho \tag{86}$$

The first term on the right-hand side is the contribution to the refractive index change due to a temperature-dependent change in the molar refractivity of the sample. This term is usually quite small relative to the density-dependent refractive index change. The temperature-dependent refractive index change may be obtained from the Lorenz–Lorentz equation, which describes the refractive index of a substance in terms of the number density and polarizability, or to the related molar refractivity expression.

The second term is the density-dependent refractive index change. This term is cast in terms of the density change but can also be related to the temperature change through the volume expansion coefficient, β. The temperature-dependent refractive index may thus be expressed as the sum of the two terms

$$\left(\frac{dn}{dT}\right) = \left(\frac{\partial n}{\partial T}\right)_\rho + \left(\frac{\partial n}{\partial \rho}\right)_T \left(\frac{\partial \rho}{\partial T}\right)_P \tag{87}$$

The relationship between refractive index and molar refractivity is given by

$$\frac{n^2 - 1}{n^2 + 2} = \frac{R_m \rho}{M} \tag{88}$$

where R_m ($m^3\,mol^{-1}$) is the molar refractivity, M ($kg\,mol^{-1}$) the molecular weight, and ρ ($kg\,m^{-3}$) the density of the medium. Both R_m and ρ are temperature dependent. The temperature-dependent index change is due to a change in the specific or molar refractivity of the sample while the second term is due to the density change. Using the molar refractivity equation yields

$$\left(\frac{\partial n}{\partial \rho}\right)_T = \frac{(n_0^2 + 2)^2}{6n_0} \frac{R_m}{M} = \frac{(n_0^2 + 2)(n_0^2 - 1)}{6n_0\rho_0} \tag{89}$$

where ρ_0 ($kg\,m^{-3}$) the density of the medium and n_0 is the refractive index. The density is linearly related to the temperature through equilibrium, isobaric thermodynamic functions; thus the second term in the temperature-dependent refractive index change is

$$\left(\frac{\partial n}{\partial \rho}\right)_T \left(\frac{\partial \rho}{\partial T}\right)_P = -\rho_0\beta \left(\frac{\partial n}{\partial \rho}\right)_T = -\beta \frac{(n_0^2 + 2)(n_2^2 - 1)}{6n_0} \tag{90}$$

For gases below a few atmospheres of pressure, the refractive index differential may be directly related to the static refractive index and the gas law (Barker and Rothem 1987). This is because low-pressure gases have static refractive indexes very close to 1. Thus for the normal pressure range of gas samples,

$$\left(\frac{\partial n}{\partial T}\right)_{T_0} \approx -\frac{1}{2}\beta(n_0^2 - 1) = -\frac{n_0^2 - 1}{2T_0} \tag{91}$$

where T_0 is the equilibrium temperature of the sample.

The temperature-dependent refractive index change that occurs at constant density is dependent on temperature-dependent changes in the molar refractivity:

$$\left(\frac{\partial n}{\partial T}\right)_\rho = \frac{\rho}{M}\frac{(n_0^2+2)^2}{6n_0}\left(\frac{\partial R_m}{\partial T}\right) \tag{92}$$

This temperature dependence, away from phase transition temperatures and in the absence of temperature-dependent equilibria or reactions, can be obtained by considering the change in Boltzmann populations of the individual states with temperature. Molar refractivity is related to molecular polarizability by

$$R_m = \frac{\langle\alpha\rangle}{3\varepsilon_0} N_A \tag{93}$$

where $\langle\alpha\rangle$ is the molecular polarizability average over all molecules and N_A (Avogadro's number). The average polarizability, in turn, may be found from

$$\langle\alpha\rangle = \frac{1}{N_{\text{tot}}} \sum_i \alpha_i N_i \tag{94}$$

where the summation is over the i states of the molecule. Each polarizability is the sum over all the spectroscopic transitions available to that level:

$$\alpha_i = \sum_j \alpha_{i,j} \tag{95}$$

where the sum is over the spectroscopic transitions and the individual components to the polarization are given by

$$\alpha_{i,j} \approx \frac{\sigma_{i,j}(v_{i,j})\lambda_{i,j}(v_{i,j}-v)}{8\pi^2\,\Delta v_{i,j}} \tag{96}$$

The $\sigma_{i,j}$ are the absorption cross sections for transitions between levels i and j, $v_{i,j}$ is the transition frequency, and v is the frequency of the polarizability, thus the refractive index determination. Neglecting any contribution due to a temperature-dependent transition line width, the only temperature-dependent terms in these equations are the level populations. The number density in a given level, N_i, may be determined from the Boltzmann population equation.

The effects of small changes in temperature on the refractive index of fluid samples are probably negligibly small for visible wavelengths. First, transitions that may be nearly resonant in the visible have excited states far above kT. There would be little change in the excited electronic-state populations for temperatures around room temperature. Second, changing rovibrational state populations should not directly affect the refractive index contribution visible because the summation is over all possible transitions. While certain terms in the summation may increase with a temperature change due to the thermal population of a lower state of a near-resonant transition, other near-resonant terms must decrease accordingly. This argument cannot be used to eliminate temperature effects but does suggest that these effects would be small.

Several dn/dT values are given in Tables 5.1 through 5.3. Refractive indexes and the changes that occur with temperature are wavelength dependent. Much of the data is compiled for the common atomic emission lines. The most common is the sodium D line at 589.3 nm, which is rarely used in photothermal spectroscopy. Dovichi points out that much of the refractive index and thermooptical coefficient data is probably inaccurate and researchers should not rely on these values as much as on their own calibrations. As a general rule, nonpolar solvents have thermal and thermooptical properties that favor large refractive index changes. The best of these nonpolar solvents is CCl_4, owing to a lack of vibrational overtone spectra in the visible and favorable thermal and thermooptical parameters.

To consider the effect of temperature on the photothermal response under normal laboratory conditions (e.g., constant pressure), one can make use of the relationship obtained from equations 70 and 71, $(dn/dT)_T \propto -\beta$. For gases, $-\beta \propto T^{-1}$, $\rho \propto T^{-1}$, and $\kappa \propto T^{1/2}$. For a given absorbed power or energy, the continuous excited photothermal signal will be proportional to $T^{3/2}$, while that for pulsed excitation is independent of T, with the approximation that the specific heat capacity is not a function of temperature. If the fraction of the absorbing analyte species in the gas sample is constant, increasing the temperature will decrease density and thus α proportional to T^{-1}. In this case low temperatures are favored for gas-phase photothermal spectroscopy using both pulsed and continuous excitation sources.

The temperature dependence of the photothermal signal for liquid samples is more difficult to predict. One must have knowledge of β, κ, and/or ρC_P, usually obtained experimentally, to determine the effect of temperature. Water is a poor solvent for photothermal spectroscopy, owing to a relatively low thermooptical coefficient, high heat capacity, and high thermal conductivity. Water is unique among these solvents for other reasons as well. Based on the density change alone, dn/dT is predicted to be zero at 4°C. In fact, measurements of the temperature-dependent refractive index would indicate that this is so (Abbate et al. 1978, Dovichi 1987). More recently, Franko and Tran (1989) have measured the temperature dependence of the photothermal lens signal in a water solvent between -9 and 80°C. They find that the temperature-dependent aqueous sample thermal lens signal can be expressed by the experimentally derived polynomial

$$\frac{S(T)}{S(20°\text{C})} = 1.2 \times 10^{-3} + 7.27 \times 10^{-2}T - 1.563 \times 10^{-3}T^2 + 2.604 \times 10^{-5}T^3 - 2.395 \times 10^{-7}T^4 + 8.891 \times 10^{-10}T^5 \quad (97)$$

where $S(T)$ is the temperature-dependent signal, $S(20°\text{C})$ is the signal at 20°C, and T is the temperature in °C. The thermal lens focal length was found to be positive below -0.1°C and negative above. The temperature-dependent signal went to zero at -0.1°C. The water remained as a supercooled liquid down to the lowest temperature studied, as evidenced by the negative relative signal magnitude below 0°C. All continuous laser-excited photothermal signal magnitudes are proportional to $dn/dT/\kappa$. Since only the dn/dT term can change sign with temperature, Franko and Tran concluded that this term must pass through zero at -0.1°C. Because the relative signal magnitude did not follow the same trend as the temperature-dependent density, it is likely that there is a significant contribution for dn/dT from $(dn/dT)_\rho$. This contribution was attributed to either a phase change into "icelike" structures or a change in the average polarizability of the water molecules near the phase transition temperature.

Index determinations from measurement of the relative photothermal lens signal strength may be more precise than direct measurements of n at various temperatures because of the differential nature of the photothermal measurement. Relatively precise expressions for n and dn/dT can be found based on the data of Franko and Tran. The temperature-dependent value of the thermal conductivity is approximately $\kappa(T°\text{C}) \approx 5.641 + 0.01817T - 6.199 \times 10^{-5}T^2$ (in mW cm^{-1} K) for water, and using accurate values of n for water (*Handbook of Chemistry and Physics* 1980) at several temperatures:

$$\frac{dn}{dT} \approx -1.221 \times 10^{-8} - 6.220 \times 10^{-6}T + 1.144 \times 10^{-7}T^2 - 1.862 \times 10^{-9}T^3 + 1.672 \times 10^{-11}T^4 - 6.003 \times 10^{-14}T^5$$

$$n \approx 1.334 - 6.220 \times 10^{-8}T - 3.110 \times 10^{-6}T^2 + 3.813 \times 10^{-8}T^3 - 4.654 \times 10^{-10}T^4 + 3.343 \times 10^{-12}T^5 - 1.000 \times 10^{-14}T^6 \quad (98)$$

for T over the temperature range -10 to 80°C.

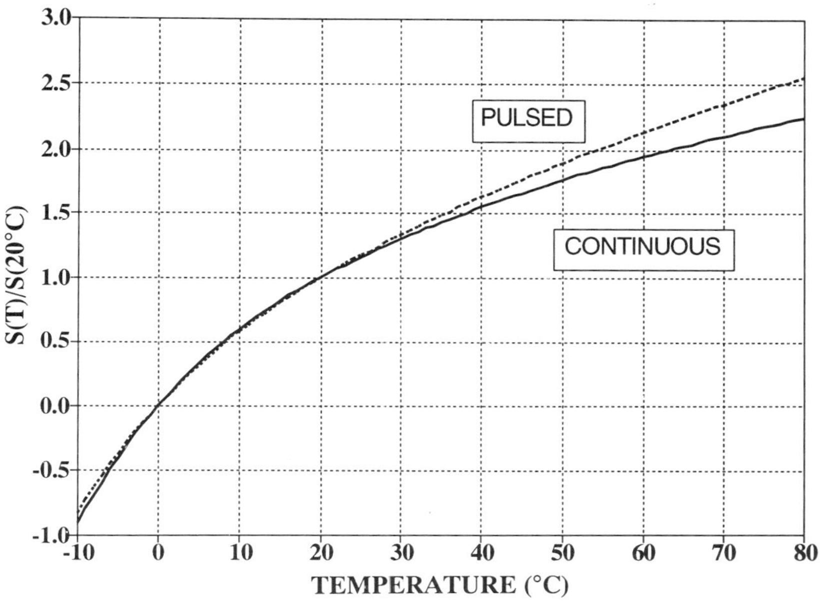

Figure 5.10. Relative temperature-dependent photothermal signal for water for continuous and pulsed excitation. Continuous data are from Franko and Tran (1989), while those for pulsed excitation are estimasted based on the continuous measurements and the thermodynamic properties of water.

Using these semiempirical relationships, the relative temperature-dependent photothermal signal can be found. The photothermal signal magnitudes relative to that at 20°C for both continuous and pulsed excitation are shown in Figure 5.10. The continuous laser excitation response shown is the equation is that given by Franko and Tran. The pulsed laser response was calculated from $S \propto (dn/dT)/\rho C_P$, which is proportional to the optical element strengths for interferometry, deflection, and lensing alike. The signal relative to that at 20°C apparently increases with temperature up to the measurement limit of 80°C and may increase to the boiling point. The approximate expression used to model the relative photothermal signal magnitude for pulsed excitation is

$$\frac{S_{\text{pulsed}}(T)}{S_{\text{pulsed}}(20°\text{C})} = 1.83 \times 10^{-3} + 6.818 \times 10^{-2}T - 1.25 \times 10^{-3}T^2 + 2.041 \times 10^{-5}T^3 - 1.832 \times 10^{-7}T^4 + 6.58 \times 10^{-10}T^5 \tag{99}$$

The slight differences between this equation and that found for the continuous excitation experiment are due to the temperature dependence of the ratio $\kappa/\rho C_P$.

These data for water illustrate a point that is often neglected in photothermal spectroscopy. It is generally assumed that the thermodynamic and optical properties of the sample are not strong functions of temperature and that the temperature change induced by the photothermal effect is small. Figure 5.10 shows that at least for water, the first assumption is not true. In this case thermodynamic and optical properties are rather strong functions of temperature, ranging to $\pm 40\%$ for a temperature differential of $\pm 10°C$ at about 20°C. Moreover, the on-axis temperature change produced by continuous excitation can be very high for long irradiation times. In theory, the on-axis temperature change increases to infinity. Thermodynamic and optical properties will change significantly for many solvents over several tens of degrees. Franko and Tran circumvented potential problems by using thermostated sample cell holders, a low excitation duty cycle, excitation times much below t_c, and low absorbance samples. Measurements performed with less attention paid to the effects of bulk temperature changes are unfortunately more common.

5.4.2. Population Dependence on Refractive Index

In addition to the thermodynamic controlled changes in refractive index, the refractive index may be influenced by index changes from nonthermal origins (Hu and Whinnery 1973). Excited-state species and photochemical products with different optical absorption characteristics will alter the sample's overall refractive index. These effects will in general have temporal characteristics which distinguish them from photothermal signals. One aspect of photothermal spectroscopy of homogeneous samples, versus heterogeneous sample experiments, where the optical probe does not pass through the excited sample itself, is that optical signals can be influenced by the excited states and products of photochemical reactions. This is because light used to monitor the refractive index change passes through the optically excited sample. Differences in the absorption coefficients of excited states and photochemical products can change the power of the light used to monitor the refractive index changes. In certain experimental configurations, changes in probe light power may be incorrectly interpreted as a photothermal signal.

The process of optical excitation can produce lower- and upper-state number densities that are far from those predicted by the temperature-dependent Boltzmann population if the rates of excited-state relaxation are slower than that for excitation. This can affect the refractive index in two ways. Consider first the effect of exciting a two-level system. The refractive index contribution due to a nearly resonant transition of the analyte is

$$n = n' + \frac{c}{4\pi}\,\alpha_0(\nu_0)\,\frac{1 - \nu/\nu_0}{\Delta\nu_0} = \frac{c}{4\pi}\,\sigma_0(\nu_0)(N_1 - N_2)\,\frac{1 - \nu/\nu_0}{\Delta\nu_0} \tag{100}$$

The population of the upper state increases relative to that of the lower state with increasing excitation irradiance. If the excitation irradiance is high enough to cause complete optical saturation, the number densities in the lower and upper levels will be equal and the contribution to the refractive index from the resonant transition will be nullified.

For organic dye molecules, saturation is not as important as optical bleaching wherein the lower singlet-state level is deleted at the expense of populating the excited triplet-state level. The excited triplet may have an optical absorption in the visible. Subsequently, optical bleaching can affect the refractive index in two ways. The first is the index change produced by ground singlet-state depletion, and the second is by population of a triplet state with visible–ultraviolet transitions. The degree to which each of these states changes the refractive index is related to the absorption coefficient and band centers relative to the wavelength of interest. Populations in the upper states of the ground singlet and excited triplet-state transitions are negligible and the population-dependent refractive index is

$$n = n' + \frac{c}{4\pi}\left[\sigma_S(\nu_S)N_S\,\frac{1 - \nu/\nu_S}{\Delta\nu_S} + \sigma_T(\nu_T)N_T\,\frac{1 - \nu/\nu_T}{\Delta\nu_T}\right] \tag{101}$$

where the subscripts S and T refer to the singlet and triplet states, respectively. For a total number density equal to the sum of singlet and triplet ground states,

$$\begin{aligned} n &= n' + \frac{c}{4\pi}N_{\text{tot}}\sigma_S(\nu_S)\,\frac{1 - \nu/\nu_S}{\Delta\nu_S} + \frac{c}{4\pi}N_T\left[\sigma_T(\nu_T)\,\frac{1 - \nu/\nu_T}{\Delta\nu_T} - \sigma_S(\nu_S)\,\frac{1 - \nu/\nu_S}{\Delta\nu_S}\right] \\ &= n_0 + \frac{c}{4\pi}N_T\left[\sigma_T(\nu_T)\,\frac{1 - \nu/\nu_T}{\Delta\nu_T} - \sigma_S(\nu_S)\,\frac{1 - \nu/\nu_S}{\Delta\nu_S}\right] \end{aligned} \tag{102}$$

The combined effects of these index changes due to population changes can result in a significant relative contribution to the photothermal spectroscopy signals. In photothermal lensing, this effect is an additive correction term and has been called a population lens (Tarazima and Hirota 1992). The effect is also commonly observed in dynamic diffraction grating experiments (Eichler et al. 1986). The number of parameter needed to model this effect are perhaps too many to result in realizable predictions. The time evolution of the triplet-state population will also control the rate of heat released to the sample and thus the temperature change. Subsequently, obtaining quantitative information regarding triplet-state energy may be difficult when the refractive index is also perturbed by the triplet.

5.4.3. Soret Effect

There are several effects that can result in refractive index changes with optical excitation. In fluids, a large temperature-dependent refractive index can occur when the identity or concentrations of molecules making up the sample change with temperature. In this case the N_i or α_i can change due to a change in the chemical identity of the species in the region being excited. Giglio and Vendramini (1974) showed that thermally induced concentration gradients can form large photothermal lens elements in binary solutions. The enhanced refractive index changes that occurred were attributed to the Soret effect. The Soret effect produces concentration gradients that are proportional to temperature gradients in solutions with more than one constituent. The steady-state concentration gradients formed produced photothermal lens signals that were about 44 times larger than those due to the thermooptical coefficient when the solution was near the critical temperature (the temperature above which the solution separates into two phases). Hardcastle and Harris (1986) used the Soret effect to enhance the photothermal lens signal produced in water. Water has particularly poor physical properties for photothermal spectroscopy. By using a solvent solution of 2,6-dimethylpyridine in water near the critical composition and temperature, they were able to enhance the signal by a factor of about 35 over that of pure water.

In the Soret effect, a steady-state concentration gradient is formed in a mixture through which a constant temperature gradient is maintained (Eichler et al. 1986). The steady-state concentration change for a sample with anisotropic temperature is given by $\delta C = -k_T \delta T/T$, where $k_T = D_T/D$ is the coefficient of thermal diffusion, D the mass diffusion coefficient, and $\delta C = \delta\rho/\rho_0$ the concentration of the heavier component (Hardcastle and Harris 1986). The concentration gradient forms to minimize the total free energy of the mixture. Free energy is reduced by accumulating the heavier component in the hotter zone of the mixture.

The dynamics of the concentration gradient is governed by a mass diffusion equation, modified to include the temperature-dependent free-energy effect:

$$\frac{\partial C}{\partial t} = D\left(\nabla^2 C + \frac{k_T}{T}\nabla^2 T\right) \tag{103}$$

The thermal and mass diffusion coefficients are typically several orders of magnitude different in condensed media. For example, the thermal diffusion coefficient for water is $1.7 \times 10^{-7}\,\mathrm{m^2\,s^{-1}}$ and the mass diffusion coefficient for aqueous benzyl alcohol is $8.21 \times 10^{-10}\,\mathrm{m^2\,s^{-1}}$. Subsequently, the time required to form or disperse the concentration gradient is $\sim 5 \times 10^{-3}$ slower than that required for the thermal gradient. On the other hand, the refractive index change with concentration can be very large, with $\partial n/\partial C \approx 0.1$ for some mixtures. So although the time required to form the concentration gradient is relatively long, the magnitude of the resulting refractive index change can be very large.

The change in the coefficient of thermal diffusion can be very large near the critical consolute point of a mixture, a point characterized by a critical concentration and temperature for a particular mixture. The critical concentration is that which has the highest temperature required for dissolution. Above the critical temperature, the two components are miscible in all proportions, while below this temperature, solutions separate into two phases. When a critical composition mixture is near critical temperature, the thermal diffusion coefficient remains nearly the same as in other parts of the phase diagram, but the mass diffusion coefficient, D, tends to zero. Thus k_T becomes very large near the critical consolute point. This enhances the Soret effect concentration gradient, and resulting photothermal signals are found to be greatest for mixtures near the critical consolute point.

The Soret effect may have an influence on photothermal signals observed for chopped laser excitation of liquid-phase samples since a quasi-steady-state temperature-change profile forms in addition to the periodic profile that is normally monitored. Substances will diffuse to minimize the free energy, but the anisotropic concentration profile will follow the quasi-steady-state temperature profile, not the temperature change produced by the chopped radiation. The quasi-steady-state concentration profile can affect the chopped photothermal signal in several ways. First, the concentration change can change the absorption coefficient in the excited region of the sample. Second, the solution composition produced by the Soret effect may affect the thermooptical coefficient and heat capacity of the solution local to the region of excitation. Gas-phase species have similar mass and thermal diffusion coefficients and, not being solutions in the respect that "solvation" does not occur, the coefficient of thermal diffusion will not change with temperature. Subsequently, the Soret effect is not expected to be as large in gases as it is in liquids.

5.4.4. Other Factors Affecting Refractive Index

Eichler et al. (1986) describe several other phenomena that can cause refractive index changes to occur when high-irradiance laser radiation is used to excite the sample. These effects are generally nonlinear and have to be considered only for high irradiances. Principal contributions to the high-irradiance refractive index changes are electrostriction, the electrocaloric effect, and the optical Kerr effect. Electrostriction occurs when the electric field of the excitation source is large enough to induce dipoles in the molecules or atoms of the sample medium. When this occurs, the induced dipoles attract one another and the medium contracts, producing an increase in the refractive index. Electrostriction was observed in early photothermal interferometry experiments.

The optical Kerr effect in liquids causes the molecules to orient due to the interaction between the electric field of the light source and the permanent or induced dipoles of the molecules. If the optical polarizability of the molecule is anisotropic, the medium becomes anisotropic and the refractive indexes change. This in turn causes birefringence. In addition, new vibrational modes can arise corresponding to molecular rocking and libration, and the vibrational energy of the molecules can redistribute among these new modes.

Expansion and contraction of an ideal fluid occur without frictional forces. With viscous fluids, frictional forces may cause heat to be generated even though there was no absorption into an excited state. Electrostriction and the optical Kerr effect can cause sample heating on expansion. This may be the cause of the electrocaloric effect. The electrocaloric effect is similar to the usual photothermal effect, with the exception that the source of heat is not that absorbed from the excitation source but rather that due to frictional forces on expansion of a medium constricted by electric field effects. The refractive index change will be related to the temperature change through the electrooptical coefficient.

A host of optically induced phenomena that alter refractive index can occur in transparent solid materials. These phenomena are complicated, owing to the presence of impurities, localized energy states, and the anisotropic nature of solids. Interested readers are referred to Eichler et al. (1986) for a summary of these effects.

5.5. LIMITATIONS

Several approximations were used to obtain theoretical temperature changes in a simple analytical form. These approximations can be unfounded in certain experimental apparatuses. The theoretical maximum temperature changes, and the resulting optical elements, probably cannot be obtained in these cases. However, the analytical solutions can often be adjusted to account for the effects neglected in the approximations. It is worth examining some typical situations to get a feeling for the accuracy of the simpler analytical results. First, the propagation mode produced from thermal expansion was neglected. For pulsed laser excitation, this omission results in erroneous density estimates during the initial perturbation. The thermal diffusion results will not correctly predict the shape or magnitude of the refractive index change that occurs during and shortly after pulsed excitations. As shown in Chapter 3, the thermal diffusion solution can be amended, with the propagating mode solution thereby yielding a more accurate description of the perturbation to the sample.

5.5.1. Excitation Beam Waist Radius Changes

It is also assumed that the excitation beam waist radius does not change significantly along the length of the sample (e.g., that the laser is collimated). As an example, consider a case where a 500-nm laser beam is focused. For a focus spot with an electric field beam waist radius of 500 μm, the confocal distance is $z_0 \sim 1.57$ m. The maximum beam wasit radius change that occurs over the length of a 1-cm-pathlength sample cell placed within one confocal distance of the focus position is only about 0.45%. This maximum occurs when the sample cell is placed $\pm z_0$ from the focus. A 0.45% change is small, so the collimated laser approximation is apparently accurate enough in this instance. If the sample cell is placed at the focus, the change in beam waist radius is negligible (about 10 ppm). On the other hand, if this laser were to be focused on a spot with a beam radius of 50 μm, the confocal distance is only $z_0 \sim 1.57$ cm. There is about a 10% change in beam radius between the focus and the window if the laser is focused into the sample cell. There would be about 44% change in the beam waist radius between the entrance and exit windows of a 1-cm sample cell placed one confocal distance from the focus. In these cases the collimated beam approximation is no valid and the average temperature-change and optical element strengths must be calculated by integrating the strengths produced by the diverging pump beam over the sample cell path.

5.5.2. Effects of Scattering and Optically Thick Samples

Another approximation used to find the theoretical temperature change was that attenuation of the excitation source was insignificant along the sample path. Optical element strengths and optical phase shift or pathlength are found by integrating the cumulative effects of thin elements along the optical path of the probe rays. Integration over the optical path of the probe rays sums the effect of these individual elements into a total element strength. For scattering and strongly absorbing samples, the absorbed power density will be a function of the distance into the sample. The power absorbed from an excitation laser beam with an irradiance $E(x, y, t)$ at the entrance of a homogeneous sample ($z = 0$), with an optical absorption coefficient and an exponential scattering loss, α_s, at any position along the z axis is

$$q_H(x, y, z, t) = \alpha E(x, y, z, t) = \alpha E(x, y, 0, t)e^{-(\alpha + \alpha_s)z} \tag{104}$$

Ignoring thermal diffusion along the z axis, which should have a small effect on the path-integrated optical element strength, the temperature change can be obtained for each thin section, dz, positioned at z into the sample. Probe ray path integration over these sections results in a factor that can be used in the element strength calculation. Assuming that optical attenuation of the excitation laser was small, probe ray path integration for parallel geometries resulted in a factor of l. For optically thick samples, parallel probe ray path integration results in a factor of $\{1 - \exp[-(\alpha + \alpha_s)l]\}/(\alpha + \alpha_s)$. The total optical element strengths are easily calculated. For example, the collinear probe-pulsed laser-excited thermal lens focal length produced in a homogeneous sample with a high optical attenuation is

$$\frac{1}{f(t)} = -\left(\frac{dn}{dT}\right)\frac{\alpha Q Y_H}{\alpha + \alpha_s}\{1 - \exp[-(\alpha_S + \alpha)l]\}\left[R'(t) * \left.\frac{d^2}{dr^2}\,\delta T_{\text{impulse}}(r, t)\right|_{r=0}\right]$$
$$= \left(\frac{dn}{dT}\right)\frac{8\alpha Q Y_H}{\pi w^4 \rho C_P(\alpha + \alpha_s)}\{1 - \exp[(\alpha_s + \alpha)l]\}R'(t) * \left(1 + \frac{2t}{t_c}\right)^{-2} \tag{105}$$

This result more accurately portrays the effects of optical absorption and scattering on the photothermal optical element strength. It does not, however, account for scattered light absorption resulting in a temperature change nor does it account for thermal diffusion along the z-axis direction.

Power and Langford (1988) used photothermal lens spectroscopy to measure optical absorbance due to dissolved organic material in unfiltered natural waters with silica scattering particles. They found no effects due to scattering for $\alpha_s l$ less than about 0.15, as evidenced by the signal strength of filtered and unfiltered samples. For the low scattering coefficients of the latter experiments, the exponential term may be approximated by $\exp[-(\alpha + \alpha_s)l] \sim 1 - (\alpha + \alpha_s)l$, and the approximated pathlength factor is l. Spear et al. (1990) have measured photothermal deflection signals in samples with much higher scattering losses. They found that the strength of the photothermal deflection signal decreases exponentially with scattering loss for coefficients as high as $\alpha_s = 3.8\ \text{cm}^{-1}$. The latter experiments used small-diameter (87-nm) latex spheres as the scattering particles, with visible excitation and probe laser sources. The scattering was within the Rayleigh limit and α_s was proportional to λ^{-4}. The latter factor should be taken into account when photothermal spectra are recorded. Yasa et al. (1982) developed a comprehensive theory to explain the effects of light scattering on transverse photothermal deflection of inhomogeneous sample which takes into account both the degree of scattering and the geometry of the scattered light. Although the theory developed is strictly applicable only to inhomogeneous samples, they find that scattering loss does not significantly affect the photothermal signal for $\alpha_s l \leqslant 0.1$ and is independent of the scattering geometry.

The effect that the scattered light irradiance has on the temperature change, and subsequently, on the optical elements, can be quantified using the diffusion theory of light scattering mentioned in Chapter 2. The diffusion theory is accurate when the point of interest is far from sources or boundaries and for $\alpha \ll \alpha_s$ (Patterson et al. 1989). According to the diffusion theory, an infinitely fast pulsed laser with an integrated irradiance $H(r) = 2Q\exp(-2r^2/w^2)/\pi w^2$ will produce a time-dependent power distribution in the sample:

$$\Phi(r, t) = \frac{2c\alpha_s Q}{\pi w^2(1 + 2t/t_s)} \exp\left[-\frac{2r^2}{w^2}\left(1 + \frac{2t}{t_s}\right)\right] \exp(-\alpha c t) \qquad (106)$$

where $t_s = w^2/4D_s c$ is the scattering time constant, $D_s = [3\alpha + 3(1-g)\alpha_s]^{-1}$ a parameter depending on the absorption and scattering coefficients and the asymmetry parameter g, and c the speed of light. The asymmetry factor is the average cosine of the particle scattering angle. For anisotropic scattering $g = 0$, while for scattering only in the forward direction, $g = 0$. The latter equation was found by convoluting the integrated irradiance with the diffusive scattering impulse-response function shown in Chapter 2. Since the result is accurate only for $\alpha \ll \alpha_s$, we will choose this limit, in which case $D_s = [3(1 - g)\alpha_s]^{-1}$. Also, assuming that α is small, the second exponential term may be neglected relative to the radial-dependent term. Since the speed of light is very much greater than thermal diffusion, the temperature change at a given time for instantaneous excited-state relaxation is

$$\delta T(r, t) = \frac{1}{\rho C_P} \int_0^t \alpha \Phi(r, t')\, dt' \approx \frac{\alpha \alpha_s c Q t_s}{\pi w^2} \left\{ E_1 \left[\frac{2r^2}{w^2(1 + 2t/t_s)} \right] - E_1 \left(\frac{2r^2}{w^2} \right) \right\} \tag{107}$$

The temperature-change distribution is that which occurs at zero time (e.g., prior to significant thermal diffusion), and the limit as t approaches infinity should give the pulsed laser-excited temperature-change distribution. The E_1 functions are not well behaved in this limit and it is perhaps easier to compare the optical elements to that obtained without scattering loss.

In the case of photothermal lens spectroscopy, the instantaneous inverse focal length due to absorption of the scattered light is found to be proportional to $(dn/dT)3\alpha\alpha_s^2(1 - g)/\pi w^2 \rho C_P$, while in the absence of scattering, the inverse focal length due to absorption of direct radiation is $(dn/dT)8\alpha/\pi w^4 \rho C_P$. Clearly scattering will not affect the inverse focal length as long as $\alpha_s^2(1 - g) \ll 8/3w^2$. The worst case is for anisotropic scattering where $\alpha_s^2 \ll 8/3w^2$ for the inverse lens element to not be affected by absorption of scattered radiation. It is apparent that scattering will not affect the inverse lens strength for most practical experiments wherein w is on the order of 1 mm or less and α_s is much less than $10\ \text{cm}^{-1}$ ($T = 4.5 \times 10^{-5}$ in a 1-cm cuvette). Even in this extreme case ($w = 1$ mm, $\alpha_s = 10\ \text{cm}^{-1}$) the scattered light lens element magnitude is less than 0.4% of the element produced from the direct light. Thus the effect of scattering on the photothermal signal can generally be accounted for by the sample path integrated loss term.

When optical absorption or scattering losses are large, thermal diffusion along the z axis might be important. Diffusion along the z axis should not affect the collinear probe photothermal spectroscopy signals. However, with transverse excitation-probe geometry, diffusion along the axis of excitation could make a significant difference in the time dependent signals. The solution to this diffusion problem is complicated by the boundary conditions at the sample cell windows but can be found using the procedures given in Carslaw and Jaeger under certain conditions. For example, if thermal diffusion through the sample cell windows is slow compared to that of the sample, as would be the case for gas phase samples, then the z-axis diffusion results in the time-dependent temperature impulse response

$$\delta T_{\text{impulse}}(z, t) = \frac{\alpha}{2} e^{\eta^2 D_T t - \eta z} \left\{ e^{2\eta z} \left[\operatorname{erf}\left(\frac{z + 2\eta D_T t}{2\sqrt{D_T t}} \right) - 1 \right] + \left[\operatorname{erf}\left(\frac{z - 2\eta D_T t}{2\sqrt{D_T t}} \right) + \right] \right\} \tag{108}$$

where $\eta = \alpha + \alpha_s$ is the total attenuation. The total thermal diffusion diffusion impulse response is the product of the radial and axial components. This result may be applied to determine the temperature change at positions far from the sample cell windows even if thermal diffusion is relatively fast through the windows (e.g., for liquid samples in quartz cuvettes). The rate of thermal diffusion is proportional to the Laplacian of the temperature-change profile. For most samples, the Laplacian along the z axis due to optical attenuation will be small compared to that due to the relatively narrow excitation laser beam. Thus thermal diffusion along the z axis should not result in a significant perturbation to the infinite extent approximation. The scale above which an accounting of the axial thermal diffusion becomes necessary is when $2\eta D_T t \sim z$.

5.5.3. Finite-Extent Sample Effects

Approximations used to reduce the thermal diffusion equation to a tractable form were that the sample is of infinite extent along the radial coordinate and that thermal diffusion through the sample cell windows and walls can be neglected. For infinite radial extent the approximation will hold up to times when the radius that the heat diffuses through is less than the minimum distance between the heat source and the cell wall. At infinite time, the thermal diffusion distance is infinite, and this approximation will fail. The approximation is probably valid for low-repetition-rate pulsed laser excitation since energy has a long time to disperse throughout the sample volume and the temperature can equilibrate with the cell and its surroundings. On the other hand, the approximation will fail for even short times when continuous lasers are used. How much the thermal conduction characteristics of the sample cell wall affect the signal depends on the relative thermal conductivity of the sample and cell materials, the position of the excitation region in the sample cell, and the geometry of the cell. Because of the different geometries of the sample cell and excitation laser (e.g., one being cylindrical, the other rectangular), solutions to this problem probably cannot be obtained in analytical form. The degree of accuracy of the infinite-extent approximation can be estimated by calculating the temperature at the cell wall both with and without thermal conduction through the wall. A fixed temperature change must be poised across the cell wall in order to have the rate of heat transferred to the sample equal the rate at which it passes through the cell wall. The heat flux is temperature-gradient dependent, given by the equation $j_E = -\kappa \nabla T$. For a standard 1-cm-pathlength quartz sample cuvette, the cell wall is approximately 1 mm thick, and the thermal conductivity is 8.5 W m^{-1} K^{-1}. We assume that the surroundings are "well mixed" and that the temperature on the outside of the cell wall is constant. This assumption is strictly valid when the sample cell is held in a temperature-controlled bath and is probably valid when the cell is in circulated air. For continuous sample excitation with a laser beam waist radius much smaller than the distance to the cell wall, the energy flux that must be dissipated at the cell wall is $j_E \approx 2\pi r \alpha Y_H \Phi$, where r is the distance from the excitation region. At equilibrium, the heat flux to the cell wall due to sample excitation is equal to that through the cell wall. Equating the two heat fluxes, $-\kappa \nabla T = 2\pi r \alpha Y_H \Phi$ for this steady-state condition to occur. Since the cell wall is thin and in equilibrium, the temperature change required to satisfy this steady-state condition is $\Delta T \approx -2\pi r \alpha Y_H \Phi \Delta x/\kappa$. For the quartz cuvette, $\Delta T \approx 3.7 \times 10^{-6} \alpha Y_H \Phi$. Even in the extreme case of 1 W m^{-1} being absorbed and converted to heat in the sample, the temperature change across the cell wall would be very small, $\Delta T \approx 4 \times 10^{-6}$ K. Thus the temperature at the cell walls is for all practical purposes that of the surroundings, and the well-mixed surrounding approximation makes the cell wall appear to be an infinite heat sink.

Another concern is the fact that the thermal diffusion equation was solved for a source of infinite extent along the axis of excitation laser propagation. The infinite-extent source approximation is valid for a sample cell with thermally insulating entrance and exit windows (i.e., a zero thermal conductivity). On the other hand, thermal diffusion through the sample cell windows will result in heat loss that will invalidate this approximation. To understand the extent of the problem, the solution for one-dimensional diffusion with a pulsed heating source can be examined (Dovichi 1987). Impulse excitation of an optically thin sample will produce heat equal to $q_H(0, z) = \alpha H Y_H$, where H is the integrated irradiance. The differential equations used to model this case are

$$\frac{\partial \delta T_w(z, t)}{\partial t} - D_{T,w} \frac{\partial^2 \delta T_w(z, t)}{\partial z^2} = 0$$
$$\frac{\partial \delta T_s(z, t)}{\partial t} - D_{T,s} \frac{\partial^2 \delta T_s(z, t)}{\partial z^2} = \frac{\alpha H Y_H \delta(t)}{\rho_s C_{P,s}} \quad (109)$$

where the subscripts s refers to the fluid sample, w to the solid window or cell wall material, and where δT is the temperature change due to thermal diffusion along the z axis. The solution to this diffusion problem for thermally thick windows (e.g., of infinite extent) is given by Carslaw and Jaeger (1986):

$$\delta T_w(z, t) = \alpha H Y_H \frac{D_{T,s}^{1/2}}{\rho_s C_{P,s} D_{T,s}^{1/2} + \rho_w C_{P,w} D_{T,w}^{1/2}} \operatorname{erfc}\left[\frac{z}{2(D_{T,w}t)^{1/2}}\right]$$

$$\delta T_s(z, t) = \alpha H Y_H \frac{D_{T,s}^{1/2}}{\rho_s C_{P,s} D_{T,s}^{1/2} + \rho_w C_{P,w} D_{T,w}^{1/2}} \left\{1 + \frac{\rho_w C_{P,w} D_{T,w}^{1/2}}{\rho_s C_{P,s} D_{T,s}^{1/2}} \operatorname{erf}\left[\frac{z}{2(D_{T,s}t)^{1/2}}\right]\right\} \tag{110}$$

where erf and erfc are the standard error functions. At zero time the temperature change in the window is zero and that of the sample is what would be expected if the window did not conduct heat. Subsequently, the zero-time, pulsed laser–excited temperature change will be exact. Note that the temperature change in the cell window is zero for a cell or window thermal diffusion coefficient of zero. When the window's thermal diffusion coefficient is zero, thermal diffusion does not occur and the temperature in the sample is stationary at $\alpha H Y_H / \rho_s C_{P,s}$.

For finite window conductivities, heat is transferred into the windows as time progresses. For continuous excitation, and assuming boundary condition such that the temperature change at the cell window is zero, thermal diffusion through the windows will effectively short the optical pathlength through the sample cell collinear excitation-probe photothermal spectroscopy experimental arrangements. The depth into which this shortening occurs can be estimated from the thermal diffusion length, $l \approx (D_T t)^{1/2}$. Counteracting this heat depletion is the absorption of heat from the excitation laser.

5.5.4. Accounting for Finite Cell Radius

The effect of using a sample cell with a finite sample chamber radius can be put on a quantitative basis. A solution to the thermal diffusion equation can be obtained for a cylindrically symmetric cell using a Bessel's series expansion of the temperature change similar to that used to obtain the propagating mode solution in Chapter 3. In this case the appropriate boundary condition is that the temperature change vanishes at the cell wail. Davis and Petuchowski (1981) have discussed the solutions to this boundary value problem for a number of different excitation beam geometries and time dependencies. We will be concerned only with the results for a Gaussian beam profile. The solution to the thermal diffusion equation can thus be cast in the form

$$\delta T(r, t) = \alpha Y_H \sum_{i=1}^{\infty} c_i(t) J_0(v_i \mathbf{r}) \tag{111}$$

where the v_i are $x_i b^{-1}$, the x_i are the roots of the zero-order Bessel function, $J_0(x) = 0$, and b is the sample cell radius. Expanding the temperature change about $J_0(v_i \mathbf{r})$ ensures that it is zero at the cell walls. We can also expand the laser irradiance in terms of equivalent Bessel functions:

$$H(r, t) = \frac{2Q}{\pi w^2} e^{-2r^2/w^2} = \sum_{i=1}^{\infty} g_i J_0(v_i \mathbf{r}) \tag{112}$$

where the g_i are the expansion coefficients. Using the integral

$$\int_0^1 r J_0(x_i r) J_0(x_j r)\, dr = \delta_{i,j} \frac{J_1^2(x_i)}{2} \tag{113}$$

where $\delta_{i,j}$ is the delta function, the integrated irradiance series can be inverted. The g_i are found to be

$$g_i = \frac{Q}{\pi b^2 [J_1(x_i)]^2} \int_0^b r e^{-2r^2/w^2} J_0(v_i r)\, dr \tag{114}$$

If the sample cell radius is much larger than the Gaussian beam waist radius,

$$g_i \approx \frac{Q}{\pi b^2 [J_1(x_i)]^2} \int_0^\infty r e^{-2r^2/w^2} J_0(v_i r)\, dr = \frac{Q}{\pi b^2 [J_1(x_i)]^2} e^{-v_i^2 w^2/8} \tag{115}$$

The time- and space-dependent temperature change is found by solving the thermal diffusion equation for pulsed excitation. Using Laplace transform techniques, the temperature change for pulsed excitation is

$$\delta T_{\text{pulsed}}(r, t) = \frac{\alpha Y_H Q}{\rho C_P} \sum_{i=1}^{\infty} g_i J_0(v_i r) e^{-v_i^2 D_T t} \tag{116}$$

Clearly, the zero-time temperature change is $H(r, 0)/\rho C_P$ and is independent of sample cell radius. If the zero-time photothermal signal is used to estimate sample absorbance, time-dependent temperature change is immaterial. The heat will diffuse as time evolves and the temperature change at any time can be found using the expression above.

When heat diffuses to the cell wall, accelerated heat transfer will perturb the temperature change within the cell. The question now is how long this heat sink allows the temperature to be calculated using the infinite extent approximation. This can be gaged from the thermal diffusion distance, $r^2/4D_T$, where D_T is that of the sample solvent. For example, using CCl_4 as the solvent, the time required for heat to undergo radial diffusion to the nearest wall in a 1-cm-diameter sample cell is about 83 s. This compares favorably with the characteristic thermal diffusion time constant for a 500-μm beam where $t_c = 0.83$ s, and even better for a 50-μm beam, for which $t_c = 0.0083$ s. In terms of the characteristic time constant, the time over which the infinite extent solution will be valid is simply $(r^2/w^2)t_c$ (i.e., is independent of thermal diffusion coefficient of the solvent or gas).

A more severe problem can arise when continuous laser sources are used to excite the sample since the heat sink at the sample cell walls may limit the maximum temperature change that can occur in the cell center. The temperature change for continuous excitation, found by integrating the pulsed temperature change over time, and by replacing the pulsed laser energy, Q, by the continuous laser power, Φ, is

$$\delta T_{\text{cw}}(r, t) = \frac{\alpha Y_H \Phi}{\kappa} \sum_{i=1}^{\infty} \frac{g_i J_0(v_i r)}{v_i^2} (1 - e^{-v_i^2 D_T t}) \tag{117}$$

For continuous excitation the maximum temperature change can be limited by thermal conduction through the cell wall if the sample cell radius is too small. This effect is illustrated in Figure 5.11 shows time-dependent on-axis temperature changes calculated for several different cell radii. The excitation beam waist radius is 100 μm and the thermal diffusion coefficient is that for argon gas. For sample cell radii less than that of the excitation laser, the g_i converge to $[x_i^3 J_0(x_i)]^{-1}$ and are not a function of w. The sum over these terms is a constant. In this limit, the maximum on-axis temperature change is proportional to $E(0)b^2$, where $E(0)$ is the on-axis irradiance. Thus the maximum temperature change will be proportional to the squared cell radius.

Using a different approach wherein the absorbed power is balanced with the power loss through the sample cell wall, Davis and Petuchowski

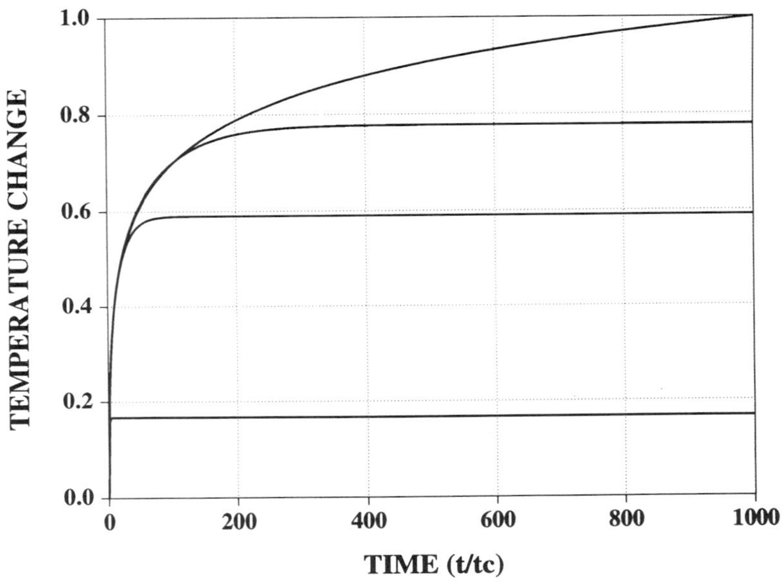

Figure 5.11. Time-dependent on-axis temperature changes for an continuous laser excitation beam waist radius of 100 μm. From top to bottom the traces are for sample cell radii of 0.5, 0.1, 0.05, and 0.01 cm. The thermal diffusion coefficient for Ar gas was used for the calculations.

(1981) were able to obtain an analytical expression for the equilibrium temperature change that would occur for a continuously irradiated sample. At equilibrium

$$-2\pi r\kappa \frac{d}{dr}\,\delta T = \Phi_0(1 - e^{-2r^2/w^2}) \tag{118}$$

which can be integrated to yield

$$\delta T_{\rm cw}(r) = \frac{\alpha \Phi_0 Y_H}{4\pi\kappa}\left[2\ln\frac{b}{r} + E_1\left(\frac{2b^2}{w^2}\right) - E_1\left(\frac{2r^2}{w^2}\right)\right] \tag{119}$$

An interesting effect, or lack of effect, predicted by the continuous laser-excited temperature change is that the photothermal lens focal length is independent of sample cell radius even when the cell radius is smaller than that of the excitation beam. The parallel excitation/probe beam inverse focal length is obtained from the second derivative of the temperature change with respect to radius, evaluated at $r = 0$:

$$\frac{1}{f_{\mathrm{cw}}(t)} = -\left(\frac{dn}{dT}\right)\left[\frac{d^2}{dr^2}\,\delta T_{\mathrm{cw}}(r, t)\right]_{r=0} = \left(\frac{dn}{dT}\right)\frac{\alpha Y_H \Phi}{\kappa}\sum_{i=1}^{\infty} g_i(1 - e^{-v_i^2 D_T t}) \quad (120)$$

The maximum lens element strength, found in the limit as time approaches infinity, is

$$\frac{1}{f_{\mathrm{cw}}(\infty)} = \left(\frac{dn}{dT}\right)\frac{\alpha Y_H \Phi}{\kappa}\sum_{i=1}^{\infty} g_i \quad (121)$$

Since the irradiance profile is given as the expansion over the Bessel functions, $E(r) = \Phi \Sigma g_i J_0(v_i r)$, the summation in the inverse focal length must be the on-axis irradiance, $E(0)$. Thus

$$\frac{1}{f_{\mathrm{cw}}(\infty)} = \left(\frac{dn}{dT}\right)\frac{\alpha Y_H E(0)}{\kappa} \quad (122)$$

This equation shows that the inverse photothermal lens focal length is independent of the sample cell radius. The maximum inverse focal length is equivalent to that obtained for the infinite cell radius approximation for $E(0) = \Phi/\pi w^2$. Several time-dependent inverse focal lengths are illustrated in Figure 5.12. The excitation beam waist is 100 μm. The inverse focal lengths for cell radii of 0.5, 0.1, and 0.05 cm are all represented by the lower trace. These traces are equivalent to that predicted by the infinite cell approximation. Apparently, the infinite cell radius approximation makes no difference in determining the magnitude of the continuous laser-excited photothermal lens element. This curious result has not been exploited in analytical applications of photothermal lens spectroscopy.

The line reaching a maximum at about $4t_c$ is for a sample cell radius of 0.01 cm. The inverse focal lengths for the smaller sample cell radii reach limiting values in less time than those for the larger cells. The response time of the sample system decreases due to the fact that the heat has less distance over which to propagate. The decreased response time of the smaller radii cells could be advantageous, in that flicker noise would be reduced in the chopped excitation signal.

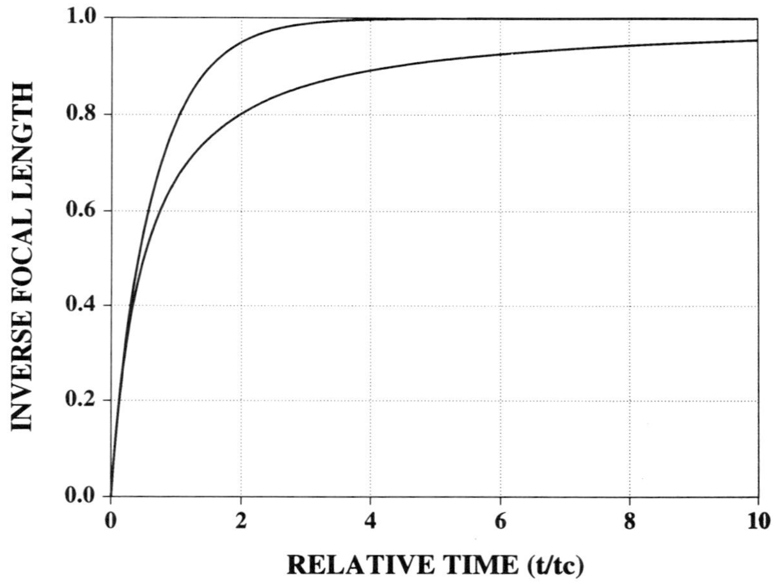

Figure 5.12. Time-dependent inverse focal length of the photothermal lens produced under the same conditions as Figure 5.11. The inverse focal length for 0.5⁻, 0.1⁻, and 0.05⁻ cm cell radii are all on the lower trace. In this time scale, these are the same as that calculated using the infinite-extent approximation. The faster response trace is for a sample cell radius of 0.01 cm. The traces are scaled to a maximum of 1. The same scale factor was used for all calculations.

REFERENCES

Abbate, G; Berninim, U.; Ragozzino, E.; and Somma, F. *J. Phys. D* **11** 1167 (1978).

Angell, C. A. In *Water: A Comprehensive Treatise*, Franks, F., ed., Vol. 7, Plenum, New York (1982).

Bailey, R. T.; Pugh, D.; and Cruickshank, F. R. *J. Chem. Soc. Faraday Trans. II* **76** 633 (1980).

Barker, J. R. and Rothem, T. *Chem. Phys.* **68** 331 (1982).

Buffett, C. E. and Morris, M. J. *Appl. Spectrosc.* **37** 455 (1983).

Carslaw, H. S. and Jaeger, J. C. *Conduction of Heat in Solids*, 2nd ed., Clarendon, Oxford (1986).

Cremers, D. A. and Keller, R. A. *Appl. Opt.* **21** 1654 (1982).

Cavis, C. C. and Petuchowski, S. J. *Appl. Opt.* **20** 2539 (1981).

Covichi, N. J. *CRC Crit. Rev. Anal. Chem.* **17** 357 (1987).

Eichler, H. J.; Günter, P.; and Pohl, D. W. *Laser-Induced Dynamic Gratings*, Springer-Verlag, New York (1986).

Fang, H. L. and Swofford, R. L. In *Ultrasensitive Laser Spectroscopy*, Kliger, D. S., ed., Academic Press, New York (1983).

Fournier, D. and Boccara, A. C. in *Photothermal Investigations in Solids and Fluids* Sell, J. A. Editor, Academic Press, Inc. New York (1989).

Franko, M. and Tran, C. D. *Anal. Chem.* **61** 1660 (1989).

Gautschi, W. and Cahil, W. F. In *Handbook of Mathematical Functions*, Abramowitz, M. and Stegun, I. A., eds., Dover, New York (1972).

Giglio, M. and Vendramini, A. *Appl. Phys. Lett.* **25** 555 (1974).

Gordon, J. P.; Leite, R. C. C.; Moore, R. S.; Porto, S. P. S.; and Whinnery, J. R. *J. Appl. Phys.* **36** 3 (1965).

Gradshteyn, I. S. and Ryzhik, I. M. *Table of Integrals, Series, and Products*, Academic Press, New York (1980).

Gupta, R. In *Photothermal Investigations in Solids and Fluids*, Sell, J. A., ed., Academic Press, New York (1989).

Handbook of Chemistry and Physics, 61st ed., Weast, R. C., Ed., CRC Press, Boca Raton, FL (1980).

Hardcastle, F. D. and Harris, J. M. *Appl. Spectrosc.* **40** 606 (1986).

Herschfelder, J. O.; Curtiss, C. F.; and Bird, R. B. *Molecular Theory of Gases and Liquids*, Wiley, New York (1954).

Hu, C. and Whinnery, J. R. *Appl. Opt.* **12** 72 (1973).

Ingle Jr., J. D. and Crouch, S. R. *Spectrochemical Analysis* Prentice Hall, Englewood Cliffs, NJ (1988).

Jackson, W. B.; Amer, N. M.; Boccara, A. C.; and Fournier, D. *Appl. Opt.* **20** 1333 (1981). *JANAF Thermochemical Tables*, 2nd ed., NSRDS-NBS 37, National Bureau of Standards, Washington, DC (1971).

Longaker, P. R. and Litvak, M. M. *J. Appl. Phys.* **40** 4033 (1969).

McGraw, D. J.; Michaelson, J. and Harris, J. M. *J. Phys. Chem.* **86** 2536 (1987).

Mori, K.; Imasaka, T.; and Ishibashi, N. *Anal. Chem.* **54** 2034 (1982).

Mori, K.; Imasaka, T.; and Ishibashi, N. *Anal. Chem.* **55** 1075 (1983).

Patterson, M. S.; Chance, B.; and Wilson, B. C. *Appl. Opt.* **28** 2331 (1989).

Power, J. F. and Langford, C. H. *Anal. Chem.* **60** 842 (1988).

Siebert, D. R.; Grabiner, F. R.; and Flynn, G. W. *J. Chem. Phys.* **60** 1564 (1974).

Siegman, A. E. *J. Opt. Soc. Am.* **67** 545 (1977).

Simo Alfonso, E. F.; Rius Revert, M. A.; Garcia Alvarez-Coque, M. C.; and Ramis Ramos, G. *Appl. Spectrosc.* **44** 1501 (1990).

Spear, J. D.; Russo, R. E.; and Silva, R. J. *Appl. Opt.* **29** 4225 (1990).

Tarazima, M. and Hirota, N. *Phys. Chem.* **96** 7147 (1992).

Twarowski, A. J. and Kliger, D. S. *Chem. Phys.* **20** 253 (1977).

Whinnery, J. R. *Acc. Chem. Res.* **7** 225 (1974).

Whinnery, J. R.; Miller, D. T.; and Dabby, F. *IEEE J. Quant. Electron.* **QE-3** 382 (1967).

Yasa, Z. A.; Jackson, W. B.; and Amer, N. M. *Appl. Opt.* **21** 21 (1982).

Zhu, Z. R. and Harris, J. M. *J. Phys. Chem.* **93** 75 (1989).

Zhu, Z. R. and Harris, J. M. *J. Opt. Soc. Am. B* **7** 796 (1990).

Zhu, Z. R.; McGraw, D. J.; and Harris, J. M. *Anal. Chem.* **64** 710A (1992).

CHAPTER

6

PHOTOTHERMAL SPECTROSCOPY IN HOMOGENEOUS SAMPLES

In this chapter, simple mathematical descriptions for signals obtained in the main types of photothermal spectroscopy used for homogeneous samples are shown. There has been steady development in the theory since the first descriptions of the photothermal lens by Gordon et al. (1965). Longaker and Litvak (1969), Hu and Whinnery (1973), Siebert et al. (1974), Twarowski and Kliger (1977a, b), Bailey et al. (1980), Davis and Petuchowski (1981), and Barker and Rothem (1987) have forwarded understanding of the physical principles underlying the photothermal spectroscopy signal-generation process in homogeneous samples. These works are sufficiently separated in time and the approaches used are different, owing to subtle differences in the area in which these researchers worked. In this chapter we consolidate this work by developing a consistent mathematical basis for signal description. These consolidated results are then used to make predictions to extend the understanding of photothermal spectroscopy. It is hoped that the consolidated theory can serve as a platform from which further studies into improving photothermal spectroscopy can be launched.

6.1. PHOTOTHERMAL INTERFEROMETRY

The most direct manner in which to monitor the refractive index change that occurs upon sample heating is by using interferometry. Almost any of the several types of interferometers could be used to measure the relative phase shift of light passing through the perturbed sample. The most widely used interferometers for photothermal interferometry are the Mach–Zehnder (e.g., Longaker and Litvak, 1969) and the Jamin (e.g., Stone, 1972, 1973) interferometers, although the Fabry–Perot has also been used (Campillo et al. 1982). The Jamin interferometer is essentially the same as the Mach–Zehnder but uses fewer optical elements and is thus more stable.

Earlier interferometer designs utilized static alignment of the optical elements. Photothermal signal detection in these devices relied on monitoring a phase shift relative to some average phase angle. Since the average

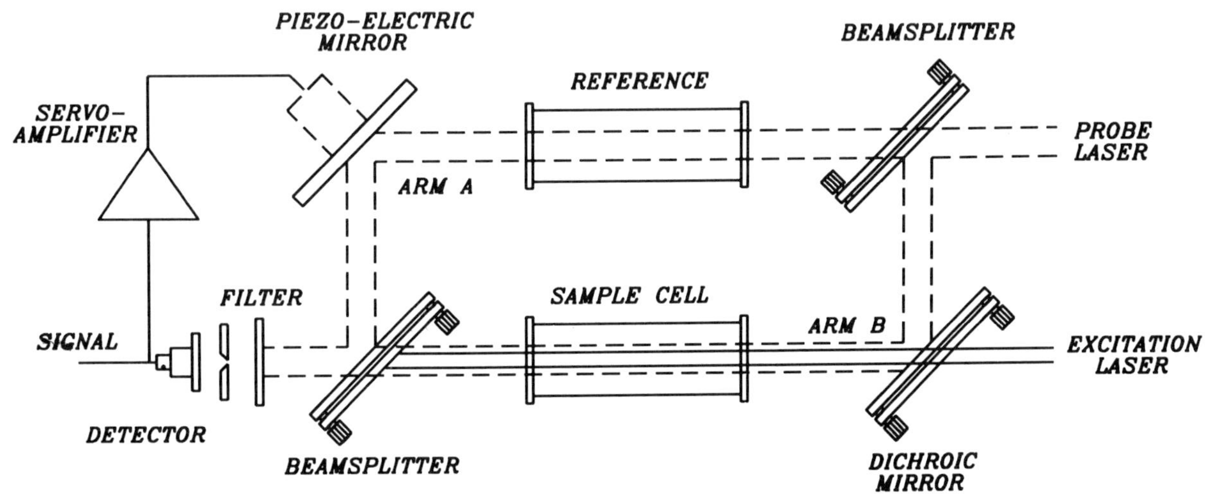

Figure 6.1. Photothermal interferometry based on the Mach–Zender interferometer. The excitation laser beam enters from the lower right and passes through the dichroic beamsplitter into the sample cell. The single-frequency probe laser beam enters from the upper right and is split into two equal components. One component passess through reference arm A, while the other passes through the sample (arm B). The two probe beams are combined at the detector. The filter eliminates excitation beam radiation. The signal is amplified and feed to piezoelectric mirror. The latter is used to control the relative phase of the probe beam passing through arm B.

phase angle would change with temperature, pressure, and so on, quantitative information was difficult to obtain with these devices. Later, interferometers were developed with components that had positions driven by piezoelectric transducers (Davis 1980, Davis and Petuchowski 1981). The piezoelectric transducer–driven elements allow dynamic compensation for probe laser frequency and environmental (e.g., temperature, air density, etc.) drifts. The basic dynamically stabilized Mach–Zehnder interferometer is illustrated in Figure 6.1. Davis (1980) coined the term *Phase fluctuation optical heterodyne* (PFLOH) *spectroscopy* for this photothermal interferometry technique, although it is actually based on a homodyne detection technique. Light from the single-frequency helium–neon probe laser is split into two beams. One beam makes up the reference arm of the interferometer, arm A, the other the sample arm, arm B. Beams from the two arms are recombined at the second beamsplitter. The optical power transmitted at this beamsplitter depends on the relative phase of the beams in arms A and B. The electric fields just prior to the combining beamsplitter are

$$\begin{aligned} \mathbf{E}_A(t) &= \mathbf{E}_{0,A} e^{i(\omega t + \phi_A)}, \\ \mathbf{E}_B(t) &= \mathbf{E}_{0,B} e^{i[\omega t + \phi_B + \delta\phi(t)]} \end{aligned} \tag{1}$$

where the phase shift, $\delta\phi$(t), is that due to the photothermal effect. Past the combining beamsplitter, the electric field is

$$\mathbf{E}_{A+B}(t) = \mathbf{E}_A(t) + \mathbf{E}_B(t) = \mathbf{E}_{0,A} e^{i(\omega t + \phi_A)} + \mathbf{E}_{0,B} e^{i[\omega t + \phi_B + \delta\phi(t)]} \tag{2}$$

and the detected power is $\Phi(t) \propto \mathbf{E}_{A+B}(t)^* \mathbf{E}_{A+B}(t)$, or

$$\Phi(t) \propto \mathbf{E}_{0,A}^2 + \mathbf{E}_{0,B}^2 + 2AB[\cos(\phi_B - \phi_A)\cos\delta\phi(t) - \sin(\phi_B - \phi_A)\sin\delta\phi(t)] \tag{3}$$

For maximum sensitivity, the static phase difference between the two arms is maintained at $(2n + 1)\pi/2$. This condition corresponds to being halfway between minimum and maximum of the sinusoidal fringe. The optical power at the detector for thermally induced phase shifts is

$$\Phi(t) \propto \mathbf{E}_{0,A}^2 + \mathbf{E}_{0,B}^2 - 2\mathbf{E}_{0,A}\mathbf{E}_{0,B}\delta\phi(t) \tag{4}$$

in this case. Because of environmental noise, it is difficult to maintain the pathlength, and therefore phase, of the two arms such as to obtain the electric field cancellation. The piezo-driven mirror is used for dynamic adjustment of the pathlength of arm *A*. This is accomplished using a servoamplifier between the detector and the piezoelectric transducer. Cremers and Keller (1982) discuss the operation of the active feedback stabilization circuit. More recently, a computer is used to monitor the photodiode current and control the piezoelectric transducer–driven mirror (Mazzoni and Davis 1991). Because of the limited translation distance of the piezoelectric transducers, analog servoamplifiers can drive the piezo-driven mirror to the end of its translation distance. When this happens the servo has to be reset so that the mirror can be positioned near the minimum of its range. With computer control, the relative mirror position can be monitored. Software is used to translate the mirror from one side of the range to the other when the limits are approached. This can be accomplished rather quickly and long-term data acquisition can be performed. The computer-controlled optical pathlength feedback apparatus, based on a modified Jamin interferometer, is illustrated in Figure 6.2.

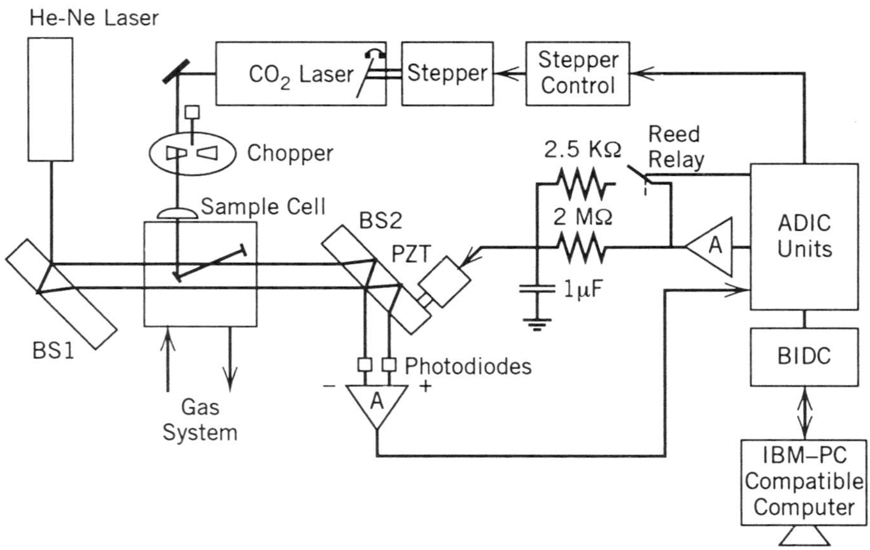

Figure 6.2. Trace gas detection system of Mazzoni and Davis (1991) based on a modified Jamin interferometer. The infrared CO_2 laser is used to excite the sample and a single-frequency helium–neon laser is used as the probe. The relative optical pathlength is controlled by a piezoelectric transducer. The computer monitors the difference photodiode signal and controls the transducer, thereby eliminating problems with long-term drift. (Reprinted with permission from Mazzoni and Davis 1991. Copyright 1991, Optical Society of America.)

There are two modes of operation of the dynamic pathlength adjustment. In the one mode, only the low-frequency or slow drift of the interferometer alignment is adjusted. In this case the higher-frequency photothermal signal is present in the detector signal, and it is ac coupled to the detection electronics. In the second mode of operation, the pathlength of arm A is adjusted for high- and low-frequency changes. In this case the pathlength is adjusted so as always be an odd multiple of $\pi/2$. The photothermal signal is present along with slower components at the voltage used to adjust the piezo-driven mirror. The latter mode of operation can be used only for chopped excitation, and a lock-in amplifier is used to recover the photothermal phase shift from the complex signal. With the low-frequency path adjustment mode, the interferometer can be used with both pulsed and chopped excitation lasers.

In addition to the computer-controlled feedback, the modified Jamin interferometer shown in Figure 6.3 incorporates several important features that compensate for problems with earlier designs. First is the elimination of an excitation laser path which passes through one of the optical elements of the interferometer. In previous designs, germanium mirrors were used to reflect the visible probe laser light while allowing transmission of the infrared excitation laser. Heating of this mirror caused the optical pathlength of arm B to change, thus limiting the ultimate absorbance limit of detection. In fact, Stone (1972) pointed out that even Pyrex sample cuvettes were inadequate because of the relatively high absorbance. Stone recommended that fused silica be used. Cremers and Keller (1982) point out that even silica cuvettes can have an absorbance high enough to cover up analyte absorbance values below 10^{-5} cm^{-1}. A second feature is the balanced mixed detection. The balanced mixer utilizes separate photodiodes to monitor the two output branches of the modified Jamin interferometer. The quadrature condition for maximum sensitivity to the photothermal phase shift is met when the signals from the two photodiodes are equal. The operational-amplifier circuit gives high rejection of common-mode amplitude noise, a feature that was lacking in earlier designs. With this design, Mazzoni and Davis (1991) were able to measure absorbance to within a factor of 3 of the theoretical shot noise limit for this type of interferometer, even though they used inferior optics.

The theoretical shot noise limit for homodyne detection is well known (Yariv 1985). Mazzoni and Davis show that the theoretical minimum phase shift gives rise to a minimum detectable absorption coefficient of

$$\alpha_{\min} = \frac{2w^2 \rho C_P \lambda T_0 f}{(n-1) l \Phi_e} \sqrt{\frac{2h\nu \, \Delta f}{\eta \Phi_0}} \tag{5}$$

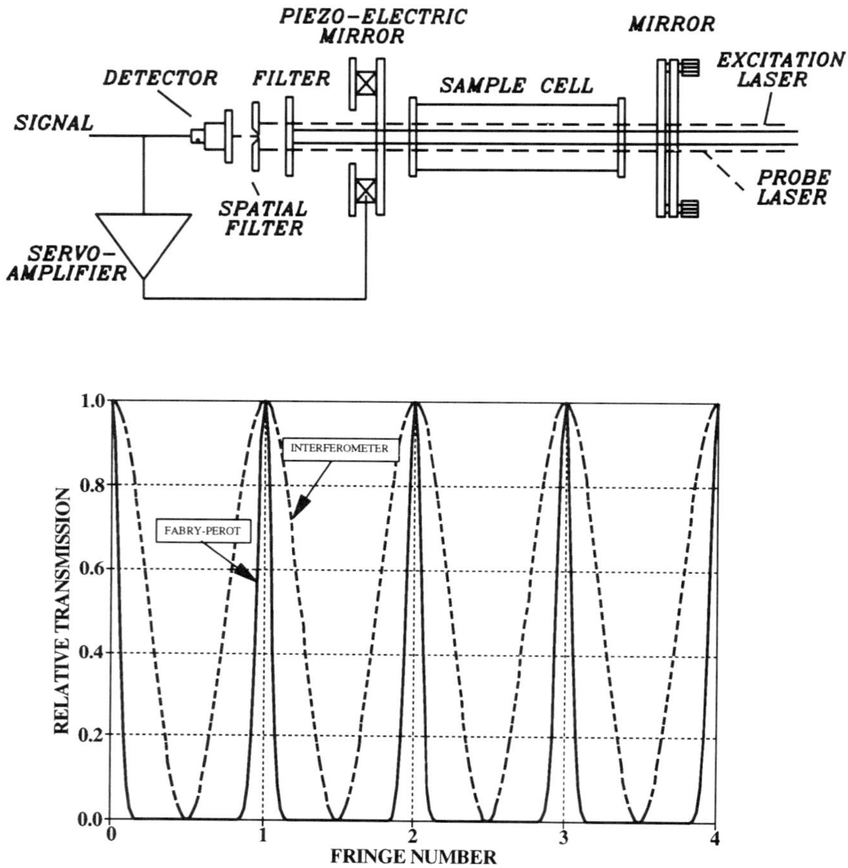

Figure 6.3. Fabry–Perot-based photothermal interferometer and the resulting relative probe laser transmission. The change in transmission with respect to a change in optical pathlength is enhanced because of the multipass character of this interferometer.

where λ and v are the wavelength and frequency of the probe laser; n, ρC_P, and T_0 the refractive index, heat capacity, and temperature of the gas sample; and Φ_e and Φ_0 the excitation and probe laser powers. The excitation laser is chopped at frequency f, and the measurement bandwidth is Δf. Using a 1-mW helium–neon probe and a line tunable 2.65-MW m^{-2} irradiance continuous CO_2 excitation laser chopped at 85.2 Hz, the theoretical minimum absorption coefficient is $\alpha_{min} \approx 1.6 \times 10^{-9}$ cm^{-1} for nitrogen buffer gas. In the experimental part of this paper they show that hydrazine compounds could be detected at ppb levels.

Photothermal interferometric detection techniques are extremely sensitive and since the early studies by Stone (1972) which showed that the absorption of pure solvents could be measured, they have been applied mostly to gas-phase analysis. However, problems associated with the solvent absorption can be overcome by placing a sample cell containing solvent in arm A of the interferometer. Both sample and reference solvent are excited. The interferometer phase shift will be the difference in phase between the heated sample and reference. Cremers and Keller (1982) describe such a system. They used a Jamin interferometer with active stabilization of the pathlength in one of the arms using a galvanometer-driven compensator plate. Fringe stabilization to 1 part in 400 allowed analyte absorbances as low as about $5 \times 10^{-6}\,\text{cm}^{-1}$ to be measured in CCl_4 solvent, which had an absorption coefficient of about 10^{-5}. $CoSO_4$ was used for the analyte and an argon-ion laser operating at 515 nm was used as the excitation source. Two sources of error limited fringe stabilization and thus the analyte absorbance limit of detection. The first was due to the excitation laser heating of the optics within the interferometer paths. Of course, this effect can be minimized with a different design [e.g., that of Mazzoni and Davis (1991)]. The second source of error was thought to be due to convection heat transfer.

Campillo et al. (1982), used a servo feedback–stabilized Fabry–Perot interferometer to enhance the photothermal fringe shift measurement. A schematic of the interferometer is shown in Figure 6.3. Fabry–Perot interferometers use a cavity instead of a single-pass interferometer design. The interference fringe structure is subsequently not $1 + \cos \delta\phi$, but rather, $\cos^{\langle n \rangle} \delta\phi$, where $\langle n \rangle$ is the average number of round trips a photon makes in the cavity. The transmission through the Fabry–Perot interferometer is given by $(1 + \delta \sin^2\phi)^{-1}$, where $\delta = (F/\pi)^2$, F being the finesse of the device. The finesse is related to the reflectivities of the partially transmitting mirrors, the flatness of the mirrors, and other factors affecting losses in the cavity (Klein 1970). Finesse values of from 10 to 50 are not uncommon using high-quality 50% transmission mirrors. The larger change in transmission with respect to phase shift allowed for more sensitive measurements to be taken. However, the higher fringe shift sensitivity did not result in lower-absorbance detection limits. The device was so sensitive that acoustic and environmental "noise" were seen to limit the minimum observable photothermal phase shift. Photothermal interferometry has been used to measure fluorescence quantum yields of dilute dye solution (Lee et al. 1985), to measure the infrared optical absorbance of gas-phase aerosols (Lin and Campillo 1985, Fluckiger et al. 1985), to monitor trace levels (ppb) of hydrazines (Mazzoni and Davis 1991), and as a gas chromatography detector (Lin et al. 1981).

6.2. PHOTOTHERMAL DEFLECTION

Photothermal deflection spectroscopy uses two optical sources: one usually a TEM_{00} laser, to excite the homogeneous sample, the other used to probe the resulting index perturbation. Laser sources are used almost exclusively for the probe source. Photothermal deflection signals are modeled by calculating the theoretical deflection angle of a ray as it passes through the homogeneous sample. There are several experimental means to obtain deflection angle information from the position of the probe laser spot after being deflected by the excited sample (we discuss these experimental means later). The photothermal deflection angle can be calculated for either pulsed or chopped excitation, and for a number of different geometries. In the sections that follows, the models used to calculate deflection angles for rays and laser beams are given.

6.2.1. Deflection Angle for Pulsed Laser Excitation

6.2.1.1. Collinear Probe Geometry

The simplest case of photothermal deflection example is that for pulsed Gaussian laser sample excitation with a probe ray propagating parallel to the excitation laser, but offset along the x and y coordinates by distances of x_0 and y_0. Using the temperature change in equation 12 of Chapter 5, the deflection angle in air is found to be

$$\theta_{x,\text{pulsed}}(x_0, y_0, t) = -\left(\frac{dn}{dT}\right)\frac{8x_0\alpha l Y_H Q}{\pi w^4 \rho C_P(1 + 2t/t_c)^2}\, e^{-2(x_0^2 + y_0^2)/w^2(1 + 2t/t_c)} \quad (6)$$

The complicated deflection angle dependence on the x- and y-axis offsets and on time needs to be examined. The relative deflection angle in the x-axis direction for Gaussian profile temperature change is illustrated in Figure 6.4. This plot was calculated for the initial, (e.g., $t = 0$) deflection angle. The x- and y-axis offsets range from $-2w$ to $2w$. For any value of y_0, the absolute value of the deflection angle is zero at the x-axis origin, positioned in the middle of the figure, and increases with x_0 offset to a maximum. Beyond the maximum, the absolute deflection angle decreases to zero.

The maximum (and/or minimum) deflection angle can be found by equating the derivative of the deflection angle with respect to x_0 to zero and solving for offset. This procedure results in

$$x_{0,\max} = \pm\frac{w}{2}\left(1 + \frac{2t}{t_c}\right)^{1/2} \quad (7)$$

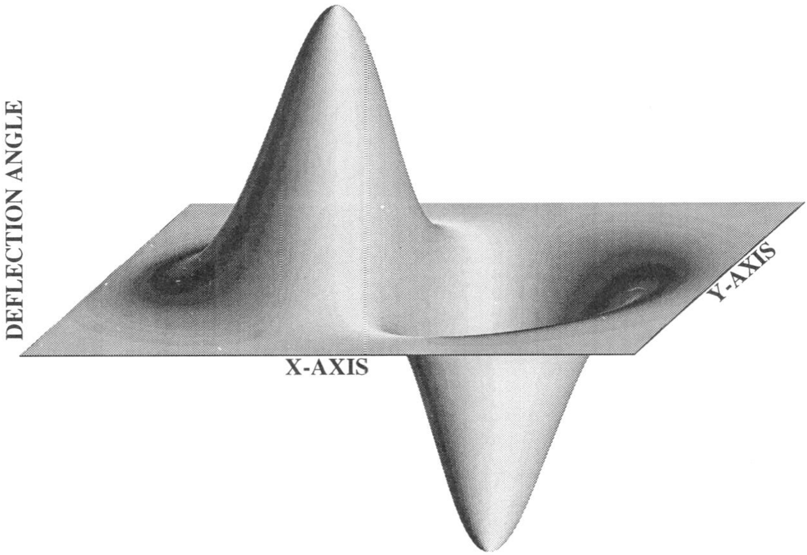

Figure 6.4. X-axis deflection angel as a function of probe ray displacement in the x and y dimensions.

Of course, a single offset must be chosen for the experiment. The maximum time-dependent deflection angle will be that which occurs for the greatest refractive index gradient. Because of the nature of thermal diffusion, tending to equalize spatial temperature variations, the maximum gradient will always occur just after the acoustic relaxation, or in the approximation used here, at $t = 0$. Thus the maximum absolute deflection angle occurs at an offset of $x_0 = w/2$. Figure 6.5 shows the deflection angle as a function of time and x-axis offset. The series of time-dependent deflection angles each show slightly different time dependencies. At the optimum x-axis offset, the deflection angle is a maximum at zero time and decreases monotonically with time. X-axis offsets different than that for the optimum zero-time deflection angle exhibit maximum angles which are less than that for the optimum. At offsets greater than that for the optimum deflection magnitude, the deflection angle may not be monotonic in time. In particular, at x-axis offsets greater than w, the signal initially increases with time, then decreases as thermal diffusion broadens the initial Gaussian temperature change distribution. The experimental implication of the different time dependencies is that time-resolved photothermal deflection spectroscopy may not be an accurate indicator of relaxation kinetics unless great care is taken to ensure that the probe offset is controlled.

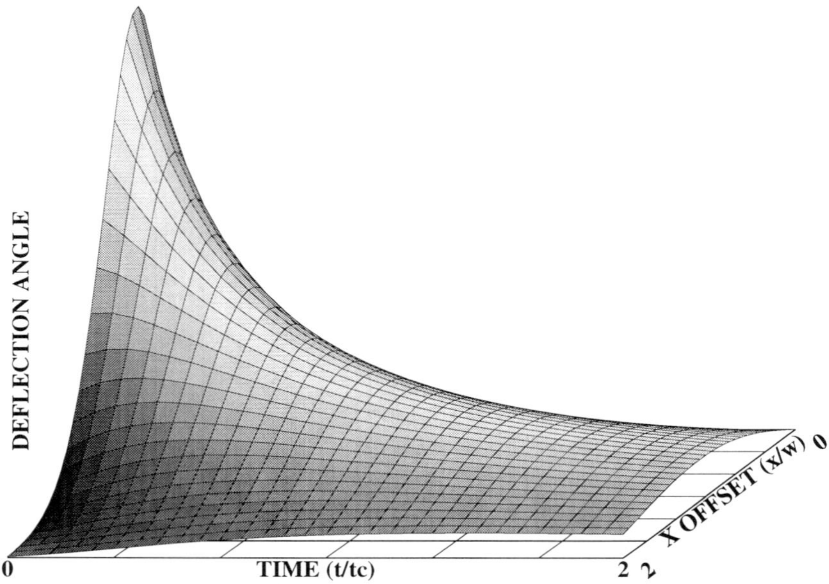

Figure 6.5. Time-dependent probe ray x-axis deflection signal as a function of x-axis offset for pulsed laser excitation. Each x-axis offset produces a signal with different time dependence, making it difficult to measure thermal diffusivities using this method.

The ray deflection angle for the optimum conditions probe ray offset is

$$\theta_{x,\text{pulsed}}(x_0 = w/2,\, y_0 = 0, t) = -\left(\frac{dn}{dT}\right)\frac{4\alpha l Y_H Q}{\pi w^3 \rho C_P (1 + 2t/t_c)^2}\, e^{-1/2(1+2t/t_c)} \tag{8}$$

This function has a maximum value at zero time

$$\theta_{x,\text{pulsed}}(x_0 = w/2,\, y_0 = 0, t = 0) = -\left(\frac{dn}{dT}\right)\frac{4\alpha l Y_H Q}{\pi w^3 \rho C_P}\, e^{-1/2} \tag{9}$$

With the approximations used to obtain this result (e.g., optically thin sample), the calculated ray deflection angle is directly proportional to α, the absorption coefficient; to l, the sample pathlength; and to $Y_H Q$, the amount of energy converted to heat. The deflection angle is also inversely proportional to w^3. This high-order dependence on the excitation laser beam waist radius indicates that the optimum photothermal deflection signals, which are proportional to the deflection angle, would be obtained using tightly focused excitation lasers. The sample's thermooptical coefficient and heat capacity will also affect the ray deflection angle magnitude. These parameters are controlled by the solvent used. The solvent may be chosen to optimize the deflection angle. However, solvent choice based solely on the thermooptic coefficient and heat capacity do not necessarily give the best detection limits, since interfering absorption can take place.

6.2.1.2. Crossed-Beam Probe Geometry

The ray deflection angle can be calculated for other excitation-probe geometries as well. For example, Fournier et al. (1980), Sell (1984), Tam et al. (1985), and Rose et al. (1986) describe experimental apparatuses in which the probe laser beam crosses the path of the excitation laser at the center of the sample cell. To calculate the x-axis deflection angle, integration over the path in equation 52 of Chapter 5 can be accomplished by substituting $y_0 = s \sin \gamma$ and a path element of $2 \cos \gamma \, ds$, where γ is the angle between the excitation beam path and probe ray. The deflection angle for a probe ray crossed at an angle γ can be shown to be

$$\theta_{x,\text{pulsed}}(x_0, t) = -\left(\frac{dn}{dT}\right) \frac{2^{5/2} x_0 \alpha Y_H Q}{\pi^{1/2} w^3 \rho C_P (1 + 2t/t_c)^{3/2} \tan \gamma} \times \operatorname{erf}\left(\frac{l \tan \gamma}{2^{1/2} w (1 + 2t/t_c)^{1/2}}\right) e^{-2x_0^2/w^2(1 + 2t/t_c)} \tag{10}$$

Path integration for a 90° excitation-probe crossing is easily accomplished by integrating over the path element dy. This results in the crossed-beam photothermal deflection ray deflection (Weimer and Dovichi 1985a, Rose et al. 1986)

$$\theta_{x,\text{pulsed}}(x_0, t) = -\left(\frac{dn}{dT}\right) \frac{2^{5/2} x_0 \alpha Y_H Q}{\pi^{1/2} w^3 \rho C_P (1 + 2t/t_c)^{3/2}} e^{-2x_0^2/w^2(1 + 2t/t_c)} \tag{11}$$

Any angle other than $\pm 90°$ will increase the magnitude of the deflection angle. The magnitude of the zero-time ray deflection angle using the optimum x_0 is

$$\theta_{x,\text{pulsed}}(x_0 = w/2, t = 0) = -\left(\frac{dn}{dT}\right) \frac{2^{3/2} \alpha Y_H Q}{\pi^{1/2} w^2 \rho C_P} e^{-1/2} \tag{12}$$

The 90° crossed deflection angle can be very much less than that of the parallel configuration. Compared to the parallel excitation-probe arrangement, the magnitude of the 90° photothermal deflection angle can be gaged by the ratio of optimum angles

$$\frac{\theta_{\parallel}}{\theta_{\perp}} = \frac{2^{1/2}}{\pi^{1/2}} \frac{l}{w} \tag{13}$$

The numerical factor is very nearly 1. The main difference is thus in the probe ray pathlength dependencies. The relative response of the parallel probe ray deflection through a 1-cm sample cell with a modest minimum beam waist radius of 100 SYMBOL 109 f 'Symbol'm is 80 times greater than that for the perpendicular probe ray.

6.2.2. Deflection Angle for Continuous and Chopped Laser Excitation

6.2.2.1. Continuous Excitation with Parallel-Probe Geometry

The ray deflection angle for continuous laser excitation can be obtained from equation 27 of Chapter 5. For a probe ray parallel to the axis of excitation, the deflection angle along the x axis is

$$\begin{aligned}\theta_{x,cw}(x_0, y_0, t) &= -\left(\frac{dn}{dT}\right)\frac{x_0\Phi_0\alpha l Y_H}{2\pi\kappa(x_0^2 + y_0^2)} \\ &\quad \times \left\{\exp\left[\frac{-2(x_0^2 + y_0^2)}{w^2(1 + 2t_1/t_c + 2\delta t/t_c)}\right] - \exp\left[\frac{-2(x_0^2 + y_0^2)}{w^2(1 + 2\delta t/t_c)}\right]\right\} \\ \delta t &= \begin{cases} t - t_1 & \text{for } t \geqslant t_1 \\ 0 & \text{for } t < t_1 \end{cases}\end{aligned} \tag{14}$$

Where use was made of the differential relationship $dE_1(z)/dz = -e^{-z}/z$. The maximum deflection angle occurs just prior to turning the excitation beam off. At this time, $\delta t = t - t_1$ and $t = t_1$. For a probe ray with $y_0 = 0$, the deflection angle is

$$\theta_{x,cw}(x_0, t) = -\left(\frac{dn}{dT}\right)\frac{\Phi_0\alpha l Y_H}{2\pi\kappa x_0}\left\{\exp\left[\frac{-2x_0^2}{w^2(1 + 2t/t_c)}\right] - \exp\left(\frac{-2x_0^2}{w^2}\right)\right\} \tag{15}$$

The probe ray offset, x_0, for the optimum deflection angle cannot be obtained in an analytical expression. Nonetheless, there exists an optimum probe ray offset for a given heating time, t_1. A plot of the optimum probe

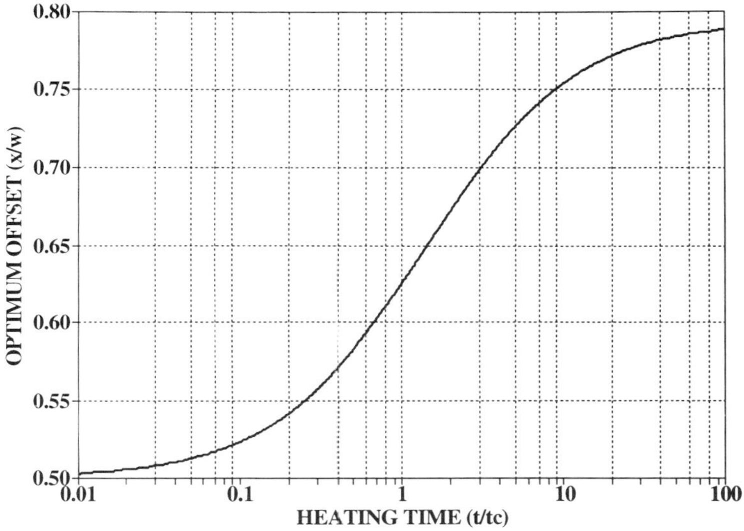

Figure 6.6. Optimum probe ray offset as a function of continuous laser excitation time. For short heating times the optimum offset is that for the pulsed laser excitation case. For longer excitation times relative to t_c, the optimum offset increases, owing to the increased thermal perturbation width.

ray offset as a function of the relative heating time, found by numerical evaluation, is shown in Figure 6.6. The short-term limiting excitation duration converges to a value of $w/2$. This is the value that was found for pulsed laser excitation. For excitation times very much greater than t_c, the optimum probe ray offset converges to a value of about $0.8w$. Relative maximum ray deflection angles at the optimum x-axis offsets are shown in Figure 6.7. The ray deflection angle, found in the limit as $t \to \infty$ of the deflection angle for heating, is

$$\lim_{t \to \infty} \theta_{x,cw}(x_0, t) = -\left(\frac{dn}{dT}\right)\frac{\Phi_0 \alpha l Y_H}{2\pi\kappa x_0}\left[1 - \exp\left(\frac{-2x_0^2}{w^2}\right)\right] \tag{16}$$

The maximum deflection angle at the optimum x-axis offset ($0.7926w$) is

$$\theta_{x,cw}(x_0 = 0.7926w,\ t = \infty) = -\ 0.9023\left(\frac{dn}{dT}\right)\frac{\Phi_0 \alpha l Y_H}{2\pi\kappa w} \tag{17}$$

The maximum relative ray deflection angle illustrated in Figure 6.7 has a value of 0.9023 (e.g., the data are scaled according to equation 17). The

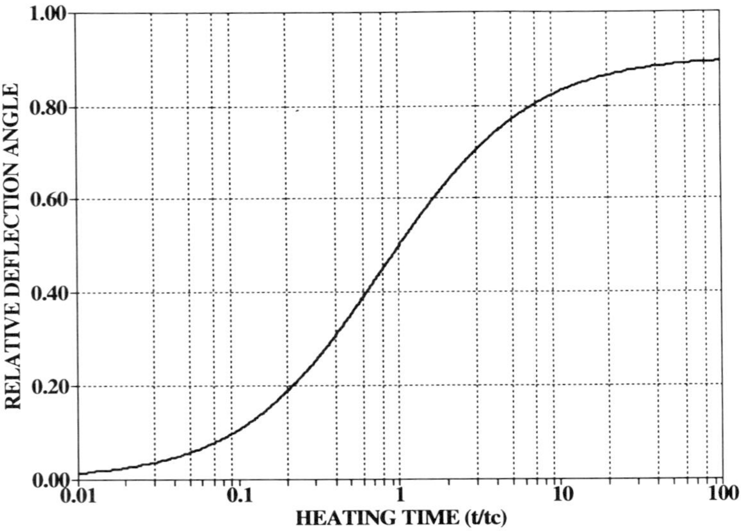

Figure 6.7. Relative photothermal deflection angle as a function of the continuous laser heating time relative to t_c. The probe ray offset is optimum for each heating time.

maximum probe ray deflection angle reaches a value that is proportional to the thermooptic coefficient, the fraction of power transferred to the sample ($\alpha l Y_H$), the excitation laser power (ϕ_0), and inversely proportional to the thermal conductivity (κ) and the excitation laser beam waist radius.

A plot of the theoretical time-dependent deflection angle at the optimum probe ray offset is shown in Figure 6.8. The time-dependent on-axis temperature changes are shown for reference. Compared to the temperature changes, the photothermal deflection angles have the temporal characteristics of a faster response time. The ray deflection angle rises and falls in times that are apparently much shorter than that of the corresponding temperature change. The reason for the apparent faster response is that the deflection angle is sensitive to the spatial derivative of the temperature change. While longer excitation results in a greater temperature change, the spatial gradient of the temperature attains a limiting value. This is most apparent in part (b) of the figure, where excitation lasts $t = 25t_c$.

6.2.2.2. *Continuous Excitation with Crossed-Probe Geometry*

The continuous laser–excited crossed-beam photothermal deflection angle can be calculated by integrating the position-dependent deflection angle along the path of the probe ray. For right-angle crossing of the excitation

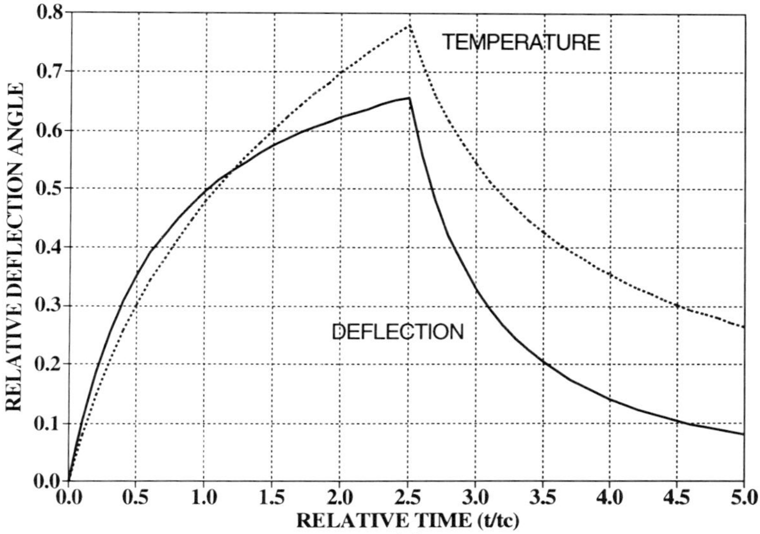

(a)

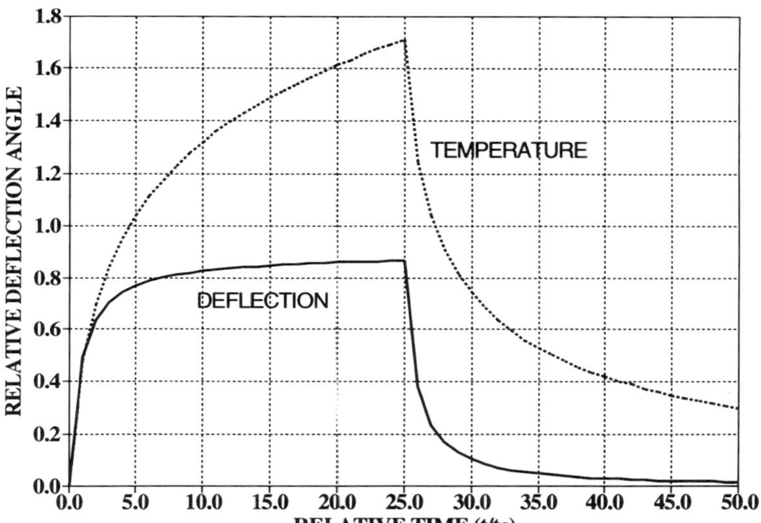

(b)

Figure 6.8. Time-dependent photothermal deflection angle and on-axis temperature change for chopped excitation: (*A*) relatively high chopper frequency; (*B*) low chopper frequency. Notice that the photothermal deflection angle has nearly reached the maximum by about $20t_c$ while the temperture continuous to increase.

and probe, path integration along the y-axis for an excitation laser propagating along the z axis results in

$$\theta_{x,cw}(x_0, t) = -\left(\frac{dn}{dT}\right)\frac{\Phi_0 \alpha Y_H}{2\kappa}\left[\operatorname{erf}\left(\frac{\sqrt{2}x_0}{w\sqrt{1 + 2\delta t/t_c}}\right) - \operatorname{erf}\left(\frac{\sqrt{2}x_0}{w\sqrt{1+2t_1/t_c+2\delta t/t_c}}\right)\right] \quad \delta t = \begin{cases} t - t_1 & \text{for } t \geqslant t_1 \\ 0 & \text{for } t < t_1 \end{cases} \tag{18}$$

The maximum signal at the end of the heating cycle is

$$\theta_{x,cw}(x_0, t) = -\left(\frac{dn}{dT}\right)\frac{\Phi_0 \alpha Y_H}{2\kappa}\left[\operatorname{erf}\left(\frac{\sqrt{2}x_0}{w}\right) - \operatorname{erf}\left(\frac{\sqrt{2}x_0}{w\sqrt{1 + 2t/t_c}}\right)\right] \tag{19}$$

The x-axis offset for maximum signal can be found in this case:

$$x_{0,\max} = \pm\frac{w}{2}\sqrt{\frac{(1 + 2t/t_c)\ln(1 + 2t/t_c)}{2t/t_c}} \tag{20}$$

As with the parallel-probe case, the optimum x-axis offset is dependent on the heating time. For short heating times, the optimum offset is the same as that for the pulsed laser-excited case, $x_0 = w/2$. However, the crossed-beam arrangement differs from the parallel-probe case in that there is no limiting optimum x-axis offset for long excitation times.

6.2.2.3. *Chopped Excitation with Parallel Probe*

The probe ray deflection angle may also be determined for chopped sample excitation. This can be performed by differentiating the temperature change in equations 36 and 37 of Chapter 5 with respect to the x dimension, integrating over the sample path, and multiplying by the thermooptical coefficient. The exact expression for the ray deflection angle during the heating cycle is

$$\begin{aligned}\theta_{x,\text{chopped}}(x_0, M\tau + \delta t) &= \left(\frac{dn}{dT}\right)\frac{\Phi_0 \alpha l Y_H}{2\pi\kappa x_0}\sum_{m=1}^{M}\left\{\exp\left[\frac{-2x_0^2}{w^2(1 + 2m\tau/t_c + 2\delta t/t_c)}\right]\right. \\ &\quad \left. - \exp\left[\frac{-2x_0^2}{w^2(1 + 2m\tau/t_c + 2\delta t/t_c - 2t_1/t_c)}\right]\right\} \\ &\quad + \left(\frac{dn}{dT}\right)\frac{\Phi_0 \alpha l Y_H}{2\pi\kappa x_0}\left\{\exp\left[\frac{-2x_0^2}{w^2(1 + 2\delta t/t_c)}\right] - \exp\left(\frac{-2x_0^2}{w^2}\right)\right\}\end{aligned} \tag{21}$$

while for cooling

$$\begin{aligned}\theta_{x,\mathrm{chopped}}(x_0, M\tau + \delta t) = \left(\frac{dn}{dT}\right)\frac{\Phi_0 \alpha l Y_H}{2\pi\kappa x_0}\sum_{m=1}^{M}\Bigg\{&\exp\left[\frac{-2x_0^2}{w^2(1 + 2m\tau/t_c + 2\delta t/t_c)}\right]\\ &- \exp\left[\frac{-2x_0^2}{w^2(1 + 2m\tau/t_c + 2\delta t/t_c - 2t_1/t_c)}\right]\Bigg\}\\ &+ \left(\frac{dn}{dT}\right)\frac{\Phi_0 \alpha l Y_H}{2\pi\kappa x_0}\Bigg\{\exp\left[\frac{-2x_0^2}{w^2(1 + 2\delta t/t_c)}\right]\\ &- \exp\left[\frac{-2x_0^2}{w^2(1 + 2\delta t/t_c - 2t_1/t_c)}\right]\Bigg\}\end{aligned} \tag{22}$$

Plots of the theoretical time-dependent ray deflection angle is illustrated in Figure 6.9. For these data the model chopper had an "on" periods, t_1, of 0.25, 2.5, and $25t_c$, and the duty cycle is 50%. Two things are apparent in comparing these plots. First, it is clear that the periodic ray deflection angle stabilizes after about $5t_c$ and is independent of the chopper period. For the $t_1 = 2.5t_c$ data, the relative deflection angle stabilizes after about three complete cycles, while the $t_1 = 0.25t_c$ data take about 10 or so periods. A peak-to-peak measure of ray deflection angle would be stable after the first few cycles periods in all three of these cases. The on-axis temperature change is also shown. The temperature change is not stable in these times scales and continues to increase. If the peak-to-peak temperature change were measured, it would decrease with increasing cycle since the average on-axis temperature increases logarithmically.

The magnitude of the deflection angle change that occurs in the times between the maximum temperature increase and decrease is often measured in a photothermal deflection experiment. Electronic band pass filters and lock-in amplifiers pass only those frequencies at or near the frequency of the chopper. The long, relatively low-frequency ray deflection angle changes occurring during the cooling part of past cycles are not detected with this type of processing. After examining the data in Figure 6.9 it is apparent that the oscillatory ray deflection angle magnitude is equal to the decaying or "laser-blocked" portion of the cycle. A quantitative equation describing the magnitude of the deflection angle change can be found from the difference between equation 21, with $\delta t = t_1$, and equation 22, with $\delta t = t_1 + t_2$. An approximate expression is found by assuming that the change in the differential ray deflection angle magnitude does not depend on the ray angle

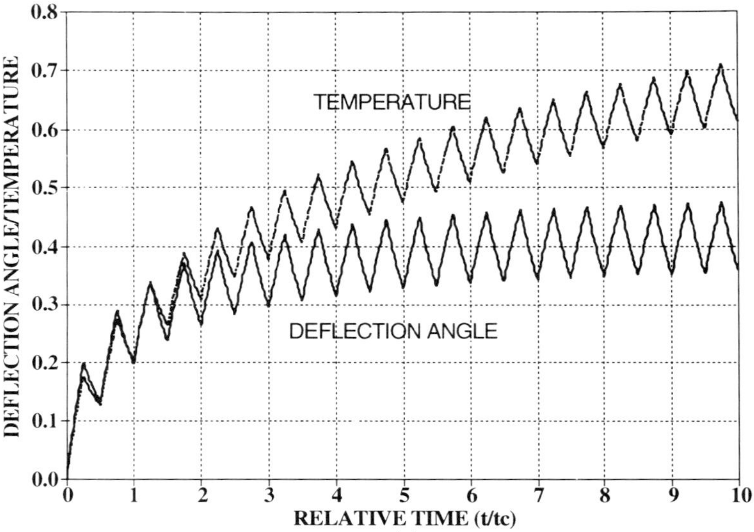

(*a*)

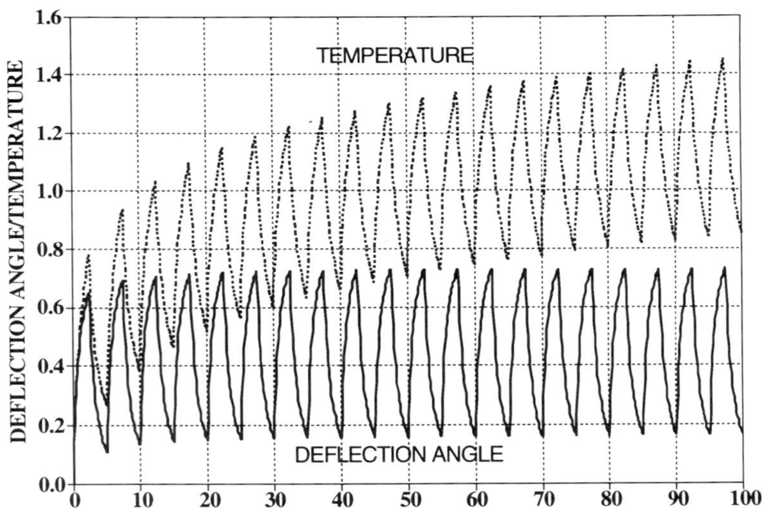

(*b*)

Figure 6.9. Multiple-cycle chopped photothermal deflection signal for several different relative heating times. The deflection angle stabilizes much sooner than does the temperature change in all cases. With the lower-frequency excitation, the deflection signal stabilizes after one or two fewer chopper cycles.

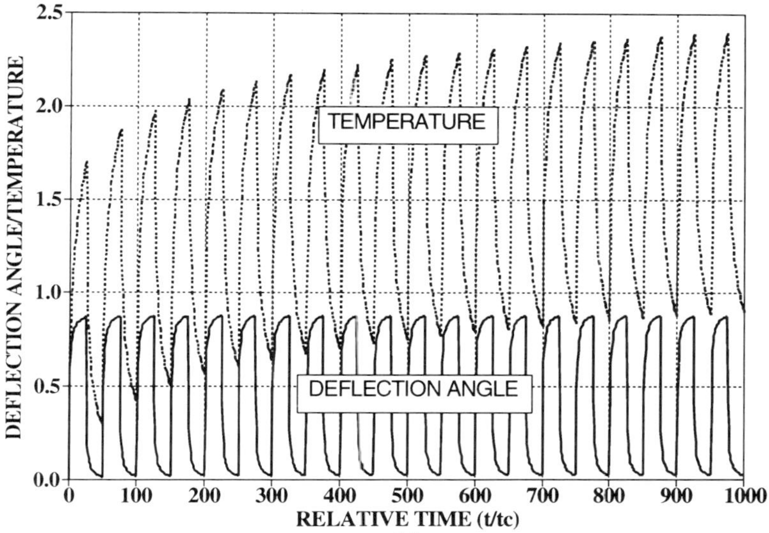

(c)

Figure 6.9. *(Continued)*

due to previous cycles. In this case the approximate magnitude is

$$\overline{\theta_{x,\text{chopped}}}(x_0) \approx \left(\frac{dn}{dT}\right) \frac{\Phi_0 \alpha l Y_H}{2\pi\kappa x_0} \left\{ 2\exp\left[\frac{-2x_0^2}{w^2(1+2t_1/t_c)}\right] - \exp\left[\frac{-2x_0^2}{w^2(1+4t_1/t_c)}\right] - \exp\left(\frac{-2x_0^2}{w^2}\right) \right\} \tag{23}$$

6.2.3. Deflection Angle Detection

Photothermal deflection spectroscopy is performed by using a probe laser to monitor the refractive index gradient produced by the excitation source. Several different schemes for obtaining an electronic signal from the probe laser beam have been used. These schemes all work satisfactorily, although each has inherent advantages and disadvantages. Most schemes are based on monitoring a change in position of the probe laser beam spot at some distance past the sample cell. Two common ways of doing this are to use a straightedge, such as a razor blade or some such aperture, followed by a detector, or to use one of the commercial position-sensing detectors with associated electronic processing modules.

6.2.3.1. *Probe Laser Beam Waist Effect*

The probe laser has a finite beam waist and so probes a finite region in the perturbed sample. Each ray comprising the probe beam will experience a slightly different deflection as a consequence of the finite sizes of the refractive index perturbation and probe beams. Bialkowski and He (1988) have considered the effect that a finite probe beam waist has on the observed photothermal deflection signal by calculating the average deflection angle. Spatial integration of the probe laser profile over the space-dependent deflection angle gives the average deflection angle that would be observed by a position-sensing detector:

$$\langle \theta_x(x_0, t) \rangle = \int_{\text{path}} \int_{-\infty}^{\infty} \int_{-\infty}^{\infty} \theta_x(x, t) E_p(x - x_0, y)\, dx\, dy\, ds \tag{24}$$

where $E_p(x - x_0, y)$ is the irradiance of the probe laser and $\theta_x(x, t)$ is the spatially dependent deflection angle. For pulsed laser excitation, the average deflection angle is

$$\begin{aligned} \langle \theta_{x,\text{pulsed}}(x_0, t) \rangle = \left(\frac{dn}{dT} \right) \frac{8x_0 \alpha l Q Y_H}{\pi \rho C_P} \frac{1}{[w^2(1 + 2t/t_c) + w_p^2]^2} \\ \times \exp\left[- \frac{2x_0^2}{w^2(1 + 2t/t_c) + w_p^2} \right] \end{aligned} \tag{25}$$

where w is the beam waist radius of the pulsed excitation laser and w_p is that of the probe laser. Note that the finite probe beam waist changes both the magnitude and the time dependence of the pulsed laser-excited deflection angle. For quantitative purposes it is sufficient to determine the effect on the initial, maximum deflection angle. The maximum average probe beam deflection angle occurs for a probe beam offset of $x_0 = \pm \frac{1}{2}(w_p^2 + w^2)^{1/2}$, not $x_0 = \pm \frac{1}{2} w$ as for the ray. The effect of the finite probe beam diameter can be seen in the ratio of the maximum average beam deflection angle to the maximum ray deflection angle:

$$\frac{\langle \theta \rangle}{\theta_{\text{ray}}} = \frac{w^3}{(w^2 + w_p^2)^{3/2}} \tag{26}$$

The average deflection angle decreases with the probe laser beam waist radius because the wider probe beam will sample more volume with relatively less ability to deflect the rays. Clearly, the ray deflection angle is optimized for both small excitation laser and probe laser beam waists. However, decreasing the excitation beam waist necessitates a decrease in the probe laser beam waist as well.

Focusing the excitation and probe lasers into the sample cell introduces a situation that is difficult to model quantitatively. When both lasers are focused into the sample cell to decrease the minimum beam waist, thereby enhancing the signal magnitude, the beam divergences on both sides of the focus will be governed by the beam waist relationship, $w^2(z) = w_0^2(1 + z^2/z_0^2)$. If the beams are focused too tightly, the collimated laser beam approximation will not be valid and divergence will decrease the signal over the value that would be expected for a given sample cell pathlength.

6.2.3.2. Straightedge Apparatus

Since first introduced (Fournier et al., 1980) the straightedge probe laser beam deflection detection scheme has been utilized extensively. This simple detection scheme does not require any equipment other than that which may be in a typical optics laboratory. The detection scheme is based on a detecting a linear probe laser beam spot displacement, through a direction orthogonal to propagation, in some plane past the sample cell. The relationship between the ray deflection angle and the beam offset in a plane containing the detector is found by considering the offset of a ray propagating through the sample, being deflected through an angle, θ, and propagating a distance of d before reaching the detector. The offset in the detector plane is $a = d \tan\theta$. For small angles, $\tan\theta \approx \theta$ and $a \approx d\theta$.

The probe beam is eclipsed with an aperture that blocks part of the beam. All of the unblocked probe laser light is then detected with a photodiode. For a probe laser with power, $(2\Phi_0/\pi w_p^2)\, exp[-2(x^2 + y^2)/w_p^2]$, the unblocked power will be

$$\frac{\Phi_{\text{detector}}}{\Phi_0} = \text{erfc}\left(\frac{\sqrt{2}a}{w_p}\right) \tag{27}$$

where erfc is the standard complementary error function (Abramowitz and Stegun 1965) and a is the position of the straightedge aperture along the x or y axes. When deflection occurs, the probe laser beam spot will move to a different position, which is equivalent to a change in the a offset. Maximum sensitivity to a change in a, found by setting the differential of the transmitted probe laser power with respect to a equal to zero, occurs for $a = 0$. This, in turn, corresponds to the straightedge blocking exactly half of the probe laser beam. The erfc function is relatively linear at this point, so the transmitted power will change linearly with a for small displacements.

Substituting the deflection angle–distance product for a and defining the measured signal, S, as the relative change in power that occurs for a given deflection angle yields

$$S(t) = \left[\frac{d(\Phi_{\text{detector}}/\Phi_0)}{d\theta}\right]_{a=0} \langle\theta_x(t)\rangle = 2\sqrt{\frac{2}{\pi}}\frac{d}{w_p}\langle\theta_x(t)\rangle \tag{28}$$

The probe laser beam is normally focused into the sample cell to maximize the deflection angle. Since the probe laser will diverge past the focus,

$$S(t) = 2\sqrt{\frac{2}{\pi(z_{0,p}^2 + d^2)}}\frac{dz_{0,p}}{w_{0,p}}\langle\theta_x(t)\rangle \tag{29}$$

For distances much shorter than the confocal distance, the signal is proportional to $d\theta_x(t)$. When the straightedge is positioned far beyond the confocal distance, the probe beam radius is $w_p = w_{0,p}(1 + d^2/z_{0,p}^2)^{1/2} \approx w_{0,p}d/z_{0,p}$, and the signal is

$$S(t) \approx 2\sqrt{\frac{2}{\pi}}\frac{z_{0,p}}{w_{0,p}}\langle\theta_x(t)\rangle = 2\sqrt{2\pi}\frac{w_{0,p}}{\lambda_p}\langle\theta_x(t)\rangle \tag{30}$$

The signal is maximized at large distances and becomes independent of distance past the sample for $d \gg z_{0,p}$. As a practical note, the probe laser power can fluctuate during the course of the measurement and must be monitored to assure that the signal measured past the straightedge aperture is proportional only to the deflection angle. This can be accomplished using a beamsplitter to direct part of the probe power onto a second photodiode. An alternative is to use one of the commercial position-sensing detectors that have processing electronics which compensate for probe laser power fluctuations.

6.2.3.3. Position-Sensing Detectors

Position-sensing detectors are more commonly used in photothermal deflection spectroscopy (e.g., Jackson et al. 1981, Fournier and Boccara 1989). Position-sensing detectors are photodiodes and photodiode arrays that are specially designed for determining beam position (Light 1986). Commercial position-sensing detectors can be obtained from many electrooptics manufacturers: Centronic, EG&G, Hamamatsu, Silicon Detector Corporation, United Detector Corporation, and others. United Detector Corporation sells processing electronic modules that convert raw electronic signals into analog signals that are proportional to the beam position. When used with the associated analog processing modules, these position sensors result in electronic signals that can be easily related to the photothermal deflection angle.

There are two main types of position sensors. The first type is the segmented detector. Segmented detectors used for photothermal deflection are made from two or four photodiodes placed in close proximity. Two-diode detectors, called bicells, are usually two rectangular diodes placed next to each other to form a square active area. The photodiode diodes are separated by a very small distance, and light missed by the inactive barrier is usually negligible relative to the active area. The two diodes share a common anode (or cathode) and have three electrical connections. These devices are used just as one would used the straightedge aperture. Quad-cells contain four photodiodes in a square pattern. This allow simultaneous determination of beam position along the two axes.

Analog processing is performed such that the difference in current between the photodiodes is divided by the sum of currents. The sum of the currents from the two diodes of the bi-cell is proportional to the total optical power while the difference is related to the position–power product. By dividing the difference by the sum, the resulting signal is independent of probe laser power. Segmented detectors have rather large frequency bandwidths, due to their PIN diode structure and photoconductive operation. Response times can be as high as about 10 ns. On the other hand, the analog processing circuitry, in particular the division circuit, limits the bandwidth and the response times of the devices. For example, the United Detector Technology Model 301DIV-30 has a bandwidth from dc to 30 kHz, limiting response times to about 3 μs. However, 30 kHz is much higher than chopper frequencies typically used in photothermal spectroscopy.

The segmented cells are designed for optical position null detection and beam-centering applications. Their response is proportional to a/w_p, so can be very sensitive to changes in position for small beam waists. When used for photothermal spectroscopy, the cells are normally placed past the confocal distance of the probe beam, and as with the straightedge aperture discussed above, their deflection angle response is independent of distance past the sample cell. However, care must be taken to ensure that their response is linear. A position change of a Gaussian laser beam will produce analog processed output proportional to $\mathrm{erfc}(\sqrt{2a/w_p})$. The erfc function is linear about the midpoint, $a = 0$. Plotting the erfc function will show that these devices should have a linear response for $a \approx \pm w_p/2$. These devices can be made very sensitive to position near the null point. Position sensitivity is such that a spot change on the order of 0.1 nm can be detected.

The other device used to detect the probe beam spot position is the lateral position-sensing detector. These devices are a single photodiode with two or four cathodes (anodes) and a common anode (cathode). The more common two cathode designs are single-axis detectors, while the four cathode devices are two-axis position detectors. The photodiode is constructed of high-resistivity silicon and the relative current at each of the cathodes is distance-weighted optical power. These devices use the same circuitry as that used by segmented photodiode detectors. Although comparable to segmented photodiodes, they are generally more noisy, due to internal resistance. On the other hand, the position response of these devices is proportional to the centroid of the spatial power distribution. This author has found that they work equally well with distorted and asymmetric beams and respond to position changes even if the entire beam is on the active area of the detector. These detectors are preferred for position-sensing-based photothermal deflection spectroscopy with highly scattering samples.

6.2.3.4. Other Methods to Detect Deflection Angle

A few other methods have been explored for detecting photothermal beam deflection angles. Glatt et al. (1984) and Horowitz et al. (1984) used moiré deflectometry to obtain images of the photothermal deflection angles in gas- and solid-phase samples. In these experiments a relatively large beam waist is used for the probe laser beam in the sample cell. Deflection is measured in different areas of the transmitted probe beam using a moiré deflectometer and a pinhole spatial filter to select the area to be monitored. Moiré deflectometry is performed by placing two Ronchi rulings, gratings with alternating transmitting and opaque strips, at positions past the sample cell. Because of the periodic structure of these gratings and the transform properties of diffracted light, diffraction of a coherent probe laser beam, expanded to approximate a plane wave, passing through the first periodic Ronchi grating structure produces an "image" of the periodic structure in subsequent planes, called Fourier planes. These Fourier planes are spaced $\Delta = p^2/\lambda$ apart, where p is the pitch or distance between the rulings. The second Ronchi with identical ruling spacing is oriented with grating rulings parallel to the first grating and is placed an integer multiple of Δ past the first grating. Since the second grating is placed in a plane where the image of the first grating is formed, the total transmission through this grating can be maximized or minimized, depending on the particular offset. When the second Ronchi grating is translated to minimize the undeflected probe laser transmission, the change in probe laser power or transmission through the moiré apparatus is given by

$$\frac{\Phi(x, t)}{\Phi_0(x)} = \frac{m2\Delta}{p}\,\theta(x, t) \tag{31}$$

where m is the Fourier plane number. The probe power and deflection angle are given as functions of the x-axis coordinate only since although this apparatus can image the deflection angle, it can be observed in only one perpendicular axis. The deflection signal is observed as a transmission through the moiré deflectometer. The apparatus can image the photothermal deflection angle. Using this apparatus, Glatt et al. (1984) were able to determine the profile of the temperature change produced by pulsed excitation of SF_6 in He buffer gas. The measured temperature-change profile was broader and distorted relative to that of the TEM_{00} excitation laser. Although the energy of the excitation laser pulse was not reported, it is likely that the SF_6 was optically bleached, thereby resulting in the broader temperature-change profile.

Another advantage is that the deflectometer can result in deflection signals with a better signal to noise ratio than that of the straight edge aperture method since the minimum probe laser transmission is much lower. Subsequently, the average shot noise will decrease due to the decreased average (non-signal) probe laser power.

Bialkowski and Long (1987) reported a different method to decrease average shot noise by using an optical setup employing two straightedge spatial filters. The first straightedge cut off half the Gaussian probe beam profile. This beam was focused into the sample cell, wherein it was deflected by a pulsed CO_2 laser-excited gas-phase sample. The probe beam exits the cell and encounters a second straightedge. This straightedge spatial filter was positioned to block the remaining half-beam. When the pulsed laser heated the gas sample, the beam angle change causes light to pass though a region not blocked by the second straightedge. With this method, the change in the probe beam power at the detector is equivalent to that of the single-straightedge method. However, a relatively small nondeflected beam power was monitored by the detector. Thus, as with the moiré deflectometer, the average shot noise was reduced and the signal contrast was dramatically improved.

Bialkowski and He (1988) describe a sensitive method that employs an analyzer etalon for angle-sensitive detection. An analyzer etalon is Fabry–Perot interferometer with transmission that is very sensitive to input angles, owing to its high finesse. In this apparatus the probe beam is focused through the sample cell and subsequently refocused onto the analyzer etalon with a lens placed behind the sample cell. The change in transmission with respect to a change in the angle at which the beam strikes the etalon was found to be given by $d(\Phi/\Phi_0)/d\theta \approx 1.3n^2F(d/\lambda)\sin\beta$, where n is the refractive index of the etalon material, F the etalon finesse, with typical values from 10 to 60, d the thickness of the etalon, λ the wavelength of the probe laser, and β the beam angle internal to the etalon. In theory, this angle detection method is about two orders of magnitude more sensitive than the position-sensing methods. The authors found an enhancement of about 40 times greater than that using the position-sensing detector or straightedge methods. This enhancement was close to the value of about 60 expected, accounting for the finite probe laser beam waist effects. The problem with this method is that it is so sensitive that slowly varying environmental effects such as air temperature and pressure drift would continuously change the sensitivity of the detection scheme by altering the average probe beam angle at the etalon. In addition, short-term environmental noise (e.g., air currents and vibrations), could easily mask the relatively weak photothermal signals. The servo feedback stabilization methods used in photothermal interferometry would help this apparatus.

An interesting method for sensitive photothermal detection was developed by Fung and Lin (1986) and by Tran (1986a, b). This method utilizes the high sensitivity that lasers have to internal cavity losses to enhance the photothermal signals. In this method a sample cell is placed inside the cavity of a helium–neon laser. The laser cavity is a Fabry–Perot interferometer. Just as minor mirror alignment perturbations change the average photon lifetime and thereby the laser output power, refractive index gradients produced inside the cavity by the photothermal effect change the output power of this laser. The laser power change is monitored by placing a detector behind one of the cavity mirrors. Samples are excited with chopped visible or infrared lasers, and the power of the helium–neon laser output is monitored. Probe laser power signals are processed with a lock-in amplifier. Laser output power is a complex function of the intracavity losses, and no attempt to describe the signal was given in these papers. In the work by Tran, the signals were not linearly proportional to the excitation power or analyte concentration, although log-log plots did appear linear in form. Fung and Lin determined how a change in output probe laser mirror angle affected the output power. The sensitivity to analyte absorbance was greatest where the output power versus mirror angle curve was linear. They adjusted the laser mirror to be within this range prior to measurement. Their measurements were found to be linear in both analyte concentration and excitation laser power. They reported 10^{-7}-cm^{-1} absorption coefficient limits of detection for gaseous analytes in nitrogen using about a 25 to

28 cm-long sample cell using a 0.8-W chopped CO_2 laser with a focused irradiance of about 10^8 W m^{-2}.

Bialkowski (1989) used an all-optical novelty filter to serve as an adaptive probe laser spatial filter for photothermal deflection. An argon-ion probe laser was used to monitor the deflection angle of a pulsed infrared CO_2 laser-excited gas sample. The probe laser writes a spatial image of its beam profile in a $BaTiO_3$ crystal. The write time constant was about 1 s. The image written in the $BaTiO_3$ crystal rejects light of the same optical frequency and spatial profile as that used to write the image. Thus, without any photothermal-induced perturbation, the probe laser is rejected from the normal ray path. When the pulsed infrared laser induces a refractive index change in the sample, the probe laser is deflected on a time scale much shorter than that required to write a new spatial rejection filter. Subsequently, the probe light is coupled out beyond the adaptive $BaTiO_3$ optical element and is monitored with a detector. This scheme enhances detection by increasing the signal contrast levels and a four- to five-order-of-magnitude signal-to-noise ratio improvement over conventional spectrophotometry was found. Kalaskar and Bialkowski (1992) laser showed that the apparatus is able to be used when the probe laser beam waist radius was much larger than that of the excitation. They describe operation of the optical novelty filter-based detector in terms of diffraction optics and showed that the resulting probe laser power should be equal to the square of the photothermal-induced refractive index perturbation. Subsequently, the detection scheme should allow for real-time imaging of optical absorptions. The relatively wide probe laser also decreased noise due to the pointing error of the excitation and probe lasers. The work of Kalaskar and Bialkowski showed that the optical detection scheme cannot strictly be characterized as photothermal deflection, but rather, is more akin to photothermal interferometry.

6.2.3.5. Differential Deflection Angle

When using ultrasensitive absorption techniques, detection limits for a particular analyte are often limited by the background absorption of the solvent. Differential absorption methods are required. Independent measurement of the analyte and blank samples separately followed by subtraction is rarely sufficient in eliminating the background absorbance in photothermal spectroscopy. This is because the excitation and probe laser sources have noise, and this noise can result in measurement errors that are proportional to the sample absorbance. There are several types of noise in laser sources that can manifest as measurement noise that is proportional error. Instantaneous power or pulse-to-pulse energy fluctuations, spatial mode noise in the laser beam profiles, and pointing noise in the beam propagation direction all contribute to the proportional error. Although the measurement errors can largely be overcome with independent measurement of the light-source characteristics followed by data correction, the shot noise in the light source and detector will ultimately limit the precision that can be obtained.

To overcome these problems, a technique that incorporates some form of optical common-mode rejection can be used. Spear et al. (1988) report a differential photothermal deflection spectrometer that used two separate beam paths for both the chopped continuous excitation and the probe laser. Analyte and blank sample cells were placed in each of these two independent beam paths. The apparatus was configured so that the probe beam defection of the analyte was in the direction opposite to that of the blank. An argon-ion pumped dye laser was used as the excitation source, and by scanning the dye laser, absorption spectra could be obtained. A polarization rotator and beamsplitter were used to adjust the power between the two beam paths to obtain an optical null. The two probe laser beams were directed on the active area of a bicell position-sensing detector. Since absorption in the analyte produced beam deflection in the direction opposite to that due to blank absorption, absorption features common to both analyte and blank samples were subtracted at the position-sensing detector. The signal past the electronics used to process the bicell current was proportional to the differential absorbance–excitation laser power product. The bicell voltage was processed with a lock-in amplifier to increase the signal-to-noise ratio. Excitation laser power was monitored with a separate photodiode, sampled, and was used to normalize the absorption spectrum data. The differential signal can be said to be generated electrooptically since it is an electronic sum of two simultaneous optical measurement.

Spear et al. were able to discriminate 8×10^{-7} M neodymium analyte absorption, with a calculated absorption coefficient detection limit of $6 \times 10^{-6}\ \mathrm{cm}^{-1}$, in aqueous samples with a background absorption coefficient of $4 \times 10^{-4}\ \mathrm{cm}^{-1}$. They also reported that probe laser beam pointing noise was noticeably reduced. This was presumably due to opposing beam spot translation at the detector for the two different beam paths. Bialkowski et al. (1992) used a single sample with two spatially separated pulsed excitation laser beams to measure a differential sample absorbance. A continuous probe laser was used to measure the differential beam deflection angle of two dye lasers, tuned to different wavelengths and pumped by a single excimer laser. The three laser beams where made parallel, but forming a triangular pattern, using beamsplitters. After passage through the sample, probe laser beam deflection in a direction parallel to the axis connecting the two excitation lasers is proportional to the differential absorbance–laser energy product, while probe beam deflection perpendicular to this axis is proportional to the sum. Difference and sum deflection signals were monitored using a quadrant cell. Rejection of signals common to both excitation wavelengths was performed optically since the beam deflection angle is proportional to the difference in absorbed energy. When combined with the sum information, the differential measurement allowed calculation of the absorbance at both wavelengths. Detection limits were not determined in this study, although pH-dependent differential absorbances of indicator dyes (bromocresol green and purple in 50:50 methanol–water solutions, and aqueous phenol red solution) could be measured quite precisely in a range from 10^{-3} to 10^{-4} absorbance unit in a standard 1-cm-pathlength cell.

6.3. THERMAL LENS FOCAL LENGTH

6.3.1. Pulsed Excitation Thermal Lens Focal Length

6.3.1.1. Time-Dependent Focal Length

The inverse focal length of the photothermal lens element is easily derived for pulsed laser excitation. For probe rays propagating collinear with the excitation beam, an instantaneous temperature change (equation 12 in Chapter 5) results in the time-dependent inverse focal length:

$$\frac{1}{f_{\text{pulsed}}(t)} = \left(\frac{dn}{dT}\right) \frac{8\alpha l Y_H Q}{\pi w^4 \rho C_P} \frac{1}{(1 + 2t/t_c)^2} \qquad \text{for } t \geqslant 0 \tag{32}$$

The maximum inverse focal length occurs immediately after sample excitation and decreases with time thereafter. The magnitude of the inverse focal length at $t = 0$ is

$$\frac{1}{f_{\text{pulsed}}(0)} = \left(\frac{dn}{dT}\right) \frac{8\alpha l Y_H Q}{\pi w^4 \rho C_P} \tag{33}$$

As with the deflection angle result, the inverse focal length is linearly proportional to the amount of energy deposited in the sample as heat, $\alpha l Y_H Q$, and the thermooptic coefficient. It is inversely proportional to the heat capacity and the excitation beam waist radius raised to the fourth power.

6.3.1.2. Sample Pathlength Limitations

The strength of the inverse focal length is directly proportional to the sample pathlength. In addition, the magnitude of the inverse focal length, and thus the optical signal, is strongly enhanced by focusing the excitation laser into the sample cell. To obtain the foregoing result it was assumed that the excitation laser was collimated through the length of the sample. Gaussian beams diverge and this approximation is valid only if the sample pathlength is much less than the confocal distance. Focusing the excitation into the sample may invalidate the constant beam waist assumption. It is simple enough to calculate the effective inverse focal length that results from focused excitation beams (Bialkowski 1986). One need only consider the thermal perturbation to be a series of thin lenses along the sample path. Focusing a pulsed excitation laser into the center of a sample cell with an optical pathlength of l, and accounting for the divergent nature of the excitation beam, results in an effective inverse focal length of

$$\frac{1}{\langle f_{\text{pulsed}}(t)\rangle} = \left(\frac{dn}{dT}\right)\frac{8\alpha Y_H Q}{\pi\rho C_P}\int_{-l/2}^{l/2}\frac{1}{[w^2(z) + 8D_T t]^2}\,dz \tag{34}$$

Substitution of $w^2 = w_0^2(1 + z^2/z_0^2)$ and integration over the pathlength results in

$$\frac{1}{\langle f_{\text{pulsed}}(t)\rangle} = \left(\frac{dn}{dT}\right)\frac{8\alpha z_0 Y_H Q}{\pi w_0^4\rho C_P}\,\frac{(X^2+1+2t/t_c)\tan^{-1}(X(1+2t/t_c)^{-12})+X(1+2t/t_c)^{1/2}}{(1 + 2t/t_c)^{3/2}(X^2 + 1 + 2t/t_c)} \tag{35}$$

where $X = l/2z_0$ is a dimensionless parameter used to describe the relative sample cell pathlength. It is assumed that thermal diffusion along the axis of laser propagation does not affect the time-dependent inverse focal length. This assumption is probably valid as long as the focus spot is not too small. The effect of sample cell pathlength may most easily seen for the initial, zero-time inverse focal length:

$$\frac{1}{\langle f_{\text{pulsed}}(0)\rangle} = \left(\frac{dn}{dT}\right)\frac{8\alpha z_0 Y_H Q}{\pi w_0^4\rho C_P}\,\frac{(X^2 + 1)\tan^{-1}(X) + X}{X^2 + 1} \tag{36}$$

The effective inverse zero-time focal length, plotted as a function of the relative sample cell length parameter, is shown in Figure 6.10. The effective

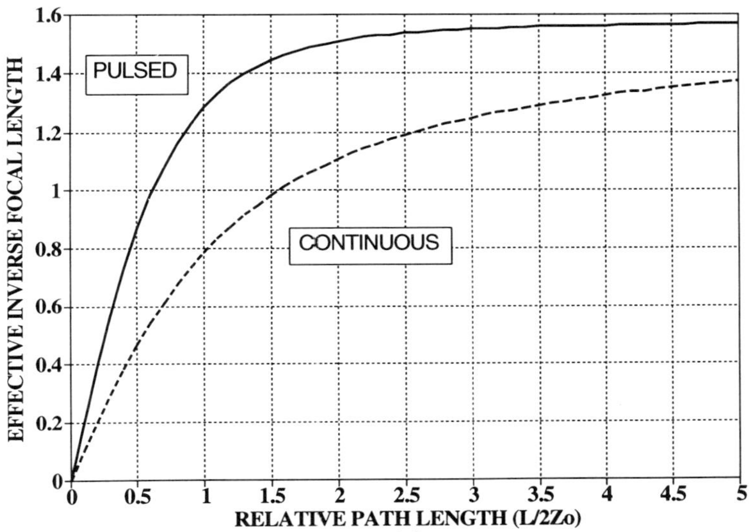

(*a*)

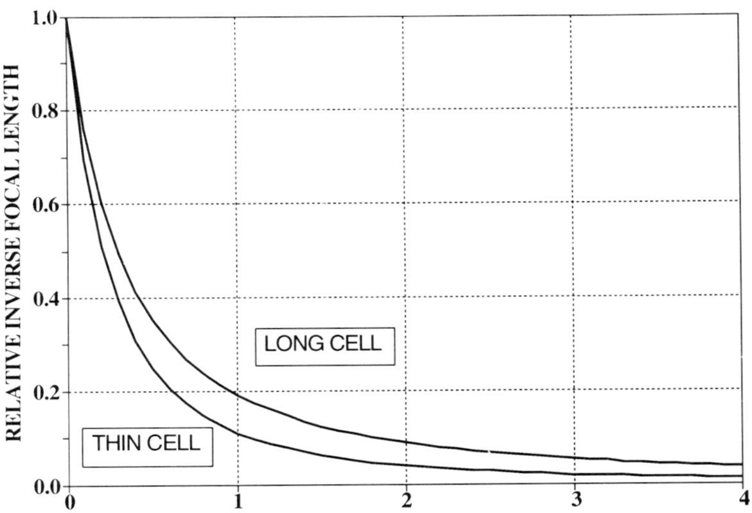

(*b*)

Figure 6.10. For long sample cells, the effective optical sample pathlength is limited by the confocal distance of the excitation laser. (*a*) The effect of increasing the relative sample pathlength on the inverse focal length is not linear. The signal is not proportional to sample pathlength with cells longer than z_0. (*b*) Effect of sample cell pathlength on decay of the pulsed laser excited photothermal lens; (*c*) Relative photothermal lens inverse focal lengths for continuous laser excitation of thin and long cells.

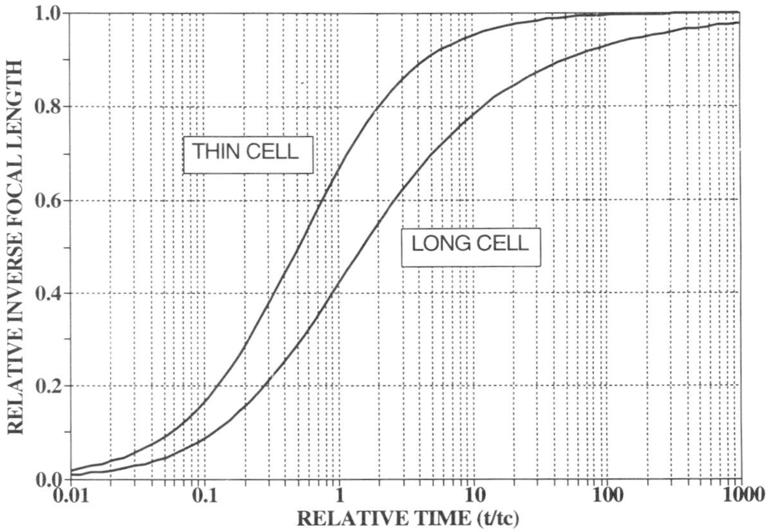

(c)

Figure 6.10. *(Continued)*

inverse focal length is seen to increase linearly with the relative sample cell length for $X < 1$, but levels off to a limiting value for long sample pathlengths. The limiting value of the effective inverse focal length is

$$\lim_{X \to 0} \frac{1}{\langle f_{\text{pulsed}}(0) \rangle} = \left(\frac{dn}{dT} \right) \frac{4\alpha z_0 Y_H Q}{w_0^4 \rho C_P} \tag{37}$$

The magnitude of the inverse focal length is limited by the confocal distance of the excitation laser. This in turn is related to the minimum beam waist radius at the focus and subsequently

$$\lim_{X \to 0} \frac{1}{\langle f_{\text{pulsed}}(0) \rangle} = \left(\frac{dn}{dT} \right) \frac{4\pi\alpha Y_H Q}{\lambda w_0^2 \rho C_P} \tag{38}$$

The effective inverse focal length of the pulsed–laser excited photothermal lens is still inversely proportional to the focused beam waist radius. The effect of a long sample cell changes not only the magnitude of the inverse focal length but also the time dependence. Again, for relatively long sample cells,

$$\lim_{X\to 0}\frac{1}{\langle f_{\text{pulsed}}(t)\rangle}=\left(\frac{dn}{dT}\right)\frac{4\pi\alpha Y_H Q}{\lambda w_0^2\rho C_P}\frac{1}{(1+2t/t_c)^{3/2}} \tag{39}$$

The inverse thermal lens decay time changes from one with a t^{-2} dependence for collimated excitation beams to one with a $t^{-3/2}$ for the focused excitation beam.

6.3.1.3. Crossed-Beam Arrangement

The inverse focal length for noncollinear excitation-probe geometries can be obtained by integrating a particular (e.g., x or y dimension) element over the path of the probe beam. Dovichi et al. (see, e.g., Weimer and Dovichi 1985a, b) have developed a photothermal lens spectroscopy technique based on crossing the excitation laser and probe beam in the sample. This technique, also called photothermal refraction, offers several advantages over the conventional collinear excitation-probe geometry. One particular advantage is that the excitation light does not interfere with the detection. This can be very useful when excitation and probe lasers are of similar wavelengths. The inverse focal length of the thermal lens for excitation and probe beams crossed at an angle is

$$\frac{1}{f_{x,\text{pulsed}}(t)}=\left(\frac{dn}{dT}\right)\frac{2^{5/2}\alpha Y_H Q}{\pi^{1/2}w^3\rho C_P(1+2t/t_c)^{3/2}\tan\gamma}\,\text{erf}\left[\frac{l\tan\gamma}{2^{1/2}w(1+2t/t_c)^{1/2}}\right] \tag{40}$$

which for 90° excitation-probe angle results in the inverse focal length:

$$\frac{1}{f_{x,\text{pulsed}}(t)}=\left(\frac{dn}{dT}\right)\frac{2^{5/2}\alpha Y_H Q}{\pi^{1/2}w^3\rho C_P(1+2t/t_c)^{3/2}} \tag{41}$$

As with the photothermal ray deflection angle, the cross-beam geometry produces a lens with an inverse focal length reduced over that for the collinear geometry. However, the same reduction is found for long sample pathlengths. Dovichi points out that crossed-beam arrangement probes a very small volume. For samples in small volume "cells" (e.g., capillary tubes), there is little if any disadvantage to using the cross-probe beam geometry. The advantages that it is easier to set up and does not require beamsplitters and excitation laser blocking filters apparently outweigh the lower-resulting signals in this case. Finally, it is interesting to note that the time dependence of the crossed-beam photothermal lens decay is the same as that obtained for the long-sample-pathlength parallel photothermal lens. The ratio of the long cell collinear inverse focal length to that of the crossed beam arrangement with the excitation laser focused into the sample cell is $\pi\sqrt{(\pi/2)}w_0/\lambda$. Apparently, the magnitude of the collinear inverse focal length is greater than that for the crossed beam arrangement for $w_0 \geqslant 4\lambda$. This beam waist radius is on the order of the smallest possible, diffraction-limited spot for a focused laser.

6.3.2. Continuous and Chopped Excitation Thermal Lens Focal Length

6.3.2.1. *Continuous Excitation*

For continuous laser excitation, the inverse focal length is found by substituting the continuous laser temperature change (equation 21 of Chapter 5) into the inverse focal length (equation 61 of Chapter 5):

$$\frac{1}{f_{\text{cw}}(t)} = \left(\frac{dn}{dT}\right) \frac{2\Phi_0 \alpha l Y_H}{\pi \kappa w^2} \frac{t_c t_1}{(t_c + 2\delta t)(t_c + 2t_1 + 2\delta t)} \qquad \delta t = \begin{cases} t - t_1 & \text{for } t \geqslant t_1 \\ 0 & \text{for } t < t_1 \end{cases} \tag{42}$$

where, again, t_1 is the time that the excitation laser is on. During the heating part of the cycle, $\delta t = 0$, $t = t_1$, and the inverse focal length develops as

$$\frac{1}{f_{\text{cw}}(t)} = \left(\frac{dn}{dT}\right) \frac{\Phi_0 \alpha l Y_H}{\pi \kappa w^2 (1 + t_c/2t)} \tag{43}$$

Although this result was obtained from the temperature-change distribution, it could also have been obtained by integrating the pulsed laser photothermal inverse lens strength over the time of irradiation. Unlike the on-axis temperature, the inverse focal length of the photothermal lens attains a maximum limiting value:

$$\frac{1}{f_{\text{cw}}(\infty)} = \left(\frac{dn}{dT}\right) \frac{\Phi_0 \alpha l Y_H}{\pi \kappa w^2} \tag{44}$$

6.3.2.2. *Sample Pathlength Limitations*

Excitation laser beam divergence also limits the sample pathlength in continuous or chopped laser-excited photothermal lens spectroscopy. Increasing the sample pathlength much over the confocal distance of the focused excitation laser will not increase the signal magnitude. This effect was first realized by Fang and Swofford (1979), who modeled the effect of the inverse focal length of the long-path photothermal lens on the probe laser using the hyperbolic *ABCD* matrix for a quadratic index GRIN lens. This effect can also be modeled by integrating the series of inverse lenses over the sample cell path (Carter and Harris 1983a). Using this model results in an effective inverse focal length that can be compared to the collimated beam results, and the temporal behavior of the lens development is more apparent. In addition to the assumption that there is no thermal diffusion along the axis of laser propagation, it must also be assumed that the thermal perturbation does not affect the propagation characteristics of the excitation laser in the sample. The later assumption is valid for small thermal perturbations (e.g., low absorption coefficients and large thermal diffusion coefficients). For samples with large thermal perturbations, the effect of the lens in each thin section of the sample must be taken into account when calculating the beam profile of the continuous excitation laser through the sample (Fang and Swofford 1979, Carter and Harris 1983a).

For small absorbance samples (e.g., $\alpha l < 10^{-3}$) the integrated sample path model should be accurate. Substitution of the beam waist radius dependence in the continuous-laser-excitation inverse-focal-length equation results in the effective focal length:

$$\frac{1}{\langle f_{\text{cw}}(t)\rangle} = \left(\frac{dn}{dT}\right)\frac{\Phi_0 \alpha Y_H}{\pi\kappa}\int_{-l/2}^{l/2}\frac{8D_T t}{w^2(z)[8D_T t + w^2(z)]}\,dz \tag{45}$$

for an excitation laser focused into the center of a sample cell with a pathlength of l. Using the Gaussian beam waist equation for $w^2(z)$ and integrating results in the effective time-dependent inverse focal length:

$$\frac{1}{\langle f_{\text{cw}}(t)\rangle} = \left(\frac{dn}{dT}\right)\frac{2\Phi_0 \alpha Y_H}{\lambda\kappa}\left[\tan^{-1}X - \frac{1}{(1+2t/t_c)^{1/2}}\tan^{-1}\frac{X}{(1+2t/t_c)^{1/2}}\right] \tag{46}$$

where again, the relatiave sample pathlength is $X = l/2z_0$. The long-path inverse focal length has a limiting value of

$$\frac{1}{\langle f_{\text{cw}}(\infty)\rangle} = \left(\frac{dn}{dT}\right)\frac{2\Phi_0 \alpha Y_H}{\lambda\kappa}\tan^{-1}X \tag{47}$$

This function is plotted in Figure 6.10. The effective inverse focal length initially increases linearly with sample pathlength for $X < 0.5$ and attains a limiting value for $X \geqslant 10$. For sample pathlengths greater than this,

$$\lim_{X\to\infty} \frac{1}{\langle f_{\mathrm{cw}}(\infty)\rangle} = \left(\frac{dn}{dT}\right)\frac{\pi\Phi_0\alpha Y_H}{\lambda\kappa} \tag{48}$$

where the definition of the confocal distance has been used. Long sample pathlengths also affect the time dependence of the lens development. For sample cell lengths much greater than the confocal distance,

$$\lim_{X\to\infty} \frac{1}{\langle f_{\mathrm{cw}}(t)\rangle} = \left(\frac{dn}{dT}\right)\frac{\pi\Phi_0\alpha Y_H}{\lambda\kappa}\left[1 - \frac{1}{(1+2t/t_c)^{1/2}}\right] \tag{49}$$

The time required to develop the thermal lens is increased over that for the collimated excitation laser. Separate excitation and probe lasers also allow the excitation laser to be focused directly into the sample, decreasing t_c and increasing the maximum inverse focal length magnitude, which scales as w^{-2}. Using a single laser, focusing into the center of the sample cell does not result in a change in the power at the center of the laser beam past the sample.

6.3.2.3. *Crossed-Beam Geometry*

The inverse focal length for the crossed-beam photothermal refraction arrangement is not limited by the pathlength of the excitation laser in the sample but rather, by the excitation laser spot size. The continuous laser excited inverse focal length for photothermal refraction is obtained either from the path integration of the inverse focal length or by integrating the pulsed laser–excited inverse-focal-length result over time. In either case, one can obtain for the heating portion of the cycle

$$\frac{1}{f_{x,\mathrm{cw}}(t)} = \frac{1}{f_{x,\mathrm{cw}}(\infty)}\left[1 - \frac{1}{(1+2t/t_c)^{1/2}}\right] \tag{50}$$

where the limiting inverse focal length that occurs at infinite time is

$$\frac{1}{f_{x,\mathrm{cw}}(\infty)} = \left(\frac{dn}{dT}\right)\frac{2^{1/2}\Phi_0\alpha Y_H}{\pi^{1/2}\kappa w} \tag{51}$$

6.3.2.4. *Chopped Excitation*

The inverse focal length for the heating portion of the Mth cycle for chopped excitation is

$$\frac{1}{f_{\text{chopped}}(M\tau + \delta t)} = \left(\frac{dn}{dT}\right)\frac{2\Phi_0 \alpha l Y_H}{\pi\kappa w^2}\sum_{m=1}^{M}\frac{t_1 t_c}{(t_c + 2m\tau + 2\delta t)(t_c + 2m\tau + 2\delta t - 2t_1)} + \left(\frac{dn}{dT}\right)\frac{2\Phi_0 \alpha l Y_H}{\pi\kappa w^2}\frac{\delta t}{t_c + 2\delta t} \tag{52}$$

while for cooling

$$\frac{1}{f_{\text{chopped}}(M\tau + \delta t)} = \left(\frac{dn}{dT}\right)\frac{2\Phi_0 \alpha l Y_H}{\pi\kappa w^2}\sum_{m=1}^{M}\frac{t_1 t_c}{(t_c + 2m\tau + 2\delta t)(t_c + 2m\tau + 2\delta t - 2t_1)} + \left(\frac{dn}{dT}\right)\frac{2\Phi_0 \alpha l Y_H}{\pi\kappa w^2}\frac{t_1 t_c}{(t_c + 2\delta t)(t_c + 2\delta t - 2t_1)} \tag{53}$$

Again, $\tau = t_1 + t_2$; t_1 is the heating time and t_2 is the cooling time of each chopper cycle, and δt is the time the sample is being irradiated after the Mth chopper cycle. The calculated inverse focal length for a chopped source, along with the corresponding on-axis temperature change, is shown in Figure 6.11. An expression for the magnitude of the inverse focal length can be obtained by calculating the difference in focal lengths heating between the maximum heating and minimum cooling times. Assuming also that the inverse-focal-length change due to previous cycles does not affect the current chopper cycle, the magnitude of the inverse-focal-length difference is

$$\frac{1}{|\delta f|} \approx \left(\frac{dn}{dT}\right)\frac{8\Phi_0 \alpha l Y_H}{\pi\kappa w^2}\frac{(t_1/t_c)^2}{(1 + 2t_1/t_c)(1 + 4t_1/t_c)} \tag{54}$$

Sample cell pathlength effects can be accounted for in the chopped laser excitation case by summing over the heating and cooling parts of the cycles. Although this calculation is straightforward, the results are very complicated and are best examined using digital simulation.

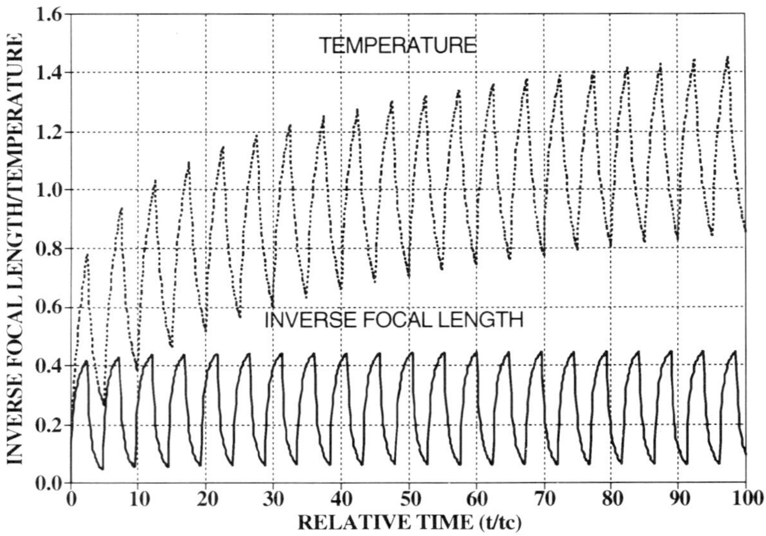

Figure 6.11. Chopped laser excited inverse focal length and temperature change.

6.3.3. Focal Length for Periodic Excitation

For periodic excitation at angular frequency, $\omega = 2\pi f$, the periodic inverse focal length is obtained from the convolution of the impulse response of the sample with a periodic excitation. The impulse response is simply that inverse focal length that results from pulsed excitation. By defining

$$f_{\text{impulse}}^{-1}(t) = \left(\frac{dn}{dT}\right)\frac{8\alpha l Y_H}{\pi w^4 \rho C_P}\frac{1}{(1 + 2t/t_c)^2} \qquad \text{for } t \geqslant 0 \tag{55}$$

and using the Laplace transform equivalent to the convolution,

$$f_{\text{osc}}^{-1}(t, \omega) = \Phi_\omega e^{i\omega t}\mathscr{L}\{f_{\text{impulse}}^{-1}(t)\}|_{\zeta = i\omega} \tag{56}$$

where Φ_ω (W m^{-2}) is the peak power of the oscillating source, the periodic inverse focal length can be found using tabulated Laplace transforms. For the case at hand,

$$\begin{aligned} f_{\text{osc}}^{-1}(t, \omega) &= i\omega\left(\frac{dn}{dT}\right)\frac{2\alpha l Y_H \Phi_\omega t_c^2}{\pi w^4 \rho C_P}\Gamma\left(-1, \frac{i\omega t_c}{2}\right)e^{i\omega(t + t_c/2)} \\ &= \frac{1}{f_{\text{cw}}(\infty)}\frac{i\omega t_c}{2}\Gamma\left(-1, \frac{i\omega t_c}{2}\right)e^{i\omega(t + t_c/2)} \end{aligned} \tag{57}$$

where $\Gamma(-1, iz)$ is the incomplete gamma function (Abramowitz and Stegun

1965) and $f_{cw}(\infty)$ is the limiting focal length for continuous excitation. The inverse focal length that would be observed with real periodic excitation with a power defined by

$$\Phi(t) = \Phi_0(1 - e^{i\omega t}) \tag{58}$$

where Φ_0 is the average power, is simply the sum of the continuous and the oscillating components:

$$f^{-1}(t) = f_{cw}^{-1}(t) + f_{osc}^{-1}(t, \omega) \tag{59}$$

The inverse focal length is thus the sum of a component that continues to increase and an oscillating component that reaches a steady-state value. The amplitude of the periodic inverse focal change is found using the relationships $\Gamma(-1, x) = \exp(-x)/x - \Gamma(0, x)$ and $\Gamma(0, ix) = i\,\mathrm{si}(x) - \mathrm{ci}(x)$ (Gradshteyn and Ryzhik 1980). The magnitude of this oscillating inverse focal length is

$$|f_{osc}^{-1}(\omega)| = \frac{1}{f_{cw}(\infty)} \sqrt{1 + \left(\frac{\omega t_c}{2}\right)^2 \left[\mathrm{ci}^2\left(\frac{\omega t_c}{2}\right) + \mathrm{si}^2\left(\frac{\omega t_c}{2}\right)\right] + \omega t_c \left[\mathrm{si}\left(\frac{\omega t_c}{2}\right)\cos\left(\frac{\omega t_c}{2}\right) - \mathrm{ci}\left(\frac{\omega t_c}{2}\right)\sin\left(\frac{\omega t_c}{2}\right)\right]} \tag{60}$$

where the vertical bars indicate the magnitude. This result is equivalent to that first found by Dovichi and Harris (1981) using Fourier transforms of the impulse response (Dovichi 1987). The inverse-focal-length magnitude is plotted as a function of relative frequency in Figure 6.12. The magnitude of the periodic inverse focal length decreases to about 0.5 of the maximum continuous excitation value for $\omega \approx 2/t_c$ and decreases rapidly for angular frequencies greater than this value. The magnitude for $\omega \approx 20/t_c$ is about 10% of the maximum.

Also plotted in Figure 6.12 is the magnitude of the frequency-dependent inverse focal length for the crossed-beam arrangement of photothermal refraction. In this case the inverse time-dependent focal length is

$$\begin{aligned} f_{x,osc}^{-1}(t, \omega) &= -e^{i\omega t}\left(\frac{dn}{dT}\right)\frac{2^{1/2}\alpha Y_H \Phi_\omega}{\pi^{1/2}\kappa w} \\ &\quad \times \left[1 - \pi^{1/2}\left(\frac{i\omega t_c}{2}\right)^{1/2} e^{i\omega t_c/2}\,\mathrm{erfc}\left(\frac{i\omega t_c}{2}\right)^{1/2}\right] \\ &= \frac{e^{i\omega t}}{f_{x,cw}(\infty)}\left[1 + i\pi^{1/2} z w(z)\right] \qquad z = \left(\frac{-i\omega t_c}{2}\right)^{1/2} \end{aligned} \tag{61}$$

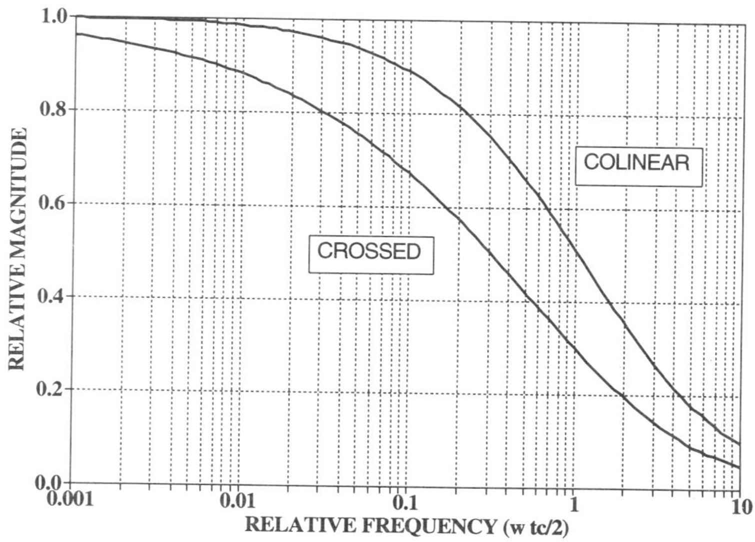

Figure 6.12. Oscillating excitation inverse focal length for collinear and crossed-beam arrangements. The magnitude of the inverse focal length is that which would be indicated by a lock-in amplifier with chopped excitation.

where $w(z) = \exp(-z^2)\,\mathrm{erfc}(-iz)$ is the complex error function (Abramowitz and Stegun 1965). The relative magnitude of the crossed-beam inverse focal length decreases slightly more rapidly with relative frequency than does the collinear inverse focal length. This is a manifestation of the inverse relationship between response time constant and frequency. Since the crossed-beam inverse focal length decays more slowly than does the collinear thin-sample focal length, the frequency response decreases more rapidly with increasing frequency. For $\omega = 2/t_c$ the inverse-focal-length magnitude has decreased to about a third of the maximum rather than half the maximum for the collinear case. Of course, if the collinear thermal lens sample pathlength is much longer than the confocal distance, the relative frequency-dependent inverse focal length will be the same as that for the crossed-beam case since the time dependencies are equivalent. The magnitude of the thermal lens focal length obtained using chopped excitation and lock-in amplifier detection is related to that which would be obtained using a periodic, oscillating excitation source. In this case the amplitude obtained by Fourier analysis of the square-wave excitation results in a fundamental sine component magnitude that is a factor of $4/\pi$ greater than that for pure sine-wave excitation.

The decrease in inverse-focal-length magnitude with increasing excitation frequency is not the only consideration in designing a photothermal lens apparatus. One should also consider the fact that flicker noise will dominate at the lower frequencies, and subsequently, the signal-to-noise ratio may actually be better at higher relative excitation frequencies. There are two ways that the experimentalist has to increase the signal-to-noise ratio of the detected photothermal lens. First, the characteristic time constant can be decreased by focusing the excitation laser to a smaller minimum beam waist. This will decrease the relative frequency for a given excitation modulation or chopper frequency. Second, the optimum excitation or chopper frequency can be found. The optimum frequency is that which maximizes the signal-to-noise ratio. Again, while higher frequencies will decrease the inverse focal length, and thus the optical signal, the flicker noise will decrease as f^{-a}.

The inverse focal length for a periodic source could be obtained only because there are no exponential time-dependent terms in the impulse-response function. This allows the Laplace transform to be found in analytical form. Analytical expressions for the periodic excitation frequency-dependent temperature change, photothermal interferometry path-length, and photothermal deflection angle cannot be obtained because they all have exponential time dependence. However, the results can be expressed in integral form and temperature change and optical elements for these parameters all behave similarly to the inverse photothermal lens focal length.

6.4. DETECTING THE THERMAL LENS

6.4.1. Signal for Symmetric Lens

The thermal lens is normally monitored by measuring the power at the center of the probe laser beam at a distance well beyond the sample cell. The relative power change is related to the inverse focal length of the photothermal lens and thus to the absorbance of the sample. The effect that the photothermal lens element has on the propagation of a probe laser can be found using the *ABCD* law for Gaussian beam propagation (Twarowski and Kliger 1977a, Swofford and Morrell 1978). Figure 6.13 shows a schematic for the optical arrangement typically used to monitor the photothermal lens. The TEM_{00} probe laser propagates along the z axis and is focused with a lens to a minimum spot electric field radius w_0. The laser is focused to a minimum spot at a distance z' before the thermal lens. The laser beam diverges past the focus and has a beam waist of w_1 on passing though the sample. The sample is assumed to have a thermal lens with a time-dependent focal length of $f(t)$. The thermal lens can be caused by the probe laser or by a separate excitation laser. After passing through the thermal

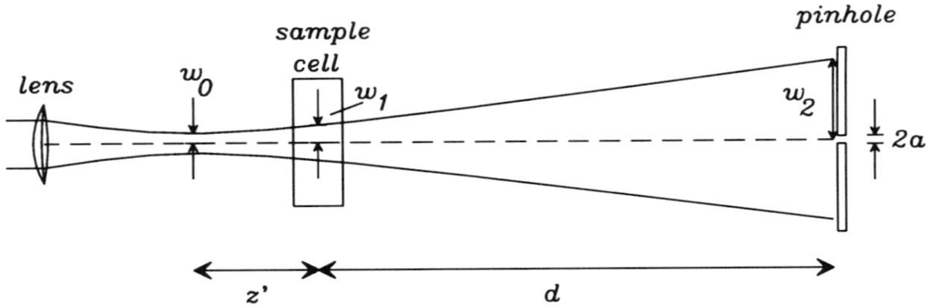

Figure 6.13. Geometry used to define the theoroetical photothermal lens signal. The probe laser enters from the left and is focused to a minimum spot radius of w_0 at a distance z before the sample cell. It has a beam waist radius of w_1 at the sample and w_2 at the pinhole aperture placed before the detector. The pinhole aperture is a distance of d after the sample cell. Probe beams focused beyond the sample are indicated by negative z.

lens, the probe laser beam propagates a distance of d before encountering the pinhole aperture. The matrix representing probe laser beam propagation past the minimum spot at the focus of the lens is

$$T = \begin{bmatrix} 1 - d/f(t) & z' + d[1 - z'/f(t)] \\ -1/f(t) & 1 - z'/f(t) \end{bmatrix} \tag{62}$$

Using the complex beam parameter at the focus of the probe laser of, $q_0 = -i/z_{0,p}$, where the probe laser confocal distance is $z_{0,p} = \pi w_0^2/\lambda_p$, the beam parameter at the pinhole aperture is found using the $ABCD$ law, $q(z) = (Aq_0 + B)/(Cq_0 + D)$. The imaginary part of $q(z)$ is $\lambda_p/\pi w_2^2$. Using the $ABCD$ law, the beam waist radius at the pinhole can be shown to be

$$w_2^2(t) = \frac{\lambda_p}{\pi z_{0,p}} \left\{ (z_{0,p}^2 + z'^2)\left[1 - \frac{d}{f(t)}\right]^2 + 2dz' + d^2\left[1 - \frac{2z'}{f(t)}\right] \right\} \tag{63}$$

The pinhole is positioned at the center of the beam and the transmitted power is

$$\Phi(t) = \frac{2\Phi_0}{\pi w_2^2(t)} \int_0^{2\pi} d\theta \int_0^a e^{-2r^2/w_2^2(t)}\, r\, dr = \Phi_0(1 - e^{-2a^2/w_2^2(t)}) \tag{64}$$

For a relatively small pinhole radius, a, the transmitted power is

$$\lim_{a/w_2(t) \to 0} \Phi(t) = \frac{2\Phi_0 a^2}{w_2^2(t)} \tag{65}$$

The signal is defined similarly for pulsed and continuous excitation. For pulsed laser excitation, the signal is defined by

$$S_{\text{pulsed}}(t) = \frac{\Phi(\infty) - \Phi(t)}{\Phi(t)} = \frac{w_2^2(t)}{w_2^2(\infty)} - 1 \tag{66}$$

while for continuous excitation

$$S_{\text{cw}}(t) = \frac{\Phi(0) - \Phi(t)}{\Phi(t)} = \frac{w_2^2(t)}{w_2^2(0)} - 1 \tag{67}$$

where the subscript cw is used to indicate continuous-wave excitation. The difference between these two definitions is due to the fact that the signal is initially a maximum for pulsed excitation, while the signal is initially zero for continuous excitation. The power and beam waist radius at the time when there is no thermal lens is the reference. Defining the signal as a ratio of powers eliminates the dependence on the probe laser beam parameters, the detector response, and the electronic devices that may be used to processing and record the probe laser power.

For either pulsed or continuous excitation, the probe laser beam waist radius in the absence of a photothermal lens is

$$w_2^2 = \frac{\lambda_p}{\pi z_{0,p}} (z_{0,p}^2 + z'^2 + 2dz' + d^2) \tag{68}$$

and the signal is

$$S(t) = -\left[\frac{2d}{f(t)}\right] \frac{z'^2 + z_{0,p}^2 + dz'}{(z' + d)^2 + z_{0,p}^2} + \left[\frac{d}{f(t)}\right]^2 \frac{z'^2 + z_{0,p}^2}{(z' + d)^2 + z_{0,p}^2} \tag{69}$$

This expression gives the exact signal that is observed with photothermal lensing. The photothermal lens signal consists of both a linear term and a second-order term. The second-order dependence on the inverse focal length results in deviations form a linear response for short-focal-length thermal lenses (e.g., large signals).

Photothermal lens apparatuses are normally set up with the pinhole aperture placed far from the sample. In addition, the requirement that the probe laser beam waist be less than that of the thermal perturbation does not allow z' to be very large. Subsequently, the optical setup is such that $d \gg z' \sim z_{0,p}$. In this case the denominator can be set to d^2 and the signal is given approximately by

$$S(t) \approx -\frac{2z'}{f(t)} - \frac{2}{f(t)} \frac{z'^2 + z_{0,p}^2}{d} + \left[\frac{1}{f(t)}\right]^2 (z'^2 + z_{0,p}^2) \tag{70}$$

and for large d,

$$S(t) \approx -\frac{2z'}{f(t)} + \frac{z'^2 + z_{0,p}^2}{f^2(t)} \tag{71}$$

Finally, for focal lengths sufficiently long that the second-order term can be neglected, the linear photothermal lens signal is

$$S(t) \approx -\frac{2z'}{f(t)} \tag{72}$$

This expression is normally assumed to be the photothermal lens signal. However, it is valid only for weak signals and when the pinhole aperture should be placed at a distance that far exceeds the confocal distance. The range over which the second-order term can be neglected may be gaged by the thermal lens focal length relative to the distance to the probe laser focus. Since the confocal distance is on the order of the z-axis offset term, z', the first- and second-order terms both vary with $z'/f(t)$. Thus when $z'/f(t) \leqslant 0.01$, the second-order term constitutes less than 1% of the signal. Since z' can be very short, on the order of 10 cm or less in many apparatuses, the thermal lens signal will be linear for most practical focal lengths. Nonetheless, very high-absorbance samples may result in focal lengths that are on the order of the confocal distance. When this is the case, the quadratic term cannot be neglected.

6.4.2 Signal for Different x and y Focal Lengths

The signal for crossed-beam or other configurations can be derived in a similar fashion. In this case there may be different focal lengths about the x and y coordinates. An elliptical Gaussian laser beam can be defined having different electric beam waist radii in the x and y dimensions (Yariv 1985):

$$E(x, y) = \frac{2\phi_0}{\pi w_x w_y} \exp\left(-\frac{2x^2}{w_x^2} - \frac{2y^2}{w_y^2}\right) \tag{73}$$

In general, elliptical beams can have different focus positions and beam divergences for the x and y dimensions. Elliptical beams have two separate q; for example,

$$\frac{1}{q_x(z)} = \frac{1}{R_x(z)} - i\,\frac{\lambda}{\pi n w_x^2(z)},$$
$$\frac{1}{q_y(z)} = \frac{1}{R_y(z)} - i\,\frac{\lambda}{\pi n w_y^2(z)} \tag{74}$$

Subsequently, the *ABCD* law can be applied to each coordinate independently.

We will consider the case where the two beam waist radii are initially equivalent, $w_{x,2} = w_{y,2} = w_2$, but then change due to a lens that has a different strength in the x and y dimensions. This situation will occur, for example, when a TEM_{00} laser is used to probe a photothermal lens in the crossed-beam geometry. The signal can be defined by

$$S_{\text{pulsed}}(t) = \frac{\Phi(\infty) - \Phi(t)}{\Phi(t)} = \frac{w_{x,2}(t) w_{y,2}(t)}{w_2^2(\infty)} - 1 \tag{75}$$

for pulsed laser excitation, with a similar definition for the continuous laser excitation case. This definition is equivalent to that of the radially symmetric lens and corresponds to the relative probe laser power detected past a pinhole aperture placed at the center of the beam. Each coordinate of the elliptical beam will have a focal-length-dependent beam waist radius, found from the ray matrix, of

$$w_{x,y,2}^2(t) = \frac{\lambda_p}{\pi z_{0,p}} \left\{ (z_{0,p}^2 + z'^2)\left[1 - \frac{d}{f_{x,y}(t)}\right]^2 + 2dz' + d^2\left[1 - \frac{2z'}{f_{x,y}(t)}\right] \right\} \tag{76}$$

The approximate signal may be found by performing a Taylor expansion of the square-root terms that result from the signal definition. For example, if only $1/f_x(t)$ is finite, the signal is

$$S(t) = \frac{\sqrt{(z_{0,p}^2 + z'^2)[1 - d/f_x(t)]^2 + 2dz' + d^2[1 - 2z'/f_x(t)]}}{\sqrt{(z' + d)^2 + z_{0,p}^2}} - 1 \tag{77}$$

Retaining the first two terms of the expansion yields the approximate signal

$$S(t) \approx -\left[\frac{d}{f_x(t)}\right] \frac{z'^2 + z_{0,p}^2 + dz'}{(z' + d)^2 + z_{0,p}^2} + \frac{1}{2}\left[\frac{d}{f_x(t)}\right]^2 \frac{d^2 z_{0,p}^2}{[(z' + d)^2 + z_{0,p}^2]^2} \tag{78}$$

and under the limits imposed by the experimental setup,

$$S(t) \approx -\frac{z'}{f_x(t)} + \frac{1}{2}\left[\frac{z_{0,p}}{f_x(t)}\right]^2 \tag{79}$$

Of course, when the thermal lens is weak, the signal is approximated by the linear term, $S(t) \approx -z'/f_x(t)$. Notice that the signal magnitude is half that obtained for the symmetric lens case.

In the general case, both $1/f_x(t)$ and $1/f_y(t)$ are finite. The signal is rather complicated in this case:

$$S(t) = \frac{\sqrt{(z_{0,p}^2 + z'^2)[1 - d/f_x(t)]^2 + 2dz' + d^2[1 - 2z'/f_x(t)]}\;\sqrt{(z_{0,p}^2 + z'^2)[1 - d/f_y(t)]^2 + 2dz' + d^2[1 - 2z'/f_y(t)]}}{(z' + d)^2 + z_{0,p}^2} - 1 \tag{80}$$

The signal is approximated by first taking the Taylor expansion about $f_x(t)$ and $f_y(t)$ equal to infinity and retaining only those terms linear in inverse focal lengths. This expansion yields the approximate signal

$$S(t) \approx -\left[\frac{d}{f_x(t)} + \frac{d}{f_y(t)}\right]\frac{z'^2 + z_{0,p}^2 + dz'}{z_{0,p}^2 + (z' + d)^2} \approx -\left[\frac{z'}{f_x(t)} + \frac{z'}{f_y(t)}\right] \tag{81}$$

When the x and y focal lengths are equal, signal equation 81 is equivalent to that for the symmetrical lens. This equation was used to calculate the pulsed-laser-excited parallel-probe laser photothermal lens signal illustrated in Figure 6.14. As the probe laser is offset from the center of the thermal distribution, the signal strength decreases to a minimum. The minimum signal is less than zero. We have used this probe laser offset-dependent behavior to “tune up” photothermal lens apparatuses. The probe laser offset can be adjusted with a mirror while watching the signal on an oscilloscope. Maximization of the signal while adjusting the orthogonal offset coordinates ensures that the excitation and probe laser beams are aligned properly.

6.4.3. Lock-in Amplifier or Pulse-Height-Detected Signal

Signals defined as the inverse probe laser power passing through a pinhole aperture are mathematically convenient. On the other hand, the arithmetic processing the raw measured power to these defined signals is tedious and often cannot be performed with the signal processing electronics. Rapid data

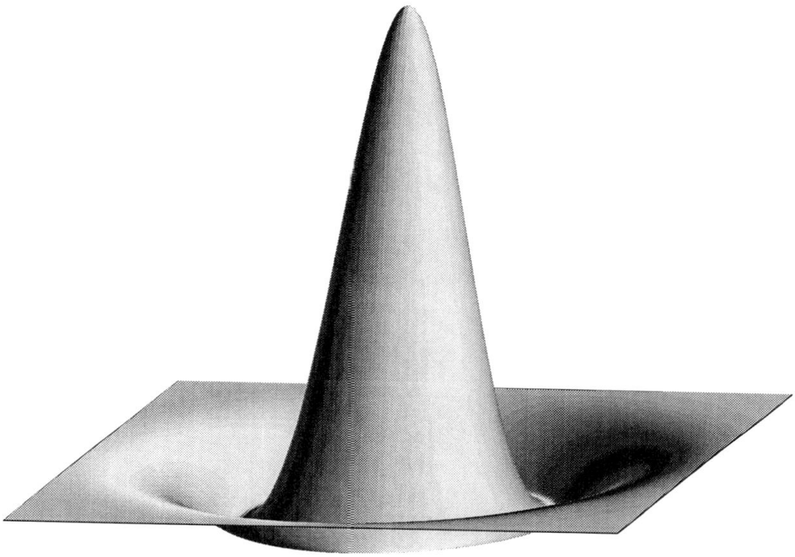

Figure 6.14. Plot of the photothermal lens strength as a function of displacement from the excitation source. The excitation source is Gaussian.

collection and processing of pulsed laser–excited photothermal lens signals is often performed using pulse height analyzers, boxcar averagers, or with sample-and-hold circuits with sampling gates set to sample the maximum change in of the time-dependent probe laser power signal. In these cases the processed signal is equal to $S'_{\text{pulsed}} \propto \Phi(\infty) - \Phi(0)$, not $[\Phi(\infty) - \Phi(t)]/\Phi(t)$ as was found convenient using the *ABCD* law. Similarly, the probe laser power changes produced with chopped excitation sources are often directly processed with lock-in amplifiers. In this case the processed signal is proportional to difference in probe laser power, $S'_{\text{chopped}} \propto \Phi(0) - \Phi(t)$, or the magnitude of the oscillating component of the signal $S'_{\text{osc}} \propto \Phi(\omega)$.

Signals generated by direct detection of the probe laser power change observed through the pinhole are proportional to both the probe laser power and the beam waist radius change that occurs upon sample excitation, Δw_2^2. The probe laser power dependence can be compensated for using independent measurement, although it is usually assumed to be constant over the course of the experiment and thus neglected. For simplicity, one can define a measured photothermal lens signal as being independent of the probe laser power:

$$S'_{\text{chopped}}(t) = \frac{\Phi(t) - \Phi(0)}{\Phi(0)} = \frac{w_2^2(0)}{w_2^2(t)} - 1 \qquad (82)$$

for chopped or periodic excitation and

$$S'_{\text{pulsed}}(t) = \frac{\Phi(t) - \Phi(\infty)}{\Phi(\infty)} = \frac{w_2^2(\infty)}{w_2^2(t)} - 1 \tag{83}$$

for pulsed excitation. The lock-in amplifier detected signal is related to the root-mean-square photodiode current produced by an oscillating or chopped source. Using a procedure equivalent to that used for error propagation calculations, the signal can be calculated from

$$S'_{\text{osc}}(\omega) = f^{-1}(\omega)\left(\frac{\partial S(t)}{\partial[1/f(t)]}\right)_{1/f(t)=0} \tag{84}$$

For chopped or periodic excitation, it is easy to show that the angular frequency-dependent signal is

$$S'_{\text{osc}}(\omega) = \frac{2d}{f(\omega)} \frac{z_{0,p}^2 + z'^2 + dz'}{z_{0,p}^2 + (z' + d)^2} \tag{85}$$

For experimental conditions where $d \gg z' \sim z_{0,p}$, $S'(\omega) \approx 2z'/f(\omega)$. The later is equivalent to that obtained using the normal signal definition.

The exact expression for a signal defined as the probe power change over that power detected in the absence of a photothermal lens is

$$S'_{\text{pulsed}}(t) = \frac{z_{0,p}^2 + z'^2 + 2dz' + d^2}{(z_{0,p}^2 + z'^2)[1 - d/f(t)]^2 + 2dz' + d^2[1 - 2z'/f(t)]} - 1 \tag{86}$$

The pulsed-laser-excited signal maximum can be found using the differential response, as shown above for periodic excitation, or by using a Taylor expansion of the exact equation, retaining only first few terms in inverse focal length. The exact result is first expressed as a ratio. Products with the inverse focal length in the denominator are set to zero. These terms are all products of the form dz^2/f and are expected to be small relative to the d^2 terms. For small inverse focal lengths and large d relative to $z' \sim z_0$, Taylor expansion yields

$$S'_{\text{pulsed}}(t) \approx \frac{2z'}{f(t)} - \frac{z_{0,p}^2 + z'^2}{f^2(t)} \tag{87}$$

which is equivalent, except in sign, to that obtained using the conventional signal definition. Of course, the sign occurs simply as a matter of the order of the signal and average power of the probe laser. It should be remembered that unlike that obtained using the measurement signal definition, this equation is an approximation. For critical measurements of the time dependence of the pulsed-laser-excited signals, the processing conforming to the usual signal definition should be used. However, it is clear that lock-in and peak height detection yields photothermal lens signal magnitudes that are approximately linear and proportional to αlQ or $\alpha l\Phi$ products. With calibration, they may be used to gage sample absorbance.

6.4.4. Signal Development with Large Apertures

The commonly used signal definitions are based on measurement of probe laser power at the center of the Gaussian beam. Pinhole apertures with diameters much less than that of the probe laser beam are usually used to satisfy the spatial beam profile sampling requirements. The requirement that the pinhole diameter be much less than the probe beam radius is dictated by approximations used in defining the photothermal lens signal. This requirement is not necessarily the same as that which would optimize the signal strength or signal-to-noise ratio (SNR).

In general, the four types of noise that must be considered in all forms of optical spectroscopy—constant thermal or Johnson noise, shot noise, source flicker noise, and interference—must also be considered in optimizing the photothermal signal. In addition, photothermal signals are influenced by the laser beam spot size and position in the sample. Subsequently, laser pointing and TEM mode instabilities constitute a noise source in these apparatuses. It is sometimes said that no one has really done photothermal spectroscopy unless the person has recognized the mode and pointing instability of the laser sources. Laser instabilities do not give rise to "noise" in the sense of a zero-mean linear additive contribution to the signal. The effect of these instabilities on the signal is usually to decrease the magnitude over that expected for the ideal case. Subsequently, the error distribution is not the normal Gaussian and propogation of errors cannot be used to determine the effects of these errors on the SNR.

Skogerboe and Yeung (1986) suggested that the SNR may be enhanced for small aperture diameters when the source is flicker-noise limited, and for large aperture diameters when the laser is limited by pointing instabilities. Erskine and Bobbitt (1988) performed a detailed study of this effect using both simulation and experimental results. The results of this study showed that the relative aperture ratio, a/w, where a is the radius of the aperture, significantly affected the SNR value for Johnson, shot, and source-flicker-noise limited conditions. For Johnson and shot-noise-limited conditions, there was an optimum relative aperture radius. For source-flicker-dominated noise, the SNR was optimized for the smallest aperture radius. The relative aperture ratio did not significantly affect the SNR value when laser pointing stability affected the measurement precision.

The optimum aperture dimension can be found from the probe laser power transmitted throught the aperture, $P(a) = 1 - \exp(-2a^2/w^2)$, and the relationship between the transmitted power and the limiting noise. Defining the signal as the change in transmitted probe laser power relative to a change in beam waist radius, the signal is simply

$$S \propto 4\,\frac{a}{w}\exp\left(\frac{-2a^2}{w^2}\right)\frac{\delta w}{w} \tag{88}$$

For constant noise, σ^2 is independent of a. In this case SNR $\propto S/\sigma$, and the optimum aperture radius found by differentiating the signal with respect to a, is $a_{\text{opt}} = 2^{-1/2}w$. For shot-noise-limited detection, the noise is proportional to $P(a)^{1/2}$ and the SNR is

$$\text{SNR}_{\text{shot}} \propto \left(\frac{a}{w}\right)^2 \frac{\exp(-2a^2/w^2)}{[1 - \exp(-2a^2/w^2)]^{1/2}} \tag{89}$$

Maximizing SNR_{shot} with respect to a results in $a_{\text{opt}} = 0.5674w$. For source flicker noise, noise is proportional to $P(a)$ and the SNR is

$$\text{SNR}_{\text{flicker}} \propto \left(\frac{a}{w}\right)^2 \frac{\exp(-2a^2/w^2)}{1 - \exp(-2a^2/w^2)} \tag{90}$$

Plotting $\text{SNR}_{\text{flicker}}$ as a function of a/w reveals that it is a maximum for $a = 0$. The SNR for a combination of these noise sources is

$$\text{SNR}_{\text{total}} \propto \left(\frac{a}{w}\right)^2 \frac{\exp(-2a^2/w^2)}{(\sigma^2 + [1 - \exp(-2a^2/w^2)] + \xi[1 - \exp(-a^2/w^2)]^2)^{1/2}} \tag{91}$$

The σ^2 and ξ terms represent the relative constant and flicker noise contributions, respectively. Erskine and Bobbitt examined $\text{SNR}_{\text{total}}$ for different relative noise contributions and compared these results to experimental data. They used a pair of acoustooptical modulators to chop the beams of two different Ar^+ lasers. The experimental apparatus allows a single laser to be used as both the excitation and probe laser (Erskine et al.

1987), and high chopper frequencies were possible. Signal and noise were analyzed with a lock-in amplifier equipped with a noise measurement feature. A water-cooled, super-Invar-based cavity laser was compared to one utilizing air-cooled construction. They concluded that the pointing instability of the water-cooled Ar^+ laser constructed with a super-Invar cavity was negligable. The noise was found to be a combination of flicker and shot noises. The flicker noise component was present even with chopper frequencies of 5 kHz. Shot noise dominated at 5 kHz. Subsequently, the optimum SNR was found for a rather large relative aperture radius. Flicker noise increased at a 35-Hz chopper frequency, and the optimum aperture was found to shift to a slightly smaller radius. On the other hand, the less expensive, air-cooled laser had significant flicker noise even with a modulation frequency of 5 kHz. The SNR was optimum for smaller aperture dimensions in this case.

6.4.5. Signal Development Based on Image Analysis and Other Optical Filters

SNR optimization does not have to be based on circular apertures. Photothermal signal generation by other spatial filtering schemes is certainly possible. Spatial filtering opposite of that produced of circular aperture has been proposed and tested (Slaby 1987, 1989). In this design, a circular blocking filter is centered on the Gaussian probe beam. This filter exhibits a decreased average transmitted power relative to circular transmission or pinhole filters. Slaby points out that decreasing the transmitted power is useful since the high background power of circular transmission apertures masks the photothermal signal, a consequence of shot noise, and measurement of the signal requires detectors with large dynamic range.

Slaby describes the theoretical optimization in terms of diffraction theory, which is more applicable in this case since the outer portions of the beam are detected. This theory is described in a section below. However, the results of the simpler refractive thermal lens thoery can be used to understand the improvements obtained with this mask. The probe laser power transmitted around a circular blocking filter of radius a is just $1 - P(a) = \exp(-2a^2/w^2)$. Using the photothermal lens signal definition used by Erskine and Bobbitt also results in a similar aperture-radius-dependent signal [e.g., $S(a) \propto (a/w)\exp(-2a^2/w^2)$]. The only difference between these two spatial filters is that the sign of the photothermal lens signal is different. Subsequently, the optimum filter radius for constant Johnson noise is equivalent to that for the pinhole aperture, $a_{\text{opt}} = 2^{-1/2}w$. However, the circular blocking filter can significantly reduce shot and flicker noise terms. The shot-noise-limited SNR is

$$\mathrm{SNR}_{\mathrm{shot}} \propto \left(\frac{a}{w}\right)^2 \exp\left(\frac{-a^2}{w^2}\right) \tag{92}$$

The optimum circular block radius is $a_{\mathrm{opt}} = w$ in this case. For flicker noise, the SNR is simply $(a/w)^2$ and the optimum a occurs for a disk of infinite extent. Comparing the SNR under optimum conditions to that obtained using the pinhole aperture reveals that the performance of the circular beam-blocking mask will be superior to the pinhole aperture for either shot- or flicker-noise-dominated measurements. In fact, the SNR ratio (SNRR) for optimum spatial filters favors the circular blocking filter by a factor of 1.5 for shot-noise-limited measurements, and $\mathrm{SNRR} = 2a^2/w^2$ for flicker noise, where a is the diameter of the blocking filter and the pinhole aperture radius is near zero. The SNRR favors the circular beam block for radiuses greater than $2^{-1/2}w$. It is apparent that the photothermal lens SNR produced using the blocking filter can equal if not surpass that produced with the pinhole aperture for all noise cases. As an aside, we have found that the circular beam block filter can be constructed by applying the seldom-used periods found in dry transfer lettering sets to a microscope slide.

A rather unique way to determine the strength of the photothermal lens is based on beam image detection and subsequent calculation of the probe laser beam waist radius. This method was first demonstrated by Miyaishi, et al. (1982) and was later used by Jansen and Harris (1985) to compare photothermal lens signals obtained using a pinhole, a photodiode array, and a transmission mask designed to compute the beam waist radius optically. Miyaishi, et al., used both continuous and chopped Ar^+ laser radiation to excite an Fe(II) complex in chloroform solution. The helium–neon probe beam was directed through the sample and onto a one-dimensional photodiode array detector. Output from the photodiode array was fed into a computer for analysis. To analyze the one-dimensional data, the background photodiode array signal was first subtracted from the beam image data, the position of the beam centroid was then calculated using the center-of-mass formula, and this in turn was used in a linear formula for determining the squared Gaussian beam waist radius. The latter calculation is the same as that used to determine variance. The resulting signal estimate was found to be about an order of magnitude better than those based only on the peak maximum (e.g., that corresponding to the pinhole spatial filter apparatus). The reason for this improvement is not certain. However, the calculations employed would have compensated for probe laser power fluctuations. Probe laser pointing error would be reduced since the centroid was calculated prior to estimating the beam waist radius. Finally, the larger

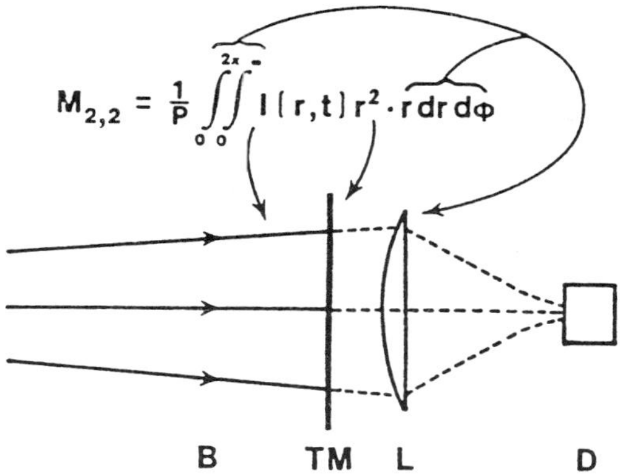

Figure 6.15. Optical probe laser beam waist radius processor of Jansen and Harris (1985). B is the laser beam in far-field, TM a parabolic transmission mask, L a collecting lens, and D a detector. The lens integrates the spatial product of the Gaussian beam with the parabolic transmission function. (Reprinted with permission from Jansen and Harris 1985. Copyright 1985, American Chemical Society.)

light detection angle could reduce effects due to light scattering by the sample.

Jansen and Harris (1985) employed a similar apparatus for their photodiode array detector experiments (Figure 6.15). Unlike the apparatus of Miyaishi et al., one laser was used as both the excitation source and the probe. Sample excitation was continuous. Their results were similar to those of Miyaishi et al. However, in addition to the factor of 2 or so precision enhancement over the pinhole spatial filter method, they found that the working curve, composed of thermal lens response versus sample absorbance, was linear over a greater range of sample absorbance for signal estimates based on computed probe laser beam waist radius. Jansen and Harris extended the concept of calculating the beam waist radius by employing an optical signal processing technique to calculate the laser beam wasit radius (Figure 6.16). After exiting the sample cell, the beam passes through a spatially dependent transmission filter. The filter mask has a radial-dependent parabolic, r^2, transmission that is a minimum at the center. The r^2 transmission filter was constructed by plotting dots onto a transparancy using a digital plotter. The dot pattern was randomized to reduce diffraction effects.

After passing through the transmission filter, the laser beam is focused onto a detector. Time-dependent signals are recorded on a computer and lens sample absorbance is later analyzed by fitting the transient signals using regression analysis. Jansen and Harris found that the absorbance detection

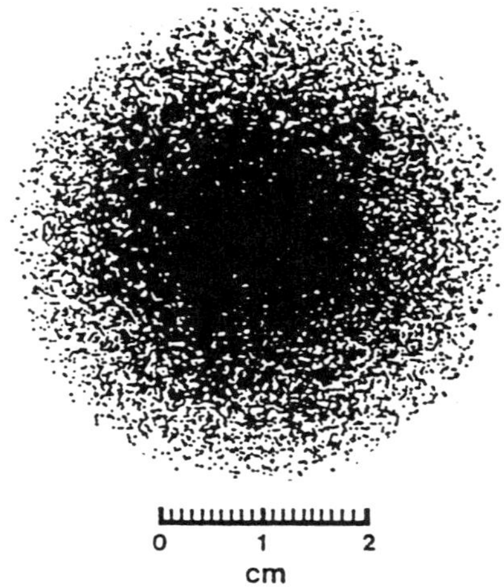

Figure 6.16. Parabolic mask used for optical processing of the photothermal lens signals. (Reprinted with permission from Jansen and Harris 1985. Copyright 1985, American Chemical Society.)

limit using optically processed beam radius determination with time-dependent signal averaging was about an order of magnitude better than that obtained using the pinhole spatial filter method. The innovative spatial optical transmission filter employed by Jansen and Harris shows that the pinhole aperture is not the most effective for photothermal lens spectroscopy. A spatially encoded optical transmission filter was used by Herrera and Bialkowski (1985) to process time-dependent photothermal deflection signals and adaptive spatial filters have been used process photothermal signals (Bialkowski 1989, Kalaskar and Bialkowski 1992, Plumb and Harris 1992). Perhaps even better innovative filter designs will arise as a consequence of the better understanding of signal processing and the techniques developed by the optical signal processing community.

6.5. TYPES OF PHOTOTHERMAL LENS APPARATUS

There are two main types of optical apparatus used in photothermal lens spectroscopy. The simplest is the single-laser apparatus. Although this apparatus is simple to construct, align, and operate, it is by no means inferior to others. The second type of apparatus commonly used is the two-laser design. In this design one beam is used to excite the sample, and a second beam is used to probe the resulting photothermal lens. Although this is called a two-laser design, it can be constructed using a single laser. The main types of photothermal apparatus can be used for both single-cell and differential measurements. Each type of apparatus is discussed below.

6.5.1. Single-Laser Apparatus

The single-laser photothermal lens apparatus is different from all other photothermal apparatus in that it uses one laser as both the excitation source and the probe. The laser is chopped at a slow rate, usually using a shutter, and lens signal is observed while the laser is exciting the sample. An expression for the single-laser photothermal lens signal is obtained by substituting the inverse focal length for heating with the linear photothermal lens signal result, $S_{cw}(t) \approx -2z'/f(t)$. Since the laser beam waist radius is $w^2 = w_0^2(1 + z'^2/z_0^2)$, the single-laser photothermal lens signal can be written

$$S_{cw}(t) \approx -\frac{2z'}{1 + z'^2/z_0^2}\left(\frac{dn}{dT}\right)\frac{\Phi_0 \alpha l Y_H}{\pi \kappa w_0^2(1 + t_c/2t)} \tag{93}$$

The optimum z′ by differential methods shows that the maximum signal occurs when $z' = \pm z_0$. For a laser focused at $\pm z_0$ relative to the center of the sample cell,

$$S_{cw}(t) \approx \mp\left(\frac{dn}{dT}\right)\frac{\Phi_0 \alpha l Y_H}{\lambda \kappa(1 + t_c/2t)} \tag{94}$$

With continuous excitation, the signal magnitude reaches a maximum as time approaches infinity. At very long times relative to t_c, the parabolic lens approximation is practical. The maximum thermal lens signal at this time is

$$S_{cw}(\infty) \approx \mp\left(\frac{dn}{dT}\right)\frac{\Phi_0 \alpha l Y_H}{\lambda \kappa} \tag{95}$$

Strictly speaking, the single-laser experiment is not accurately modeled using the parabolic lens approximation. This is especially true during initial heating when the beam waist radius of the "probe laser" will be identical to the temperature-change profile. With continued irradiation, the width of the temperature-change profile increases and the parabolic approximation was originally thought to be valid for times much greater than t_c (see Figure 5.1). By careful measurement of the signal as a function of the relative sample and laser focus positions, it was later found that this approximation was not valid (Sheldon et al. 1982; Carter and Harris 1984; Berthoud et al. 1985). Because the thermal perturbation is not parabolic, the signal cannot be found from Gaussian beam propagation through a simple lens and diffraction theory must be used to obtain the proper result (Sheldon et al. 1982). This subject is taken up later in this chapter. The difference between the beam geometries predicted using the parabolic approximation and that using diffraction theory is significant, a factor of 1 versus $3^{1/2}$. However, for the purpose of calculating relative beam geometries and enhancement factors, the validity of the parabolic lens approximation is assumed in this chapter.

Dovichi and Harris (1979) make the argument that since the signal measured in conventional absorption spectrophotometry is $S_{\text{abs}} = \alpha l$ for low-absorbance samples, the ratio of the photothermal lens signal to that produced by absorption should indicate the enhancement in signal magnitude obtained using photothermal lens spectroscopy (Figure 6.17). They defined the enhancement factor as the ratio, $E = S_{\text{tls}}/S_{\text{abs}}$. The theoretical enhancement assuming that the limiting noise is equivalent in both spectroscopies is

$$E_{\text{cw}} = \left(\frac{dn}{dT}\right)\frac{\Phi_0 Y_H}{\lambda\kappa} \tag{96}$$

The heat yield, Y_H, is included here although it does not appear in the original expression. The theoretical enhancement can be made arbitrarily large simply by increasing the power of the laser. It is also dependent on the thermooptical characteristics of the sample. Sample solvents with large dn/dT value and low thermal conductivities favor photothermal lens spectroscopy and photothermal spectroscopy in general.

Carter and Harris (1983a) examined the effect of sample pathlength on the single-laser photothermal lens signal. They treated this problem using both the ray transfer matrix formalism and by integrating the far-field thermal lens signal over the sample pathlength. The later approach is valid

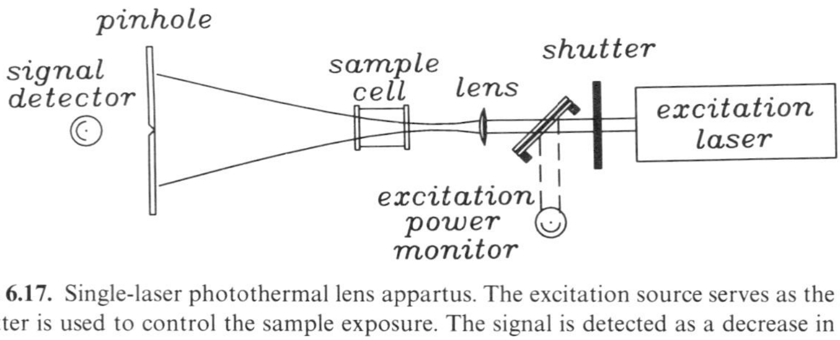

Figure 6.17. Single-laser photothermal lens appartus. The excitation source serves as the probe. A shutter is used to control the sample exposure. The signal is detected as a decrease in power at the beam center.

for small absorbance samples and is illustrated here. Starting with the infinite time signal for small absorption,

$$S_{\mathrm{cw}}(\infty) \approx -\frac{2z'}{1+z'^2/z_0^2}\left(\frac{dn}{dT}\right)\frac{\Phi_0 \alpha l Y_H}{\pi \kappa w_0^2} \tag{97}$$

an integral over the sample path is substituted for the pathlength. Integrating over the path from z_a to z_b yields the signal

$$\begin{aligned} S_{\mathrm{cw}}(\infty) &\approx -\left(\frac{dn}{dT}\right)\frac{2\Phi_0 \alpha Y_H}{\pi \kappa w_0^2}\int_{z_a}^{z_b}\frac{z'}{1+z'^2/z_0^2}\,dz' \\ &= \left(\frac{dn}{dT}\right)\frac{z_0^2 \Phi_0 \alpha Y_H}{\pi \kappa w_0^2}\ln\frac{z_a^2+z_0^2}{z_b^2+z_0^2} \end{aligned} \tag{98}$$

Expression 98 does not lend itself to analysis for optimum sample placement since the sample pathlength is $l = z_b - z_a$. Carter and Harris found that the optimum sample position was, in fact, a function of sample pathlength. In addition to the corrections to optimum sample placement, the signal was found to be a nonlinear function of sample pathlength. In general, it was found that the signal was less than expected when the sample pathlength was greather that z_0.

Although the single-laser apparatus is limited in many respects, the advantage is in its experimental simplicity. Using a single laser, lens, pinhole, and detector, sensitive absorption measurements can be obtained. The disadvantage is that signal can be collected only during the excitation part of the heating–cooling cycle, and that the excitation and probe beams are not optimized for maximum signal.

6.5.2. Differential Single-Laser Apparatus

Dovichi and Harris (1980a) used the fact that the signal changes sign depending on whether the sample is in place before or after the laser beam focus to develop a differential photothermal lens apparatus (Figure 6.18). In principle, with analyte and blank samples placed $\pm \boldsymbol{z}_0$ from the beam focus, the apparatus should produce a response of

$$S_{\mathrm{cw}}(\infty) = \pm\left(\frac{dn}{dT}\right)\frac{\Phi_0}{\lambda \kappa}\left[(\alpha l Y_H)_s - (\alpha l Y_H)_b\right] \tag{99}$$

In practice, reflections from the windows and beam divergence due to the first cell will affect the power at the second. This can be compensated for by

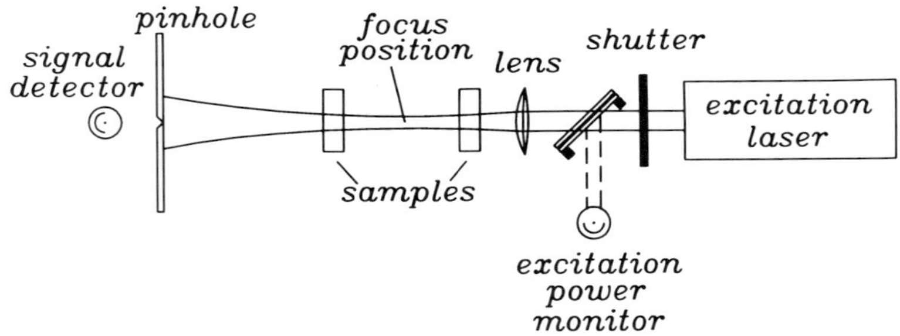

Figure 6.18. Differential photothermal lens apparatus similar to that first used by Dovichi and Harris (1980a). The beam of this single-laser apparatus is focused to a position between the two sample cells. The lens created in the first sample cause the power past the pinhole aperture to increase while that in the second causes it to decrease. The net result in a signal that is proportional to the difference absorbance of the two samples.

translating the first cell to a position that is slightly less than optimum. Differences in characteristic time constants between the two samples do not affect the long-term limited signal. Using this apparatus, differential absorption coefficients of $\delta\alpha \approx 10^{-6}\,\text{cm}^{-1}$ were obtained for I_2 in CCl_4, and coefficients of $\delta\alpha \approx 2 \times 10^{-6}\,\text{cm}^{-1}$ were obtained for aqueous iron 1,10-phenanthroline complex in 1-cm quartz sample cells (Dovichi and Harris 1980b).

The unique characteristic of this apparatus is that the differential signal is obtained optically, not using an electronic difference past the detector. If the relative standard deviation of the laser power is the main source of error in the measurement, the error in the differential measurement would be $\sigma_d = (\sigma_\Phi/\Phi_0)|\alpha_a - \alpha_b|$, where the α are analyte and blank absorption coefficients. On the other hand, sequential measurement with electronic difference generation would produce a standard deviation of $\sigma_s = (\sigma_\Phi/\Phi_0)(\alpha_\alpha^2 + \alpha_b^2)^{1/2}$. This optical common-mode-rejection feature is unique to differential photothermal apparatuses and is clearly advantageous when analyte absorption is small compared to the blank.

6.5.3. Two-Laser Apparatus

The two-laser apparatus is based on using one laser to create the photothermal lens and a second laser to probe the resulting lens element (Figure 6.19). The two-laser apparatus is obviously required for pulsed or oscillating excitation since signals are derived from the thermal lens formed both during and after excitation. There is an advantage for chopped excitation as well. Both the heating part, when the lens is being formed, and the cooling

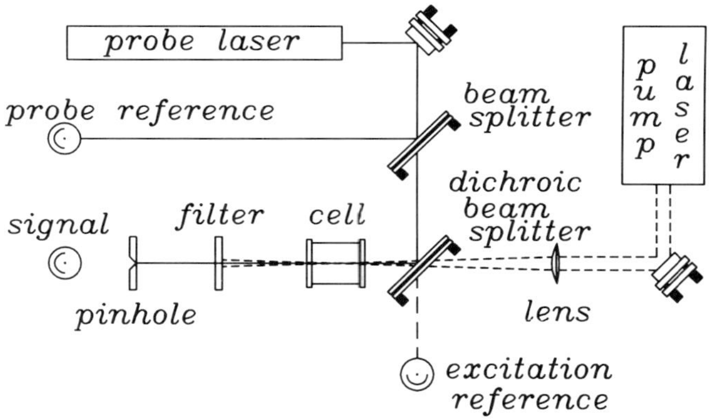

Figure 6.19. Two-laser photothermal lens apparatus. Separate excitation (pump) and probe lasers are used. The filter stops the continuous, chopped, or pulsed excitation source from being detected.

portion of the cycle, when the lens dissipates, can be monitored. With the two-laser experiment, the width of the temperature change profile expands with time for the cooling part of the cycle. So the restrictions on the beam waist radius of the separate probe laser are not so critical during cooling.

The two-laser apparatus is more flexible than the single-laser one since the excitation and probe lasers propagate independent of one another. This flexibility allows for crossed and collinear excitation-probe beam arrangements, and an infinite number of excitation-probe focus position geometries. In general, the excitation laser is focused into the sample, and with the restriction that the pump laser beam radius be smaller than that of the focused excitation laser to maintain the validity of the parabolic lens approximation, the probe laser can be focused anywhere along the axis of laser propagation.

The flexibility allowed by using separate excitation and probe lasers is of some concern since the signal magnitude changes with the beam geometry of the probe laser. On one hand, it is useful to determine theoretical optimum geometric conditions and then to design the apparatus to meet these conditions. On the other hand, these conditions can be difficult to satisfy and perhaps more difficult to reproduce. Because of this difficulty, it may be better to use a reproducible excitation-probe laser beam geometry for analytical measurements. Most, if not all, of the literature reports of detection limits and the like using two-laser-based photothermal lens measurements do not give enough information about the beam geometry used in the apparatus to allow comparison. Subsequently, the reasons for the reported detection limits are not clear. Reproducible geometry in apparatus design would allow for reproducible and cross-comparable measurements.

There has been much discussion over the correct excitation and probe beam geometry to use in the two-laser apparatus (see Dovichi 1987). The terms *mode matched* and *mode mismatched* have been used to describe the relationship between the excitation and probe laser beam waists at the sample. When the excitation beam waist and divergence is equal to that of the probe, the apparatus is said to be mode matched. In this case both beams propagate with the same propagation mode. If this is not the case, the beams are mode mismatched. There is no reason why the two laser beams should be mode matched in a photothermal lens apparatus.

With mode-matched continuous excitation and probe laser beams, the photothermal lens signal will be optimum for the same conditions as that for the single-laser apparatus. In fact, in their detailed study of single- and dual-beam configurations for thermal lens spectrometry, Carter and Harris (1983b) showed that there is little advantage to the two-laser apparatus for mode-matched conditions. Different provisions are required for cases where the collinear excitation and probe lasers are focused at different positions along the axis of propagation. The simplest approach to determining the effect of the excitation and probe laser beam focus positions is to use the far-field pinhole aperture placement result of Twarowski and Kliger, equation 72.

With the pinhole aperture placed far from the sample cell, and with the excitation and probe lasers focused at distances, z and z', respectively, in front of the sample cell, the photothermal lens signal is

$$S_{\mathrm{cw}}(t) = -\left(\frac{dn}{dT}\right)\frac{\Phi_0 \alpha l Y_H}{\pi \kappa w_0^2(1 + t_c/2t)}\frac{z'}{1 + z^2/z_0^2} \tag{100}$$

The factor of $1 + z^2/z_0^2$ in the denominator is due to the divergence of the excitation laser, while the z' in the numerator is from the thermal lens signal equation, $2z'/f(t)$. It is easier to see the effect of the two laser focuses on the signal by defining a distance between excitation and probe laser focus positions, $\delta z = z' - z$. In this case

$$S_{\mathrm{cw}}(t) = -\left(\frac{dn}{dT}\right)\frac{\Phi_0 \alpha l Y_H}{\pi \kappa w_0^2(1 + t_c/2t)}\frac{z'}{1 + (z' - \delta z)^2/z_0^2} \tag{101}$$

The two-laser photothermal lens apparatus is normally operated with the excitation laser focused into the sample cell and with the probe laser focus position set to maximize the resulting signal. The small focused excitation beam waist increases the strength of the photothermal lens and decreases the characteristic time constant; both effects are beneficial to the measurement. The optimum probe beam offset is $z'_{\text{opt}} = \pm(z_0^2 + \delta z^2)^{1/2}$. With the excitation laser focused into the sample cell, $\delta z = z'$ and the simple result predicted from $S_{\text{cw}}(t) = 2z'/f(t)$ describes the resulting signal.

Mori et al. (1982) were among the first to use a pulsed excitation laser for analytical measurements. They derive a description of the signal magnitude using mode-mismatched excitation and probe lasers and calculated the theoretical enhancement factor for pulsed laser excitation. They started with the weak lens signal equation, $S(t) = -2z'/f(t)$, and then included the implicit dependence on the excitation beam waist radius, $w^2 = w_0^2(1 + z^2/z_0^2)$, where z is the sample cell-to-excitation laser focus distance. Assuming that the excitation beam waist radius does not change along the path through the sample cell,

$$S_{\text{pulsed}}(t) \approx -\left(\frac{dn}{dT}\right)\frac{16\alpha l Y_H Q}{\pi\rho C_p(1 + 2t/t_c)^2 w_0^4}\frac{z'}{(1 + z^2/z_0^2)^2} \tag{102}$$

Using the definition for distance between the excitation and probe laser focus positions as $\delta z = z' - z$, we obtain

$$S_{\text{pulsed}}(t) \approx -\left(\frac{dn}{dT}\right)\frac{16\alpha l Y_H Q}{\pi\rho C_p(1 + 2t/t_c)^2 w_0^4}\frac{z'}{[1 + (z' - \delta z)^2/z_0^2]^2} \tag{103}$$

The maximum signal for a given excitation-probe offset is

$$z' = \frac{2\delta z \pm \sqrt{4\delta z^2 + 3(\delta z^2 + z_0^2)}}{3} \tag{104}$$

The signals are generally greater for mode-mismatched conditions, $\delta z \neq 0$. However, the range of z is not without bounds. The parabolic lens approximation is not valid for large distances between the probe laser focus and sample cell because of divergence in the probe laser beam.

Mode matching the excitation and probe lasers requires that $\delta z = 0$. In this case the optimum laser focus distance to the sample cell is $z' = \pm 3^{1/2} z_0$ and photothermal lens signal is

$$S_{\text{pulsed}}(t) \approx \mp z_0\left(\frac{dn}{dT}\right)\frac{3^{3/2}\alpha l Y_H Q}{\pi w_0^4 \rho C_p(1 + 2t/t_c)^2} \tag{105}$$

The maximum signal at zero time found by substituting the confocal distance is

$$S_{\text{pulsed}}(t) \approx \mp \left(\frac{dn}{dT}\right) \frac{3^{3/2}\alpha l Y_H Q}{\lambda w_0^2 \rho C_P} \tag{106}$$

where the wavelength is that of the excitation laser. Defining the enhancement factor in a fashion equivalent to that for continuous excitation, $E_{\text{pulsed}} = S_{\text{pulsed}}/S_{\text{abs}}$, the pulsed laser enhancement factor is

$$E_{\text{pulsed}} = \pm \left(\frac{dn}{dT}\right) \frac{3^{3/2} Y_H Q}{\lambda w_0^2 \rho C_P} \tag{107}$$

As was found for the continuous laser enhancement, the thermal lens signal is enhanced by large dn/dT values and short wavelengths. In contrast to the continuous excitation enhancement, the pulsed laser enhancement is independent of the thermal conductivity but instead, is inversely dependent on the heat capacity of the sample. The pulsed laser enhancement is also inversely dependent on excitation beam waist radius and is thus maximized for smaller beams. Nickolaisen and Bialkowski (1986) compared pulsed to continuous-laser-excited photothermal spectroscopy based on the relative enhancement factors. The relative enhancement is simply

$$\frac{E_{\text{pulsed}}}{E_{\text{cw}}} = \frac{3^{3/2}}{4} \frac{Q}{t_c \Phi_0} \tag{108}$$

Under conditions where t_c is small (e.g., small excitation laser beam waists or large thermal diffusion constants), pulsed laser excitation is favored. Pulsed laser excitation is also favored is flowing samples where the effective time constant is limited by the rate a material transport out of the excited region.

The treatment of Mori et al. assumed that the photothermal lens signal could be modeled by first neglecting the effects of probe laser beam waist, focus position and detector position and then reintroducing the excitation and probe beam-to-cell offsets in the beam waist radius equation. The optimum probe beam mode for the two-laser photothermal lens apparatus may best be found by using the unsimplified form of the signal definition. Long and Bialkowski (1984) made the simple argument that excitation and probe laser propagation modes should be optimized independently. The photothermal lens inverse focal length magnitude is maximum when the excitation laser is focused into the sample. Subsequently, this condition was used to optimize the excitation laser beam geometry. Using the unsimplified form of the signal definition, equation 71, relationships between the optimum probe laser focus position and the pinhole aperture distance, d, where found. For pulsed laser excitation, they found that the optimum pinhole distance was $d_{\text{opt}} = -(z'^2 + z_{0,p}^2)/(z' \pm z_{0,p})$. For any given pinhole aperture distance, the optimum probe laser focus was found to occur at $z'_{\text{opt}} = -(d \pm z_{0,p})$. Since both $z_{0,p}$ and d are positive, and $d > z_{0,p}$, the optimum probe laser focus position must be past the sample cell.

The optimum condition can be seen in the plot of the photothermal lens signal as a function of pinhole aperture and probe laser focus distances shown in Figure 6.20. In this figure the photothermal lens signal is that given by equation 72. The signal is plotted as a function of the relative detection plane distance, $d/z_{0,p}$, and the probe laser-to-sample cell position, $z'/z_{0,p}$. Positive relative focus positions are for a probe laser focus prior to the sample cell, and negative relative focus positions are for probe laser focus positions between the sample cell and the pinhole. A positive signal corresponds to a decrease, while a negative signal corresponds to an increase in probe power at the beam center. Probe laser focuses between the sample cell and the pinhole aperture can result in positive or negative signals for a negative lens element. The maximum and minimum values in the signal as a function of pinhole aperture distance seen in this figure correspond to the optimum conditions of Long and Bialkowski.

Dovichi (1987) argued that although almost any probe laser focus position can be used, the simplest experimental optics arrangement is one wherein both the excitation and probe lasers are focused directly into the sample cell (e.g., $z' = 0$). This is called a confocal arrangement, to indicate the common focus position of the lasers. This experimental arrangement may be the easiest to reproduce since the focus positions of the excitation and probe lasers are rather easy to detect and to change by moving the focusing optics positions. The detector pinhole distance behavior of this configuration is clearly visible in Figure 6.20. It is apparent that there is an optimum detection plane distance. The optimum detection plane or pinhole aperture distance can be found from the beam propagation equation. With this geometry the photothermal lens signal in equation 72 becomes

$$S(t) = -\left[\frac{2d}{f(t)}\right]\frac{z_{0,p}^2}{d^2 + z_{0,p}^2} + \left[\frac{d}{f(t)}\right]^2 \frac{z_{0,p}^2}{d^2 + z_{0,p}^2} \tag{109}$$

The optimum detector distance found from differentiating the signal is

$$d_{\text{opt}} = \left|\frac{z_{0,p}}{2f(t)}\left[\sqrt{z_{0,p}^2 + 4f^2(t)} - z_{0,p}\right]\right| \tag{110}$$

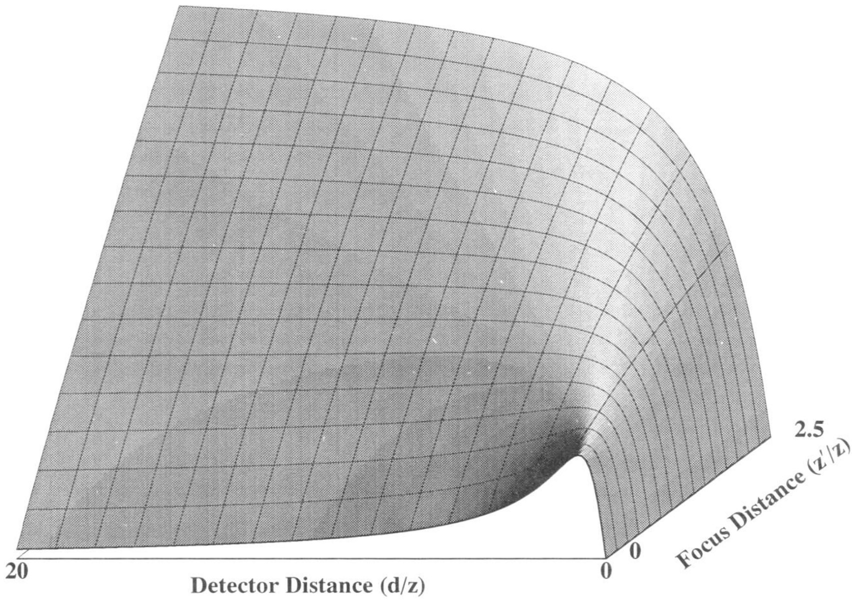

(a)

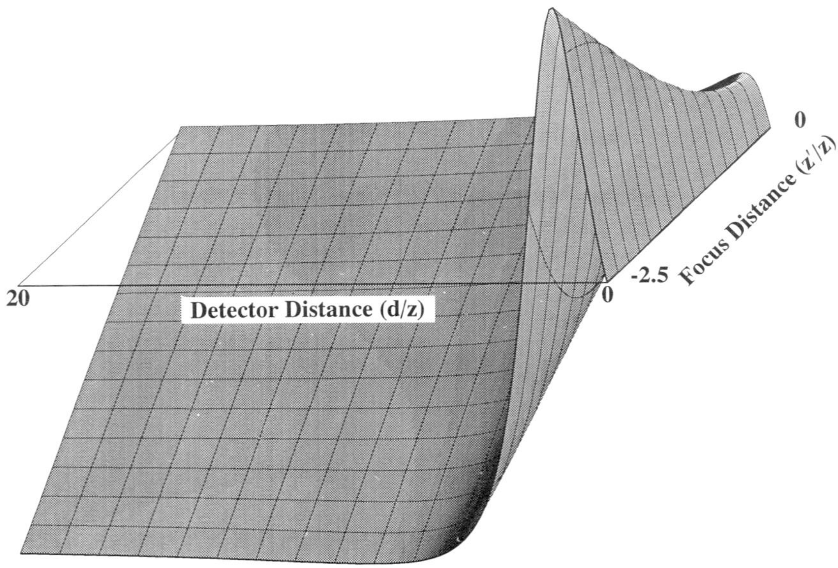

(b)

Figure 6.20. Photothermal lens signal as a function of probe laser beam geometry for two-laser experiments: (*a*) for the probe laser focused before the sample; (*b*) for the probe focused after the sample, between the sample cell and the pinhole aperture. These results are predicted from refractive optics and do not account for the finite excitation beam waist radius.

which in the limit of a small inverse focal length results in $d_{\text{opt}} = z_{0,p}$. The signal produced with this geometry is

$$S(t) = -\frac{z_{0,p}}{f(t)} \tag{111}$$

Because the maximum probe focus offset, z', is usually limited by the requirement that the probe laser beam waist be less than or equal to that of the excitation source, the confocal optical arrangement will produce photothermal lens signals that are not that different from the theoretical maximum. For example, if z' is constrained to be $\pm 2z_{0,p}$ in order to satisfy the parabolic lens approximation conditions, the $z' = 0$ arrangement is only a factor of 5 less sensitive than the optimum.

Berthoud et al. (1985) performed a detailed experimental study designed to find the optimum probe laser focus position. In this study the probe laser beam was not constrained to have a beam waist radius less than or equal to that for the excitation beam in the sample cell. The experimental details

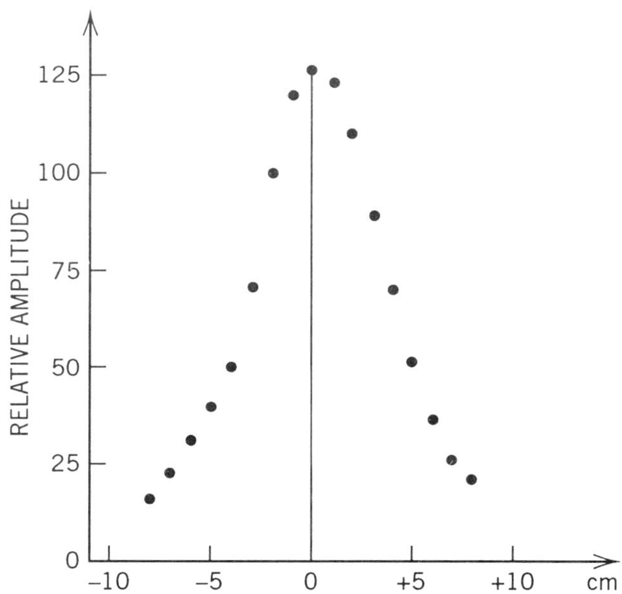

Figure 6.21. Experimental measure of the effect of excitation laser beam waist size. The relative amplitude of the photothermal lens signal is plotted as a function of the excitation laser beam focus position. A focus position of zero corresponds to the minimum spot of the excitation laser being in the sample cell. The probe laser was focused 10 cm before the sample cell. (Reprinted with permission from Berthoud et al. 1985. Copyright 1985, American Chemical Society.)

are as follows. A chopped argon-ion laser pumped tunable dye laser is used to excite aqueous neodymium solutions, and a helium–neon laser serves as the probe. The minimum beam waist radius of the excitation laser is $w_0 = 0.024$ mm, that for the probe is $w_{0,p} = 0.110$ mm. The pinhole aperture of 1 mm diameter is placed 2.5 m after the sample cell. The results of these measurements are shown in Figures 6.21 and 6.22. Figure 6.21 shows the thermal lens signal as a function of the excitation laser focus position with the probe laser focus fixed at 10 cm before the sample cell. As expected, the thermal lens signal is maximized when the excitation laser is focused directly into the sample. Figure 6.22 shows the signal as a function of the probe laser focus position. The signal is similar to that found for the single laser apparatus however, it does not conform to the theory in that the signal is seen to decrease for probe focus distances greater than about ± 10 cm. These results cannot be interpreted in terms of the parabolic lens approximation since the probe laser beam waist is not less than that of the excitation source.

The reason for the decrease in photothermal lens signal with increasing the z' was interpreted in terms of the relatively large probe laser beam waist radius. Berthoud et al. suggest that this effect may be due to diffraction since the maxima occur at about $\pm 3^{1/2} z_{0,p}$, as predicted by diffraction theory for

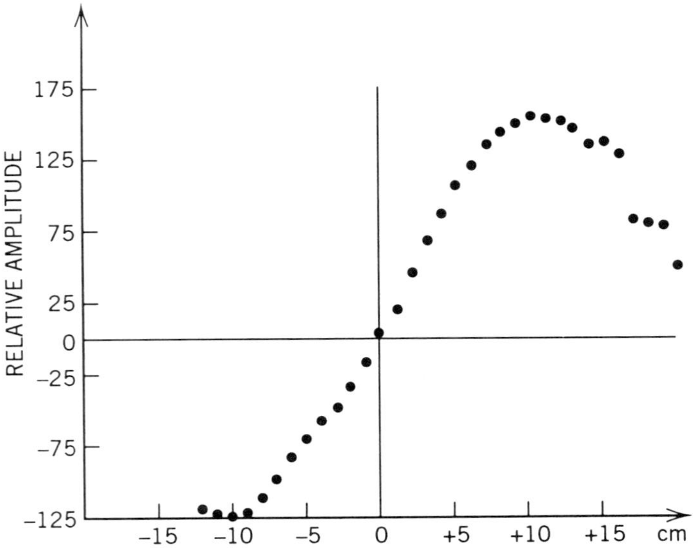

Figure 6.22. Experimental measure of the relative photothermal lens signal as a function of the z parameter. The excitation laser was focused into the sample cell. (Reprinted with permission from Berthoud et al. 1985. Copyright 1985, American Chemical Society.)

the single-beam apparatus (Sheldon et al. 1982, Carter and Harris 1983a). However, the latter theory was derived for mode-matched conditions and the excitation beam waist radius changes with changing focus offset from the sample cell.

6.6. TWO-LASER PHOTOTHERMAL LENS SPECTROSCOPY

Separate excitation and probe lasers allow the use of tunable sources for photothermal lens spectroscopy. Tuning the wavelength of the excitation lasers while monitoring the photothermal lens signal allows photothermal absorption spectra to be obtained. Tunable dye lasers were employed successfully in some of the earliest applications of photothermal lens spectroscopy. Continuous ion laser pumped dye lasers have been employed for the excitation source (Long et al. 1976, Swofford and Morrell 1978). Earlier dye lasers were restricted in the wavelength range, operating primarily at wavelengths of 500 to 1100. Current technology has extended this range. Present ultraviolet Ar^+ laser pumped dye lasers are tunable to about 370 nm in the ultraviolet, and intracavity frequency doubling can produce continuous wavelengths as short as 270 nm. The power is low in the ultraviolet. A typical photothermal spectroscopy design is to use a chopped, tunable dye laser excitation source with a continuous helium–neon (632.8 nm) probe laser to monitor the resulting photothermal lens. Colored glass or interference filters, or even dispersion devices, are used to remove light from the excitation laser prior to pinhole aperture used to develop the lens signal.

Twarowski and Kliger (1977b) used a pulsed tunable dye laser excitation source with a continuous probe laser to measure the multiphoton absorption spectrum of benzene vapor. Common pump sources for pulsed dye lasers include N_2, excimer, YAG, and flashlamps. Pulse durations vary from a few picoseconds up to microseconds. This technique is not restricted to visible wavelengths since adequate energy is produced through a wavelength range from 340 to 1100 nm in the fundamental. The wavelength range can be extended to about 220 nm using frequency doubling or nonlinear wave mixing, although at about one-tenth the energy of the fundamental dye laser wavelength.

Higashi et al. (1984) used a tunable infrared CO_2 gas laser to measure infrared absorption spectra of gas-phase species. The carbon dioxide laser is not continuously tunable but produces a series of about 70 lines, corresponding to specific rovibrational transitions of the CO_2 molecule. Nonetheless, enough information was obtained to allow species identification. There is no reason why other gas lasers could not be used for infrared photothermal lens spectroscopy. However, the CO_2 laser is preferred due to the atmospheric transmission window at 10 μm.

In all this apparatus, the excitation laser was focused into the sample cell while the probe laser focus was set to some other position, presumably to maximize the photothermal lens signal as described above. Excitation laser power or energy is monitored and the photothermal lens signal is corrected for tunable laser output variations with wavelength. This arrangement allows one to obtain a photothermal absorption spectrum. As discussed in the next section, the photothermal absorption spectrum is different from that which would be obtained using a conventional spectrophotometer. Even accounting for the wavelength-dependent response of the detector used to monitor the excitation laser energy, there is a wavelength-dependent focus spot size that must be accounted for to obtain accurate sample absorbance information.

6.6.1. Excitation Wavelength Dependence in Two-Laser Photothermal Spectroscopy

Twarowski and Kliger (1977b) and Fang and Swofford (1983) point out that the photothermal lens absorption spectrum obtained by scanning the excitation dye laser wavelength will be slightly different from that obtained using conventional absorption spectrophotometry. This is because the laser beam waist radius changes with wavelength. They apply the basic theory of laser operation discussed by Yariv (1985) to deduce how the beam waist radius changes with wavelength.

The cavity confocal distance, z_c, is a fundamental and constant parameter of the laser cavity. When the operating wavelength is changed, the cavity's confocal distance remains the same, but the minimum electric field beam waist radius varies according to the relationship $w_c^2 = \lambda z_c/\pi$, where w_c is the electric field beam waist radius in the cavity. The position of the minimum beam waist in the laser does not change with wavelength since it is a function only of the laser cavity resonator design (Yariv 1985). Focusing the laser beam with a lens placed a distance z from the minimum beam waist will produce a spot with a minimum radius of (see Chapter 4, Equation 88)

$$\frac{\pi w_0^2}{\lambda} = z_c \frac{f^2}{(z - f)^2 + z_c^2} \tag{112}$$

Since all terms on the right-hand side of this equation, including z_c, are constant, the focus spot size will vary with wavelength as $w_0 \propto \lambda^{1/2}$. It is also important to note that focus position will not change if an achromatic lens or mirror focusing element is used (see Chapter 4, Equation 87). Scanning the wavelength of the excitation laser will alter the photothermal signal only through the change in beam in beam waist radius. The beam waist radius affects both the strength of the photothermal lens and the characteristic time constant, t_c. The strength of the maximum photothermal lens varies as λ^{-1} for continuous laser excitation, and λ^{-2} for pulsed laser excitation, owing to the respective inverse beam waist dependence of the inverse focal lengths. For chopped or oscillating laser excitation with lock-in detection at the fundamental frequency, an additional wavelength dependence, as indicated by Equation (66), must be compensated for to obtain accurate absorption spectra.

Although the above were derived with photothermal lens spectroscopy in mind, similar relationships will result for photothermal deflection and even photothermal interferometric measurements. The signal dependence on excitation wavelength for these are also related directly to the changes in focused beam waist radius. For photothermal deflection, a change in excitation laser beam waist radius will affect the magnitude of the thermal-induced refractive index gradient and also the relative excitation to probe laser beam offset. In photothermal interferometry, a change in the excitation laser beam radius will affect the irradiance and thus the temperature change.

6.7. DIFFERENTIAL TWO-LASER APPARATUSES

Experimental configurations for measurement of differential photothermal lens signals can be employed for both chopped and pulsed laser excitation sources. Two main types of apparatus have been employed. The first type generates the difference signal optically, in much the same fashion as the differential apparatus first reported by Dovichi and Harris (1980a). This type of apparatus is useful for measuring the differential absorbance between blank and analyte solutions. The second type of apparatus generates the difference electronically. There are two main variations of the electronic-based differential apparatuses. In one variation, the difference in absorbance between two samples is measured. In the other variation, the differential absorbance at two wavelengths or two different polarizations can be measured.

Berthoud and Delorme (1987) describe two-cell optically generated differential absorbance apparatus that utilizes excitation and probe laser beam geometry and is very similar to the single-laser differential apparatus. This apparatus uses excitation and probe lasers focused at a position that lies between the two sample cells. The excitation and probe lasers are mode matched in the respect that the focuses are at the same position. A chopped

Ar^+ laser pumped dye laser is used for sample excitation, and a continuous helium–neon laser probes the resulting differential lens response. Lock-in amplifier signal processing is used to enhance the differential signal. Optical nulling of the optically generated differential response was performed by placing one of two sample cells on a translation stage. The sample cell could be translated to compensate for a change in beam focus position due to strong photothermal lens. Berthoud and Delmore could obtain differential absorbance spectra of lanthanide (Nd^{3+} and Pr^{3+}) ions in aqueous solutions by scanning the wavelength of the excitation laser and were able to detect Nd^{3+} in $HClO_4$ solutions to about 5×10^{-6} $\underline{M}$.

As discussed above, absorption spectra obtained using photothermal spectroscopy will, in general, not be the same as those obtained with a conventional spectrophotometer because of the wavelength-dependent focused beam waist radius of the excitation laser. Berthoud and Delmore point out that this effect is compounded in the differential two-cell apparatus since a change in excitation laser focus position occurs with wavelength due to changing analyte or blank absorption. Subsequently, the differential spectra are relative and though related to the conventional optical absorption spectra, will be equivalent to the latter only with compensation for excitation beam waist changes and in the limit of small absorbance in the sample cell placed first in the optical apparatus.

One limitation of the foregoing design is that the excitation laser beam waist radius is larger than the minimum possible. This physical restriction limits the possible magnitudes and thermal time constants that can be obtained. Erskine and Bobbitt (1989) recognized that the excitation beam could be split and focused into two sample cells independently. In their experiment, a continuous Ar^+ laser excitation source first passed through a polarization rotator and was then split into two beams with a polarization beamsplitter. By rotating the polarization angle, the power in the two excitation beams could be adjusted for optical null. These excitation beams were independently focused and the beam paths directed to cross the probe laser path at an oblique angle. Differential thermal lens response was generated by focusing a helium–neon probe laser between the two sample cells. The helium–neon laser was chopped to move the measurement band out of the flicker noise. In their experiment, the sample cell were connected to the effluent of a liquid chromatograph. The reduced time constant afforded by the smaller pumped laser beam waists allowed for sensitive detection even in the presence of mass transfer.

Probe laser power was monitored through a pinhole aperture with a photodiode, and a lock-in amplifier was used to process the signal. The two cells were connected in series so that the effluent first passed through one cell, then the other. This in turn resulted in a time-derivative response of the detector to the LC effluent. Integration of the time-differential response results in the typical chromatogram "peaks." The integration also serves to smooth the data. This innovative detection scheme resulted in measurements of differential absorbance at an impressive 2×10^{-7} AU level. This same technique was later found to be applicable to HPLC effluent measurement at the 2×10^{-7} AU level, with a background or common absorbance of 0.02 AU (Rice et al. 1991). The apparatus was also demonstrated to be useful for circular dichroism measurements in the latter.

Teramae and Winefordner (1987) have reported on the second type of apparatus for reference sample subtraction. This method generates the difference photothermal lens signal electronically. This two-cell photothermal lens spectrometer uses a chopped Ar^+ excitation laser. The excitation laser beam is split and focused into two separate sample cells. The output of a helium–neon probe laser is also split into two beams and used to probe the photothermal lens formed in the two sample cells. Two pinhole apertures and two photodiode detectors are used to detect the resulting signals. Essentially, two separate photothermal lens apparatuses were constructed with common excitation and probe lasers. The photodiode detectors are connected anode to cathode. Thus the total electronic current is the difference of the two photodiode currents. The electronically generated difference signal was processed with a lock-in amplifier. Detection limits (5×10^{-6} AU) were not as good as those of Erskin and Bobbitt but are nonetheless impressive.

Apparatus based on a similar concept was described by Yang and Hall (1988). This apparatus uses a single laser and technically is of single-laser design. However, the laser beam is split into analyte and reference beams and chopped synchronously such that the reference and analyte are excited at different times. The two beams are recombined using polarization rotation optics, and the common beam is directed through a single pinhole aperture and a detector. A lock-in amplifier is used to process the resulting signal. Since the lock-in amplifier detects only that oscillating component of the signal that is synchronous with the chopper, the output is proportional to the differential photothermal lens signal. The experimental concept is essentially the same as that used in conventional two-beam spectrophotometers. Detection limits reported by Yang and Hall compared favorably with those reported by Teramae and Winefordner for the duplicate two-laser two-cell photothermal lens design.

A single sample cell is used in the second variation of the electronically generated differential absorption apparatus. In this case the excitation laser is modulated to excite the sample at either two or more wavelengths or at different polarizations. Wavelength modulation allows in situ solvent absorption subtraction as long as the solvent absorbance is the same at all wavelengths. Polarization modulation is used to obtain circular dichroism measurements. Development work for this method has been pioneered by Tran and co-workers. In all of these papers, a modulated Ar^+ laser is used for sample excitation and a helium–neon laser is used as the probe. The usual thermal lens setup is used to detect the signal. For wavelength modulation studies, the Ar^+ laser is operated with all lines. In one apparatus (Franko and Tran 1988), a prism is used to separate the various wavelengths. These beams are asymmetrically chopped and recombined in the sample cell. More recently, an acoustooptical bandpass filter is used to select two or more Ar^+ wavelengths from 514.5 to 457.9 nm (Tran and Simianu 1992). Detector signals are recorded digitally and analyzed on a computer. Computer data analysis allows detailed analysis of the transient signals. However, this processing technique limits the dynamic reserve of the apparatus due to the absence of common-mode rejection and limited resolution of analog-to-digital converters. This limitation was overcome in later designs (Xu and Tran 1990a) by using a lock-in amplifier. Circular dichroism measurements were made by modulating the excitation laser from right to left circular polarization using either a Fresnel rhomb/chopper combination (Tran and Xu 1989) or a Pockels cell (Xu and Tran 1990a,b).

6.8. DIFFRACTION EFFECTS

Although the mathematics is somewhat tedious, it is relatively straightforward to determine the effect of the real (i.e., nonparabolic) refractive index perturbation on the propagation of the Gaussian probe laser beam. As was shown in Chapter 4, there are two equivalent methods to solve for Gaussian beam propagation when there is a complex transmission. The first method is to use Fresnel diffraction integrals. The second method is performed by adding a phase shift in the Gaussian beam electric field equation. Results obtained using either of these two methods should be equivalent; the choice depends on which is simplest for the case at hand. Sheldon et al. (1982) used a diffraction integral approach to solve for the signal obtained for single-laser photothermal lens. Approximations made in the derivations are similar to those of Fresnel or even Fruanhofer diffraction, and results are valid only for far-field detection plane or pinhole aperture placement. The thermal refractive index perturbation was approximated as a linear phase shift. The results were later analyzed and compared to those using parabolic lens approximation by Higashi et al. (1983) and Carter and Harris (1984). The phase-shift modified Gaussian beam propagation equation method has been used to find both the radial power distribution for Gaussian refractive index perturbations under mode-matched conditions (Weaire et al. 1979) and the resulting radial probe power and photothermal lens signal for mode-mismatched pulsed-laser excitation (Bialkowski 1985, Sheik-Bahae et al., 1989, Castillo et al. 1994). The Fresnel diffraction integral method has been used to find the pulsed laser excited photothermal lens signal (Slaby 1986) and the probe laser radial power distribution (Power 1990). More recently, Shen et al. (1992) used the Fresnel diffraction approach to calculate the photothermal lens signal obtained with the two-laser mode-mismatched optical apparatus for continuous excitation. As in the work of Sheldon et al., they approximated the exponential phase shift in linear terms. However, the latter is the most general of the continuous excitation laser results since it can be used to predict the optimum beam conditions for the two-laser apparatus.

All of these reports use the same general procedure. First, the complex transmission is determined from the temperature change and the resulting refractive index change. Second, the effect that the passage of the Gaussian probe beam through a sample is found by multiplying by the complex transmission function. Implicit here is the assumption that the sample cell is thin enough that the probe beam waist radius does not change significantly on passing through the sample. In diffraction, this assumption is valid when the distance between the sample cell and the detection plane is much greater than the optical pathlength through the cell. Third, the perturbed probe beam field in the detection plane is found using either a diffraction integral or the Gaussian beam propagation equations.

6.8.1. Probe Laser Diffraction Effects for Pulsed Excitation

We examine first the case of a pulsed laser excited sample probed with a continuous laser. The geometry of Twarowski and Kliger will be used to facilitate comparison between the diffraction and refraction (i.e., simple lens results). In pulsed TEM_{00} laser excitation, the refractive index perturbation is of a radially symmetric Gaussian form. The time-dependent complex transmission is given by the exponential phase shift. The phase shift, in turn, is simply the time-dependent temperature change, multiplied by the thermooptical coefficient and the sample pathlength, l. Thus

$$t(r) = \exp[i\delta\phi(r)] \approx \exp\left[ikl\left(\frac{dn}{dT}\right)\delta T(r, t)\right]$$

$$t(r) \approx \exp\left\{k\left(\frac{dn}{dT}\right)\frac{2\alpha lQY_H}{\pi w^2(t)\rho C_P}\exp\left[-\frac{2r^2}{w^2(t)}\right]\right\} \tag{113}$$

where $\delta\phi(r)$ is the phase shift, $k = 2\pi/\lambda_p$ is the probe laser wave vector, and

$w^2(t) = w^2(1 + 2t/t_c)$ is the time-dependent Gaussian radius of the temperature change. The equation describing electric field of a probe laser, focused a distance, z', before the sample cell is the Gaussian beam equation at a distance z' from the minimum beam waist position. Without the longitudinal phase terms, the electric field of a probe laser is

$$\mathbf{E}(r, z') = \mathbf{E}_0 \frac{q(0)}{q(z')} e^{-ikr^2/2q(z')} \tag{114}$$

The longitudinal or z-axis phase terms do not affect the probe beam power and are neglected here in favor of simplicity. $q(z') = iz_{0,p} + z'$ is the complex beam parameter normally analyzed by the inverse relationship $1/q(z') = 1/R_p(z') - i2/kw_p^2(z')$ where a subscript p is used to indicate parameters for the probe laser. After passing through the sample, the electric field will be $\mathbf{E}(r, z')t(r')$.

In diffraction theory, the electric field in the detection plane, placed a distance, d, past the sample, is found by performing the integrations or Fourier transforms required in Fresnel diffraction calculations. For the rather complicated form of the phase-shift term produced by the Gaussian temperature change, integration requires some type of simplification. In all cases reported, this simplification is accomplished by expanding the exponential transmission in a series as shown by Weaire et al. (1979):

$$\begin{aligned} t(r) &= \exp[i\delta\phi(r)] = \sum_{m=0}^{\infty} \frac{[i\phi(t)]^m}{m!} e^{-2mr^2/w^2(t)} \\ \phi(t) &= -k\left(\frac{dn}{dT}\right)\frac{2\alpha l Q Y_H}{\pi\rho C_p w^2(t)} \end{aligned} \tag{115}$$

With this definition, the negative photothermal lens generated in most samples results in positive ϕ. Since the series expansion is a sum over terms with a Gaussian radial functionality, and since the product of two Gaussians is also a Gaussian, the electric field in the detector plane can be found by simply modifying the complex beam parameter of each Gaussian beam in the series using the ABCD method for Gaussian beam propagation discussed in Chapter 4. The derivation used herein will use this method. As was shown in Chapter 4, results obtained using Fresnel diffraction are equivalent to those of the ABCD method for Gaussian beams.

Multiplying the electric field by the transmission, the electric field just past the sample cell is the sum

$$\mathbf{E}(r, z') = \mathbf{E}_0 \frac{q(0)}{q(z')} \sum_{m=0}^{\infty} \frac{[i\phi(t)]^m}{m!} e^{-ikr^2/2q_m(t)} \tag{116}$$

The complex beam parameter, $q_m(t)$, is a function of both excitation and probe laser beam waists,

$$\begin{aligned} \frac{1}{q_m(t)} &= \frac{1}{R_p(z')} - i\,\frac{2}{kw_m^2} \\ \frac{1}{w_m^2} &= \frac{1}{w_p^2(z')} + \frac{2m}{w^2(t)} \end{aligned} \tag{117}$$

where $R_p(z') = (z_0^2 + z'^2)/z'$ and $kw_p^2(z')/2 = (z_0^2 + z'^2)/z_0$ are the Gaussian beam parameter definitions of the unperturbed probe laser. It may also be shown that the complex beam parameter is $1/q_m(t) = 1/q(z') - i[4m/kw^2(t)]$. Since the effect of a lens on the propagation of a Gaussian beam is found from $1/q - 1/f$, the Gaussian probe laser beam apparently experiences a series of complex lenses of focal lengths $f = -ikw^2(t)/4m$. Using the *ABCD* method, the complex beam parameter describing the electric field in the detection plane some distance, d, past the sample cell is $q_m(t) + d$. The resulting electric field in the detection plane is

$$\mathbf{E}(r, d) = \mathbf{E}_0 \frac{q(0)}{q(z')} \sum_{m=0}^{\infty} \frac{[i\phi(t)]^m}{m!} \left[\frac{q_m(t)}{q_m(t) + d}\right] e^{-ikr^2/2(q_m(t)+d)} \tag{118}$$

For Gaussian beam propagation of a single beam, only the real part of the inverse complex beam parameter is retained. Here the complex electric field amplitude is retained for each Gaussian so that the electric fields may cancel in the superposition. This result shows that the electric field in the detection plane is a series of Gaussian beams, each with a different beam waist radius and phase. The first term in the series describes the unperturbed probe laser beam. Subsequent terms reflect the corrections due to the photothermal perturbation. The series will converge rapidly for small $\phi(t)$. Often, retaining only the first two terms is sufficient for describing the photothermal lens signal.

This electric field result is nearly identical to that obtained based on Fresnel diffraction (Slaby 1986, Power 1990). Besides trivial differences due to parameter definitions, the only real difference between the Fresnel diffraction and a Gaussian beam results is the linear longitudinal phase terms. These linear longitudinal phase shift terms could have been included. They were left out of equation 118 because they cancel in the probe power calculation. Equation (118) is amended easily to include this term by multiplying by $\exp[-i(z' + d)]$.

This result must be put into perspective with those obtained by Weaire et al. (1979) and Bialkowski (1985). Recall that there are actually two parts to the longitudinal phase shift, a linear term depending only on z-axis distance, and a tangent term that depends on z-axis distance and also the confocal distance, z_0. Weaire et al. and Bialkowski included both longitudinal phase-shift terms in keeping with the Gaussian beam propagation solution to Maxwell's equations. With the tangent phase term, the real part of the preexponential term, $\mathscr{R}\{[1 + d/q_m(t)]^{-1}\}$, is used. Strictly speaking, the results of Weaire et al. and Bialkowski should be more accurate than those obtained using Fresnel diffraction since the small-angle approximation implied in Fresnel diffraction is not imposed in the Gaussian beam propagation equation. In the end analysis, Fresnel diffraction, Gaussian *ABCD*, and Gaussian beam results should yield equivalently accurate predictions for far-field detection plane optical configurations.

Although exact to within the approximations of Fresnel diffraction, this result does not lend itself to easy interpretation. In addition to the dependence on the probe focus, z', and detection plane, d, positions, and the time-dependent photothermal perturbation strength, $\phi(t)$, the diffraction result is a function of the radial offset in the detection plane, r, and excitation and probe laser beam waist radiuses, $w(t)$ and $w_{0,p}$. There is no simple way to analyze these data. The probe laser power in the detection plane is calculated by first performing the sum over the electric field component, then taking the complex square of the field. Performing this calculation for all possible configurations is time consuming at best. The amount of information generated would be enormous. The discussions below will be restricted to the results obtained using a pinhole aperture to monitor the thermal lens signal. Diffraction results using other types of detection plane masks can be found in Slaby (1986, 1987, 1989). The probe laser radial power distribution in the detection plane has been discussed by Power (1990).

Of primary concern are the effects of the relative distances of the probe laser focus and the detection plane, and the probe laser beam waist radius. First, photothermal lens signals calculated for an excitation beam waist greater than about five probe beam waists are identical to that predicted by the refraction equation. Only when the probe laser beam waist becomes on the order of, or greater than, the pump waist does diffraction theory need to be used. Shown in Figure 6.23 is the result of a diffraction calculation for the relative photothermal lens signal for such a case. In this plot the pulsed laser excitation beam radius is $w = 20\ \mu$m and a probe laser beam waist is $w_{0,p} = 100\ \mu$m. The probe laser wavelength is 632.8 nm and the confocal distance is $z_{0,p} = 5$ cm. A photothermal phase shift of $\phi(t) = 10^{-5}$ is used since most samples have negative thermooptical coefficients. For the plots shown here, the series was summed up to the 20-power term. This is not usually necessary since convergence typically occurs with two or three terms.

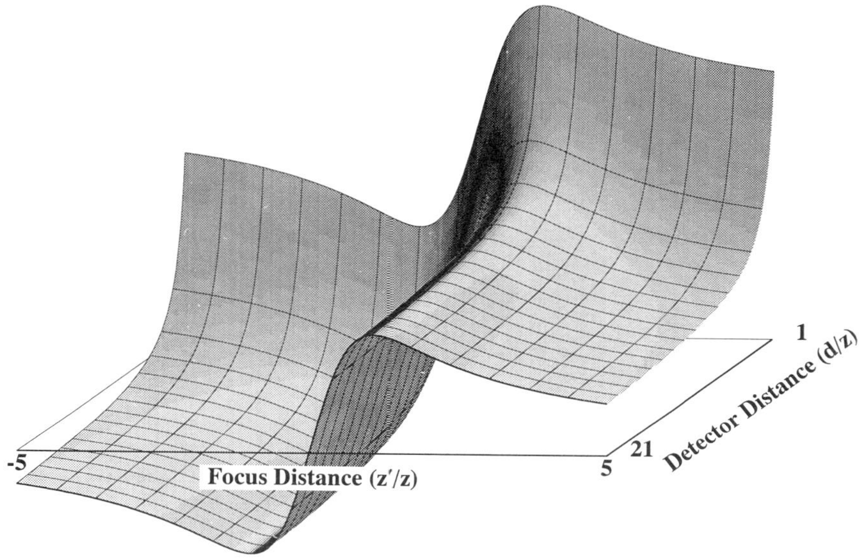

Figure 6.23. Pulsed laser excited photothermal lens signal predicted from diffraction theory, as a function of the probe laser beam geometry. The excitation laser beam waist was 20 μm in the sample. the minimum 632.8-nm probe laser beam radius was 100 μm. The photothermal perturbation was small and the signal was defined on the usual fashion.

The plot is oriented with the detection plane distance increasing toward the viewer. The probe laser focus position varies across the surface. The $d = 0$ point was not plotted. The theoretical signal is found to be zero at this point, independent of z' distance. In addition, near-field points with $d \leqslant w_p$ are of questionable value due to the small-angle approximation of Fresnel diffraction. These near-field points should be calculated using the procedures of Weaire et al. and Bialkowski.

The signal behavior predicted by this example calculation is clearly different from that using the refractive lens theory. In refraction theory, the signal "levels off" with increasing detection plane distance. In contrast, the signal calculated using the more accurate diffraction theory decreases with detection plane distance. With the detection plane near the sample cell, the signal is initially positive for negative z', indicating a decrease in probe laser power with formation of the photothermal lens. This is due only to the fact that the focus position of the probe beam is beyond the detection plane. In fact, the same behavior can be seen in Figure 6.22 for refractive lens theory. For detection plane distances far from the sample cell, the signal exhibits a sigmoidal dependence with probe laser beam focus position. This is in contrast to refraction theory, wherein the signal is approximately proportional to z' at large d. The sigmoidal dependence is similar to that observed by Berthoud et al. (1985), although they used a chopped excitation laser. However, the maximum calculated diffraction signal occurs at $z' \approx z_0$, not $\sim\sqrt{3z_0}$, for this set of calculation parameters.

Figure 6.24 illustrates the effect of relative probe laser beam waist on the theoretical pulsed laser excited photothermal lens signal. In this case the detection plane is fixed at a distance of 10 m (i.e., farfield) and the probe laser focus position is prior to the sample. The relative minimum probe laser beam waist radius varies logarithmically from a factor of 0.1 to 10 of that of the excitation beam. The excitation beam waist radius is 100 μm, the probe laser wavelength is 632.8 nm, and the phase shift was 10^{-5}. The figure spans a range from that adequately predicted by refraction theory to that wherein diffraction must be used. It is interesting to note that the diffraction result reproduced the trends predicted from refractive optics in the appropriate region. A maximum signal strength value appears for probe beam

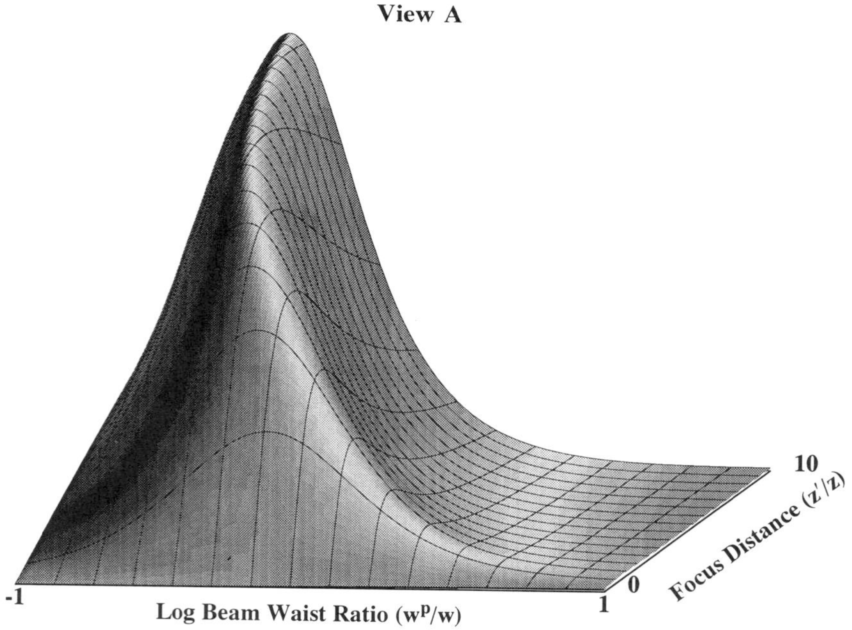

Figure 6.24. Far-field diffraction theory predictions for pulsed laser excited photothermal lens signal as a function of the relative probe laser beam waist radius and focus position. The detector plane is at $d = 10$ m, the excitation beam radius was 100 μm in the sample, and the perturbation was small (10^{-5}). Two views are given to allow inspection of the predicted surface. In view B, the line on the right-hand side is equivalent to that predicted from refractive optics.

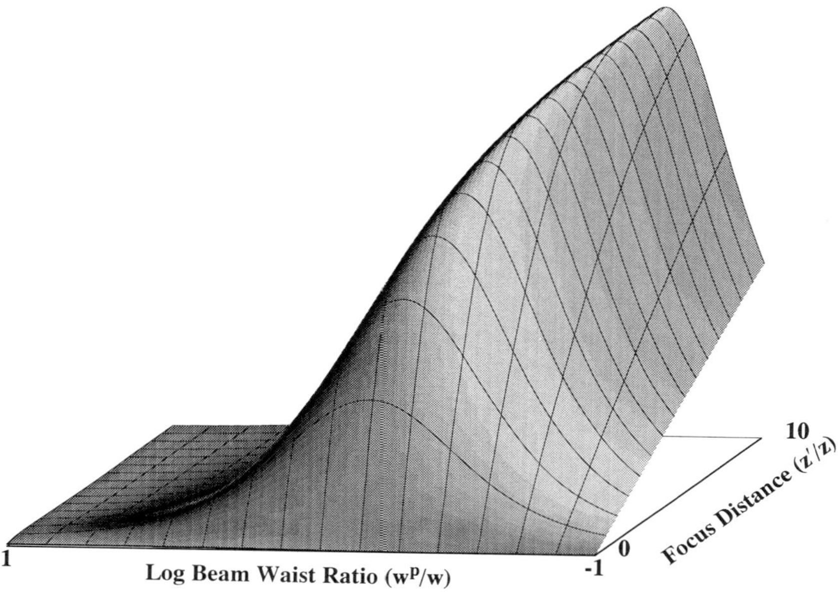

Figure 6.24. Continued

waist radiuses slightly less than that of the excitation source. Whether or not this indicates a trend can be addressed only by examining the effects of detection plane position. This figure also shows that there is no single probe laser beam focus position that optimizes the signal. There apparently is no "best" z'/z_0 value for pulsed laser excited photothermal lens signals predicted by diffraction.

Figure 6.25 shows the photothermal lens signal response calculated for the near-field condition of $d = 5$ cm. Of interest here is the apparent flat signal region for large probe beam waist. Taken in combination with the trends shown in Figure 6.23, wherein it is shown that the signal increases with decreasing sample-to-detection plane distance, it would seem that a good optical configuration for pulsed laser excited photothermal lensing is one where a relatively large probe laser is focused several $z_{0,p}$ values prior to the sample and the detection plane is close to the sample. The latter can be accomplished by placing the unfocused probe laser close to the sample. This extremely simple optical configuration should be tested.

Diffraction affects the time-dependent decay as well as the of linearity of the photothermal lens signals. These effects are discussed by Bialkowski (1985). In general, these effects vary in a complex fashion with the relative

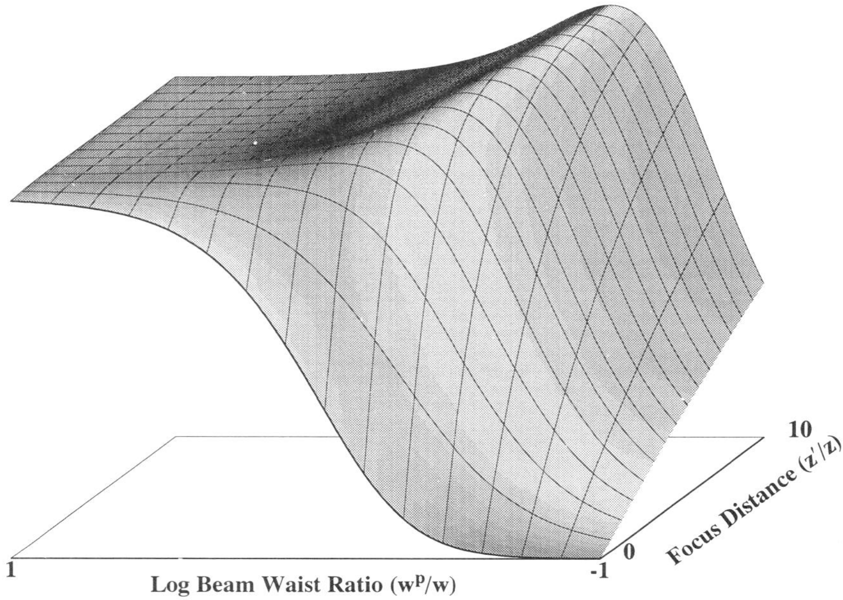

Figure 6.25. Near-field diffraction theory predictions for pulsed laser excited photothermal lens signal as a function of the relative probe laser beam waist radius and focus position. The detection plane is at $d = 5$ cm in this case. All other parameters are the same as in Figure 6.24.

beam waist radiuses and the optical geometry of the probe and detection plane positions. It is generally difficult to measure accurately the optical parameters required to model the signal. Bialkowski suggests that to obtain accurate estimations, each experimental setup must be calibrated in terms of the temporal response and the concentration linearity.

Insight into how experimental geometry affects the time-dependent signal can be gained through an approximate analytical expression for the diffractive photothermal lens signal. Sheldon et al. (1982) used only the first two terms in the series (e.g., $m = 0$ *and* $m = 1$) to approximate the probe laser power passing through the pinhole aperture in the detection plane. This is equivalent to approximating the exponential phase shift by, $\exp(i\delta\phi) \approx 1 + i\delta\phi$. The higher-order expansion terms are necessary only for accurate signal prediction in highly absorbing samples and/or using high-power excitation sources. Sheldon, et al. argue that since the phase shift induced in most photothermal lens experiments is very much less than one, higher-order series approximations are not necessary. Using this approximation allows an analytical expression for the probe laser power transmitted through the pinhole aperture. Using only the first two terms in the series and retaining terms only up to first order in the photothermal phase shift, the probe laser power ratio is

$$\frac{\Phi(t)}{\Phi(\infty)} \approx 1 - \phi(t) \frac{8dkw^2(t)(z_{0,p}^2 + z'^2 + dz')}{k^2w^4(t)(z_{0,p}^2 + z'^2 + 2dz' + d^2) + 8kw^2(t)d^2z_{0,p} + 16d^2(z_{0,p}^2 + z'^2)} \tag{119}$$

If the detector–pinhole plane is in the far field,

$$\frac{\Phi(t)}{\Phi(\infty)} \approx 1 - \phi(t) \frac{8kw^2(t)z'}{k^2w^4(t) + 8kw^2(t)z_{0,p} + 16(z_{0,p}^2 + z'^2)} \tag{120}$$

The photothermal lens signal is calculated from the power ratio. A simple expression for the time-dependent signal is found by assuming that $\phi z' \ll z_{0,p}$ and $z'^2 \ll z_{0,p}^2$ in the denominator of the signal expression. With these assumptions

$$S_{\text{pulsed}}(t) \approx \frac{2z'}{f_{\text{pulsed}}(0)} \frac{1}{(1 + 2w_{0,p}^2/w^2 + 2t/t_c)^2} \tag{121}$$

where $1/f_{\text{pulsed}}(0) = 4\phi(0)/kw^2$ is the definition for the initial pulsed laser photothermal lens focal length predicted from refractive optics theory. The result of this simplification is strikingly similar to that predicted using refraction optics under the same conditions. In fact, when the probe laser beam waist, $w_{0,p}$, is much smaller than that of the excitation source, the diffraction and refraction methods yield equivalent results. For finite probe beam radiuses, the maximum ($t = 0$) signal will be a factor of $w^2/(w^2 + 2w_{0,p}^2)$ smaller than that predicted by refraction.

The exact probe laser beam focus position resulting in the optimum signal past a pinhole placed in the far field can be obtained from the far-field results in equation 120. Solving for the initial photothermal lens signal, then differentiating with respect to z', results in the optimum focus position, $z'_{\text{opt}} = \pm z_{0,p}(1 + w_0^2/2w_{0,p}^2)$. Using this result, the optimum initial lens signal can be found. Substituting into the signal equation and eliminating terms nonlinear in ϕ, the optimum photothermal lens signal is

$$\begin{aligned} S_{\text{pulsed,opt}}(t) &= \frac{\Phi(t) - \Phi(\infty)}{\Phi(t)} \\ &= \pm\left(\frac{dn}{dT}\right) \frac{8\alpha lQY_H}{\lambda_p \rho C_P} \frac{2w_{0,p}^2 + w_0^2}{[2w_{0,p}^2 + w_0^2(1 + 2t/t_c)]^2 + (2w_{0,p}^2 + w_0^2)^2} \end{aligned} \tag{122}$$

and the zero-time (maximum) signal is

$$S_{\text{pulsed,opt}}(0) = \frac{\Phi(0) - \Phi(\infty)}{\Phi(0)} = \pm\left(\frac{dn}{dT}\right)\frac{4\alpha l Q Y_H}{\lambda_p \rho C_P(2w_{0,p}^2 + w_0^2)} \tag{123}$$

As with the approximate expression 121, this result explicitly shows how the probe laser beam waist radius affects the photothermal lens signal.

6.8.2. Probe Laser Diffraction Effects for Continuous Excitation

The probe laser transmission function for the continuous laser excitation case can be obtained from the time integral of the pulsed laser temperature change in the exponent of the transmission function. Time-integration for a Gaussian source yields the temperature change produced for continuous laser excitation. In this case, the transmission is

$$t(r) = \exp\left[i\int_0^t \delta\phi(r, t')dt'\right] = \sum_{m=0}^{\infty}\frac{i^m}{m!}\left[\int_0^t \phi(t')\exp\left(\frac{-2r^2}{w^2(t')}\right)dt'\right]^m \tag{124}$$

Sheldon et al. (1982) and Shen et al. (1992) use the phase shift approximation, $\exp(i\delta\phi) = 1 + i\delta\phi$, retaining terms only up to first order in the photothermal phase shift. This linear approximation allows analytical solutions for the diffraction result. The electric field at the detector plane is

$$\mathbf{E}(r, d) \approx \mathbf{E}_0\frac{q(0)}{q(z'+d)} + \mathbf{E}_0\frac{q(0)}{q(z')}\int_0^t i\phi(t')\left(\frac{q_1(t')}{q_1(t')+d}\right)\exp\left(\frac{-ikr^2}{2(q_1(t')+d)}\right)dt' \tag{125}$$

For the pinhole detection scheme used in photothermal lens spectroscopy, the integral is tractable since the radius is zero and thus the exponential term is unity. With a little algebra, the integral can be rewritten in terms of the integration variable, $T = 1 + 2t/t_c$. It is simple enough to calculate the probe laser power prior to performing the integration over time:

$$\frac{\Phi(t)}{\Phi(0)} - 1 \approx \frac{t_c\phi}{2}\int_1^{1+2t/t_c}\left\{\frac{iq(z'+d)}{q(z')}T^{-1}\frac{q(z')T - i4/kw^2}{[q(z')+d]T - i4/kw^2} + \text{C.C.}\right\}dT \tag{126}$$

where C.C. is the complex conjugate of the first term. The probe laser power ratio for continuous excitation is

$$\frac{\Phi(t)}{\Phi(0)} \approx 1 + t_c\phi \tan^{-1}\left\{\frac{kw^2[z_{0,p}^2 + (z' + d)^2] + 4d^2 z_{0,p}}{4d(z_{0,p}^2 + z'^2 + dz')}\right\}$$
$$- t_c\phi \tan^{-1}\left\{\frac{kw^2(1 + 2t/t_c)[z_{0,p}^2 + (z' + d)^2] + 4d^2 z_{0,p}}{4d(z_{0,p}^2 + z'^2 + dz')}\right\} \quad (127)$$

Unlike the results of Sheldon et al., who calculate only the far-field result, the result above gives the relative power change for any set of experimental parameters. In continuous laser excited photothermal lens spectrometry, the maximum signal is for long irradiation time. Taking the limit as time approaches infinity and calculating the signal in the usual fashion results in the signal expression

$$S_{cw}(\infty) = \frac{\Phi(0) - \Phi(\infty)}{\Phi(\infty)} \approx \frac{1}{1 - t_c\phi \tan^{-1}\left\{\dfrac{4d(z_{0,p}^2 + z'^2 + dz')}{kw^2[z_{0,p}^2 + (z' + d)^2] + 4d^2 z_{0,p}}\right\}} - 1 \quad (128)$$

The signal magnitude is a function of the $t_c\phi$ product. This product can be represented in several forms:

$$-t_c\phi = \frac{w^2 \rho C_P}{4\kappa} k \left(\frac{dn}{dT}\right) \frac{2\alpha l \Phi_0 Y_H}{\pi \rho C_P w^2} = \left(\frac{dn}{dT}\right) \frac{\alpha l \Phi_0 Y_H}{\lambda_p \kappa} = \frac{\lambda}{\lambda_p} \frac{z_0}{f_{cw}(\infty)} \quad (129)$$

The relationship 129 shows the relationship to the inverse focal length calculated from refractive optics theory. This connection may be used to make adjustments to the result, for example, in the case where the effective sample pathlength is limited by the beam divergence of the excitation laser.

Equation 128 is plotted in Figure 6.26 for a case where the probe laser beam waist is much larger than that of the excitation source. Parameters used to generate this plot are the same as those of Figure 6.23, with the exception that the magnitude term is here $t_c\phi = -(dn/dT)\alpha l \Phi_0 Y_H/\lambda\kappa$. Signals above the plane correspond to a decrease in probe power and those below correspond to an increase in power. For detector plane positions near the sample, only positive signals result. This is because the negative focal length of the thermal lens causes blooming of the probe beam. As the detector plane is moved away from the sample, the signal can be positive or

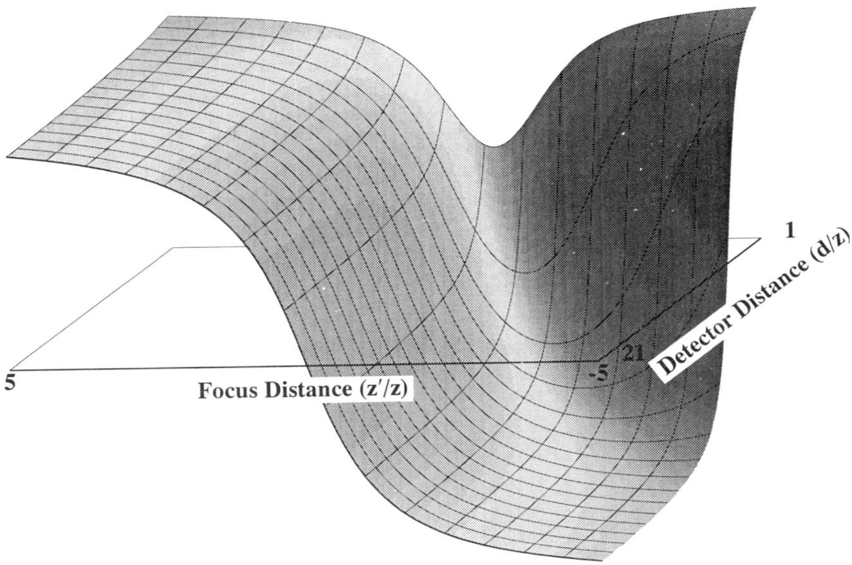

Figure 6.26. Continuous laser excited photothermal lens signal predicted from diffraction theory. The signal magnitude is defined by $\Phi(\infty)\,\Phi(0)/\Phi(0)$. The excitation laser beam waist is 20 μm and the 632.8 nm probe laser waist is 100 μm. The signal is apparently maximum at small detection plane distances.

negative, depending on whether the probe laser was focused prior to or beyond the sample. Within the range of parameters used in this model calculation, a maximum signal occurs for near-field conditions (e.g., $d = z_{0,p}$, and for $z' = \pm 5z_{0,p}$). On comparing this plot to that for pulsed laser excitation, it is apparent that the signal varies more slowly with z' in the far field. In fact, the behavior is similar to that predicted by refraction theory for probe focus positions within $\pm z_{0,p}$. The reason for this may be due to the fact that the continuous laser-excited photothermal perturbation is broader than the initial perturbation generated using a pulsed source. One final feature of this plot that should be noticed is the apparent lack of a signal maximum with respect to z' position. This is confirmed by solving the differential of equation 127 with respect to z'. Apparently, the continuous laser-excited photothermal lens signal is optimized for large probe-beam focus positions, but there is no one best probe laser focus position.

The relative power resulting from probe-beam diffraction may also be compared to the previous studies of diffraction effects in two-laser continuous-excitation photothermal lens spectroscopy. To compare this with previous results, the detection plane is taken to be far from the sample cell. For this geometry the resulting power is

$$\lim_{d\to 0} \frac{\Phi(t)}{\Phi(0)} \approx 1 + t_c\phi \tan^{-1}\frac{kw^2 + 4z_{0,p}}{4z'} - t_c\phi \tan^{-1}\frac{kw^2(1 + 2t/t_c) + 4z_{0,p}}{4z'} \tag{130}$$

This result also leads to a compact expression for the maximum probe laser power change that would occur for infinitely long excitation laser irradiation times. In this case

$$\frac{\Phi(\infty)}{\Phi(0)} - 1 \approx -t_c\phi \tan^{-1}\left[\frac{4z'}{k(w^2 + 2w_{0,p}^2)}\right] \tag{131}$$

It is interesting that the two-laser apparatus appears to be linear in absorbed energy, $t_c\phi$, when the photothermal lens signal is defined by $[\Phi(\infty) - \Phi(0)]/\Phi(0)$. This expression also shows the signal dependence on the excitation and probe laser beam waist radiuses. Decreasing either of these will enhance the photothermal lens signal detected in the far field for a given z', with the caveat that decreasing the excitation laser beam waist may decrease the effective pathlength throughout the sample.

6.8.3. Diffraction Effects for Single-Laser Photothermal Lens

Diffraction results for a single-laser photothermal apparatus can be found from the continuous-excitation two-laser case by simply equating excitation and probe laser beam parameters. Sheldon et al. introduced a different way to define the continuous-laser-excited photothermal lens signal that yields a more compact result for diffraction theory. They defined the time-dependent signal as a ratio of the power change to that which occurs at infinite time. Using this definition, the time-dependent photothermal lens signal obtained in the far field is

$$S_{cw}(t) = \frac{\Phi(t) - \Phi(\infty)}{\Phi(\infty)} = \frac{t_c\phi \tan^{-1}\{4z'/[kw^2(1 + 2t/t_c) + 4z_0]\}}{1 - t_c\phi \tan^{-1}[4z'/(kw^2 + 4z_0)]} \tag{132}$$

We use z_0 here since there is no distinction between the excitation and probe laser in the single-laser photothermal lens apparatus. For the mode-matched conditions of a single-laser experiment, w and z_0 are related. By first substituting $w_0^2(1 + z'^2/z_0^2)$ for w^2, then using the definition for $z_0 = kw_0^2/2$, the far-field signal for the single-laser apparatus is

$$S_{\text{cw}}(t) = \frac{t_c\phi \tan^{-1}\left\{\dfrac{2z'/z_0}{(z'/z_0)^2 + 3 + 2[(z'/z_0)^2 + 1]t/t_c}\right\}}{1 - t_c\phi \tan^{-1}\{(2z'/z_0)/[(z'/z_0)^2 + 3]\}} \tag{133}$$

With this signal definition, the maximum power change is

$$S_{\text{cw}}(0) = \frac{\Phi(0) - \Phi(\infty)}{\Phi(\infty)} = \frac{1}{1 - t_c\phi \tan^{-1}\{(2z'/z_0)/[(z'/z_0)^2 + 3]\}} - 1 \tag{134}$$

It is straightforward to show by derivative methods that the maximum signal will occur when $z' = \pm\sqrt{3}z_0$. Finally Sheldon et al. point out that the time-dependent signal measured with this optimum optical geometry will be given by the signal equation evaluated at the optimum focus-sample offset. In this case

$$S_{\text{cw,opt}}(t) = \frac{t_c\phi \tan^{-1}[3^{1/2}/(3 + 4t/t_c)]}{1 - t_c\phi(\pi/6)} \tag{135}$$

Aside from difference in mathematical structure and notation, these results are equivalent to those obtained by Sheldon et al. Minor difference result from the fact that Sheldon et al. cast their results in terms of the ratio t_c/t and that t_c was included in the signal strength term, ϕ.

6.8.4. Effect of Diffraction on the Thermal Lens Enhancement Factor

The diffraction optics results for pulsed laser–excited photothermal lens spectroscopy yields an approximate signal which clearly shows the effect of the probe laser beam waist radius on the resulting signal:

$$S_{\text{pulsed}}(t) = \frac{\Phi(t) - \Phi(\infty)}{\Phi(t)} \approx \frac{2z'}{f_{\text{pulsed}}(0)} \frac{1}{(1 + 2w_{0,p}^2/w^2 + 2t/t_c)^2} \tag{136}$$

where the maximum inverse thermal lens focal length occurs at zero time and is given by

$$f^{-1}_{\text{pulsed}}(0) = \left(\frac{dn}{dT}\right)\frac{8\alpha l Y_H Q}{\pi w^4 \rho C_p} \tag{137}$$

where w (m) is the radius of the excitation laser beam at the sample. We will use this approximate result to determine the theoretical photothermal lens enhancement factor for mode-matched conditions. Using the method for determining the pulsed excitation laser photothermal lens enhancement (Section 6.5.3) for the condition where the excitation and probe lasers are focused at the same position along the z-axis, it can be shown that the enhancement is given by

$$E_{\text{pulsed}} \approx \left|\left(\frac{dn}{dT}\right)\right| \frac{3^{3/2} Y_H Q}{\lambda \rho C_P (w_0^2 + 2w_{0,p}^2)} \tag{138}$$

A similar expression can be found using the optimum pulsed laser photothermal lens signal in equation 123. However, in the later case, a numerical factor of 4, versus $3^{3/2}$, occurs in the numerator and the wavelength is that of the probe laser. In the enhancement factor above, w_0 and λ are the minimum beam waist radius and wavelength of the excitation laser. The implication here is that the excitation laser is focused into the sample. The distance to the optimum excitation and probe laser beam focus is $z = z' = z_0$, where z_0 is the confocal distance of the excitation laser. This expression shows that the theoretical enhancement is a function of both excitation and probe laser beam waists, a feature missing from refractive optics theory. However, if the minimum probe laser beam waist radius is much smaller than that of the excitation source, the theoretical enhancement factor is equal to that obtained using refractive optics. If, on the other hand, the optics of the apparatus are further restrained to have excitation and probe laser beam waists equal at the focus (i.e., mode matched), the enhancement is

$$E_{\text{pulsed}} \approx \left|\left(\frac{dn}{dT}\right)\right| \frac{3^{1/2} Y_H Q}{\lambda w_0^2 \rho C_P} \tag{139}$$

Compared to the theoretical enhancement obtained using refraction optics for mode-matched beams, the diffraction result is a factor of $\frac{1}{3}$ less.

The enhancement for single continuous laser excitation is easier to predict since the beam geometry for optimum signal is fixed. According to the diffraction theory description of the single-laser photothermal lens signal shown in section 6.8.3 the optimum signal for the detector plane placed in the far field is

$$S_{\text{cw,max}} = \frac{\Phi(0) - \Phi(\infty)}{\Phi(\infty)} = \frac{(\pi/6)t_c\phi}{1 - (\pi/6)t_c\phi} \tag{140}$$

For small thermal perturbations, the term $\pi t_c \phi/6$ is much less than 1 and can be neglected in the denominator. The continuous, single-laser photothermal lens enhancement factor, E_{cw}, defined as the ratio of the photothermal lens signal to that obtained using spectrophotometry, is then

$$E_{cw} = \frac{(\pi/6)t_c\phi}{\alpha l} = \left|\frac{dn}{dT}\right| \frac{\pi\Phi_0 Y_H}{\lambda\kappa} \tag{141}$$

which is a factor of $\pi/6 \approx 0.523$ of the enhancement factor obtained from refractive optics theory. Before the development of diffraction theory, experimental determination of the enhancement factor fell short of that predicted by refraction. The diffraction theory result explains why this was observed (Higashi et al. 1983).

REFERENCES

Abramowitz, M., and Stegun, I. A., eds. *Handbook of Mathematical Functions*, Dover, New York (1965).

Bailey, R. T.; Pugh, D.; and Cruickshank, F. R. *J. Chem. Soc. Faraday Trans. 2* **76** 633 (1980).

Barker, J. R. and Rothem, T. *Chem. Phys.* **68** 331 (1982).

Berthoud, T. and Delorme, N. *Appl. Spectros.* **41** 15 (1987).

Berthoud, T.; Delorme, N.; and Mauchien, P. *Anal. Chem.* **57** 1216 (1985).

Bialkowski, S. E. *Appl. Opt.* **24** 2792 (1985).

Bialkowski, S. E. *Anal. Chem.* **58** 1706 (1986).

Bialkowski, S. E. *Opt. Lett.* **14** 1020 (1989).

Bialkowski, S. E. and Herrera, S. *Anal. Chem.* **60** 1586 (1988).

Bialkowski, S. E. and He, Z. F. *Anal. Chem.* **60** 2674 (1988).

Bialkowski, S. E. and Long, G. R. *Anal. Chem.* **59** 873 (1987).

Bialkowski, S. E.; Gu, X.; Poston, P. E.; and Powers, J. S. *Appl. Spectrosc.* **46** 1335 (1992).

Campillo, A. J.; Petuchowski, S. J.; Davis, C. C.; and Lin, H.-B. *Appl. Phys. Lett.* **41** 327 (1982).

Carter, C. A. and Harris, J. M. *Appl. Spectrosc.* **37** 166 (1983a).

Carter, C. A. and Harris, J. M. *Anal. Chem.* **55** 1256 (1983b).

Carter, C. A. and Harris, J. M. *Appl. Opt.* **23** 476 (1984).

Castillo, J.; Kozich, V. P.; and Marcano O. A. *Opt. Lett.* **19** 171 (1994).

Cremers, D. A. and Keller, R. A. *Appl. Opt.* **21** 1654 (1982).

Davis, C. C. *Appl. Phys. Lett.* **36** 515 (1980).

Davis, C. C. and Petuchowski, S. J. *Appl. Opt.* **20** 2539 (1981).

Dovichi, N. J. *CRC Crit. Rev. Anal. Chem.* **17** 357 (1987).

Dovichi, N. J. and Harris, J. M. *Anal. Chem.* **51** 728 (1979).

Dovichi, N. J. and Harris, J. M. *Anal. Chem.* **52** 2338 (1980a).

Dovichi, N. J. and Harris, J. M. *Anal. Chem.* **52** 695A (1980b).

Dovichi, N. J. and Harris, J. M. *Proc. SPIE* **288** 372 (1981).

Erskine, S. R. and Bobbitt, D. R. *Appl. Spectros.* **42** 331 (1988).

Erskine. S. R.; Foley, E. S.; and Bobbitt, D. R. *Appl. Spectrosc.* **41** 1189 (1987).

Erskine, S. R. and Bobbitt, D.R. *Appl. Spectrosc.* **43** 668 (1989).

Fang, H. L. and Swofford, R. L. *J. Appl. Phys.* **50** 6609 (1979).

Fang, H. L. and Swofford, R. L. In *Ultrasensitive Laser Spectroscopy*, Kliger, D. S., ed., Academic Press, New York (1983).

Fluckgier, D. U.; Lin H.-B.; and Marlow, W. H. *Appl. Opt.* **24** 1668 (1985).

Fournier, D. and Boccara, A. C. In *Photothermal Investigations in Solids and Fluids*, Sell, J. A., ed., Academic Press, New York (1989).

Fournier, D.; Boccara, A. C.; Amer, N. M.; and Gerlach, R. *Appl. Phys. Lett.* **37** 519 (1980).

Franko, M. and Tran, C. D. *Anal. Chem.* **60** 1925 (1988).

Fung, K. H. and Lin, H.-B. *Appl. Opt.* **25** 749 (1986).

Glatt, I.; Karny, Z.; and Kafri, O. *Appl. Opt.* **23** 274 (1984).

Gordon, J. P.; Leite, R. C. C.; Moore, R. S.; Porto, S. P. S.; and Whinnery, J. R. *J. Appl. Phys.* **36** 3 (1965).

Gradshteyn, I. S. and Ryzhik, I. M. *Table of Integrals, Series, and Products*, Academic Press, New York (1980).

Higashi, T.; Imasaka, T.; and Ishibashi, N. *Anal. Chem.* **55** 1907 (1983).

Higashi, T.; Imasaka, T.; and Ishibashi, N. *Anal. Chem.* **56** 2010 (1984).

Horowitz, L.; Band, Y. B.; Kafri, O.; and Heller, D. F. *Appl. Opt.* **23** 2229 (1984).

Hu, C. and Whinnery, J. R. *Appl. Opt.* **12** 72 (1973).

Jackson, W. B.; Amer, N. M.; Boccara, A. C.; and Fournier, D. *Appl. Opt.* **20** 1333 (1981).

Jansen, K. L. and Harris, J. M. *Anal. Chem.* **57** 1698 (1985).

Kalaskar, S. D. and Bialkowski, S. E. *Anal. Chem.* **64** 1824 (1992).

Klein, M. V. *Optics*, Wiley, New York (1970).

Lee, W.-K.; Güngör, A.; Ho, P-T.; and Davis, C. C. *Appl. Phys. Lett.* **47** 916 (1985).

Light, B. *Laser & Applications* April, p. 75 (1986).

Lin, H.-B. and Campillo, A. J. *Appl. Opt.* **24** 442 (1985).

Lin, H.-B.; Gaffney, J. S.; and Campillo, A. J. *J. Chromatography* **206** 205 (1981).

Long, G. R. and Bialkowski, S. E. *Anal. Chem.* **56** 2806 (1984).

Long, M. E.; Swofford, R. L.; and Albrecht, A. C. *Science* **191** 183 (1976).

Longaker, P. R. and Litvak, M. M. *J. Appl. Phys.* **40** 4033 (1969).

Mazzoni, D. L. and Davis, C. C. *Appl. Opt.* **30** 756 (1991).

Miyaishi, K.; Imasaka, T.; and Ishibashi, N. *Anal. Chem.* **54** 2039 (1982).

Mori, K.; Imasaka, T.; and Ishibashi, N. *Anal. Chem.* **54** 2034 (1982).
Nickolaisen, S. L. and Bialkowski, S. E. *Anal. Chem.* **58** 215 (1986).
Plumb, D. M. and Harris, J. M. *Appl. Spectrosc.* **46** 1346 (1992).
Power, J. F. *Appl. Opt.* **29**, 52 (1990).
Rice, P. D.; Thorne, J. B.; and Bobbitt, D. R. *SPIE Proc.* **1435** 104 (1991).
Rose, A.; Vyas, R.; and Gupta, R. *Appl. Opt.* **25** 4626 (1986).
Sell, J. A. *Appl. Opt.* **23** 1586 (1984).
Sheik-Bahae, M.; Said, A. A.; and Van Stryland, E. W. *Opt. Lett.* **14** 955 (1989).
Sheldon, S. J.; Knight, L. V.; and Thorne, J. M. *Appl. Opt.* **21** 1663 (1982).
Shen, J.; Lowe, R. D.; Snook, R. D. *Chem. Phys.* **165**, 385 (1992).
Siebert, D. R.; Grabiner, F. R.; Flynn, G. W. *J. Chem. Phys.* **60** 1564 (1974).
Skogerboe, K. J. and Yeung, E. S. *Anal. Chem.* **58** 1014 (1986).
Slaby, J. *Opt. Comm.* **60**, 133 (1986).
Slaby, J. *Opt. Commun.* **64** 89 (1987).
Slaby, J. *Anal. Chem.* **61** 2496 (1989).
Spear, J. D.; Russo, R. D.; and Silva, R. J. *Appl. Spectrosc.* **42** 1103 (1988).
Stone, J. *J. Opt. Soc. Am.* **62** 327 (1972).
Stone, J. *Appl. Opt.* **12** 1828 (1973).
Swofford, R. L. and Morrell, J. A. *J. Appl. Phys.* **49** 3667 (1978).
Tam, A. C; Sontag, H.; and Hess, P. *Chem. Phys. Lett.* **120** 280 (1985).
Teramae, N. and Winefordner, J. D. *Appl. Spectrosc.* **41** 164 (1987).
Tran, C. D. *Anal. Chem.* **58** 1714 (1986a).
Tran, C. D. *Appl. Spectros.* **40** 1108 (1986b).
Tran, C. D. and Simianu, V. *Anal. Chem.* **64** 1419 (1992).
Tran, C. D. and Xu, M. *Rev. Sci. Instrum.* **60** 3207 (1989).
Twarowski, A. J. and Kliger, D. S. *Chem. Phys.* **20** 253 (1977a).
Twarowski, A. J. and Kliger, D. S. *Chem. Phys.* **20** 259 (1977b).
Weaire, D.; Wherrett, B. S.; Miller, D. A. B.; and Smith, S. D. *Opt. Lett* **4** 331 (1979).
Weimer, W. A. and Dovichi, N. J. *Appl. Opt.* **24** 2981 (1985a).
Weimer, W. A. and Dovichi, N. J. *Appl. Spectrosc.* **39** 1009 (1985b).
Xu, M. and Tran, C. D. *Anal. Chem.* **62** 2467 (1990a).
Xu, M. and Tran, C. D. *Appl. Spectrosc.* **44** 962 (1990b).
Yang, Y. and Hall, S. C. *Appl. Spectrosc.* **42** 72 (1988).
Yariv, A. *Optical Electronics*, 3rd ed., Holt, Rinehart and Winston, New York (1985).

CHAPTER

7

ANALYTICAL MEASUREMENT AND DATA PROCESSING CONSIDERATIONS

Special considerations are required for successful application of photothermal spectroscopy for analytical measurement. In addition to the usual considerations for sampling, sample preparation, separations, reagent and solvent purity, and the like, one must also consider the thermal and optical properties of the solvents and optical materials used in the apparatus. Because of the extreme sensitivity to absorbance and the spatial-dependent nature of the measurement, one must also be aware of how the environmental affects the measurement. Vibrations, temperature fluctuations, air currents, and even absorption of airborne species can have deleterious effects on measurement quality.

Photothermal spectroscopy measurements are less than routine. Consider the reported abilities of photothermal lens spectroscopy since it is perhaps the best known and most used of all the photothermal spectroscopy methods for analytical applications. The relative simplicity of the single-laser apparatus coupled with the low-solution absorption detection limits, on the order of 10^{-7} cm^{-1}, make it highly attractive for trace analysis applications. But although the opportunity for developing low-absorbance-measurement-based analysis methods is at hand, applications have not developed as fast as might be expected. Part of the reason for this may that although absorbance detection limits are low, even lower than that of the solvent in some cases, it is difficult to obtain absolute quantitative information on the sample absorbance. The photothermal lens signal is a function of sample absorbance as well as sample matrix composition, optical geometry, and excitation laser power. Accurate qualitative information (e.g., in the form of optical absorbance spectra) is even more difficult to obtain for trace-level analytes. This is primarily an optical source problem. Continuous laser excitation sources typically used in solution-phase photothermal spectroscopy do not have a wide wavelength tuning range. Pulsed laser sources have a wide tuning range, but the pulse-to-pulse energy and beam profile reproducibility is generally low for these sources. Even with stable tunable sources and compensation for excitation-laser wavelength dependencies, the ability to discriminate a trace analyte absorbance feature on the background absorbance due to vibrational overtones is extremely difficult since measurement precision obtained using photothermal spectroscopy is modest.

Despite these shortcomings, there are situations where photothermal spectroscopy can outperform conventional approaches to trace analysis. However, given the negative aspects, at least some attempt to justify the use of photothermal spectroscopy is apparently in order. The theoretical basis for photothermal spectroscopy is discussed in terms of the requirements for analytical application in this chapter. Conditions where the use of photothermal spectroscopy is advantageous are explored and the limitations as well as the virtues are probed.

7.1. SENSITIVITY OF PHOTOTHERMAL SPECTROSCOPY

For analytis of low-absorbance samples, it is often necessary to choose a technique that maximizes the sensitivity (i.e., instrument response per absorbance unit change) in order to have signals that are large compared to the noise of the measurement system. Without the required sensitivity, measurements cannot be made. The relative merit of using photothermal spectroscopy for trace measurements is encompassed in the theoretical enhancement factor. This factor is the ratio of the signal produced by photothermal lens spectroscopy to that produced by absorption spectrophotometry. Thus it can be used to gage the relative sensitivity of photothermal lens spectroscopy. Although strictly applicable only to photothermal lens spectrometry using pinhole-aperture-based apparatus, the enhancement factor gives the relative figure of merit for all linear photothermal spectroscopy techniques since the signals produced by the various apparatus are related. By definition (Dovichi and Harris 1979) the enhancement factor is the ratio of the photothermal signal to that obtained using transmission spectrophotometry under the assumption that both measurements have the same limiting noise. Measurements obtained using photothermal spectroscopy can be compared to absorption spectrophotometry through the enhancement factor. It is a very useful figure of merit when comparing the relative capabilities of the two techniques. It is also useful for designing analytical procedures since the theoretical enhancement factor can be estimated for a variety of experimental conditions.

7.1.1. Thermal Lens Enhancement Factors

There are two ways to calculate the theoretical enhancement for photothermal lens spectroscopy. One is based on the refractive theory of optics and the other on diffraction optics. The results of these two theories are nearly the same for most experimental conditions. As shown in Chapter 6, the diffraction description of the single-laser photothermal lens signal yields a theoretical photothermal lens enhancement factor for a far-field detector plane of

$$E_{cw} = \left|\left(\frac{dn}{dT}\right)\right| \frac{\pi\Phi_o Y_H}{6\lambda\kappa} \tag{1}$$

Clearly, the photothermal lens signal can be enhanced by choice of solvent since the thermooptical (dn/dT) and thermal conductivity (κ) parameters are strictly solvent dependent in trace analysis. In addition, the power of the excitation laser, Φ_0(W), is an experimentally controlled parameter. In fact, the enhancement can be made arbitrarily large by increasing the excitation power. To a lesser extent, the heat yield, Y_H, can be controlled. A less than unit heat yield implies either trapping of the molecule in some excited state or energy loss through fluorescence and phosphorescence. To the extent that the addition of excited-state quenchers can diminish these effects, the heat yield can be enhanced. For example, Chartier et al. (1990a, b, 1991) have used solution pH, solution oxygen depletion, and an external quencher (iodide) to affect the heat yield for a number of dye molecules that have high triplet-state yields.

For weakly absorbing samples, the single-laser photothermal lens signal is equal to the enhancement factor–optical absorbance product. Thus although the strength of the photothermal lens element increases with decreasing excitation laser beam waist radius, the single-laser photothermal lens signal is independent of beam waist radius. Carter and Harris (1984) rightly predicted that since signals produced in continuous laser photothermal lens spectroscopy do not depend on the beam waist of the excitation source, the absorbance detection limit should be independent of optical pathlength. They performed a study wherein the photothermal lens signals for 1-cm and 1-mm-pathlength samples were compared. Shorter confocal distances were used for the short-path sample, presumably to have a smaller excitation laser beam waist. They found that absorbance detection limits were nearly equivalent for the two sample cells, although slightly better for the short-path sample. They attributed the small improvement in the short-path sample to reduced effects of pointing noise in the laser source. They concluded that continuous-laser-excited photothermal lens spectrometry is suited for absorbance determination of short-optical-path low-volume samples.

Diffraction results for pulsed laser–excited photothermal lens spectroscopy yields a time-dependent signal that exhibits an effect due to the probe laser beam waist radius. As with the time-dependent signal, the enhancement factor will be a maximum at zero time. The enhancement factor resulting from an optimum probe laser beam focus position of $z_{\text{opt}} = \pm z_{0,p}(1 + w_0^2/2w_{0,p}^2)$ (Chapter 6 Equation 121) is

$$E_{\text{pulsed}} = \left|\left(\frac{dn}{dT}\right)\right| \frac{4Y_H Q}{\lambda_p \rho C_P (w_0^2 + 2w_{0,p}^2)} \tag{2}$$

where w_0 and $w_{0,p}$ are the minimum radiuses of the excitation and probe lasers, and λ_p is the wavelength of the probe laser. The pulsed laser enhanement clearly depends on both excitation and probe laser beam radiuses at the sample in this expression. As a general rule, the greater the time-integrated excitation irradiance, the greater the enhancement. However, if the probe laser beam radius is too large, the enhancement factor is proportional only to the pulse energy, Q (J). It is difficult to calculate a general enhancement factor for the pulsed-laser case because of the two beam waists. However, restraining the optics to having the excitation and probe laser beam waists equal at their respective focuses, we have

$$E_{\text{pulsed}} = \left|\left(\frac{dn}{dT}\right)\right| \frac{4Y_H Q}{3\lambda_p w_0^2 \rho C_P} \tag{3}$$

There are a few differences between the pulsed and continuous single-laser enhancements. One difference is that pulsed laser photothermal spectroscopy does not depend on the thermal conductivity but is inversely proportional to the specific heat of the sample. Another difference is that the pulsed laser photothermal spectroscopy enhancement is inversely proportional to the excitation laser beam waist radius. Subsequently, pulsed laser photothermal lens enhancement increases with increasing integrated irradiance of the excitation source, while that for the continuous laser experiment increases with laser power, independent of the beam waist at the sample.

The single- and pulsed laser–excited photothermal lens enhancement factors for a number of different gases and solvents are given in Tables 7.1 and 7.2. The enhancement are calculated using the diffraction optics formulas for unit heat yields. Excitation laser powers and energies are consistent with SI radiometric units (e.g., continuous laser powers of 1 W and pulsed laser integrated irradiances of J m^{-2}). Also listed in these tables are the minimum continuous laser powers and pulsed laser integrated irradiances needed to obtain an enhancement of 1. These can be used as a gage to determine whether a particular laser is suited for use in photothermal spectroscopy. For most liquids, the minimum continuous laser powers are less than 1 mW continuous and less than 1.4 nJ pulsed energy focused to

Table 7.1. Photothermal Properties of Gases

Substance	κ^a ($W m^{-1} K^{-1}$)	C_P ($J kg^{-1} K^{-1}$)	ρ ($kg m^{-3}$)	$10^6 \times dn/dT$ (K^{-1})	E_{cw} [b]	E_{pulsed} [c]	$\Phi_{cw,E=1}$ [d] (mW)	$H_{pulsed,E=1}$ [e] ($J m^{-2}$)	t_c [f] ($ns\ \mu m^{-2}$)
Helium	0.1500	5320	0.16	−0.11	1	0.00034	1300	2900	1.42
Hydrogen	0.1815	14400	0.08	−0.42	2	0.00097	410	1000	1.59
Air	0.0261	1010	1.17	−0.88	35	0.0020	28	500	11.3
Nitrogen	0.0260	1040	1.14	−0.90	36	0.0020	28	490	11.4
Oxygen	0.0267	920	1.30	−0.82	32	0.0018	31	560	11.2
Argon	0.0177	420	1.62	−0.86	51	0.0027	20	370	11.9
Carbon dioxide	0.0166	840	1.79	−1.38	87	0.0024	11	410	22.6
Methane	0.0343	2240	0.65	−1.35	41	0.0025	24	400	10.6
Ethane	0.0218	1760	1.22	−2.35	110	0.0029	9	340	24.6

[a] κ, C_P, ρ, and dn/dT data from Dovichi (1987).
[b] Calculated for a 500-nm 1-W laser.
[c] Calculated for a 500-nm 1-J/m^2 excitation laser.
[d] Continuous 500-nm laser power required for $E_{cw} = 1$.
[e] Pulsed 500-nm laser integrated irradiance for $E_{pulsed} = 1$.
[f] Characteristic time constant for 1-μm beam radius.

Table 7.2. Photothermal Properties of Some Liquids

Substance	κ^a ($W\ m^{-1}\ K^{-1}$)	C_P ($J\ kg^{-1}\ K^{-1}$)	ρ ($kg\ m^{-3}$)	$10^4 \times dn/dT$ (K^{-1})	$E_{cw}{}^b$	$E_{pulsed}{}^c$	$\Phi_{cw}{}^d$ (mW)	$H_{pulsed}{}^e$ ($J\ m^{-2}$)	$t_c{}^f$ ($\mu s\ \mu m^{-2}$)
n-Pentane	0.118	2290	630	−5.5	4880	0.0010	0.20	980	3.1
n-Hexane	0.123	2230	660	−5.2	4430	0.00094	0.23	1100	3.0
n-Heptane	0.127	2210	670	−4.9	4040	0.00088	0.25	1100	2.9
n-Octane	0.132	2190	710	−4.7	3730	0.00081	0.27	1200	2.9
n-Nonane	0.136	2180	720	−4.5	3460	0.00076	0.29	1300	2.9
n-Decane	0.140	2210	730	−4.2	3140	0.00070	0.32	1400	2.9
Cyclohexane	0.123	1810	780	−5.56	4730	0.0010	0.21	950	2.9
Carbon disulfide	0.150	1020	1260	−8.09	5650	0.0017	0.18	600	2.1
Carbon tetrachloride	0.103	850	1600	−6.12	6220	0.0012	0.16	830	3.3
Chloroform	0.117	950	1520	−6.03	5400	0.0011	0.19	900	3.1
Methylene chloride	0.122	1080	1330	−5.50	4720	0.0010	0.21	980	2.9
Benzene	0.137	1710	880	−6.52	4980	0.0012	0.20	860	2.8
Toluene	0.135	1690	870	−5.68	4410	0.0010	0.23	970	2.7
o-Xylene	0.143	1770	880	−5.00	3660	0.00086	0.27	1200	2.7
Ethyl ether	0.137	2310	720	−6.06	4630	0.00097	0.22	1000	3.0
Acetone	0.190	2180	800	−5.42	3000	0.00083	0.33	1200	2.3
Ethyl acetate	0.149	1960	910	−4.90	3440	0.00073	0.29	1400	3.0
Methanol	0.202	2460	790	−3.94	2040	0.00054	0.49	1800	2.4
Ethanol	0.167	2360	800	−4.00	2510	0.00056	0.40	1800	2.8
n-Propanol	0.156	2220	800	−3.70	2480	0.00056	0.40	1800	2.8
Acetic acid	0.172	2050	1040	−3.90	2374	0.00048	0.42	2000	3.1
Water (0°C)	0.564	4210	1000	−0.00	0	0	—	—	1.9
Water (20°C)	0.598	4180	1000	−0.91	159	0.00006	6.28	17000	1.8
Water (80°C)	0.670	4200	970	−2.31	361	0.00015	2.77	6600	1.5

[a]κ, C_P, ρ, and dn/dT data from Dovichi (1987). [b]Calculated for a 500-nm 1-W laser. [c]Calculated for a 500-nm 1-J/m^2 excitation laser. [d]Continuous 500-nm laser power required for $E_{cw} = 1$. [e]Pulsed 500-nm laser integrated irradiance for $E_{pulsed} = 1$. [f]Characteristic time constant for 1-μm beam radius.

1 μm^2. For gases the continuous laser powers needed to favor photothermal spectroscopy are a few tens of milliwatts, while the pulsed laser irradiances required to break even are less than 1 nJ/μm^2 for all but He. Finally, the characteristic thermal diffusion times are given for a 1-μm beam radius in the sample. To scale these to other excitation beam radiuses, the t_c values in the tables are multiplied by the square of the beam radius in μm. It is notable that the t_c for gas samples is on the order of the acoustic limited rise time for the 1-μm excitation beam. Subsequently, the photothermal lens signal is not expected to follow the "thermal diffusion only" approximate theory but must be calculated using the methods described in Chapter 3.

Enhancement factors for other photothermal apparatuses can be derived from those of the pulsed laser apparatus. For example, although diffraction theory does not predict an optimum probe laser focus geometry for the continuous-excitation two-laser photothermal lens apparatus, an enhancement can be obtained by time integration of the pulsed excitation laser result. The optimum probe laser focus offset from the sample for pulsed laser excitation was found to be $z'_{\text{opt}} = \pm z_{0,p}(1 + w^2/2w_{0,p}^2)$, where z' is the probe laser focus to sample distance, $z_{0,p}$ and $w_{0,p}$ are the confocal distance and minimum beam waist of the probe laser, and w is the beam waist radius of the excitation laser at the sample. The continuous laser photothermal lens signal can be found by integration of the optimum time-dependent thermal lens signal resulting from far-field placement of the pinhole

$$S_{\text{cw,opt}}(t) = \pm\left(\frac{dn}{dT}\right)\frac{8\alpha l Q Y_H}{\lambda_p \rho C_P}\int_0^t \frac{2w_{0,p}^2 + w^2}{[2w_{0,p}^2 + w^2(1 + 2t'/t_c)]^2 + (2w_{0,p}^2 + w^2)^2}\,dt'$$

$$S_{\text{cw,opt}}(t) = \pm\left(\frac{dn}{dT}\right)\frac{\alpha l \Phi_0 Y_H}{\lambda_p \kappa}\left[\tan^{-1}\left(1 + \frac{2w^2}{w^2 + 2w_{0,p}^2}\frac{t}{t_c}\right) - \frac{\pi}{4}\right] \tag{4}$$

The maximum signal occurs when the argument of the $\tan^{-1}$ term is maximum. This occurs for minimum excitation and probe beam waist radiuses (the factor t_c is proportional to w^2) and for excitation times much greater than t_c. Subsequently, for any finite irradiation time, the excitation laser should be focused into the sample cell ($w = w_0$) and the probe laser focused to the smallest possible beam size for maximum signal. The photothermal lens enhancement is obtained by dividing the infinite-time signal by αl:

$$E_{\text{cw,opt}} = \left|\left(\frac{dn}{dT}\right)\right|\frac{\pi\Phi_0 Y_H}{4\lambda_p \kappa} \tag{5}$$

This enhancement, independent of the excitation and probe laser beam waists, is similar to, though a factor of $\frac{3}{2}$ larger than, that obtained using the single-laser apparatus.

Enhancement factors that include other effects can be found by recasting the enhancement in terms of the inverse focal length of the resulting photothermal lens element. In this form, the continuous-excitation two-laser enhancement is

$$E_{\mathrm{cw,opt}} = \frac{1}{\alpha l} \frac{\pi}{4\lambda_p} \frac{\pi w_0^2}{f_{\mathrm{cw}}(\infty)} \tag{6}$$

An apparatus commonly used in photothermal lens spectroscopy uses a chopped excitation laser with a lock-in amplifier. The enhancement factor for this case can be found simply by substituting the magnitude of the oscillating thermal lens for that of the continuous lens:

$$E_{\mathrm{osc,opt}} = \left|\left(\frac{dn}{dT}\right)\right| \frac{\pi \Phi_0 Y_H}{4\lambda_p \kappa} \times \sqrt{1+(\omega t_c/2)^2[\mathrm{ci}^2(\omega t_c/2)+\mathrm{si}^2(\omega t_c/2)]+t_c[\mathrm{si}(\omega t_c/2)\cos(\omega t_c/2)-\mathrm{ci}(\omega t_c/2)\sin(\omega t_c/2)]} \tag{7}$$

where si and ci are the sine and cosine integrals as defined in Gradshteyn and Ryzhik (1980). The relative magnitude of this enhancement, as a function of frequency, ω, is equivalent to that plotted in Figure 6.12. The magnitude decreases to about half maximum for $\omega \approx 2/t_c$, and decreases rapidly for angular frequencies greater than this value.

The realizable enhancement for tightly focused excitation beams will be less than that predicted from the equations above because the Gaussian laser beams diverge at a greater rate when focused to smaller spots. It was shown in Chapter 6 that when the sample cell pathlength was longer than about 10 confocal distances of the excitation laser, the effective inverse photothermal lens focal length is given by

$$\frac{1}{\langle f_{\mathrm{cw}}(\infty)\rangle} = \left(\frac{dn}{dT}\right) \frac{\pi \alpha \Phi_0 Y_H}{\lambda \kappa} \tag{8}$$

Subsequently, the continuous, two-laser photothermal lens enhancement factor for this case is

$$E_{\mathrm{cw,long\,cell}} = \left|\left(\frac{dn}{dT}\right)\right| \frac{\pi^2 z_0 \Phi_0 Y_H}{4 l \lambda_p \kappa} \tag{9}$$

In the long sample cell limit, the enhancement is proportional to the confocal distance of the excitation laser ($z_0 = \pi w_0^2/\lambda$), indicating that the excitation laser beam waist radius should be large enough to result in long confocal distances. In the latter case, the photothermal lens element requires longer times to reach steady state, but since the final photothermal lens signal is independent of beam waist radius, the enhancement can still be realized. The inverse sample pathlength dependence is due to the fact that the sample pathlength dependence of the inverse focal length is lost when integration over the hourglass-shaped excitation beam is performed. The end result is that when the excitation laser is focused to decrease the time constant to increase the signal for a given chopper frequency, photothermal spectroscopy is favored over absorption spectrophotometry for short path samples.

Similar calculations for the pulsed laser excitation case yields a long-sample-cell enhancement:

$$E_{\text{pulsed,long cell}} = \left|\left(\frac{dn}{dT}\right)\right| \frac{3\pi z_0 Q Y_H}{l\lambda_p w_0^2 \rho C_p} \tag{10}$$

and since the confocal distance is proportional to w_0^2,

$$E_{\text{pulsed,long cell}} = \left|\left(\frac{dn}{dT}\right)\right| \frac{3\pi^2 Q Y_H}{l\lambda_p \lambda \rho C_p} \tag{11}$$

In this case the theoretical enhancement is independent of excitation beam waist radius. However, it is apparent that a pulsed-laser excited photothermal lens, and deflection for that matter, is favored in sample cells with optical pathlengths that are short relative to the confocal distance of the excitation laser. These small-focus-spot-size laser beam effects also explain several anomalies observed in previous and subsequent studies. First, the absorbance limits of detection do not increase with the continuous laser apparatuses as predicted from the irradiance dependence of the inverse focal length. Second, using a tight-excitation-focus cross-beam photothermal lens arrangement, Nolan and Dovichi (1986) found that an absorption coefficient detection limit of about 10^{-5} cm^{-1} was obtained. However, the optical pathlength was only 200 μm. So the absorbance detection limit was $A = 2.303\alpha l = 5 \times 10^{-7}$. Considering the fact that previous measurements had been made in 1-cm cuvettes, the absorbance sensitivities of the limited-volume cross-beam apparatus are the same as those of the 1-cm-pathlength mesurements, about $A = 10^{-7}$. Third, two-laser photothermal lens spectroscopy exhibits an apparent absence of interference due to optical absorption by the sample cell windows when short-focal-length lenses are used to focus the excitation beam into the sample cell. The excitation laser beam is usually focused at the center of the sample cell and the waist is larger at the windows. Since the photothermal lens signal scales as w^{-2}, window absorption effects can be insignificant. The trends shown in the equations above explain these observations. It is also interesting that under the proper conditions, photothermal lens spectroscopy may be used to obtain absorption coefficient measurements without any knowledge of the sample pathlength.

7.1.2. Relative Sensitivity of Photothermal Lens and Deflection Spectroscopies

Photothermal lens and deflection spectroscopies are commonly used for analytical measurement in homogeneous samples. The relative sensitivity of photothermal lens and photothermal deflection spectroscopy of homogeneous samples has not been addressed in the past, but it can be derived using the results presented in Chapter 6. One problem with the comparison is the difference in the way signals are generated and calculated from subsequent time-dependent laser power. In photothermal lens spectroscopy, a pinhole aperture is used to develop the optical signal, and the analytical signal is calculated from this. In deflection, a straightedge aperture can be used to develop the optical signal. The ratio of this signal to the probe laser power is used as the analytical signal.

It was shown in Chapter 6 that the photothermal deflection signal produced by a straightedge aperture placed at the center of a TEM_{00} probe laser beam is given by

$$S_{\mathrm{pds}}(t) \approx 2\sqrt{2\pi}\,\frac{w_{0,p}}{\lambda_p}\langle\theta_x(t)\rangle \tag{12}$$

where $\langle\theta_x(t)\rangle$ is the spatially averaged probe laser beam deflection angle at the aperture and the aperture is placed far beyond the probes laser confocal distance. Substitution of the maximum pulsed laser deflection angle results in the signal

$$S_{\mathrm{pds,pulsed}} = \sqrt{\frac{2}{\pi}}\left|\left(\frac{dn}{dT}\right)\right|\frac{8e^{-1/2}\alpha l Y_H Q}{\lambda_p \rho C_P (w_0^2 + w_{0,p}^2)^{3/2}} \tag{13}$$

The ratio of this signal to that derived for pulsed-laser excited photothermal lens spectroscopy gives an indication of the relative sensitivity of the two techniques as well as the relative enhancement factors:

$$\frac{S_{\mathrm{tls,pulsed}}}{S_{\mathrm{pds,pulsed}}} = \frac{E_{\mathrm{tls,pulsed}}}{E_{\mathrm{pds,pulsed}}} = \frac{e^{1/2}}{2w_{0,p}} \sqrt{\frac{\pi}{2}} \frac{(w_0^2 + w_{0,p}^2)^{3/2}}{w_0^2 + 2w_{0,p}^2} \tag{14}$$

This expression may be used to relate photothermal lens enhancement factors, E_{tls}, more precisely to those expected from photothermal deflection, E_{pds}, for the case of pulsed laser excitation. For most practical pulsed laser excitation experimental apparatuses, this ratio is close to 1. There is no apparent advantage of one technique over the other. The ratio for the continuous-laser-excitation case can be found by substitution of the time-dependent deflection angle and taking the ratio to the continuous photothermal lens signal. The complicated reult indicates that the relative enhancement factor for photothermal deflection degrades relative to that for photothermal lens spectroscopy with continuous excitation.

7.1.3. Relating Photothermal Signals to Absorbance and Enhancement

In theory, the photothermal signal can be related to optical absorbance through the equations that predict the signal. For example, single-continuous-laser photothermal lens apparatus utilizing a short-optical-path sample cell has a signal

$$S_{\mathrm{cw}}(\infty) = E_{\mathrm{cw}}\alpha l = \left|\left(\frac{dn}{dT}\right)\right| \frac{\alpha l \pi \Phi_0 Y_H}{6\lambda\kappa} \tag{15}$$

Note that the signal can be expressed in terms of the product of the theoretical enhancement factor with the αl product. The sample absorbance is given as $A = \log_{10}(T)$, where T is transmission. Since $T = \exp(-\alpha l)$, $A = 0.4343\alpha l$. The photothermal lens signal can then be expressed in terms of the theoretical enhancement factor and the sample absorbance

$$S_{\mathrm{cw}}(\infty) = 2.303 E_{\mathrm{cw}} A \tag{16}$$

Since photothermal spectroscopy is a relative measurement that depends on thermal and optical properties of the sample and on the specific geometry of the excitation laser and, if used, probe laser, it is rarely practical to have quantitative knowledge of all parameters necessary to calculate sample absorbance accurately directly from the photothermal signal. First, the thermal (e.g., heat capacity and thermal conductivity) and optical (e.g., temperature-dependent refractive index, beam waists, focus geometries) are not accurately known for most solvents. The thermal and thermooptical parameters of mixed solvents usually do not follow linear relationships. All of this is not to say that there is no value in knowledge of the approximate parameters required to predict an experimental result. Even though not accurate, the relative parameters can be used to optimize the experiment.

7.1.4. Intrinsic Enhancement of Two-Laser Methods

The fundamental limitation of all spectroscopy methods is due to the shot-noise-like absorption process and the shot noise associated with the detection. The precision of an absorbance calculations is directly proportional to the precision of the measurements of the photothermal signal and the excitation laser energy or power. In addition, Poisson statistics of the absorption process will result in a variation in the number of excitation laser photons absorbed. In turn, this will result in a variation in the probe laser deflection angle or photothermal lens strength. The variance in the number of photons is equal to the average number absorbed. The intrinsic variance is easily calculated and applied to the relative precision of photothermal spectroscopy for pulsed excitation sources (Bialkowski et al. 1992). These results are summarized here.

The average number of absorbed photons, $\langle p \rangle$, for pulsed laser excitation is

$$\langle p \rangle = \frac{Q(1 - T)}{h\nu} \tag{17}$$

where $h\nu$ is the energy per absorbed quantum, $1 - T$ is the loss in sample transmission due to absorption, and $Q(1 - T)$ is the total absorbed energy. The heat absorbed in the volume of the excitation source is

$$Q' = \langle p \rangle h\nu \tag{18}$$

and the variance in Q' is

$$\sigma_{Q'}^2 = (h\nu)^2 \sigma_p^2 \tag{19}$$

where the variance in the number of photons is $\sigma_p^2 = \langle p \rangle$. In a two-laser photothermal spectroscopy, the optical element produced is proportional to the absorbed heat, CQ'. The proportionality constant, C, is a function of thermooptical parameters of the sample and geometric factors of the pump and probe lasers. The measured signal variance due to difference in the absorbed heat is

$$\sigma_S^2 = C^2\sigma_{Q'}^2 \tag{20}$$

and the relative variance is thus

$$\frac{\sigma_S^2}{S^2} = \frac{h\nu}{Q(1 - T)} \tag{21}$$

To obtain the absorption coefficient from a measure of the photothermal signal, and independent measurement of the excitation pulse energy must be obtained. The variance in measured transmission is easiest to calculate.

$$\sigma_T^2 = (1 - T)^2 \left(\frac{\sigma_S^2}{S^2} + \frac{\sigma_Q^2}{Q^2}\right) \tag{22}$$

The relative variance for shot-noise-limited detection of Q is

$$\frac{\sigma_Q^2}{Q^2} = \frac{h\nu}{Q} \tag{23}$$

The variance in sample absorbance is equal to the relative variance in transmission (i.e., $\sigma_{\alpha l}^2 = \sigma_T^2 T^{-2}$) and the limiting signal-to-noise ratio of the optical absorption coefficient is

$$\mathrm{SNR} = \frac{\alpha l}{\sigma_{\alpha l}} = T\alpha l\sqrt{\frac{Q}{h\nu}\frac{1}{(1 - T) + (1 - T)^2}} \tag{24}$$

Using similar arguments, the signal-to-noise ratio for transmission-based spectrophotometry is

$$\mathrm{SNR} = \alpha l\sqrt{\frac{Q}{h\nu}\frac{T}{1 + T}} \tag{25}$$

A comparison of the relative signal-to-noise ratios for pulsed laser–excited photothermal spectroscopy and transmission spectrophotometry is illustrated in Figure 7.1. The pulsed laser–excited photothermal spectroscopy signal has a shot-noise-limited signal-to-noise ratio that exceeds that of transmission spectrophotometry for low-absorbance samples. Of course, the comparison is somewhat artificial in that only the Poisson character of the

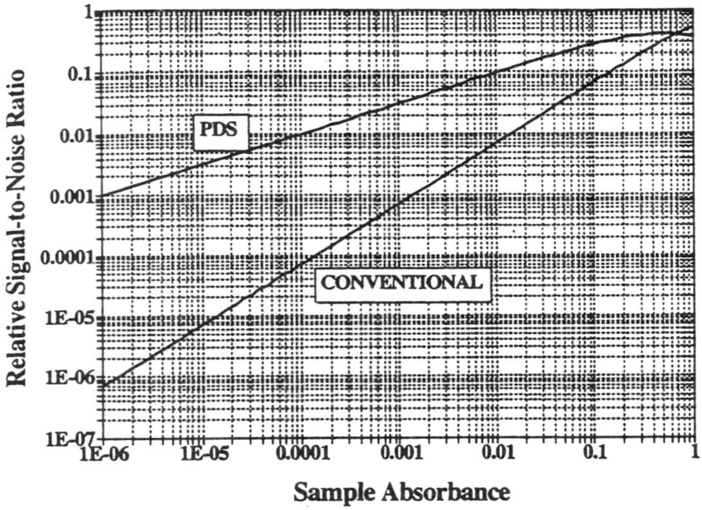

Figure 7.1. Comparison of the theoretical signal-to-noise ratios for transmission (conventional) and pulsed laser photothermal deflection (PDS) spectroscopies. The measurements are both assumed to be shot-noise limited in the theoretical calculations. (Reprinted with permission from Bialkowski et al. 1992. Copyright 1992, Society for Applied Spectroscopy.)

absorption and detection was taken into account. The photothermal spectroscopy measurements also have a shot-noise term due to the probe laser, but since the probe laser power can be arbitrarily large, this contribution may be neglected.

The ratio of these two signal-to-noise results illustrates the advantage of photothermal spectroscopy in much the same fashion as does the theoretical single-laser enhancement factor. Under the assumption that spectrophotometric transmission and photothermal experiments are both limited by shot noise, the enhancement factor is defined as the sensitivity ratio. The enhancement factor, E, defined as the ratio of two signal-to-noise ratios derived under shot-noise-limiting conditions, is

$$E = \frac{\mathrm{SNR}_{\mathrm{photothermal}}}{\mathrm{SNR}_{\mathrm{transmission}}} = \sqrt{\frac{(1 - T) + (1 - T)^2}{T(1 + T)}} \tag{26}$$

This enhancement is plotted in Figure 7.2. The enhancement is linear for sample absorbances less than about 0.1. For an optically thin sample, T

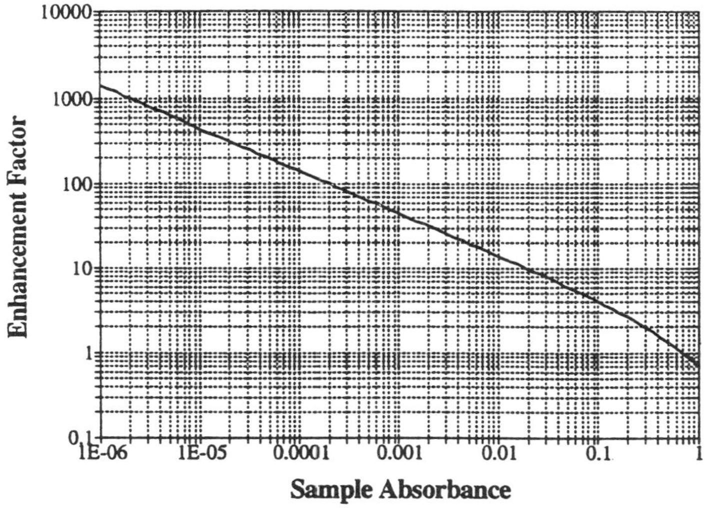

Figure 7.2. Ratio of signal-to-noise ratios resulting in the theoretical, shot-noise-limited enhancement factor for pulsed laser–excited photothermal spectroscopy. (Reprinted with permission from Bialkowski et al. 1992. Copyright 1992, Society for Applied Spectroscopy.)

approaches unity while $1 - T$ approaches αl, and the result above reduces to

$$\lim_{T \to 0} E = \sqrt{\frac{2}{\alpha l}} \tag{27}$$

Above $\alpha l \sim 0.1$, the enhancement is nonlinear, due to significant loss in light energy in the sample. The point where the photothermal signals are better than those obtained using transmission spectrophotometry occurs for a sample that transmits more than 50% of the incident light. An interesting feature of these enhancement factors is that they are independent of the physical and thermooptic parameters of the sample.

7.1.5. Enhancement Limitations

There are limitation to the photothermal enhancement factor in addition to those produced by finite optical pathlength and other beam geometric considerations. The basic principle of photothermal spectroscopy predicts that signal magnitudes can be made arbitrarily large by increasing the excitation energy or power. However, at high power levels, and in particular with relatively high absorbance samples, one must be concerned over convective heat transfer. Convection occurs when too much power is deposited in a small volume. With convection, the photothermal optical element can be destroyed (Buffett and Morris 1983, Alfonso et al. 1990). Approximate limits on the power or energy able to be absorbed by the sample was given in Chapter 5. In addition, multiphoton or multiple-photon absorption and optical saturation and ground-state bleaching can occur at the high irradiance. Since all of these effects are nonlinear in irradiance, the resulting temperature change will be distorted over that of the Gaussian profile of the excitation source. Multiphoton absorption enhances the photothermal spectroscopy signal by narrowing the Gaussian temperature change distribution. The net effect of multiphoton absorption is to increase photothermal signal due to a combination of increased absorbed energy and decreased temperature perturbation width (Twarowski and Kliger 1977).

At high irradiance, optical saturation or ground-state bleaching can occur, thereby limiting the minimum absorbance that may be detected using absorption spectrophotometry. Photothermal spectoscopy has distinct advantages over absorption spectrophotometry when transitions are easily saturated (Long and Bialkowski 1984, 1985). Photothermal spectroscopy is an indirect measurement technique wherein the amount of energy absorbed by the sample is monitored. The sample absorbs energy even when optical saturation or bleaching occurs. The absorbed energy results in a measurable signal, although the signal may not be linear, due to the spatial distortion particular to a given phenomenon. On the contrary, optical saturation results in an effective decrease in optical absorbance in conventional absorption spectrophotometry. Photothermal spectroscopy may thus overcome one of the fundamental limitations of absorption spectrophotometry. It is also suited as a technique for studying nonlinear absorption since the signals are finite even at irradiances far in excess of those required for optical saturation.

In addition to increasing the irradiance, the magnitude of photothermal signals may also be increased by choosing solvents that have large dn/dT values. The thermooptical coefficient is enhanced near concentration and temperature critical points for partially miscible solvents (Giglio and Vendramini 1974). Supercritical fluids have also been used for solvents (Leach and Harris 1984). Near the critical point, supercritical fluids should enhance the photothermal signals. The main drawback to using solvents near their critical points is in the inherent instability of the system. Long-range-order correlation can occur, causing large turbulent changes in the refractive indexes (Debenedetti and Reid 1986).

7.2. CONSIDERATIONS FOR TRACE ANALYSIS

Trace analysis requires that one be cognizant both of the chemistry involved in sample preparation and development into measurable form and of the limiting background signals of the instrumentation. Because photothermal spectroscopy is an extremely sensitive technique for measuring optical absorbance, great care must be taken in the chemical design of the method and in sample preparation. Harris and Williams (1985), and more recently Kitamori and Sawada (1991), point out that instrumental methods capable of measuring low absorbances are now common, but they may not be useful in trace analysis applications. The reason for this is that many samples have background absorbance which affects the measurement of the analyte absorbance. Background absorbance due to the solvent, impurities in the solvent, concommitants, and free ligands used in analysis where colored metal complexes are formed all contribute the total absorbance of the sample. Although the minimum detectable absorbance is a useful figure of merit for characterizing the potential of photothermal and other ultrasensitive methods, the relative precision is more appropriate when discussing trace analysis.

In photothermal spectroscopy, one must prepare the sample in such a fashion as to reduce background absorbances. This is done by choosing a pure, low-absorbance solvent and by choosing chelating or other color-developing reagents that have minimum absorbance and minimum side reactions that produce absorbing products. Solution-phase solvents should be chosen from those that do not have strong vibrational overtone absorptions. In general, the solvent should have relatively weak infrared absorptions and should not contain isolated high-frequency vibrations. In particular, it should not be aromatic or have isolated high-transition-energy infrared absorptions that could result in relatively strong vibration overtone absorption in the visible. Gas-phase samples are easier to prepare since permanent gases do not have many optical transitions and can be obtained in high purity. In direct atmospheric sampling, the air can be cleaned up prior to measurement using molecular traps (e.g., silica gels, molecular sieves). These practical experimental measures do not increase the precision of the measurement but simply reduce interference. Nonetheless, sample preparation and handling are the most important steps in trace analysis, regardless of the method used to quantify the analyte.

Even with the most carefully planned and executed procedure, finite residual absorption will limit the ultimate detection since absorbance differences must be taken to determine the analyte absorbance. Differential absorbance can be obtained by consecutive or simultaneous measurement of different samples, a given sample at different wavelengths, or a flowing sample at different times. In all cases the same experimental procedure must be followed for a working sample set consisting of the analytical samples, blanks, and calibration samples or standards. The fact that government agency trace measurement protocols require sets of measurements to be taken implies that reproducibility of both the measurement and the chemical preparation procedure must be high.

Mho and Yeung (1985) introduced the dynamic reserve figure of merit to evaluate trace methods in terms of their ability to detect small signal changes in the presence of a background signal. The dynamic reserve is defined as the ratio of the background signal to the minimum detectable signal change. The dynamic reserve is an effective way to report relative precision and can be used to compare different methods used for trace analysis. It is most important in the context of fluctuating excitation sources and when sample matrix absorbance is greater than that of the analyte. Since trace analysis often results in analyte absorbances that are less than that of the sample matrix, the dynamic reserve of a method or apparatus should be increased to the greatest possible extent of useful application. High-dynamic-reserve methods mostly rely on differential absorption measurements. In particular, with simultaneous differential absorbance measurement techniques, fluctuations in excitation power and environmental factors cancel.

There are three main techniques used to increase the dynamic reserve of photothermal spectroscopy for trace analysis: (1) differential sample-blank measurements, (2) wavelength discrimination, and (3) time- or space-series measurements. Differential absorbance methods are based on simultaneous measurement of two samples, one sample containing the analyte, the other the blank. This method is commonly used in absorption spectrophotometers. The second technique is to obtain spectroscopic absorbances at two or more wavelengths. In a two-component system, spectroscopic discrimination of the analyte can be derived from independent estimates of the absorbances at two wavelengths. This procedure works if the absorbances of the two components are different at the wavelengths selected. Even when using separation techniques, samples can have more than one coeluting species, or the solvent itself can exhibit a relatively high absorbance. With complexes, ligands will always be present in a relative concentration determined by the complex formation constant. Therefore, spectroscopic discrimination of the analyte over that of the solvent, impurity, or free ligands should always be performed by taking measurements at more than one wavelength. A time series can be used to measure sequentially optical absorbances due to the analyte and the solvent. This is performed in flow injection analysis and in chromatography. If the sample is heterogeneous, a series of measurements that vary over space can be used to discriminate absorptions of the analyte from that of the background. This results in images of the spatially dependent absorbance or the spatially dependent thermal properties of the sample. With the exception of the first technique, these all rely on the precision and reproduction accuracy of the individual measurements. Only the first technique truly increases the dynamic reserve. Increasing the dynamic reserve has been the subject of most instrumental developments in analytical applications of photothermal spectroscopy.

7.3. OPTICAL INSTRUMENTATION FOR ANALYSIS

7.3.1. Differential Measurement

Although photothermal spectroscopy is very sensitive, it does not have an intrinsically high dynamic reserve. In transmission spectrophotometers, the analyte absorbance is found by measuring the ratio of transmissions between the sample and the blank. Perhaps the most important invention used for photothermal spectroscopy of trace-level species was the differential photothermal lens spectrometer first reported by Dovichi and Harris (1980). In this device the signal due to the blank sample can be optically subtracted from that of the analytical sample. The two sample cells are arranged so that one cell results in a positive signal, the other in a negative signal. Differential absorbance is measured by placing the sample in one cell and the blank in the other. This concept is similar to that applied to conventional spectrophotometry. The distinction is that the difference signal is generated optically, whereas in the conventional spectrometers, difference signals are generated electronically.

Some of limitations of the two-cell differential photothermal lens apparatus that might arise are (1) sample cell placement errors, (2) changes in laser propagation characteristics of the beam entering the second sample cell due to the photothermal lens element formed in the first sample cell, (3) changes in the laser minimum beam position with times, and (4) differences in sample temperatures that would influence the thermooptical characteristics of the two samples. These potential limitations can be overcome using stationary cells with sample fluid injection. Elution chromatography and flow injection analysis allows measurement of sample and blank absorbance as a function of time. With this method, measurements are taken in a single sample cell so that sample placement errors are eliminated. Changes in the laser mode and in propagation characteristics due to first-sample-cell lens formation are more difficult to overcome. Placing the laser in the far field will help reduce errors due to subtle changes in the mode. Dovichi and Harris translated one of the sample cells along the laser beam axis to obtain an optical null even when the background absorbance was high. Errors due to photothermal lens formation in the first sample cell can be reduced, if not eliminated, by optically nulling sample blanks prior to the analysis with the analyte sample placed in the second sample cell and blank solution placed in the first sample cell.

A different approach to background absorbance correction based on spectroscopic discrimination has been pursued by Xu and Tran (1990a,b). They developed photothermal lens techniques for differential two-color and differential circular dichroism measurements. These techniques are based on the modulation of either the wavelength or the polarization of a continuous-wave Ar^+ laser. Since wavelength or dichroism was used as the basis for discrimination, they used a single sample cell. The modulated optical signal was detected and processed with a lock-in amplifier. This modulation and processing scheme is essentially that used in conventional spectrophotometry. The only advantage is that due to the signal enhancement inherent in photothermal spectroscopy. As such, it cannot increase the precision of the differential measurement for samples with high-absorbance interferant. Of course, static sample differential methods using continuous-laser excitation sources will all suffer from convective heat transfer effects. In general, methods that attempt to subtract signals due to solvent and interferant absorbance using twin cell designs will always be limited ultimately by convection heat transfer. Tran and Simianu (1992) extended the two-color approach to several wavelengths. In this apparatus an Ar^+ laser is operated with a simple cavity that produces a multiple-wavelength output. The polychromatic output is filtered with an electronically tunable acoustooptical bandpass filter. This filter, in combination with the multiple-line output of the Ar^+ laser, allows flexibility in selecting wavelengths for sample excitation. Time-dependent filter modulation produces a series of wavelengths multiplexed in time. The photothermal lens signal is collected.

Bialkowski et al. (1992) demonstrated differential absorption photothermal deflection spectroscopy by using two simultaneous pulsed excitation lasers tuned to different wavelenghs. This technique combines wavelength discrimination with simultaneous background subtraction. As with the two-cell design of Dovichi and Harris, differences are produced optically. Sample cell placement did not produce errors because the measurement is taken in a single stationary sample cell. Injection is used to introduce the samples into a stationary sample cell. The differential signal works within certain limits of the relative solvent absorption. Although technique does increase the dynamic reserve of the measurements, instabilities in the pulsed dye lasers limited the measurement precision.

Rosenzweig and Yeung (1993) recently used the electronic circuit described by Haller and Hobbs (1991) to reduce laser-source flicker noise associated with single-laser photothermal lens spectroscopy. This work may constitute the most accurate and direct comparison of photothermal spectroscopy to conventional absorption. The circuit is designed to allow shot-noise-limited detection when using laser light sources. Equivalent light source, detectors, and signal processing circuits were used to measure both photothermal and absorption response. With shot-noise-limited detection, the greater the optical power or rate of photon detection, the greater will be the signal-to-noise ratio. Laser sources have high optical power and thus promise to enhance spectroscopy in general. However, their power is not constant, and shot-noise-limited detection of the time-dependent signals has presented a problem. Although the potential for highly accurate measurements through the use of laser light sources is not new, there has been a general lack of detectors able to monitor high powers without saturating, and the circuits used to process the current-producing quantum detectors often added much excess noise. The photothermal lens signal is proportional to the excitation laser power. For example, a 1% variation in laser power, which is typical of a gas discharge laser, will result in a 1% measurement error.

The circuit used produces a signal that is proportional to the log ratio of two photodiode signals. By splitting the laser beam prior to sample excitation, directing one beam to the reference detector and the other to through the sample, pinhole spatial filter, and finally onto the signal detector, Rosenzweig and Yeung effectively created a single-laser photothermal lens spectrometer that accounted for flicker noise in the source. They applied this technique to the analysis of 2′-Cl-4-dimethylaminobenzene using HPLC with isocratic methanol mobile phase. Their system demonstrated a noise-equivalent detection of 2×10^{-7} AU using a 34-mW argon ion laser. They found that this detection limit was an order of magnitude better than optical absorption spectroscopy using the same laser and detector circuit, and constituted a two-order-of-magnitude improvement over absorbance detectors using more conventional light sources.

7.3.2. Spectroscopic Measurement

Data obtained using two-color methods do not contain enough information to perform multiple-component analysis, which requires a spectrum measurement. There have been a few reports on apparatus designed to discriminate the analyte from the sample matrix by measurement of the photothermal excitation spectra. In particular, Jackson et al. (1981) were able to observe the C—H vibrational overtone of benzene in CCl_4 solvent by scanning the wavelength of the pulsed excitation laser. Fournier et al. (1980), using photothermal deflection, and Higashi et al. (1984), using photothermal lensing, recorded differences in the photothermal signals obtained using different wavelengths of the infrared CO_2 molecular laser. However, spectroscopic discrimination and analyte quantitation in mixtures are difficult. Even though the absorption coefficient limits of detection for gas-phase species is 10^{-7} cm^{-1} or better in the 10-μm infrared transmission window of the atmosphere, the background absorption due to water vapor is at least 10^{-7} cm^{-1} in this region and can be orders of magnitude higher, depending on the relative humidity. The signal magnitude precision and reproducibility of laser-based photothermal spectroscopy measurements are not high enough to allow detection of analytes with absorption coefficients less than those of the water vapor. These analysis are perhaps easier performed using sample separation techniques.

It is worth noting that much, if not all, of the poor precision performance of the photothermal apparatus used for spectroscopic measurements is due to probe-laser flicker noise, excess noise in the electronic circuits used for laboratory prototype apparatus, or a lack of accuracy in measuring the excitation laser power or energy. This performance limitation could be largely overcome if the analytical spectroscopy community were to use low-noise electronic photodiode current-processing circuits such as that described by Haller and Hobbs (1991) for both excitation and probe lasers.

7.3.3. Fiber Optics

Instrumental artifacts often ultimately limit the precision of photothermal measurement. In homogeneous samples, the development and detection of the photothermal signal is based on the spatial characteristic of the excitation and probe laser sources. Laser pointing and TEM mode noise can thus strongly affect the precision of the measurement. Spatial filtering can be used to reduce the TEM mode noise, but pointing noise (i.e., random changes in the direction of the laser beam) is not reduced by this technique. Although pointing noise can be reduced to a great extent by proper placement of the lenses used to focus the excitation and probe laser beams, a much more satisfactory means of eliminating both pointing and spatial mode noise is to use single-mode optical fibers. Optical fibers may also reduce the complexity of the apparatus and make for a more compact design.

Several authors have used fiber optics in place of the pinhole aperture in photothermal lens spectroscopy (Buffett and Morris 1982, Sepaniak et al. 1984, Yang 1984, Pang and Morris 1985). The main advantage here is that the detector may be placed far from the experiment, and thus radio-frequency interference from the laser sources may be reduced. However, although the use of optical fibers to improve the deterministic noise of the excitation laser is apparent, there have been relatively few reports utilizing them for photothermal spectroscopy. Bialkowski (1986) used a multimode fused-silica optical fiber to improve on pointing instability and transverse or TEM mode noise of a pulsed nitrogen laser source used for photothermal lens spectroscopy. Although measurement precision improvements were realized, the use of solid-core optical fibers for short-pulse-duration lasers is limited because of the damage that occurs at relatively modest irradiances (Allison et al. 1985). Nakanishi et al. (1987) discuss the characteristics of single- and multiple-mode optical fibers as they apply to the transmission of the excitation source in photothermal lens spectrometry. Imasaka et al. (1987) discuss an application of single-mode optical fibers for source delivery and focusing of the laser beam for single-laser photothermal lens spectroscopy. The later constituted an important improvement to the practicality of single-laser photothermal lens spectroscopy that has not been subsequently exploited.

Discussions of the operating characteristics can be found in many texts devoted to this subject, as well as in modern optics texts. However, Nakanishi et al. (1987) have summarized the important characteristics of optical fibers as they pertain to photothermal spectroscopy. Optical fibers consist of a high-refractive-index material called the core, surrounded by a transparent medium of lower refractive index called the cladding. Light propagating in the core region at an angle relative to the wall normal greater than the critical angle is totally reflected from the cladding of the fiber. The core and cladding refractive indexes can change abruptly, as in the step-index fibers, or may change gradually, as in graded-index fibers. The two important physical features of an optical fiber are its core diameter and the core refractive index. If the relative core diameter is small enough, only a single TEM mode can propagate along the length of the fiber. In fact, optical fibers may allow either a single or multiple TEM modes to propagate along the fiber, depending on whether the core diameter, d, is small or large. The coupling efficiency of multimode fibers (e.g., the fraction of laser light introduced into the fiber) is generally higher than that of the single-mode fibers due to their greater cross-sectional area. Nakanishi et al. found that the poor spatial profile of the exiting beam lowers the photothermal lens enhancement factor for multimode fibers and recommend that they not be used in photothermal lens spectroscopy. In addition, the spatial profile of beams exiting from multimode fibers are sensitive to the orientation of the fibers. Small changes in the fiber position (e.g., bending and the like) can dramatically change the spatial structure of the output. In a word, single-mode fibers are preferred for photothermal spectroscopy. For single-mode fibers, the exiting beam waist radius is related to the core diameter by $2w_0 = 0.72d$, where d is the core diameter.

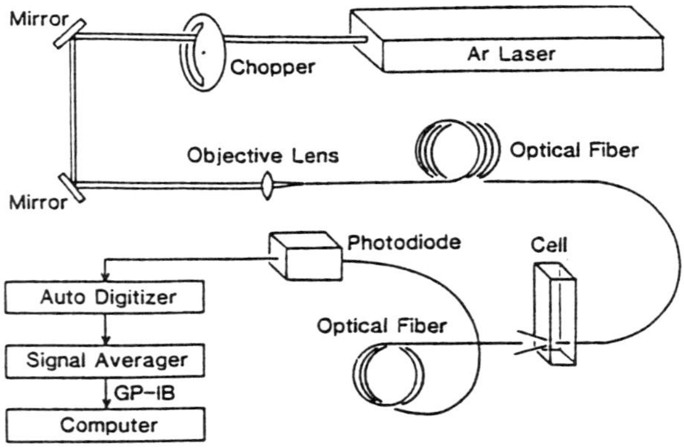

Figure 7.3. Experimental schematic diagram of the photothermal lens apparatus based on fiber optics. One fiber is used to deliver the excitation source to the sample and a second fiber is used to monitor the beam passing through the sample. The monitor fiber serves an equivalent function to the pinhole spatial filter used in most other apparatuses. The autodigitizer and signal average perform ensemble averaging. (Reprinted with permission from Imasuka et al. 1987. Copyright 1987, American Chemical Society.)

In a companion article, Imasaka et al. (1987) designed and tested apparatus that utilized single-mode optical fibers placed inside the cell for sample excitation. A second fiber, coupled to a detector, was used to detect the resulting single-laser photothermal lens signal. A schematic of their experiment is shown in Figure 7.3. A chopped 488-nm argon ion laser was introduced into a 6- or 10-μm core single-mode optical fiber through a $10\times$ microscope objective. The chopper rate was slow enough to allow the thermal lens signal to develop fully. The single-mode fiber was inserted into the sample cell so that light was introduced directly. Sample-pathlength-dependence studies where performed by adjusting the distance between the end of the fiber and the exit window of the sample cell. The photodiode coupling fiber position could also be adjusted. Imasaka et al. studied the effects of detection fiber placement distance and found that the signal did not vary for distances between the two fibers greater than about 8 mm. The photodiode signal was digitized and averaged over several chopper cycles to improve the signal-to-noise ratio. The resulting data were analyzed by inspection of the transient signal and calculation of the photothermal lens signal using the usual definition, $S = (\Phi_0 - \Phi_\infty)/\Phi_0$. The minimum detectable absorbance was comparable to other photothermal lens apparatus, being 7.6×10^{-5} AU for 130-mW 488-nm excitation power in CCl_4 solvent. The analyte used to determine detection limits was I_2.

Experimental results for the effect of optical pathlength are illustrated in Figure 7.4. The fact that the signal did not increase proportional to

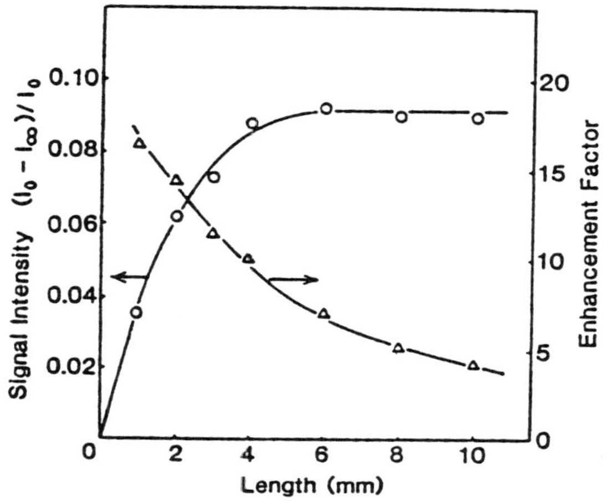

Figure 7.4. Sample pathlength dependence of the optical fiber–based photothermal lens apparatus shown in Figure 7.3. The signal initally increases, then levels off to a stationary value. The enhancement decreases due to a less than linear pathlength-dependent signal magnitude. (Reprinted with permission from Imasuka et al. 1987. Copyright 1987, American Chemical Society.)

increasing sample length is not surprising. However, the leveling off of signal magnitude at sample pathlengths greater than about 5-mm signal is somewhat contrary to that predicted using the integrated pathlength approach to signal calculation. Using the integrated pathlength equation developed by Carter and Harris (1983) and shown in Chapter 6, the pathlength-dependent signal is, in this case,

$$S_{\text{cw}}(t) \approx -\int_{\text{path}} \frac{2z'}{f_{\text{cw}}(t)}\, ds \approx \left(\frac{dn}{dT}\right) \frac{z_0^2 \Phi_0 \alpha Y_H}{\pi \kappa w_0^2} \ln \frac{z_0^2 + l^2/(1 + 2t/t_c)}{z_0^2 + l^2} \tag{28}$$

where z' is the path integration variable, t_c the characteristic time constant for the minimum beam waist at the fiber output, $t_c = w_0^2/8D_T$, and l is the pathlength or the distance between the excitation fiber and the sample cell window. It is assumed here that the minimum beam waist is at the output of the excitation fiber and that the refractive index change occurring at the sample cell–air interface does not affect the signal magnitude. The maximum signal occurring at infinite time is

$$S_{\text{cw}}(\infty) \approx \left(\frac{dn}{dT}\right) \frac{z_0^2 \Phi_0 \alpha Y_H}{\pi \kappa w_0^2} \ln \frac{z_0^2}{z_0^2 + l^2} \tag{29}$$

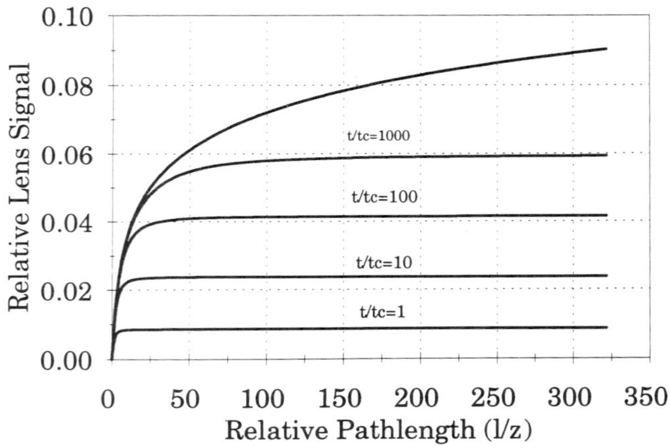

Figure 7.5. Theoretical pathlength-dependent optical fiber–based photothermal lens signal magnitude for different relative irradiance times. The upper curve is for continuous excitation. The decrease in the plateau signal with faster irradiance times is described in the text.

The z_0^2/w_0^2 dependence apparently favors larger core diameter fibers (e.g., larger minimum beam waist radius). This is due to the geometry particular to this type of experiment.

The minimum beam waist radius, w_0, at the excitation fiber was only 2.5 μm and the confocal distance, z_0, at 488 nm was 31 μm in these experiments. At the maximum sample pathlengths used, the relative path, l/z_0, was about 325. A plot of theoretical signal over this range for several t/t_c ratios is illustrated in Figure 7.5. The t_c value is that based on the minimum beam waist at the fiber. The uppermost curve is that for infinite irradiation time. The signal leveling for finite t/t_c ratios is understandable in terms of the expanding beam waist radius as the light exits the fiber and propagates through the sample since times required to develop the thermal lens increase with increasing beam waist. Comparison to the data in Figure 7.4 shows that although the general trend is the same for theory and experiment, the experimental signal data increase more slowly with pathlength and level off at a relative pathlength of about 200. The leveling-off behavior observed in the experiment is understandable since finite irradiation times were used. In fact, the shape of the $t/t_c = 1000$ curve is similar to that observed by Imasaka et al. The leveling off of the experimental signal suggests that the chopper rate was too fast to allow complete development of the signal. The relative pathlength where this leveling off occurs can be accounted for only if the minimum beam waist radius is greater than that predicted based on the core diameter of the fiber. A twofold increase in minimum beam waist could account for the observed effect. Imasaka et al.

suggest that these effects may be due either to excitation light propagating through the fiber cladding, increasing the effective minimum beam waist radius, or to thermal diffusion along the axis of propagation. Both of these effects would increase the radial extent of the thermal perturbation, thereby decreasing the relative pathlength, l/z_0.

Imasaka et al. also found a background signal component when the fiber was used. They ruled out the possibility of adsorption on the surface of the fiber since the contribution to the lens signal from this point is negligible. They proposed that this signal may be due to absorption in the optical fiber itself and point out that absorption measurements in optical fibers are extremely difficult to measure. Transmission loss in fibers is due mostly to scattering loss. Since photothermal spectroscopy is not directly dependent on scattering, this method may prove valuable in characterization of optical fibers.

7.4. PROCESSING PHOTOTHERMAL SIGNALS

Signal processing techniques are used to increase the precision and accuracy of the photothermal spectroscopy signal measurement. The signal, of course, is derived from an intensity or power of either the laser used in single-laser photothermal lens spectroscopy or from the probe laser used in the two-laser techniques: photothermal lens, deflection, diffraction, or interferometry. For the case of interferometry, the signal may be derived from the mirror positioning voltage required to produce a given phase shift. Independent of the way the signal is generated, all of these forms of photothermal spectroscopy generate a current or voltage that contains noise. Practically speaking, better estimates of the analytical signal in the presence of this noise will produce more precise estimates of the absorbance, and it follows that lower detection limits will be obtained. It has been shown repeatedly that the means for signal processing can make significant differences in the measurement precision and detection limits produced from the same photothermal spectroscopy apparatus (Dovichi and Harris 1981; Nickolaisen and Bialkowski 1986; Bialkowski 1986, 1987).

Two principal approaches that have been used to process the photothermal signal response. The first is to measure the detector signal at the beginning and the end of the transient time-dependent photothermal signal. These two measurements are then used to determine the detector response through the equations that govern the signal. Equivalently, the amplitude of the signal may be measured when periodic sample excitation is used. With knowledge of the excitation laser power or energy, requiring an additional measurement for the two-laser apparatus case, and the calibration factor relating thermal response to analyte absorbance, the sample absorbance may be estimated using the equations that describe the physical processes of signal generation. This approach is based on the conviction that the resulting signal should be linear in the sample absorbance–excitation laser strength product, and that the thermal and optical properties of the sample are known well enough to allow accurate calculation. The second approach is to take measurements over the course of the time-dependent signal. The entire transient signal estimate is then used to determine the parameters that affect the analytical signal. This may be done by regression of the data to an appropriate model. The latter approach is generally more precise, owing to the redundancy of information in the extensive data set and to the fact that the models can account for signal nonlinearity. In either case the detector response for a series of excitation laser powers or energies, and calibration solutions covering the range of expected measurements, should be determined. Detector response to sample absorbance and excitation laser power or energy is nonlinear unless the range of variation is small. In any event, the absorbance estimate can be determined from the resulting calibration curves if they exhibit a positive sensitivity.

There are two primary means to process electronic signals. The first is based on analog processing of the voltages or currents produced by the optical or piezoelectric transducers. The later is based on the use of passive frequency filters, active amplifier-based filters, integrators, and differentiators, readout devices such as voltmeters and oscilloscopes, and even more sophisticated instrumentation such as lock-in amplifiers, gated integrators, and boxcar averagers. The second way to process the signals is to use digital electronics. In digital processing the analog signals produced by the transducers are converted to digital signals. The digitized signals can then be stored for indefinite periods of time. They can be processed either with the use of simple spreadsheet programs or with sophisticated filters and computer algorithms. In most modern instrumentation, even analog-processed signals can be digitized and transferred to a computer for analysis and archival storage. In the following sections we address the ways that signals produced in photothermal spectroscopy have been processed and used to estimate properties of the sample.

It is important to keep in mind the types of noise encountered in performing optical measurements using laser light sources. Although it is beyond the scope of this book to address these noise sources in detail, a short summary is given here. Detailed analysis of these noise sources can be found in the book by Ingle and Crouch (1988). According to Ingle and Crouch, noise can be classified as being fundamental or nonfundamental. Fundamental noise is that due to the nature of the light and the electrons produced by the optical transducers. There is nothing that can be done to reduce fundamental noise. Nonfundamental, or excess, noise is measurement error that occurs in addition to the random processes giving rise to fundamental noise. It is due to, for example, inferior optical transducers, electronic circuits, and interference. Excess noise can often be reduced by using higher-quality or more appropriate components or by better experimental design.

There are three primary types of noise produced in electronic circuits: Johnson, quantum, and flicker. Each of these has associated with it a different random physical process that produces the noise. Johnson or thermal noise is produced by the random motion of electrons in resistive electronic elements. The magnitude of this noise is given by $\sigma_J^2 = 4kRT\,\Delta f$, where σ_J^2 (V^2) is the voltage variance due to Johnson noise, k ($J\,K^{-1}$) is Boltzmann's constant, R (Ω) is the element resistance, T (K) is the temperature, and Δf (Hz) is the measurement bandwidth. This noise has a white-noise power spectrum, meaning that the noise magnitude is independent of frequency. The noise magnitude is proportional to the measurement bandwidth. Johnson noise is usually independent of the fundamental signal magnitude. The primary means used to reduce the relative content of Johnson noise are to increase the signal magnitude, to cool the circuit or component, and to reduce the measurement bandwidth. Quantum noise, also called shot or Schottky noise, is a fundamental noise produced by the nature of particle counting. The binomial distribution applies to the mean number fluctuations that occur when counting noninteracting particles. The magnitude of quantum noise is directly proportional to the number of particles counted. In fact, $\sigma_N^2 = N$, where N is the number of electrons, photons, ions, and so on, counted. In terms of electronic signals, the noise is given by the shot noise variance, $\sigma_S^2 = 2ei\,\Delta f$, where σ_S^2 (A^2) is the squared noise current, e (C) is electron charge, i (A) is the average current in the signal, and Δf (Hz) is the measurement bandwidth. Like Johnson noise, this noise has a white-noise power spectrum. The noise power is proportional to the average current, and the relative noise contribution, or SNR, can be reduced by increasing the signal magnitude. Another means is simply to count more particles. In effect, increasing the power of the excitation or probe laser will increase the SNR. What is in effect a Johnson noise source can also arise due to quantum noise produced by the detection of stray radiation. This noise can be minimized by using light-tight detector mounts with laser-line optical filters.

Flicker noise is actually a broad category of different noise sources arising from both fundamental and determinant processes. It is characterized as having a colored-noise power spectrum, with more noise power at the lower frequencies. Mathematically, the noise is given by $\sigma_F^2 = K_F f^{\alpha}\,\Delta f$, where K_F is a scale factor and α is a frequency exponent, typically between 0.8 and 1.2.

The power spectrum is often referred to as "pink" to indicate the weighting toward the lower frequencies. The fundamental processes giving rise to flicker noise may arise as a consequence of chaotic system behavior, such as those that occur in hydrodynamics and even electrodynamics (Arecchi 1988). Better understanding of these processes is currently emerging with progress in the mathematical descriptions of chaos. In fact, an understanding of the source of this type of noise in lasers has resulted in stabilization of these sources (Roy et al. 1992). Major flicker noise components in photothermal spectroscopy are that produced by the excitation and probe laser sources, that resulting from the turbulence produced by sample convection at higher excitation laser powers and/or sample absorbances, refractive index gradients produced by flowing the analyte through the sample cell, and the dynamic quality of light scattering in scattering samples. Flicker noise sources will manifest as noise proportional to the photothermal signals. Increases in signal will result in a proportional increase in noise.

Perhaps more pertinent to photothermal spectroscopy are the particular nonfundamental noise sources that occur when using laser light sources. Since most forms of photothermal spectroscopy make use of the spatial integrity of the light sources, spatial variations (e.g., mode and pointing noise) also constitutes a form of flicker noise in that the noise produced from these fluctuations will be directly proportional to the photothermal signal strength. Temporal noise (e.g., power drift in continuous lasers or pulse energy variation in pulsed lasers) may be overcome by monitoring the energy or power of the source. Proper correction for this temporal noise requires that the bandwidths of all processing electronics be the same. This is particularly true for the noise produced by longitudinal mode beating, wherein the output power may vary rather substantially (up to 50% at certain times) in the MHz frequency range.

The power spectra of these laser-unique noise sources have not been examined in detail for a wide range of lasers. This type of data must, in general, be obtained in individual research laboratories. Although laser manufacturers are becoming more aware of the need for this type of data and are starting to report pointing stability and other spatial noise figures of merit, the method for measurement and reporting these data are not consistent. On the positive side, Reticon delivers software for reporting pointing noise with their laser beam profilers. With the advent of commercial measurement devices capable of reporting mode and pointing noise, one can hope that the data needed to make educated choices among the lasers will become available in the near future. There have been only a few studies addressing spatial noise in laser and other sources. Pawliszyn (1987) compared the pointing noise in helium–neon and diode lasers to that obtained using a light-emitting diode (LED). The pointing noise for all these sources was most significant below about 10 Hz. The helium–neon laser (Uniphase model 1103P) exhibited from one to two orders of magnitude more noise than that for the LEDs, while the laser diodes tested (Sharp models LT022MC and LT023MC) were about one order of magnitude worse than the LEDs. By themselves, LEDs are not as useful as the laser sources for photothermal spectroscopy, due to their relatively high divergence. However, Pawliszyn points out that this limitation may be overcome by coupling the LED output to single-mode optical fibers. In another study, Johnson and Voigtman (1990) examined the temporal noise and noise power spectra produced with a pulsed excimer laser. In this study the pulse-to-pulse energy variation of the excimer laser output was examined as a function of laser repetition rate. Pulse-to-pulse energy variation can cause problems when the laser is used as an excitation source for nonlinear optical processes. Minimization of this variation can be an effective way of improving measurement precision in these experiments. They described a method for measuring and analyzing the data, and found that the pulse-to-pulse variation reached a minimum at a 40-Hz repetition rate for their Questek model 2210 excimer laser.

The primary problem with spatial noise is that the error probability distributions are not Gaussian (normal). Subsequently, compensating for the errors produced by spatial noise is not a straightforward task. While it is simple enough to measure the laser pulse energy or time-dependent power, and to correct the data subsequently for the variation, measurement, analysis, and prediction of the result of spatial beam variations is not simple. The technology for performing this analysis and variation compensation apparently does not exist. Perhaps the best way is to eliminate the error source. If one were not lucky enough to have received a laser with low spatiotemporal variation, the only real way to overcome spatial noise would be to spatial filter the excitation and probe laser beams and to either control or monitor and compensate for fluctuations in the laser powers used in the experiment. Interestingly enough, this is rarely done in photothermal and other laser-based spectroscopies, although it is commonly performed in absorption spectrophotometry using conventional light sources.

In addition to the fundamental noise sources, interference and distortion can often lower the SNR and alter the signal. Interference such as the 60-Hz signals due to ground loops and electromagnetic radiation picked up by unshielded wires, particularly in high-impedance circuits, can often overcome the relatively small signals in photothermal spectroscopy. Sources of interference include the 60-Hz, 120-Hz, and higher harmonics of the ac line, ground loops produced by components powered from various ac outlets, and high-frequency interference produced by switching power supplies. These can be reduced to a great extent by careful experiment design and adherence to the basic rules of grounding and shielding. Distortion can also alter the experimental signals. Distortion occurs when, for instance, the irradiance of a laser source exceeds the linear working range of a detector. Distortion is difficult to detect by its manifestation in the signal. The easiest way to prevent distortion is to check that the laser source is within the linear working range of the detector. This can be accomplished simply by using calibrated neutral density filters. If the transducer signal is attenuated by an amount less than that expected by the neutral density filtered, the detector is probably saturated. Detector saturation can be circumvented by decreasing the light power at the detector, but this increases the relative quantum noise. Saturation may also be diminished by decreasing the load impedance used to terminate the photodiode detector. This decreases the forward bias produced by the *IR* drop at the load. Transimpedance amplifiers are particularly useful for this purpose in that they effectively hold the photodiode anode at ground potential.

7.4.1. Analog Signal Processing

The simplest and perhaps most straightforward means to process signals is by direct measurement of the voltage or current produced by the photodetector while the sample is being irradiated. This is often accomplished by watching the deflection of the trace on an oscilloscope or voltmeter. The precision of this method is not very high, and with the exception of oscilloscopes with signal-averaging capabilities, the SNR can be improved only through the use of electronic low-pass filters. Nonetheless, this technique was used in many of the earlier reports of photothermal lens spectroscopy, and absorbance as low as about 10^{-5} was measured in this fashion.

Gated sampling with exponential averaging, also called gated averaging or integration, can be used to enhance the SNR in repetitive signals. With gated sampling, the signal is monitored during a time period that is generally much less than the time required to repeat the signal. During this period, the signal is amplified and used to charge a capacitor. The charge delivered to the capacitor is proportional to the integral of the signal over the time of the gate period. When the gate signal is synchronous with the experiment repetition, several measurements cycles are sampled. The potential across the capacitor represents the average value of the potential or current present during the gate after several cycles have been sampled. A large-value resistor is usually placed across the capacitor. This allows the capacitor to discharge slightly between the charging part of the cycle. The capacitor potential reaches a steady state when the rate of charging due to the gated signal sampling equals the *RC* discharge rate.

The advantage of this type of analog averaging is that it can be used to measure fast, repetitive signals. It is particularly useful in averaging low-duty-cycle signal: for example, when the signal is present for only a small fraction of the total time required to repeat the cycle, such as those produced in low-repetition-rate pulsed-laser excited photothermal spectroscopy. The gated averager can also be used to average and measure the temporal signal. In this case, called boxcar averaging, the gate timing is delayed relative to the experiment cycle by a small amount with each cycle. By measuring the gated average signal as a function of the experiment start time to sample gate delay, the temporal profile of the signal can be recorded. Gated sampling and boxcar averaging can produce SNR improvements approximately equal to the square root of the number of experimental cycles averaged. However, it is important that the magnitude of the signal at any given experimental start delay time does not change significantly from cycle to cycle. The latter condition is often difficult to meet with pulsed-laser sources. The noise produced by lasers with poor pulse-to-pulse energy or mode structure can be overcome somewhat by using longer RC time constants in the capacitor integrator circuit. Gated averaging and boxcar integration can be performed using digital electronics. The sample-and-hold circuits used on the fast analog-to-digital converters are functionally equivalent to the gated integrators, and infinite impulse-response digital filters can be used to perform the same function as the averaging capacitors.

When the duty cycle of a periodic signal is near 50%, lock-in amplifiers may be the best choice for data processing. Lock-in amplifiers have variable-frequency bandpass filters that are dynamically tuned to the experimental repetition frequency. The bandpass filter frequency is locked in to the experimental repetition frequency by supplying the lock-in with an external reference signal. This reference signal is usually derived from the controller of the light beam chopper or acoustooptic modulator used to modulate the excitation laser beam. Amplification occurs only for signals that are within a small-frequency-bandwidth band and centered at the frequency of the reference. The main SNR improvement comes from the decreased bandwidth of the measurement, but most, if not all, lock-in amplifiers also have a low-pass filter that averages the potential at the reference frequency. The lock-in amplifier is not only sensitive to the amplitude of the signal but also to the phase of the signal relative to that of the reference. There are, in fact, two modes in which lock-in amplifiers can be used. In the first mode, the lock-in produces signals proportional to the magnitude of that part of the signal oscillating at a frequency equal to that of the reference, and the phase of the signal's sine wave, again relative to the reference. This mode of operation is useful when the phase shift between the excitation and signal is to be measured or when the relative phase of the signal is not known. However, out-of-phase noise will be introduced into the magnitude estimate. With phase-sensitive detection, the lock-in amplifier produces two signals, proportional to the amplitudes of the sine and cosine components of the input signal. An internally generated phase shift is used to shift the relative phase of the reference signal to match any phase shift produced by the sample excitation-measurement process. This latter mode of operation yields better SNR, but the phase relationship between the source and the signal must be known and fixed. This is often the case for photothermal spectroscopy.

Lock-in amplification can generally be used only for two-laser photothermal spectroscopies, where the signal is proportional to the magnitude of the change on detected probe laser power or analog-computed probe beam position. The two laser beams do not have to be from separate lasers; in fact, schemes for using a single laser as both pump and probe based on polarization encoding have been reported (Yang 1984, Yang and Hairell 1984, Pang and Morris 1985). This method may reduce some of the pointing noise produced as a consequence of using two separate lasers in a properly designed optical system (e.g., one where both excitation and probe beam optics image the laser into the sample cell and where the pinhole aperture is large). A method for using lock-in detection for single-laser photothermal lens spectroscopy was developed by Jansen and Harris (1985). In their apparatus, a chopped Ar^+ laser beam was split into two beams. One beam was used in a conventional single-laser photothermal lens arrangement. The other beam was directed through a variable neutral density filter and onto a second, reference photodiode detector. Signals from the thermal lens and reference photodiode detectors were subtracted by wiring them anode to cathode prior to the transimpedance amplifier. Photodiode balance was accomplished by changing the attenuation of the reference beam with the neutral density filter. The resulting difference signal was processed with a lock-in amplifier. This apparatus demonstrated short-term (ca. 1-min) SNR performance about twice that obtained with digital processing though the long-term (10-min) performance was inferior to digitally processed signals.

7.4.2. Digital Signal Processing

Digital signal processing allows for a maximum flexibility in the processing functions (Bialkowski 1988a,b). As the name implies, the data must be in digital format in order to use these methods. The raw data produced by the detector are in analog form and must be converted to digital form. This conversion is performed with an analog-to-digital converter. There are several manufacturers of analog-to-digital converters that utilize personal computer hosts. A personal computer host equipped with one of these converters serves as a convenient and cost-effective platform to collect and process photothermal data. Converters range in their acquisition speeds, typically specified as in terms of the maximum conversion frequency, conversion precision, typically specified in the maximum number of bits that the digital data possess, accuracy, the ability to adjust the input range of the analog data, thereby effectively extending measurement precision, and in their ability to sample the analog signal on trigger command. The specific converter needed for effective digital processing of photothermal spectroscopy depends on the frequency of the particular type of data. In general, converters with external trigger capacity, sample-and-hold or track-and-hold circuits, and programmable gain amplifiers prior to the analog-to-digital converter, and from 12- to 16-bit digital precision, are needed for continuous-laser-based methods. Digital data requirements for analysis of pulsed laser based methods usually imply maximum sampling rates of 1 MHz or greater. Measurement precision is often traded for relatively high sampling frequency. This is because the high measurement bandwidths admit more noise and the digital measurement precision requirement is subsequently lower. Boards sampling at rates up to 50 MHz with 8-bit precision are available for under $2000 due to the technological developments in video digital processing. The precision lost by using 8-bit converters can be recovered by oversampling the time-dependent data. Twelve-bit boards are also available but at substantially higher cost. With high conversion rates, data are often stored temporarily on the converter board itself during conversion. After conversion, the data can either be directly accessed by the microprocessor or transferred to main memory for complex processing.

Simple real-time digital filter processing schemes used to perform filtering, smoothing, and transient signal magnitude estimation have been described by Bialkowski (1988a,b). These functions are often similar to analog processing, although the digital filters are much more flexible and easily adapted to different signal-processing requirements. In addition, functions that are extremely difficult to execute with analog circuits, such as interference orthogonal matched filtering, are sometimes easier to perform as the digital smoothing equivalents to analog circuits. Although the apparent advantage to digital signal processing is the ease with which filtering characteristics and parameters can be changed, the main advantage is that optimum signal processing can be performed. In addition, because of the intrinsic ability of computers to process data to more interpretable forms, digital signal processing can result in more facile data interpretation. Finally, optimum adaptive signal processing schemes can be employed. These schemes yield the theoretical maximum data precision (Bialkowski 1987).

One of the simplest and most effective digital data processing schemes is multichannel averaging. Multichannel averaging is most often used to process time-dependent data. It is applicable whenever the experiment is reproduced periodically. Signals corresponding to a particular time element beyond the relative experiment start time are averaged into a particular data location in digital storage. Each data element is called a channel. Sequential time elements are summed into sequential channels. Thus the resulting data constitute a time series with times relative to the experiment start time. The contribution that the photothermal signal makes to the total summed measurement in a given channel will increase by a factor of N, the number of experiments summed. When the noise is not correlated to the signal, the noise contribution to the total signal at any given time delay will vary randomly from experiment to experiment. By multichannel averaging random, additive noise, the relative noise contribution to the summed measurement, increases as $N^{1/2}$. Taken together, the signal and noise contributions predict a signal-to-noise ratio improvement that is equal to a factor of $N^{1/2}$. This simple but effective processing scheme is extremely useful for processing low signal-to-noise ratio data when the time dependence of the signal is sought. It can also be used to improve SNR for data magnitude (e.g., peak height) determinations by inspecting the resulting data. Single-channel averaging can be used in this case, or better still, regression analysis of the multichannel data averaged data can be used to better estimate signal magnitudes and other parameters of the measurement.

One way to model a repeatable experiment is to specify an impulse-response signal, $s(t)$, with unit amplitude, and an experiment cycle-dependent amplitude, a_i, such that the signal for the ith experiment cycle is $a_i s(t)$. Notice that in this sense, repeatable means only that the form of the impulse response be the same; the amplitude can vary from cycle to cycle. This condition is indicative of a linear system. The measurement, $x(t)$, is a sum of the signal and noise components, $x(t) = a_i s(t) + v(t)$. For random noise that is not correlated to the experiment start time, this noise is not a function of measurement time, and $v(t) = \sigma$, σ being a random variable. Multichannel averaging produces an average measurement, $\hat{x}(t) = \hat{a}s(t) + \sigma/N^{1/2}$, where $\hat{a} = \Sigma a_i/N$. This is often interpreted as being the signal estimate, $\hat{x}(t) = \hat{s}(t)$. The reason for the inverse-square-root N dependence of the noise term is that the noise averages as the noise power. The signal estimate is thus the signal impulse response that has an amplitude scale of $\hat{a}$, the average amplitude over the course of the experiment cycles, and that has an additive noise contribution of $\sigma/N^{1/2}$. It is important to remember that signals produced in photothermal spectroscopy are nonlinear for strong thermal perturbations. If signal amplitudes vary too widely, the linear system approximation may not be valid, and the ensemble-averaged result will be averaged over the amplitude-weighted impulse response.

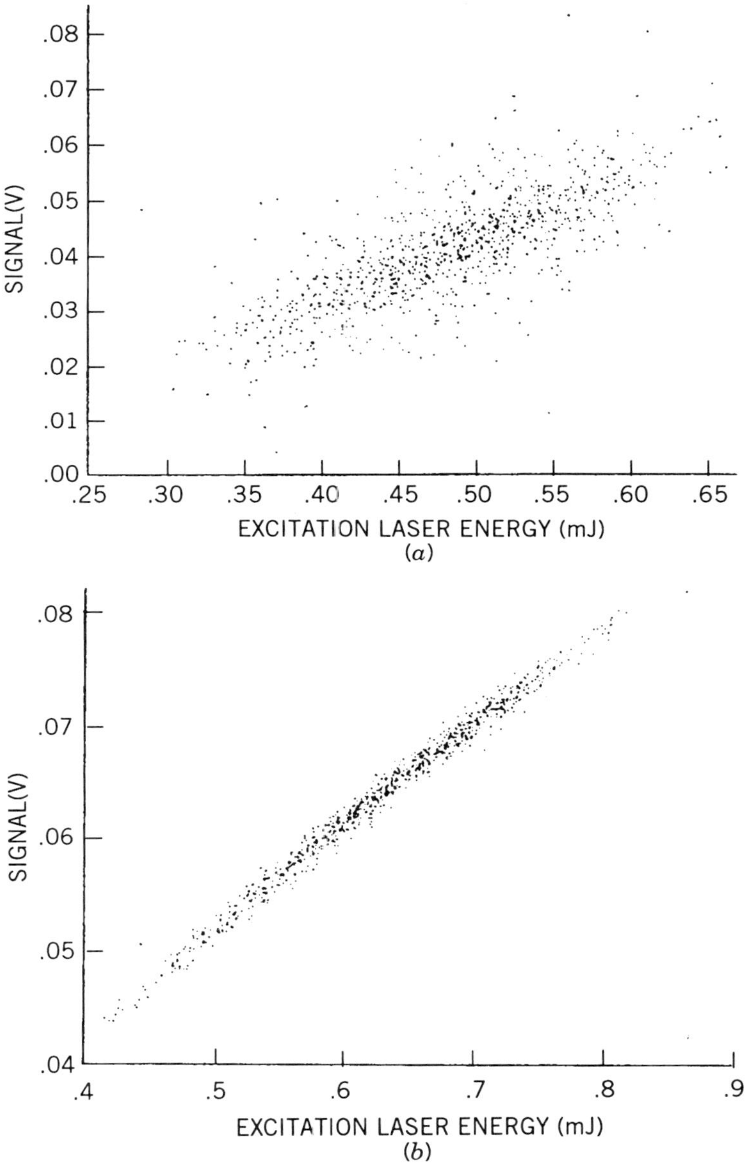

Figure 7.6. Scatter plots produced by measuring the excitation energy and pulsed-infrared-laser-excited photothermal signal for two digital signal processing schemes. Plot (*a*) used gated integration sampling of the transient signal; plot (*b*) used matched filtering with an interference–orthogonal innovations filter. The improvement signal-to-noise ratio is apparent from the spread of data in the signal axis direction. (Reprinted with permission from Bialkowski 1987. Copyright 1987, American Physical Society.)

For pulsed laser–excited experiments performed under equivalent conditions (e.g., where the thermal, optical, and characteristic thermal decay time constants are the same), the amplitudes of the multichannel averaged signals can be used to compare the relative strength of the photothermal response (Figure 7.6). Estimates of signal amplitude are normally based on the difference between the baseline estimate and that of the signal maximum. These estimates are not very precise. The noise in each is $\sigma/N^{1/2}$. A better way to estimate the signal amplitude is by using a matched filter (Nickolaisen and Bialkowski 1986; Bialkowski 1986, 1987). The matched filter developed by Nickolaisen and Bialkowski used a well-averaged signal as an estimate of the impulse response. The impulse response is made zero mean by subtracting the vector average from each vector element and then is normalized in power. This results in the estimation vector $h(t) = (s(t) - \hat{s})/[(s(t) - \hat{s}) \cdot (s(t) - \hat{s})]$. Upon converting each signal transient, an estimate of the signal amplitude is obtained from the vector dot product, $a_i = x_i(t) \cdot h(t)$. Averaging is performed by averaging the individual signal amplitude estimates.

A more sophisticated method, which compensates for interference produced by the excitation source, is to use an innovations vector in place of $h(t)$ (Bialkowski 1986, 1987, 1988a). The innovations vector is the matched filter vector made orthogonal to coherent interference (cyclostationary noise) using a Gram–Schmidt factorization procedure. The components to

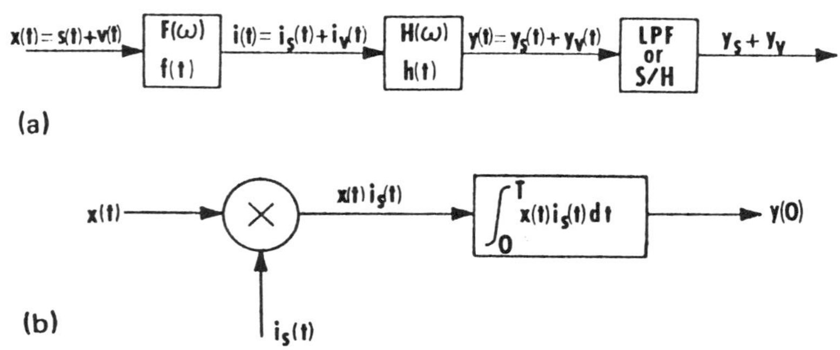

Figure 7.7. Schematic representation of the real-time matched filter in both frequency and time domains. Filter (*a*) is composed of three stages: the whitening filter $F(\omega)$ with impulse response $f(t)$, the matched filter $H(\omega)$ with impulse response $h(t)$, and the low-pass or sample-and-hold filter. The input, $\mathbf{x}(t)$, composed of signal, $\mathbf{s}(t)$, and noise, $\mathbf{v}(t)$, is first "whitened," resulting in the innovation $\mathbf{i}(t)$. The $\mathbf{i}(t)$ is then filtered, resulting in a signal estimate, $\mathbf{y}(t)$. The signal estimate is sampled at the time of maximum signal-to-noise ratio. Filter process (*b*) shows an implementation of (*a*) for the case where the form of $\mathbf{s}(t)$ and $\mathbf{v}(t)$ are known prior to measurement. (Reprinted with permission from Bialkowski 1987. Copyright 1987, American Physical Society.)

innovations filter vector construction are illustrated in Figure 7.7. Construction of this vector requires high-quality estimates of both the signal and cyclostationary noise impulse responses. In certain cases the extra time required to set up the experiment pays off. Bialkowski (1986) showed that the SNR improvement ratios for this signal processing scheme was about an order of magnitude better than that obtained using gated integration or simple matched filtering.

For equivalent sample thermal and optical constants, but where the characteristic time constant may vary due to optical alignment, excitation laser mode, or wavelength changes, the multichannel averaged signal transient may best be processed using regression to the equations that describe the transient signal. In the case of pulsed laser excitation, the maximum signal, that at zero time, is a function of the thermal and optical parameters of the sample matrix as well as optical parameters (e.g., w_0). Proper reduction of the thermal response to sample absorbance requires a determination of the excitation beam waist radius. This could be determined using regression. Regression analysis would automatically determine both the signal magnitude and the characteristic thermal decay time constant required to estimate sample absorbance.

This approach was used by Dovichi and Harris (1981) to determine sample absorbance from transient signals produced in a single-laser photothermal lens apparatus. To obtain sample absorbance, they first multichannel averaged 100 transients of the photodiode detector response. The ensemble-averaged transients were then fit to the equation

$$\Phi(t) = \Phi(0)\left[1 + \frac{2.303EA}{1 + t_c/2t} + \frac{1}{2}\left(\frac{2.303EA}{1 + t_c/2t}\right)^2\right]^{-1} \tag{30}$$

where E is the photothermal lens enhancement factor and A is sample absorbance. A nonlinear regression (Marquardt) algorithm was used to determine $\Phi(0)$, EA, and t_c. Although the nonlinear regression analysis is time consuming, the results obtained are substantially better than that obtained with the two-point signal estimation procedures. The minimum detectable absorbance using this approach was found to be 7×10^{-8} AU in CCl_4 solution using a 160-mW Ar^+ laser. Moreover, the absorbance range of the technique was extended to higher absorbances by fitting only the early part of the transient signal. Dynamic range expansion was possible because the $\Phi(t = \infty)$ point is not required to estimate $2.303EA$. This avoids problems associated with diffractive image aberration, which occurs for strong thermal lens element, and convection heat loss and turbulence when the temperature change induced in the sample is too high.

7.5. PHOTOTHERMAL DATA PROCESSING

Signal processing is used to generate data of a measurable quantity that can be related to the sample absorbance, excitation laser energy or power, optical design parameters, and so on. The process of making this relationship falls within the realm of data processing. Before discussing data processing methods it is important to realize that few, if any, photothermal spectroscopy signals are strictly linear in the product of the sample absorbance with excitation laser power or energy. The photothermal lens signal is intrinsically nonlinear due to the optics of probe beam propagation through the thermal lens. In addition, the analytical signal is defined in a complex fashion in relationship to the measured signal (e.g., photodiode response), to allow more facile interpretation in terms of sample absorbance. Data processing is performed to convert between measured and analytical signals. Photothermal deflection signals, estimated from changes in the probe laser power passing a straightedge aperture, are nonlinear for large deflection angles. A commercial position-sensing detector greatly improves and simplifies the photothermal deflection experiment by compensating for probe laser power fluctuations. But these, too, can exhibit nonlinear response, particularly when using bicell detectors. Photothermal diffraction and certain types of interferometry produce probe laser irradiance changes that are proportional to the square of the optical absorbance-excitation energy or power product. In photothermal interferometry, the voltage used to drive the piezoelectric mirror to an interferometric optical null may be used to estimate the optical absorbance. Although the change in probe laser power transmitted through the interferometer can produce a linear signal for photothermal-induced small phase shifts, it is generally nonlinear for larger photothermal perturbations.

7.5.1. Calibration

Photothermal spectroscopy is not a direct method for measuring sample absorbance. The photothermal signal depends on the thermal and optical parameters of the sample and the geometry of the particular apparatus. It is difficult to calculate sample absorbance from the signal strength and estimates of these parameters alone. The finite accuracy to which these parameters are known does not allow accurate calculation. Because of uncertainties in the optical apparatus and thermal and optical parameters of the sample, absolute absorbance determinations can only be obtained by calibrating the instrumental response with a series of solutions of known absorbance. Clearly, calibration solutions must be made from the same solvent as that of the analyte since the theoretical enhancement factors depend on solvent. In addition, care must be taken to ensure that there are no alignment changes or sample cell placement changes between measurements. Calibration is usually performed by measuring the instrumental response to samples at several concentrations. The graphical representation of the instrumental responses to sample concentration is the working curve. The working curve is used to calibrate the instrument response to unknown sample absorbance. Most protocols for ultrasensitive or trace environmental analysis require this standardization procedure, independent of the particular method used.

Of primary importance then in quantitative analysis using photothermal spectroscopy is the use of calibration procedures to generate working curves. The working curve is the line that best describes the instrumental response to the photothermal perturbation strength. For instrument calibration purposes, the woking curve must be generated in equivalent samples. It should extend over several orders of magnitude of analyte absorbance and/or excitation laser energy or power. Photothermal signals are intrinsically nonlinear in sample absorbance and excitation laser energy or power. Subsequently, one should not expect a linear working curve, independent of the methods used for signal processing. The working curve is normally produced by measuring the photothermal response for a series of solutions made from the same solvent but with varying analyte concentration. The raw signals [e.g., $\Phi(t)$] are measured, the data are processed [e.g., using $S = \Phi(0)/\Phi(\infty) - 1$], and the processed data are plotted versus analyte concentration. Regression is typically used to determine the mathematical relationship between analyte concentration and signal. The regression to a power series model, wherein the processed signals are estimated by

$$S(C) = \sum_{i=0}^{M} a_i C^i \tag{31}$$

where C is the analyte concentration and the a_i are the coefficients determined from the regression, is linear and thus easily performed. The series is usually terminated at $M = 1$ (linear) or $M = 2$ (quadratic). The quadratic model is obviously applicable to photothermal thermal lens spectroscopy. Extending the series to third ($M = 3$) or even fourth ($M = 4$) power can account for signal nonlinearity due to diffraction and other effects. Inversion of the series yields the concentration as a function of processed signal, $C = f(S, a_i)$; the inverse function is complex for series with $M \geqslant 2$. For the quadratic model, inversion of the regression equation is obtained from the quadratic formula, $a_0 + a_1 C + a_2 C^2$:

$$C = \frac{\sqrt{4a_2(S - a_0) + a_1^2} - a_1}{2a_2} \tag{32}$$

The concentration estimate standard error can be obtained using error propagation and the series coefficient variance estimates produced by many linear least-square regression routines. The concentration error in the case of a quadratic calibration curve is

$$\sigma_C^2 = \frac{1}{(a_1 + 2a_2C)^2}\,(\sigma_S^2 + \sigma_{a_0}^2 + C^2\sigma_{a_1}^2 + C^4\sigma_{a_2}^2) \tag{33}$$

The variance in the concentration estimate will increase with concentration due to both the high-order concentration dependence in the numerator and the fact that a_2 can be of a sign opposite that of a_1, reducing the value of the denominator term.

It follows from maximum likelihood principles that if the variances of the data used in a regression estimate are not equal, each datum should be weighted inversely proportional to its variance. Often, the signal variance increases with increasing concentration (or signal). Proportional errors arise from excitation or probe laser power variations, mode noise, and so on. This flicker noise results in concentration standard deviations that are proportional to the concentration. In this case weighted linear regression should be used to find the a_i coefficients, and the concentration variance will also depend on the signal variance. Weighted linear regression programs are not common in commercial software. However, there exists several useful texts on the subject, some of which have the source code for performing the analysis (e.g., Bevington 1969, Carnahan et al. 1969).

Application of weighted or unweighted linear regression to the calibration data presupposes that errors are confined only to the measurement process and that these errrors are normally distributed. In many cases neither of these two assumptions are valid. Errors in preparing the standard solutions can easily be larger than the measurement errors. In addition, excitation flicker noise sources are typically not normally distributed. For example, pointing noise in the lasers used in the photothermal apparatus will produce an error distribution that is exponential-like. Since the photothermal signal magnitude is a maximum when there is no pointing error, any change in the beam angle will produce signals that are of lower magnitude. Thus random Gaussian pointing noise in the excitation and probe lasers will result in a measurement probability or frequency distribution that is maximum at the maximum measurement value and that will decrease with decreasing measurement magnitude. A similar distribution should be found for spatial mode noise that is time dependent since the spatial noise is focused off axis to the Gaussian laser beam. In both of these cases the correct or accurate value is the most probable measurement, but the distribution is so skewed that common statistical means of correcting the measurement error distribution fail. There is currently no satisfactory way to find accurate calibration parameters for these situations.

7.5.2. Excitation Irradiance Curves

Obtaining the data required to produce the working curve from a series of solutions at different analyte concentrations can take a long time. However, the instrumental linearity may also be measured by varying the excitation laser power at a fixed standard sample absorbance. The later technique relies on the fact that the photothermal optical element is proportional to a power series in αlQ, for pulsed, or $\alpha l\Phi$ for continuous laser excitation. In the absence of nonlinear photodynamics, the calibration curve obtained by varying the excitation source strength is equivalent to that obtained using standard solutions.

Bialkowski and He (1992, 1994) used this method to linearize the instrumental response of a pulsed laser–excited photothermal lens apparatus. They needed linearized response plots to assess accurately the effect of optical bleaching on pulsed-infrared-laser-excited photothermal signals. Since optical bleaching also produced nonlinear excitation energy-dependent signal response, it would be too difficult to model the raw data if there were also instrumental nonlinearities. High-precision calibrations of the instrument response to excitation energy were generated by changing the excitation energy continuously, pulse by pulse, while measuring the magnitude of the photothermal lens signal. The calibration data were modeled using a second-order polynomial in excitation energy:

$$S(\alpha lQ) = a_0 + a_1(\alpha lQ) + a_2(\alpha lQ)^2 \tag{34}$$

where αl is the analyte absorption, Q is the excitation energy, and the a_i were determined using weighted regression. Inversion of the quadratic model equation yielded the linear response

$$\alpha = \frac{\sqrt{4a_2(S - a_0) + a_1^2} - a_1}{2lQa_2} \tag{35}$$

This equation was used to linearize the raw photothermal lens data. The resulting linearized data were also calibrated since the calibrations were obtained using samples of known αl using calibration species known not to have nonlinear photobleaching in the range of excitation energies used. Subsequent data sets processed using this linearization step revealed the nonlinear excitation-energy-dependent absorption coefficient, $\alpha(Q)$.

This method can also be used to model the instrument response for analytical purposes. It is much easier to vary the excitation power or energy than it is to prepare several solutions of known concentration. However, excitation calibration will not be able to compensate for optical attenuation of highly absorbing samples. It is a valid means of apparatus calibration only if solvent absorption is negligibly small compared to the analyte and analyte concentrations are such that the minimum transmission is greater than about 90%.

7.5.3. Limits of Detection and Quantitation

Limits of detection (LODs) are normally defined as that analyte concentration or absorbance which produces a signal that is three times the standard deviation of the blank measurement, s_b, while the limit of quantitation (LOQ) is that concentration producing a signal that is $10s_b$ (Katemen and Buydens 1993). The blank measurements are based on samples that do not contain the analyte. Sample and blank samples should be prepared in the same fashion, using the same solvents, and measurements should be taken under equivalent conditions. The LOD is often reported in photothermal spectroscopy papers. However, reported LOD do not necessarily follow the $3s_b$ convention. Very often, the LOD is taken to be three times the standard deviation of the noise produced by the measurement apparatus, or the blank standard deviation is determined from the noise in the signal. Although this method is useful in reporting a relative figure of merit for the apparatus, it is not, strictly speaking, a LOD. This method is valid only when the solvent does not have finite absorbance (e.g., in gas-phase analysis).

An interesting problem arises when the solvent has a finite absorbance. This is particulary troublesome for aqueous samples since the absorption coefficient for water is relatively high for a photothermal measurement ($\alpha \sim 10^{-2}$ to $10^{-1}\,\text{m}^{-1}$) in the visible portion of the spectrum (Tam and Patel 1979). The high solvent absorbance can dramatically lower the LOD of the technique, especially when the excitation source has time-dependent energy or power that is not compensated for in the calculations. Again, this problem can be overcome only by using a differential technique with high dynamic reserve. Taking the difference between measurements obtained using separate sample and blank measurements will generally not have the precision necessary to measure small absorbances in high-absorbance solvents.

A problem with reporting the relative merit of a photothermal apparatus arises when there is no apparent blank signal. Bialkowski and He (1988) found that the signal-to-noise ratios of data obtained using pulsed-infrared-laser-excited photothermal deflection of dichlorodifluoromethane in ultrapure argon could always be improved upon by increasing the number of signal transients averaged. Ensemble averaging decreased the intrinsic noise due to the electronic circuits and the shot noise in the probe laser beam of their etalon-based photothermal deflection apparatus. The LOD was difficult to define since it was a proportional to the root number of averages, $N^{1/2}$. As a consequence, they reported an LOD of 0.7 ppm(v/v) on a per-excitation-pulse basis. Since the LOD is improved with increased number averaged, a more correct way to report the results would be in units of ppm(v/v) pulse$^{-1/2}$ to account for the signal-to-noise ratio improvement that occurs upon signal avering.

Practically speaking, experimental measurement times are usually limited by the times that the analyte is available, as in a chromatography experiment, or by flicker noise. Bialkowski and He found that flicker noise caused by changes in the laboratory environment limited the times over which data could be averaged. However, this could be overcome with dynamic stabilization of the optical apparatus. With stabilization, there is apparently no limit to the LOD for gas samples that do not have interferent species present. This type of proportional flicker noise does not affect the blank or background signal, so although the signal magnitude changes over time, the absence of appreciable blank signal does not lower the limit of detection, although it does affect the accuracy of the estimates.

7.6. TRACKING DOWN AND REDUCING NOISE

Assembling a photothermal apparatus is relatively straightforward compared to the task of finding the sources of measurement errors, noise, or interference, and correcting them. However, there are several tools that can be used to aid in the endeavor of optimizing an experimental apparatus. Long and Bialkowski (1986) used a combination of noise power spectra analysis and measurement scatter plot analysis to determine the source of and so subsequently reduce measurement errors found in pulsed infrared laser–excited photothermal deflection spectroscopy of gas-phase analytes.

Noise power spectra estimates can be obtained using spectrum analyzers, modern digital oscilloscopes, or by averaging fast Fourier–transformed time-series data. The later digital signal processing technique is easily programmed on modern PC laboratory processors, although commercial data analysis software rarely has the flexibility to do this type of processing. The signal voltage data are collected as a time series using an analog-to-digital converter or a transient waveform recorder. The data are transformed with a fast Fourier transform routine and the resulting complex data are converted to either a power or a magnitude spectrum. The signal power spectrum data are averaged and displayed for analysis. We use exponential averaging to allow the spectra to change as the apparatus is inspected. After a number of data collection, transform, and averaging steps, the power spectrum is

$$P_N(\omega) = (1 - a)P_{N-1}(\omega) + a|\mathscr{F}\{s_N(t)\}|^2 \tag{36}$$

where the P are the power spectra, s_N is the time-series signal (or noise) measurement, and a $(0 < a < 1)$ is the exponential average weighting factor (Bialkowski 1988a).

The power spectra are useful in determining the relative frequency-dependent content of the signal and the noise. For pulsed laser experiments, the time series are obtained and displayed synchronous with the excitation. The synchronous data are useful in designing analog filters. Power spectra data obtained with and without the signal reveals the frequency components of signal and noise. The optimum frequency filter $H_{\text{opt}}(\omega)$ is one that maximizes the signal-to-noise ratio in the filtered result. Using the signal transform, $S(\omega)$, and noise magnitude, $|N(\omega)|$, spectra, the optimum filter can be found from (Papoulis 1984, Bialkowski 1987)

$$H_{\text{opt}}(\omega) = k \frac{S^*(\omega)}{|N(\omega)|} e^{-i\omega t_0} \tag{37}$$

where k is an arbitrary scaling factor and the exponential is a phase term compensating for the time delay, t_0. This is the frequency-domain representation of the matched filter. For noncausal data (e.g., those obtained from continuous excitation sources), and when the signal and noise are uncorrelated (orthogonal), a Wiener filter, defined by

$$H(\omega) = \frac{|S(\omega)|}{|N(\omega)| + |S(\omega)|} \tag{38}$$

may be used. In the experiments of Long and Bialkowski (1986), signal and noise power spectra were used to design the low-frequency cutoff and slope of an active analog high-pass frequency filter. Their apparatus showed a predominance of low-frequency flicker-type noise. The low-pass filter was essentially an implementation of the noncausal Weiner filter.

Noise power spectra can also help in locating and eliminating specific interferences. Interference at specific frequencies appears as peaks in the noise power spectrum. Experimental spectra normally exhibit peaks at 60 Hz and harmonics due to pickup of the radiated energy from the power lines in every laboratory. In photothermal experiments, peaks are often found at other low frequencies. These peaks may initially appear to be due to power-line pickup, but on closer inspection it is often found that they are at the wrong frequencies and their magnitudes change over time. We have found that these peaks are due to the vibration of optical elements used to construct the apparatus. Spatially sensitive detection used in photothermal spectroscopy is particularly sensitive to vibrations in the optical elements that correspond to changes in beam position. The optical elements responsible for the excess noise can be located by lightly tapping the elements while observing the noise power spectra estimates. Once located, the interference can often be reduced by remounting the element. In severe cases the apparatus can be reconfigured and tested with the aid of the resulting noise power spectra.

Long and Bialkowski also used scatter plots to deduce the sources of error in their pulsed laser photothermal deflection experiments. The scatter plots, obtained by plotting the signal versus excitation pulse energy for a series of laser cavity configurations, were used to deduce that signals obtained with spatial filtered TEM_{00} exciton beams had signal-to-noise ratios superior to the unfiltered multimode excitation beams.

REFERENCES

Alfonso, E. F. S.; Revert, M. A. R.; Alvarez-Coque, M. C. G.; and Ramos, G. R. *Appl. Spectrosc.* **44** 1501 (1990).

Allison, S. W.; Gillies, G. T.; Magnuson, D. W.; and Pagano, T. S. *Appl. Opt.* **24** 3140 (1985).

Arecchi, F. T. In *Order and Chaos in Nonlinear Physical Systems*, Lundqvist, S.; March, N. H.; and Tosi, M. P., eds., Plenum, New York (1988).

Bevington, P. R. *Data Reduction and Error Analysis for the Physical Sciences*, McGraw-Hill, New York (1969).

Bialkowski, S. E. *Anal. Chem.* **58** 1706 (1986).

Bialkowski, S. E. *Rev. Sci. Instrum.* **58** 687 (1987).

Bialkowski, S. E. and He, Z.-F. *Analytical Chemistry* **60** 2674 (1988).

Bialkowski, S. E. *Anal. Chem.* **60** 355A (1988a).

Bialkowski, S. E. *Anal. Chem.* **60** 403A (1988b).

Bialkowski, S. E. and He, Z.-F. *Proc. SPIE* **1637** 134 (1992).

Bialkowski, S. E. and He, Z.-F. *Proc. SPIE* **2138** 140 (1994).

Bialkowski, S. E.; Gu, X.; Poston, P. E.; and Powers, L. S. *Appl. Spectrosc.* **46** 1335 (1992).

Buffet, C. E. and Morris, M. D. *Anal. Chem.* **54** 1824 (1982).

Buffet, C. E. and Morris, M. D. *Appl. Spectrosc.* **5** 455 (1983).

Carnahan, B.; Luther, H. A.; and Wilkes, J. O. *Applied Numerical Methods*, Wiley, New York (1969).

Carter, C. A. and Harris, J. M. *Anal. Chem.* **55** 1256 (1983).

Carter, C. A. and Harris, J. M. *Anal. Chem.* **56** 922 (1984).

Chartier, A.; Georges, J.; and Mermet, J. M. *Chem. Phys. Lett.* **171** 347 (1990a).

Chartier, A.; Georges, J.; and Mermet, J. M. *Spectrochim. Acta* **46A** 1737 (1990b).

Chartier, A.; Georges, J.; and Mermet, J. M. *Spectrochim. Acta* **47A** 1505 (1991).

Debenedetti, P. G. and Reid, R. C. *J. Amer. Inst. Chem. Eng.* **32** 2034 (1986).

Dovichi, N. J. and Harris, J. M. *Anal. Chem.* **51** 728 (1979).

Dovichi, N. J. and Harris, J. M. *Anal. Chem.* **52** 2338 (1980).

Dovichi, N. J. and Harris, J. M. *Anal. Chem.* **53** 106 (1981).

Dovichi, N. J. *CRC Critical Reviews in Analytical Chemistry* **17** 357 (1987).

Fournier, D.; Boccara, A. C.; Amer, N. M.; and Gerlach, R. *Appl. Phys. Lett.* **37** 519 (1980).

Giglio, M. and Vendramimi, A. *Appl. Phys. Lett.* **25** 555 (1974).

Gradshteyn, I. S. and Ryzhik, I. M. *Table of Integrals, Series, and Products*, Academic Press, New York (1980).

Haller, K. L. and Hobbs, P. C. D. *Proc. SPIE* **1435** 289 (1991).

Harris, T. D. and Williams, A. W. *Appl. Spectrosc.* **39** 28 (1985).

Higashi, T.; Imasaka, T.; and Ishibashi, N. *Anal. Chem.* **56** 2010 (1984).

Imasaka, T.; Nakanishi, K.; and Ishibashi, N. *Anal. Chem.* **59** 1554 (1987).

Ingle, J. D. and Crouch, S. R. *Spectrochemical Analysis* Prentice Hall, Englewood Cliffs, NJ (1988).

Jackson, W. B.; Amer, N. M.; Boccara, A. C.; and Fournier, D. *Appl. Opt.* **20** 1333 (1981).

Jansen, K. L. and Harris, J. M. *Anal. Chem.* **57** 1698 (1985).

Johnson, M. E. and Voigtman, E. *Appl. Spectrosc.* **44** 958 (1990).

Katemen, G. and Buydens, L. *Quality Control in Chemical Analysis*, 2nd ed., Wiley, New York, (1993).

Kitamori, T. and Sawada, T. *Spectrochem. Acta. Rev.* **14** 275 (1991).

Long, G. R. and Bialkowski, S. E. *Anal. Chem.* **56** 2806 (1984).

Long, G. R. and Bialkowski, S. E. *Anal. Chem.* **57** 1079 (1985).

Long, G. R. and Bialkowski, S. E. *Anal. Chem.* **58** 80 (1986).

Leach, R. L. and Harris, J. M. *Anal. Chem.* **56** 1481 (1984).

Mho, S. and Yeung, E. S. *Anal. Chem.* **57** 2253 (1985).

Nakanishi, K.; Imasaka, T.; and Ishibashi, N. *Anal. Chem.* **59** 1550 (1987).

Nickolaisen, S. L. and Bialkowski, S. E. *J. Chem. Inf. Comput. Sci.* **26** 57 (1986).

Nolan, T. G. and Dovichi, N. J. *IEEE Circuits Devices.* **2** 54 (1986).

Pang, T. J. and Morris, M. D. *Anal. Chem.* **57** 2153 (1985).

Papoulis, A. *Probability, Random Variables, and Stochastic Processes*, 2nd ed., McGraw-Hill, New York (1984).

Pawliszyn, J. *Rev. Sci. Instrum.* **58** 245 (1987).

Rosenzweig, Z. and Yeung, E. S. *Appl. Spectrosc.* **47** 1175 (1993).

Roy, R.; Murphy, T. W.; Maier, T. D.; Gills, Z.; and Hunt, E. R. *Phys. Rev. Lett.* **68** 1259 (1992).

Sepaniak, M. J.; Vargo, J. D.; Ketter, C. N.; and Maskarinec, M. P. *Anal. Chem.* **56** 1252 (1984).

Tam, A. C. and Patel, C. K. N. *Appl. Opt.* **18** 3348 (1979).

Tran, C. D. and Simianu, V. *Anal. Chem.* **64** 1419 (1992).

Twarowski, A. J. and Kliger, D. S. *Chem. Phys.* **20** 253 (1977).

Xu, M. and Tran, C. D. *Appl. Spectrosc.* **44** 962 (1990a).

Xu, M. and Tran, C. D. *Anal. Chem.* **62** 2467 (1990b).

Yang, Y. *Anal. Chem.* **56** 2153 (1984).

Yang, Y. and Hairell, R. E. *Anal. Chem.* **56** 3002 (1984).

CHAPTER

8

ANALYTICAL APPLICATIONS

In the eyes of the analytical chemist, applications of photothermal spectroscopy fall into two categories: measurement of analyte concentration, and measurement of physical properties of the analyte or sample. This chapter provides an overview of applications of photothermal spectroscopy to chemical analysis when the sample has been prepared in homogeneous form. The emphasis is on chemical concentration measurement, although some measurements of analyte and sample properties are discussed.

8.1. AREAS OF ANALYTICAL APPLICATION

Photothermal spectrometry has been applied to a variety of chemical analysis problems where the measurements are made on homogeneous samples. One normally thinks of photothermal spectroscopy as an ultrasensitive technique wherein optical absorbances much smaller than those able to be measured using transmission spectrophotometry can be obtained. Although this may be true, it is not the only utility of the technique. It has also been used to measure thermal and optical properties of samples, excited-state relaxation and chemical reaction rates, and photophysical parameters such as excited-state absorption cross-section and triplet-state quantum yields. Although the later measurements are not analytical in nature, they do constitute important applications of the technique.

In chemical analysis, photothermal spectroscopy has been most often applied to the measurement of concentrations of species in dilute solutions. The linear quantitative aspect of photothermal spectroscopy is strictly maintained in dilute solutions making data analysis easier. At higher concentrations, nonlinearities due to changes in the thermal and optical parameters, as well as to optical effects, can occur. The secondary influence of analyte concentration on signal magnitude does not circumvent the application of photothermal spectroscopy but does limit the linear working range of concentration. The highly sensitive nature of photothermal spectroscopy can be advantageous in analyte concentration measurement. Because of higher sensitivity, preconcentration steps may be eliminated from an analytical procedure. When sample preconcentration is required for conventional transmission analysis, application of photothermal spectroscopy can result in reduced sample preparation time and greater measurement precision. Subsequently, photothermal spectroscopy has most often been applied or targeted for trace analysis.

There are two primary applications of photothermal spectroscopy to chemical analysis of low-absorption analytes in homogeneous samples. The first is analysis of stationary samples. In this application, analysis samples are introduced into the photothermal apparatus and the photothermal signal is subsequently measured. The measurement is related to the amount of analyte with instrument calibration. This application is most useful for trace or ultratrace analysis. The second application for measurement of optical absorption in dynamic samples. Dynamic samples exhibit a change in analyte absorbance as a function of time. Flowing samples, such as found in flow injection analysis, column chromatography, and electrophoreses, and time-dependent stationary samples, such as systems undergoing chemical reaction or physical change, are dynamic in nature. In these cases the photothermal spectroscopy signals must either be obtained on time scales short compared to the characteristic thermal diffusion time, or the apparatus must be designed to account for the dynamic behavior of the system.

Although most analytical studies have focused on the ultrasensitive nature of photothermal spectrometry, there is no reason why photothermal spectrometry cannot be used to quantitate any species that could otherwise be measured using conventional spectrophotometry. In fact, statistical analysis of the limiting noise source in spectrophotometry reveals that it is advantageous to use photothermal spectrometry for samples with absorbances up to about 0.1 absorbance unit (Bialkowski et al. 1992). However, the photothermal spectrometer must be designed to take into account the excitation attenuation along the optical path. Weimer and Dovichi (1988) used a unique transverse-probe photothermal lens spectrometer to probe the optical attenuation along the path of the excitation laser. They were able to obtain the absolute optical attenuation by fitting the space-dependent thermal lens to Beer's law. Bicanic et al. (1989) used a continuous infrared CO_2 laser to measure optical absorption in neat CH_3OH in a novel photothermal deflection spectrometer. The refractive index gradient was generated by the attenuation of the excitation beam as it passed into the sample. As the excitation beam passes through the sample, optical attenuation produced a spatially inhomogeneous gradient that followed the Beer's law absorption. The probe laser passed at a right angle relative to the excitation beam and within the sample.

8.2. PHOTOTHERMAL SPECTROSCOPY ANALYSIS OF STATIONARY HOMOGENEOUS SAMPLES

8.2.1. Measurement Techniques

Most of the photothermal spectroscopy apparatus discussed earlier have been used for quantitative measurements. No particular apparatus seems to perform exceptionally better than the others. However, each apparatus has unique features that could be used to some advantage in certain measurement situations. Photothermal deflection apparatuses that are based on position-sensing detectors have the advantage of being immune to transient absorption and scattering attenuation. Differential photothermal lens and deflection techniques have the advantage of optically compensating for solvent and matrix absorption.

Both continuous and pulsed laser excitation sources have been used in analytical studies. Several reports have shown that absorption errors can occur when using pulsed laser light sources with organic dye molecules (Terazima et al. 1989, Chartier et al. 1990a, b, Chartier and Georges 1991, Ecker et al. 1992). These experiments all showed that the energy recovered from pulsed excitation was often less than that expected from the optical absorbance and theoretical heat yield. The source of this error is not clear. Terazima suggests that energy was lost through coupling electronic energy to vibrational modes in the solvent. In this author's opinion, the effect is probably due to optical bleaching of the ground singlet state followed by triplet-state absorption under the high-irradiance conditions of pulsed excitation. Further experiments definitely need to be performed here to elucidate the mechanism for heat recovery of less than unity. Problems associated with incomplete heat recovery from certain analytes (accuracy problems) combined with effort required to overcome the pulse-to-pulse variation of the pulsed lasers (precision problems) apparently have led many to think them less desirable as excitation sources for analytical measurements.

As points in favor of pulsed laser excitation, Imasaka, et al. (1991) have examined the relative utility of pulsed versus continuous laser–excited photothermal spectroscopy. This study concluded that pulsed laser photothermal spectroscopy is sensitive for certain analysis, in particular gas phase. Waldron and Dovichi (1992) have applied pulsed laser–excited photothermal spectroscopy as a detector for capillary electrophoresis. The absorbance detection limit was about 2×10^{-6} AU in aqueous buffer, much lower than has been obtained using continuous excitation apparatuses. One could conclude that pulsed laser–based photothermal spectroscopy apparatus is

Table 8.1. Applications of Photothermal Spectroscopy to Gas-Phase Analysis

Technique	Excitation Source	Sample Pathlength or Volume	Signal Processing	Analyte	Matrix	Detection Limits	Ref.
Photothermal lens	N_2/dye laser, 420 nm, 20 μJ	20 cm	Signal averaging	NO_2	Air	0.8 ppm	1
Photothermal lens	Ar^+ laser, 488 nm, 700 mW	1 m	Lock-in amplifier	NO_2	Air	5 ppb, 10^{-6} m^{-1}	2
Photothermal lens	Excimer/dye laser, 463 nm, 1.5 mJ	1 m	Signal averaging	NO_2	N_2	4 ppb, 10^{-6} m^{-1}	3
Photothermal deflection	TEA-CO_2 laser, 10.72 μm, 100 mJ	7 cm	Signal averaging	CHF_2Cl	Ar	1 ppb, 10^{-6} m^{-1}	4
Photothermal deflection	CO_2 laser, 9.2 and 10.6 μm, 10 mJ	7 cm	Position sensor/ signal averaging	CHF_2Cl, CF_2Cl_2	Ar	170 ppt, 270 ppt, 10^{-7} m^{-1}	5
Photothermal deflection	TEA-CO_2 laser, 9.62 and 9.23 μm, 10 mJ	7cm	Photodiode/ signal averaging	1,3-Butadiene, CHF_2Cl, CF_2Cl_2	Ar	<10 ppm	6
Photothermal deflection	CO_2 laser, 10.72 μm, 1 mJ	7cm	Photodiode/ signal averaging	CF_2Cl_2	Ar	0.7 ppm	7
Photothermal interferometry	TEA-CO_2 laser, 10.6 and 9.67 μm, 10 mJ	10 cm	Oscilloscope	SF_6, CH_3OH	Air He	10 ppb, 40 ppb	8
Photothermal interferometry	CO_2 laser, 10.53 and 10.72 μm, 1.60 W/cm^2	10 cm	Lock-in amplifier	SF_6, CF_2Cl_2	Air Air	16 ppb, 4 ppm	9

Photothermal deflection	Nd:YAG/dye laser, 571 nm, 1 μJ	Molecular beam, ~180 μm	Avalanche photodiode/ boxcar averager	C_2H_2	He	20%	10
Photothermal deflection	CO_2 laser, 10.4 μm, 1 W	20 cm	Photodiode/lock-in amplifier	C_2H_4	N_2	5 ppb, 10^{-5} m^{-1}	11
Photothermal interferometry	CO_2 laser, 9.17–10.97 μm, 100 W/cm^2	20 cm	Lock-in amplifier	C_2H_4, $(NH_4)_2SO_4$ aerosols	N_2	300 ppt, 10^{-6} m^{-1} 100 μg m^{-3}	12
Photothermal interferometry	CO_2 laser, 10.6 μm	15 cm	Lock-in amplifier	NH_3	Air	5 ppb	13
Photothermal interferometry	CO_2 laser, 9.1–10.9 μm, 2.2 W	1 cm	Balanced mixer/ signal processing	N_2H_4, $CH_3N_2H_3$, $(CH_3)_2C_6H_5N_2H$	N_2	30 ppb, 40 ppb, 60 ppb	14
Photothermal lens	CO_2 laser, 9.1–10.9 μm, 20 W	100 cm	Lock-in amplifier	Aromatic hydrocarbons, C_2H_5OH, CH_3OH	N_2	4 ppb, 19 ppb, 12 ppb, 6×10^{-5} m^{-1}	15

References:

1. Mori, K.; Imasaka, T.; and Ishibashi, N. *Anal. Chem.* **55** 1075 (1983).
2. Higashi, T.; Imasaka, T.; and Ishibashi, N. *Anal. Chem.* **55** 1907 (1983).
3. Imasaka, T.; Higashi, T.; and Ishibashi, N. *Anal. Chim. Acta* **245** 191 (1991).
4. Bialkowski, S. E. and Long, G. R. *Anal. Chem.* **56** 2806 (1984).
5. Long, G. R. and Bialkowski S. E. *Anal. Chem.* **57** 1079 (1985).
6. Bialkowski, S. E. and Long, G. R. *Anal. Chem.* **59** 873 (1987).
7. Bialkowski, S. E. and He, Z.-F. *Anal. Chem.* **60** 2674 (1988).
8. Davis, C. C. *Appl. Phys. Lett.*, **36** 515 (1980).
9. Weston, N. D.; Sakthivel, P.; and Mukherjee, P. *Appl. Opt.* **32** 828, (1993).
10. Hineman, M. F.; Rodriguez, R. G.; and Nibler, J. W. *J. Chem. Phys.* **89** 2630 (1988).
11. Fournier, D.; Boccara, A. C.; Amer, N. M.; and Gerlach, R. *Appl. Phys. Lett.* **37** 519 (1980).
12. Lin, H.-B. and Campillo, A. J. *Appl. Opt.* **24** 422 (1985).
13. Campillo, A. J.; Lin, H. B.; Dodge, C. J.; and Davis, C. C. *Opt. Lett.* **5** 424 (1980).
14. Mazzoni, D. L.; and Davis, C. C. *Appl. Opt.* **30** 756 (1991).
15. Higashi, T.; Imasaka, T.; and Ishibashi, N. *Anal. Chem.* **56** 2010 (1984).

more sensitive than continuous laser excitation apparatus, but that the apparatus construction and problems associated with relaxation dynamics require that more care be taken in analyzing the results.

8.2.2. Applications

Several studies have applied photothermal spectroscopy to analysis of stationary samples. Many of these studies are listed in Tables 8.1 and 8.2. These tables illustrates the broad range of techniques and applications that have been explored. A more detailed description of a few of these studies is given below. The tables give the technique, the excitation source, wavelength, and power or energy. The average power is reported for continuous or chopped excitation, while the energy is reported for pulsed excitation. In papers that use or report more than one power or energy, the maximum or that which produced the best signal is reported. Often the detection and signal processing was not specified. These cases simply report signal averaging as the processing method. One problem in making these tables was reporting the detection limits in a consistent fashion. For the gas-phase studies, detection limits have been put in common units of parts per million or parts per billion. If an absorption coefficient was specified, this is listed below the common concentration units in units of m^{-1}. A quick survey of the latter indicates that the photothermal techniques have absorption coefficient detection limits from 10^{-5} to 10^{-7} m^{-1}. There does not seem to be any advantage of one technique over the other when measuring absorbance, although some are more instrumentally intensive. The minimum detected absorption coefficient may often be improved with better signal processing.

Many different units have been reported for detection limits in the liquid-phase studies. I finally gave up attempting to put the reported figures of merit in a common unit. Report in absorbance units (AU) are perhaps the most useful for comparison purposes since they are independent of path length. However, absorption coefficients, concentrations, and even moles have been reported. For circular dichroism work, the minimum detectable change in absorbance is given. Absorbance detection limits range from 10^{-4} to 10^{-7} AU. Better detection limits are apparent with higher-power lasers, higher-quality lasers, and for the differential techniques wherein the solvent absorbance is compensated for. It is of some concern that real detection limits are almost never reported. Detection limits are most often extrapolated from signals at some high concentration and baseline noise.

8.2.2.1. Gas Phase

Gas-phase photothermal lens absorbance measurements based on excitation of a molecular electronic transition in NO_2 have been reported by Higashi

Table 8.2. **Applications of Photothermal Spectroscopy to Liquid-Phase Analysis**

Technique	Excitation Source	Sample Pathlength or Volume	Signal Processing	Analyte	Solvent	Detection Limits	Ref.
Photothermal lens	Ar^+ laser, 514.5 nm, 160 mW	1 cm	Signal averaging	I_2	CCl_4	7×10^{-8} AU	1
Photothermal interferometry	Ar^+ laser, 515 nm, 100 mW	5 cm	Lock-in amplifier	I_2	CCl_4	5×10^{-6} cm^{-1}	2
Photothermal lens	Ar^+ laser, 514 nm, 150 mW	1 cm	Lock-in amplifier	I_2	CCl_4	5×10^{-7} AU	3
Photothermal lens	Ar^+ laser, 514.5 nm, 30 mW	1 cm	Lock-in amplifier	I_2, Co complex	CCl_4, H_2O, CH_3OH	1×10^{-8} AU, 2×10^{-6} AU, 4×10^{-7} AU	4
Photothermal deflection	He–Ne laser, 632.8 nm, 10 mW	200 μm	Signal averaging	Bromocresol green	H_2O, dimethyl-pyridine	Observation of spinodal decomposition	5
Photothermal lens	He–Ne laser, 632.8 nm, 2 mW	1 cm	Lock-in amplifier	Cu:EDTA, bromophenol blue, LD-690	H_2O (pH 7), C_2H_5OH, $CHCl_3$	2.8×10^{-3} AU, 2.5×10^{-6} AU, 7.8×10^{-4} AU	6
Photothermal deflection	Excimer/dye laser, 440–610 nm	1 cm	Position-sensitive detector/signal averaging	Bromocresol dyes, phenol red	$H_2O:CH_3OH$(1:1), H_2O	pK_a determination	7
Photothermal lens	Kr^+ laser, 647.1 nm, 10–100 mW	1 mm	Photomultiplier tube	Methylene blue	H_2O	Equilibrium constant determination	8

Table 8.2. (Continued)

Technique	Excitation Source	Sample Pathlength or Volume	Signal Processing	Analyte	Solvent	Detection Limits	Ref.
Photothermal lens	Ar^+ laser, 514 nm, 40 mW, N_2/dye laser, 514 nm, 50 μJ	1 cm	Boxcar averager, lock-in amplifier	Erythrosine Cobalt nitrate	H_2O (pH 10)	Photothermal lens signal yield determination	9
Photothermal refraction	N_2/dye laser, 480 nm, 0.38 mJ		Photodiode	Amaranth	CH_3OH	9×10^{-5} cm^{-1}	10
Photothermal refraction	Ar^+ laser, 488 nm, 200 mW	1 cm	Lock-in amplifier	Amaranth	CH_3OH	400 pM, 5×10^{-6} cm^{-1}	11
Photothermal lens	He–Ne laser, 3.39 μm, 6.2 mW	1 cm	Pb-Se detector/ Signal averaging	Hydrocarbons	CCl_4	5×10^{-4} AU	12
Photothermal lens	Ar^+/dye laser, 600 nm, 150 mW	1 cm	Photodiode array/ signal averaging	Formaldehyde	H_2O	15 n$\underline{M}$ 1×10^{-4} cm^{-1}	13
Photothermal deflection	CO_2 laser, 10.59 μm, 10.26 μm, 5 W	5 cm	Position sensitive detector/lock-in amplifier	Methanol	CH_3OH	0.122×10^5 m^{-1} 0.162×10^5 m^{-1}	14
Photothermal lens	Kr^+/F-center laser, 2.73–2.95 μm	100 μm	Lock-in amplifier	Water	D_2O Tetrahydrofuran	3.7×10^{-4}% (w/w), 1.2×10^{-4}% (w/w)	15
Photothermal lens	Kr^+ 647.1 nm, 100 mW	1 cm	Chart recorder	Al: chrome azurol *S*-cetylpyridinium chloride	H_2O	0.17 μg L^{-1}	16
Photothermal lens	Ar^+ laser, 488 nm, 600 mW	1 cm	Photodiode array/ signal averaging	Fe(II): batho phenanthroline, disulfonic acid	H_2O $CHCl_3$	500 p$\underline{M}$, 10^{-5} AU, 30 p$\underline{M}$ 6×10^{-7} AU	17

Photothermal refraction	He–Cd laser, 441.6 nm, 4 mW	25 pL	Lock-in amplifier	Fe(II) : 1,10-phenanthroline	CH_3OH/H_2O (1/1)	300 n$\underline{M}$	18
Photothermal refraction	Nd : YAG, 532 nm, 0.6 mJ	1 cm	CCD array/signal averaging	Fe(II) : 1,10-phenantroline	CH_3OH		19
Photothermal lens	Ar^+/dye laser, 458 nm	1 cm	Signal averaging	Colloidal $Fe_2O_3(H_2O)_x$ Cu : $Fe_2O_3(H_2O)_x$	H_2O	Kinetics and scattering effect measurements	20
Photothermal lens	GaAlAs laser, 780 nm, 15 mW	1 cm	Signal averaging	Fe(II) 2-Nitroso-5-diethylaminophenol	$CHCl_3$	8.1 nM	21
Photothermal refraction	Ar^+ laser, 575 nm, 25 mW	8 μL	Signal averaging	Co(III) ethylene diamine	H_2O	180 ng, 10^{-4} cm^{-1}	22
Photothermal lens	Ar^+ laser, 488 nm, 1.5 W	1 cm	Signal averaging	Bi, Co, Fe, Mo, and Ni complexes	H_2O, $CHCl_3$, MEK, C_2H_5OH/H_2O (4/1)	Bi:10 pg mL^{-1}, Co: 0.17 pg mL^{-1}, Fe: 50 pg mL^{-1}, Mo: 10 pg mL^{-1}, Ni: 9 pg mL^{-1}	23
Photothermal lens	He-Ne laser, 632.8 nm, 4 mW	1 cm	Signal averaging	Cu(II) EDTA	H_2O, acetone : H_2O 1:1 acetone : H_2O 3:1	5.1 m$\underline{M}$, 3.5 m$\underline{M}$, 1 m$\underline{M}$	24
Photothermal lens	N_2/dye laser, 417 nm, 10 μJ	1 cm	Signal averaging	Cu(II) mesotetraphenyl-porphinetrisulfonic acid sulfate	C_6H_6, $CHCl_3$	1 n$\underline{M}$	25
Photothermal lens	Ar^+ laser, 514.5 nm, 22 W	1 cm	Lock-in amplifier	Ca^{2+} dithizone	CCl_4	8 pg mL^{-1}	26
Photothermal lens	Ar^+/dye laser, 570 to 600 nm, 5 W	1 cm	Lock-in amplifier	Nd^{3+}	0.1 m$\underline{M}$ $HClO_4$ in CO_3^{2-} solution	5×10^{-6} $\underline{M}$	27

Table 8.2. (Continued)

Technique	Excitation Source	Sample Pathlength or Volume	Signal Processing	Analyte	Solvent	Detection Limits	Ref.
Photothermal deflection	Ar^+/dye laser, 575 nm, 40 mW	1 cm	Position sensitive detector/lock-in amplifier	Nd^{3+}	H_2O	8×10^{-7} M	28
Photothermcal deflection	Ar^+/dye laser, 575 nm, 35 mW	1 cm	Position sensitive detector/lock-in amplifier	1.3 nM Nd^{3+}, 0.087-μm latex spheres	0.4 mM HCl + surfactant	Light-scattering effect	29
Photothermal lens	Ar^+/dye laser, 575 nm, 25 mW	1 cm	Signal averaging	Nd^{3+}	H_2O	7×10^{-6} M 1.04×10^{-4} cm^{-1}	30
Photothermal lens	Diode laser, 794.6 nm, 10 mW	1 cm	Signal averaging	Nd^{3+}	H_2O	8×10^{-5} M	31
Photothermal lens	Ar^+ laser, 514.5 and 457.9 nm, 20 mW	?	Signal averaging	Pr^{3+}	H_2O (pH 6.2)	4.7×10^{-5} M	32
Photothermal lens	Ar^+ laser, 20 mW	1 cm	Signal averaging	Lanthanides	Organic solvent, crown ether	Extracted ion-pair stochiometry	33
Photothermal lens	Nd:YAG laser, 520 nm; Nd:YAG/dye laser, 690 nm, 1 μJ	4 mm, 2.5 mm	Boxcar averager	U(IV) alkoxides, Nd^{3+}	Methylcyclohexane, 77 K, 0.1 M HCl	10^{-5} AU 4×10^{-5} AU	34
Photothermal lens	Excimer/dye laser, 450 nm, 1 mJ	?	Boxcar averager	U(VI) carbonates	0.5 M $NaClO_4$	Equilibrium constant determination	35

Photothermal lens and deflection	He-Cd laser, 442 nm, 4 mW	1 cm	Photomultiplier tube, position-sensitive detector	UO_2^{2+}	1 M $HClO_4$, carboxylic acids	Observation of photodecarboxylation	36
Photothermal lens	He–Cd laser, 441.6 nm, 100 mW	5 mm	Photodiode/voltmeter	PO_4^{3-}	Isobutanol : hexane 1 : 2	0.6 ng mL^{-1}	37
Photothermal lens	Diode laser, 823.9 nm, 10 mW	1 cm	Signal averaging	KHP	H_2O, 2-butanol $CHCl_3$	0.35 ppb, 0.024 ppb, 0.014 ppb	38
Photothermal lens	Kr^+ laser, 647.1 nm, 70 mW		Photodiode/signal averaging	KHP	NaCl, isobutyl alcohol	Study matrix salinity	39
Photothermal lens	Ar^+/dye laser, 600–458 nm, 60 mW	1 cm	Signal averaging	Fulvic acid	H_2O	40 $\mu g\ L^{-1}$ humic material <300 $\mu g\ L^{-1}$, DOC for aquatic materials 3×10^{-5} AU	40
Photothermal lens	Ar^+ laser, 514.5–457.9 nm, 4.6–115 mW		Signal averaging	1,2-diamino-anthraquinone	CH_3OH	100 pM	41
Photothermal lens	Ar^+ laser, 488 nm, 100 mW	8 mm^3	Signal averaging	Catecholamines	$H_2O : C_2H_5OH$, 1 : 1	1 ng mL^{-1}	42
Photothermal diffraction	Ar^+ laser, 514.5 nm, 440 mW	1-cm cell, 170 nL	Signal processing	I_2	CCl_4	$7 \times 10^{-4}\ cm^{-1}$	43
Photothermal diffraction	Nd : YAG, 532 nm, 460 μJ	1-cm cell, 38 nL	Signal processing	Co^{2+}	15 mM H_2SO_4, aqueous	$6 \times 10^{-4}\ cm^{-1}$	44

Table 8.2. (Continued)

Technique	Excitation Source	Sample Pathlength or Volume	Signal Processing	Analyte	Solvent	Detection Limits	Ref.
Photothermal diffraction	Ar^+ laser, 488 and 514.5 nm, 150 mW	0.5-mm cell, 140 nL	Signal processing	Eosin B	C_2H_5OH	2.9×10^{-18} mole, 21 p$\underline{M}$	45
Photothermal diffraction	Ar^+ laser, 488 nm, 250 mW	0.1-mm cell, 98 pL	Digital signal processing	$\pm Co(en)_3^{3+}$	$H_2O:C_2H_5OH$, 1:1	29 $\mu\underline{M}$, $\Delta A = 5.5 \times 10^{-7}$ (CD)	46

References:

1. Dovichi, N. J. and Harris, J. M. *Anal. Chem.* **53**, 106 (1981).
2. Cremers, A. D. and Keller, R. A. *Appl. Opt.* **21** 1654 (1982).
3. Yang, Y. *Anal. Chem.* **58** 1420 (1986).
4. Yang, Y. and Hall, S. C. *Appl. Spectrosc.* **42** 72 (1988).
5. Hardcastle, F. D. and Harris, J. M. *Appl. Spectrosc.* **40** 606 (1986).
6. Yang, Y. *Anal. Chem.* **56** 2336 (1984).
7. Bialkowski, S. E.; Gu, X.; Poston, P. E.; and Powers, L. S. *Appl. Spectrosc.* **46** 1335 (1992).
8. Lowe, R. D. and Snook, R. D. *Analyst* **118** 613 (1993).
9. Ecker, A.; Georges, J.; and Mermet, J. M. *Appl. Spectrosc.* **46** 577 (1992).
10. Teramae, N.; Voigtman, E.; Lanauze, J.; and Winefordner, J. D. *Anal. Chem.* **58** 761 (1986).
11. Teramae, N. and Winefordner, J. D. *Appl. Spectrosc.* **41** 164 (1987).
12. Carter, C. A.; Brady, J. M.; and Harris, J. M. *Appl Spectrosc.* **36** 3 (1982).
13. Alfheim, J. A. and Langford, C. H. *Anal. Chem.* **57** 861 (1985).
14. Bicanic, D.; Kruger, S.; Torfs, P.; Bein, B.; and Harren, F. *Appl. Spectrosc.* **43** 148 (1989).
15. Tran, C. D. and Grishko, V. I. *Appl. Spectrosc.* **48** 96 (1994).
16. Lowe, R. D. and Shook, R. D. *Anal. Chim. Acta* **250** 95 (1991).
17. Miyaishi, K.; Imasaka, T.; and Ishibashi, N. *Anal. Chem.* **54** 2039 (1982).

18. Nolan, T. G.; Weimer, W. A.; and Dovichi, N. J. *Anal. Chem.* **56** 1704 (1984).
19. Klatt, L. N. *Appl. Spectrosc.* **46** 1583 (1992).
20. Gutzman, D. W. and Langford, C. H. *Anal. Chim. Acta* **283** 773 (1993).
21. Imasaka, T.; Sakaki, K.; and Ishibashi, N. *Anal. Chim. Acta* **243** 109 (1993).
22. Tran, C. D. and Xu, M. *Rev. Sci. Instrum.* **60** 3207 (1989).
23. Abroskin, A. G.; Belayeva, T. V.; Filichkina, V. A.; Ivanova, E. K.; Proscurnin, M. A.; Savostina, V. M.; and Barbalat, Y. A. *Analyst* **117** 1957 (1992).
24. Dovichi, N. J. and Harris, J. M. *Anal. Chem.* **51** 728 (1979).
25. Mori, K.; Imasaka, T.; and Ishibashi, N. *Anal. Chem.* **54** 2034 (1982).
26. Ramis Ramos, G.; Garcia Alvarez-Coque, M. C.; Smith, B. W.; Omenetto, N.; and Winefordner, J. D. *Appl. Spectrosc.* **42** 341 (1988).
27. Berthoud, T. and Delorme, N. *Appl. Spectrosc.* **41** 15 (1987).
28. Spear, J. D.; Russo, R. E.; and Silva, R. J. *Appl. Spectrosc.* **42** 1103 (1988).
29. Spear, J. D.; Russo, R. E.; and Silva, R. J. *Appl. Opt.* **29** 4225 (1990).
30. Rojas, D.; Silva, R. J.; and Russo, R. E. *Anal. Chem.* **63** 1927 (1991).
31. Rojas, D.; Silva, R. J.; and Russo, R. E. *Rev. Sci. Instrum.* **63** 2989 (1992).
32. Franko, M. and Tran, C. D. *Anal. Chem.* **60** 1925 (1988).
33. Tran, C. D. and Zhang, W. *Anal. Chem.* **62** 830 (1990).
34. Berg, J. M.; Morris, D. E.; Clark, D. L.; Woodruff, W. H.; and Van Der Sluys, W. G. *SPIE* **1435** 331 (1991).
35. Bidoglio, G.; Cavalli, P.; Grenthe, I.; Omenetto, N.; Pan, Q. I.; and Tanet, G. *Talanta* **38** 443 (1991).
36. Roach, R. J. and Snook, R. D. *Anal. Chim. Acta* **262** 231 (1992).
37. Grishko, V. I.; Yudelevich, I. G.; and Grishko, V. P. *Anal. Chim. Acta* **176** 51 (1985).
38. Nakanishi, K.; Imasaka, T.; Ishibashi, N. *Anal. Chem.* **57** 1219 (1985).
39. Philips, C. M.; Crouch, S. R.; and Leroi, G. E. *Anal. Chem.* **58** 1710, (1986).
40. Power, J. F. and Langford, C. H. *Anal. Chem.* **60** 842 (1988).
41. Tran, C. D. and Simianu, V. *Anal. Chem.* **64** 1419 (1992).
42. Sanchis-Mallols, J. M.; Villanueva-Camanas, R. M.; and Ramis-Ramos, G. *Analyst* **117** 1367 (1992).
43. Pelletier, M. J.; Thorsheim, H. R.; and Harris, J. M. *Anal. Chem.* **54** 239 (1982).
44. Pelletier, M. J. and Harris, J. M. *Anal. Chem.* **54** 1537 (1983).
45. Wu, Z. and Tong, W. G. *Anal. Chem.* **61** 998 (1989).
46. Nunes, J. A. and Tong, W. G. *Anal. Chem.* **65** 2990 (1993).

et al. (1983) and Mori et al. (1983) using continuous and pulsed laser excitation two-laser apparatuses, respectively. Detection limits of 5 ppb (v/v) NO_2 in air were found using a 700-mW 488-nm Ar^+ laser. This corresponds to an absorbance detection limit of 1.3×10^{-6} m^{-1} in the 1-m sample cell. Higher detection limits, 800 ppb (v/v) NO_2 in air, were found using pulsed excitation, owing to the low pulse dye laser energy (20 to 40 μJ) utilized. Equivalent or better detection limits were predicted for the pulsed laser experiment using 1-mJ pulses if the pulsed laser photothermal lens signals scale proportional to pulsed laser energy. Detection limits found for photothermal spectroscopy compared favorably to those of conventional visible absorption and phototacoustic spectroscopies. However, they are were about an order of magnitude higher than those obtained in long-path transmission spectrophotometry using a tunable *IR* lead salt diode laser source.

An interesting application of tunable pulsed visible laser–excited photothermal lens spectroscopy of gas-phase samples was reported by Hineman et al. (1988). This study utilized a pulsed visible laser excitation source to excite a vibrational overtone band of acetylene. The most interesting part is that measurements were performed in an expanding supersonic beam. Helium was used as the expansion gas. The dye laser was scanned over the vibrational transition and the rotational structure was apparent. The laser was of sufficiently narrow linewidth to allow both rotational and translational temperatures to be estimated. It was found that the acetylene had a rotational temperature of 125 K and a translational temperature of only 70 K. The measurement time was set by the spatial displacement of the excitation and probe laser beams in the translating (expanding) sample. It was estimated that only about 100 collisions took place over the measurement time set. This time was thought to be sufficient for complete rotational relaxation, but only partial vibrational relaxation took place since the ν_5 to ground-state relaxation bottleneck requires about 1000 collisions.

With the exception of a few gaseous compounds (e.g., NO_2, O_3, and SO_2) very few gas-phase species absorb visible and near-ultraviolet radiation. Subsequently, most gas-phase analysis using photothermal spectrometry have been performed with infrared excitation sources. The most utilized source for these studies has been the CO_2 laser. This robust infrared source can produce high powers or pulse energies and has extremely good "wall plug" power conversion efficiency. The source is line tunable throughout the 10-μm infrared atmospheric transmission window. The 10-μm region is the fingerprint region of the infrared; hence most species have finite absorbances accessible by this laser.

The applicability of photothermal spectroscopy for high-sensitivity infrared measurements was first demonstrated by Fournier et al. (1980) and

Davis (1980). Fournier et al. used continuous CO_2 laser excitation in a photothermal deflection apparatus and obtained absorption coefficient detection limits of about 10^{-5} m^{-1}, corresponding to 5 ppb (v/v) ethylene in air over the 10-cm interaction pathlength of their acute-angle crossed-beam apparatus. Davis used the PFLOH interferometric technique to detect phase shifts induced by a pulsed CO_2 laser. Concentration detection limits of about 10 ppb for SF_6 in air were reported. A subsequent report (Davis and Petuchowski 1981) predicted that absorption coefficient detection limits from 10^{-5} m^{-1}, acoustic (environmental) noise limited, to 10^{-7} m^{-1}, electronic noise limited, were possible. Even lower absorption coefficients, on the order of 10^{-8} m^{-1}, were thought possible using Stark modulation of the absorbing species in a 10-W continuous infrared laser–excited mode. By using Stark or possibly other means of optical absorption coefficient modulation, signals due to analyte absorption may be observed even in the presence of strong interferant absorption.

After comparing several different modes of gas-phase infrared absorption detection, we have found that the etalon-based pulsed infrared laser–excited photothermal deflection apparatus gives the best signal-to-noise ratio for a given set of detectors and optical elements (Bialkowski and He 1988). Measurements obtained on primary gas standard CFC−12 mixtures yielded a per-pulse absorption coefficient detection limit of about 10^{-3} m^{-1} for 1-mJ pulses at 933 cm^{-1}. The detection limits do not appear to be as good as previous results due to differences in the way they were calculated. This reported detection limit was obtained without signal averaging and was based on the measurement reproducibility (i.e., the standard deviation of the blank signal). In fact, many detection limits reported on in photothermal, and other new instrumental techniques for that matter, are figures of merit extrapolated from signals obtained at higher concentrations and the electronic instrumentation noise. This means for reporting detection limits often does not take into account the effect of blank absorption. Further, pulsed laser photothermal signals increase with increasing excitation energy. We used a standard 1-mJ pulse to report the detection limit. In fact, a 10-mJ excitation laser was used in these experiments and 10^{-4}-m^{-1} absorbance detection was demonstrated on a per-pulse basis. This corresponds to roughly 70 ppb(v/v) CFC−12 in air, still three orders of magnitude away from what can be measured with an ECD detector. However, signal-to-noise ratio improvement by ensemble averaging was apparently unlimited since background absorbance in the Ar matrix gas samples was essentially zero. Detection limits were improved by several orders of magnitude by signal averaging, although measurement reproducibility was severely limited by changes in the laboratory environment and acoustic (environmental) noise.

These gas-phase analysis studies illustrate several important aspects of gas-phase photothermal spectroscopy. It is apparent that extremely sensitive absorbance estimates can be obtained on gas-phase samples. Because of the relatively high thermal conductivity of gases, measurements can be made on a short time scale. The method is amenable to application for gaseous sample analysis due to the low solvent matrix absorption. Since the photothermal signal depends on the spatial overlap of excitation and probe sources and on some form of spatial filtering of the probe laser, the measurements are susceptible to, and often limited by, acoustic or environmental noise, as well as the spatial noise in the excitation and probe lasers.

8.2.2.2. Liquid Phase

The primarily analytical use of solution-phase photothermal spectrometry has been for metal ion and metal ion complex detection and for the detection of organic substances. The first solution-phase analytical application of photothermal spectroscopy was the trace-level determination of Cu(II) with an EDTA complex reported by Dovichi and Harris (1979) using a single-laser photothermal lens apparatus. Cu(II):EDTA is not a particularly good complex, having a molar absorption coefficient of only 47 $\underline{M}^{-1}\,cm^{-1}$ at 632.8 nm, and the molar detection limit of 1.6×10^{-5} $\underline{M}$ is no better than that which could be obtained using modern absorption spectrophotometry instruments. Since that initial study, there have been several refinements to the experimental design that have resulted in much lower detection limits. In particular, the use of metal chelators with high molar absorption coefficients, high excitation laser powers or pulse energies, and more favorable solvents have been exploited to enhance the detection limits (for examples, see Dovichi 1987).

High optical absorbance metal ion complexes formed using organic chelators are particularly advantageous for detection of trace levels of metal cations. The high molar optical absorbance of the chelate, coupled with low absorbance detection limits of photothermal spectroscopy results in trace metal sensitivity. Analyte molar detection limits as low as 5×10^{-10} $\underline{M}$ for iron(II):bathophenantholine in aqueous solution and 3×10^{-11} $\underline{M}$ in chloroform have been reported for photothermal lens spectroscopy using a 488-nm Ar^+ laser excitation source and beam profile signal processing (Miyaishi et al. 1982). To put this into the context of other techniques, 5×10^{-10} $\underline{M}$ aqueous iron(II) complex corresponds to mass concentration of about 30 ppt (w/w) aqueous. Although this detection limit is much better than that which can be obtained using ICP atomic emission or flame atomic absorption on the most sensitive lines, it is about an order of magnitude less sensitive than electrothermal atomic absorption spectrophotometry.

There is still some margin for improving molar detection limits for metal complexes. Several papers pressing the sensitivity of photothermal spectroscopy indicate that the practical absorption coefficient detection limit is on the order of 10^{-7}. The strongest molar absorption coefficients, those for charge transfer complexes, are in the range 10^4 to 10^5 $\underline{M}^{-1}$ cm^{-1}. Using the strongest absorbing charge transfer complex will result in a best molar detection limit from 10^{-11} to 10^{-12}, or in the case of iron, from 0.3 to 3.0 ppt (w/w). Iron(II):bathophenantholine detection limits were significantly better in chloroform solution. Organic complexes can sometimes be extracted from aqueous phase to an organic phase with favorable thermo-optical properties. Miyaishi et al. tried extraction of the aqueous iron(II):bathophenantholine complex into chloroform to improve limits of detection. In this case, however, solvent extraction did not improve detection limits. This was attributed to increased absorbance due to extracted bathophenantholine–disulfonate reagent. Nonetheless, the application of photothermal spectroscopy for metal ion detection was demonstrated to have detection limits rivaling that of the most sensitive atomic spectroscopy methods. The current limits can probably be improved upon with diligent attention paid to experiment design and careful choice of calorimetric reagent.

In the case of organic substances, mass detection limits of 42 attomole for glycine as the 4-dimethylaminoazobenzene-4-sulfonyl chloride (DABSYL) derivative has been found using 130 mW of a 458-nm argon ion excitation laser and a helium–neon probe in a crossed-beam photothermal refraction apparatus laser (Yu and Dovichi 1988). These low mass detection limits were obtained for on-column detection in a capillary zone electrophoresis apparatus. The solvent was a 50% acetonitrile mixture with aqueous buffer. The acetonitrile–water mixture improved the photothermal response over that of aqueous solvent.

Solution-phase molecular absorbance measurements are not constrained to molecules with electronic transitions. Carter et al. (1982) reported on an application of photothermal spectroscopy to measure molecular absorbance in the mid-infrared region. The 3.39-μm line of a continuous helium–neon laser was used to detect saturated hydrocarbons in CCl_4 solvent using single-laser photothermal lens spectrometry. Using a 6.2-mW excitation source and a PbSe photoconductive detector, the minimum sample absorbance was found to be 5×10^{-4}, corresponding to 8 ng of 2,2,4-trimethylpentane. The numbers could be improved on with better equipment (e.g., better infrared detector and higher power laser). Nonetheless, the study demonstrated that mid-infrared wavelength radiation could be used to measure a variety of hydrocarbons. The wavelength used overlapped fundamental vibrations of the methyl group and could thus be used to measure any saturated hydrocarbon.

8.2.3. A Novel Cross-Beam Apparatus for Photothermal Lens Spectroscopy

Through my review of the literature on photothermal spectroscopy, one apparatus struck me as being quite unique in its conception and design. This apparatus has potential for application to the measurement of higher absorbance samples. It apparently has not been exploited. Weimer and Dovichi (1988) reported on a novel pulsed laser–excited photothermal lens apparatus that utilized a photodiode array to detect simultaneously the spatially dependent photothermal lens elements produced as the excitation beam traverses a relatively high absorbance sample. A pulsed nitrogen laser pumped dye laser was focused through a 1-cm cuvette. The probe laser was defocused to a size that would probe the length of the sample cell, directed through the sample at right angles to the excitation beam. After passing through the sample, the probe beam was focused onto the photodiode array with a cylindrical lens. The small pixel size of the photodiode array presumably also served as the spatial filter in this case. Each element of the photodiode array corresponded to a different position within the sample. The photodiode array was read out with each pulse of the excitation laser and averaged to improve the signal-to-noise ratio. The probe laser intensity recorded by each element of the array corresponded to a different position along the length of the sample. Examined in total, the photodiode array signal showed the effect of pump laser beam attenuation as it traversed the sample. The spatially dependent attenuation of the excitation source was observed as a decrease in the spatially dependent photothermal lens signal. Absorption of the excitation source was obtained simply by observing the spatially dependent photothermal lens strength.

8.3. PHOTOTHERMAL SPECTROSCOPY DETECTION IN CHROMATOGRAPHY

One of the problems with using photothermal spectroscopy to detect flowing samples is that the heat produced by the photothermal effect is lost by mass transfer in the flowing sample. Dovichi and Harris (1981) reported substantial signal magnitude decreases with increasing flow rate with turbulent flow conditions. Weimer and Dovichi (1985a, b, c) found that flow decreases single-laser photothermal lens signals even for laminar flows. The reason for diminished signal with sample flow is due simply to heat loss through mass transfer out of the region irradiated by the excitation source. Pulsed laser excitation may be more suitable because the signal is maximum at short times, versus continuous excitation, where the signal increases with time. In fact, Nickolaisen and Bialkowski found the photothermal signal magnitude to be essentially unaffected by the flow rate for pulsed laser excitation of gas (Nickolaisen and Bialkowski 1985) and liquid (Nickolaisen and Bialkowski 1986a) samples. The effective characteristic decay time constant of the observed transient signals was observed to decrease with increasing sample flow rate in both gases and liquids. In the gas-phase study, the effect of sample flow was not easily modeled because the flow profile was turbulent. This was due to the particular sample cell design. The liquid cell used by Nickolaisen and Bialkowski allowed for a laminar flow profile, which, in turn, is easily modeled because of coherent fluid motion.

8.3.1 Temperature Change in Flowing Samples

The effect of flowing samples on the temperature change and subsequent photothermal signals has been addressed by several researchers. Sell (1984, 1985, 1989) developed a theory to account for photothermal deflection signals observed in flowing gas samples and applied the theory to the measurement of gas velocities. Sontag and Tam (1985) describe a model useful for the determination of flow velocities and species concentrations using a time-resolved pulsed laser–excited photothermal lens experiment. In the later, the probe laser is set downstream from the region excited by the pulsed laser. Rose et al. (1986) Vyas and Gupta (1988), Vyas et al. (1988), and Gupta (1989) describe a comprehensive model for photothermal spectroscopy in flowing media. Gupta (1989) extended the model to include temperature-dependent effects that might be observed in flames. Nickolaisen and Bialkowski (1986a) developed a model to account for their flowing liquid sample observations using pulsed excitation laser and accounting for finite excited-state relaxation time. Weimer and Dovichi (1985a–c) developed a model describing the time-dependent photothermal lens signal obtained for continuous laser excitation in both collinear and crossed-beam apparatuses.

The temperature change occurring with pulsed laser sample excitation in a flowing medium is described simply by accounting for the mass transfer of the sample. The thermal diffusion does not change with mass flow. However, the center of the temperature-change distribution moves along with the flowing sample. Consider a liquid sample flowing at a fixed linear velocity perpendicular to the direction of excitation laser beam propagation. For a rapid pulsed Gaussian excitation laser, the resulting space- and time-dependent temperature change for sample flowing transverse to the direction of excitation is simply

$$\delta T_{\text{impulse}}(x, y, t) = \frac{2}{\pi w^2(t)\rho C_P}\, e^{-2[x^2+(y_0-v_y t)^2]/w^2(t)} \tag{1}$$

where the flow is along the y-axis direction at a linear velocity of v_y (m s^{-1}), and $w^2(t) = w^2(1 + 2t/t_c)$ is the time-dependent radius of the Gaussian temperature change. This impulse response is the temperature change expected to occur for unit pulse energy, Q, absorption coefficient, α, and heat yield, Y_H. The temperature change for rapid pulsed laser–excited samples with negligible excited-state relaxation time is $Q\alpha Y_H \delta T(x, y, t)$. The effect that sample flow has on the normal time-dependent temperature change distribution is to translate it along the y-axis direction. This fact has led some authors to refer to the photothermal lens technique wherein the thermal lens element is observed downstream from the excitation source as the traveling thermal lens (Sontag and Tam 1985).

The impulse-response equations above can serve as a starting point to develop the temperature change and photothermal signals produced under other conditions. For example, with the very small focused beam waist radiuses able to be obtained with visible lasers, even modest flow rates may move the temperature perturbation out of the excitation region over the time required for excited-state thermalization. Nickolaisen and Bialkowski (1986a) found that the maximum photothermal lens signal magnitude decreased by about 5% at the higher flow rates (0.2 m s^{-1}), even though the measured thermal relaxation time was rather fast, 10 μs for the aqueous 2-mercaptopyridine sample used in that study. In cases where the rate of mass transfer, v_y/w, competes with the thermalization rate, the temperature change can be found by convolution of the rate of heat production with the impulse response

$$\delta T(x, y, t) = R'(t) * \delta T_{\text{impulse}}(x, y, t) \tag{2}$$

where $R'(t)$ is the rate of heat production. The deflection angles and inverse focal lengths can be found by similar convolutions.

In laminar flow, the linear flow rate is a function of distance from the center of the sample cell. The flow profile is different for different cell geometries but is generally parabolic. For example, the laminar flow profile for a square-cross-section sample cell is (Weimer and Dovichi 1985a)

$$v_y(x, z) = \frac{9V}{16b^2}\left(1 - \frac{x^2}{b^2}\right)\left(1 - \frac{z^2}{b^2}\right) \tag{3}$$

where V (m^3 s^{-1}) is the volume flow rate and b (m) is one-half the cell width. Of course, equation 1 was derived assuming that the cell was of infinite extent. In fact, this approximation may be maintained for thermal diffusion along the x dimension if the excitation beam waist radius is much smaller than the cell dimensions. It will not in general be true for the z dimension. Laminar flow causes the temperature change to move fastest in the center of the sample cell. It is stationary near the cell wall. After excitation and as time progresses, the Gaussian temperature change distribution will arch from the entrance to the exit window of the sample cell, and thermal diffusion along the z axis may become important.

The effect of turbulent flow on spatial distributions has not been addressed specifically in connection to photothermal spectroscopy. It would be difficult, at best, to model exactly. Turbulence is a chaotic effect. Although turbulence may homogeneously destroy the spatial integrity of the temperature change profile, this happens only if the size of the turbulent feature (e.g., eddies) are smaller than the size of the spatial distribution. The dimensions of focused lasers are often smaller than the size of the turbulent features. Thus the spatial distribution will decay over time scales that change from measurement to measurement. This would be particularly true when the sample flow rate is about the onset of turbulent flow. If turbulent temperature distribution destruction follows the frequency dependence typically found for chaotic systems (i.e., f^{-a}, where $a \sim 1$), the spatial profile of the time-dependent temperature change would decay with the scale $(t/\tau)^{-a}$, where τ(s) is a time-scale factor. This time dependence is equivalent to that found for thermal diffusion. In fact, turbulent heat transfer can be modeled as an average effect using the semi-empirical mass and heat transfer coefficients. Cussler (1984) casts the heat transfer equation in a form containing the thermal diffusion term plus a turbulent mixing term, $-q = (\kappa + b'v)\nabla T$. The $b'v$ term is deduced by fitting data to semiempirical models. For example, in fluid flow through cylindrical tubing, the turbulent term is

$$b'v = 0.027\kappa \left(\frac{vd\rho}{\eta_b}\right)^{0.8} \left(\frac{\eta_b C_P}{\kappa}\right)^{0.33} \tag{4}$$

where d is the tubing diameter. The two terms in parentheses are the Reynolds and Prandtl numbers, respectively. The odd powers that these numbers are raised to and the multiplicative factor are accepted values found from fitting large volumes of heat transfer data (Cussler 1984). The average thermal diffusion coefficient for a sample flowing through a tube with known fluid velocity, v, and tube diameter, d, can be modeled as being the sum of the heat transfer and turbulent terms. Thus D_T may be replaced by $D_T + b'v/\rho C_P$.

As a final comment on turbulent flow, it is worth noting that turbulence really only results in mass transfer beyond that expected from mass diffusion. Mass transfer in turn couples to the heat transfer (i.e., the heat transfer coefficient is a combination of thermal diffusion and mass transfer terms). Thus the heat, mass, and momentum transfer coefficients of turbulence are due to the same effect: the chaotic mixing produced by the eddies. The effect of this combined mass transfer is to make the flow profile appear to be isotropic, albeit noisy, across the cross section of the flow cell. On the average, turbulent flow may be modeled as being isotropic with a smaller effective thermal diffusion time constant.

8.3.2. Deflection Angles and Inverse Focal Lengths in Flowing Samples

8.3.2.1. Isotropic and Turbulent Flow

Photothermal deflection and lens signals for pulsed laser excitation is found using methods described in Chapter 6. In the case of collinear photothermal deflection, the two nonequivalent time-dependent deflection angles are

$$\begin{aligned}\theta_{x,\mathrm{impulse}}(x_0, y_0, t) &= -\left(\frac{dn}{dT}\right)\frac{8x_0 l}{\pi w^4(t)\rho C_p}\, e^{-2[x_0^2+(y_0-v_y t)^2]/w^2(t)} \\ \theta_{y,\mathrm{impulse}}(x_0, y_0, t) &= -\left(\frac{dn}{dT}\right)\frac{8(y_0-v_y t)l}{\pi w^4(t)\rho C_p}\, e^{-2[x_0^2+(y_0-v_y t)^2]/w^2(t)}\end{aligned} \tag{5}$$

where x_0 and y_0 are used to notate the displacement from the excitation laser beam center position to that of the probe laser beam. The maximum deflection angles occur at zero time and for $x_0 = \pm 2^{-1/2}w$ or $y_0 = \pm 2^{-1/2}w$. As might be expected, sample flow has no effect on the maximum instantaneous signal. The time-dependent x- and y-axis probe laser beam position displacements for a maximum signal for x-axis deflection are $x_0 = \pm 2^{-1/2} w(t)$, $y_0 = 0$, and for y-axis deflection are $x_0 = 0$, $y_0 = \pm 2^{-1/2} w(t) - v_y t$.

The collinear photothermal lens element will, in general, have two nonequivalent inverse focal lengths. The x- and y-axis inverse focal lengths are, respectively,

$$\begin{aligned}1/f_{x,\mathrm{impulse}}(x_0, y_0, \mathrm{t}) &= -\left(\frac{dn}{dT}\right)\frac{8l}{\pi w^4(t)\rho C_p}\left[\frac{4x_0^2}{w^2(t)} - 1\right] e^{-2[x_0^2+(y_0-v_y t)^2]/w^2(t)} \\ 1/f_{y,\mathrm{impulse}}(x_0, y_0, t) &= -\left(\frac{dn}{dT}\right)\frac{8l}{\pi w^4(t)\rho C_p}\left[\frac{4(y_0-v_y t)^2}{w^2(t)} - 1\right] e^{-2[x_0^2+(y_0-v_y t)^2]/w^2(t)}\end{aligned} \tag{6}$$

While the effect of the time-dependent x- and y-axis inverse focal lengths can be observed using a slit aperture at the detection plane, a pinhole aperture is normally used in photothermal spectroscopy. In this case the inverse focal length monitored will be the average of the x- and y-axis inverse focal lengths

$$1/f_{r,\text{impulse}}(x_0, y_0, t) = -\left(\frac{dn}{dT}\right)\frac{8l}{\pi w^4(t)\rho C_P}\left\{\frac{4[x_0^2 + (y_0 - v_y t)^2]}{w^2(t)} - 1\right\} e^{-2[x_0^2+(y_0-v_y t)^2]/w^2(t)} \tag{7}$$

As with the collinear deflection signal, the maximum instantaneous lens signal is equivalent to that in stationary samples. In addition, the maximum time-dependent inverse focal length will always occur for $x_0 = 0$ since flow along the y-axis direction does not affect the x-coordinate thermal transfer. The maximum probe laser displacement along the y-axis direction is simply the distance that the lens has traveled, $y_0 = v_y t$.

The transverse photothermal deflection and lens signals are obtained by directing the probe laser into the sample cell at right angles relative to the excitation beam. In this case the optical path probed by the laser is limited by the radius of the temperature-change distribution. Although there are three possible sample flow directions, flow in the same direction as the excitation beam propagation is inconsequential in the infinite cell approximation. Sample flow perpendicular to the excitation beam propagation direction can be probed by either collinear or perpendicular geometry. The perpendicular geometry (i.e., where the mass flow, excitation beam, and probe beam propagation directions make up the axes of the Cartesian coordinates), the transverse photothermal deflection and lens signals are observed. The temperature change is symmetrical along the z axis in the case of bulk or turbulent flow. The transverse photothermal deflection signal impulse response obtained by integration along the x axis is

$$\theta_{\perp,\text{impulse}}(y_0, t) = -\left(\frac{dn}{dT}\right)\frac{8(y_0 - v_y t)}{\sqrt{2\pi} w^3(t)\rho C_P} e^{-2(y_0-v_y t)^2/w^2(t)} \tag{8}$$

The maximum time-dependent deflection angle occurs at a probe laser beam displacement of $y_0 = \pm w(t) + v_y t$. The photothermal lens inverse focal length for the transverse geometry with homogeneous fluid motion is due only to probe divergence in the y dimension:

$$1/f_{\perp,\text{impulse}}(y_0, t) = -\left(\frac{dn}{dT}\right)\frac{4}{\sqrt{2\pi}w^3(t)\rho C_P}\left[\frac{4(y_0 - v_y t)^2}{w^2(t)} - 1\right]e^{-2(y_0 - v_y t)^2/w^2(t)} \tag{9}$$

For continuous excitation, the rate of heat production is constant or square periodic in the case of chopped beams. The temperature change during the heating cycle can be found from

$$\delta T_{\text{cw}}(x, y, t) = \alpha Y_H \Phi(t) * \delta T_{\text{impulse}}(x, y, t)$$

$$\delta T_{\text{cw}}(x, y, t) = \alpha Y_H \int_0^t \Phi(t - t')\, \delta T_{\text{impulse}}(x, y, t')\, dt' \tag{10}$$

$$\delta T_{\text{cw}}(x, y, t) = \frac{2\Phi_0 \alpha Y_H}{\pi \rho C_P} \int_0^t \frac{1}{w^2(t')} e^{-2[x^2 + (y^2 - v_y t')^2]/w^2(t')}\, dt'$$

This integral does not lend itself to closed-form solutions and must be solved numerically. The deflection angles and inverse photothermal lens focal length are obtained by a similar convolution.

Temperature changes and thermal lens signals for continuous sample excitation under a variety of flow conditions have been calculated by Vyas et al. (1988), Vyas and Gupta (1988), and Gupta (1989). Figure 8.1 illustrates the effect of sample flow on the temperature change. For this figure the flow direction was along the x axis and the temperature change is for $y_0 = 0$. The excitation laser beam waist radius was 0.5 mm. Two main effects may be observed: (1) as the flow velocity increases, the maximum temperature change diminishes, and (2) the temperature-change distribution becomes more asymmetrical with increasing flow velocity. The temperature change extends farther downstream.

8.3.2.2. Laminar Flow

Under laminar flow conditions, the characteristic time constant is just that of the static sample. On the other hand, the time-dependent temperature change is not symmetrical along the z axis in this case. This is because redistribution of the photothermal optical elements along the z axis due to the parabolic flow rate profile will induce optical elements. For the square cell, the velocity is a function of the x and z coordinates, as shown in

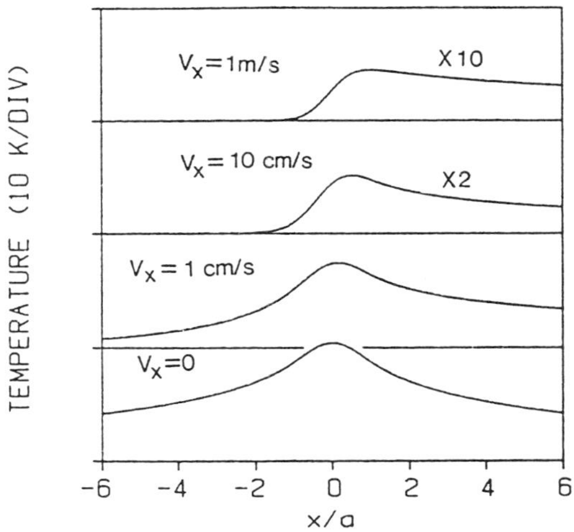

Figure 8.1. Cross section of the temperature-change distribution produced by continuous laser excitation of stationary and flowing (bulk or turbulant) samples. The excitation laser beam waist radius (w) is 500 μ. Temperature changes for stationary and three different flow velocities (v_x) are plotted as a function of the relative displacement from the excitation beam center. The temperature changes were computed using numeric integration. The vertical scales on the upper two flow rate velocity data have been expanded. (Reprinted with permission from Vyas et al. 1988. Copyright 1988, Optical Society of America.)

equation 2. The impulse-response temperature change for z axis excitation and y axis flow from equations 1 and 2 is

$$\delta T_{\text{impulse}}(x, y, z, t) = \frac{2}{\pi w^2(t)\rho C_P} \exp\left[-\frac{2}{w^2(t)}\left\{x^2 + \left[y_0 - \frac{9Vt}{16b^2}\left(1 - \frac{x^2}{b^2}\right)\left(1 - \frac{z^2}{b^2}\right)\right]\right\}\right] \tag{11}$$

for a square-cross-section flow cell. For collinear probe laser geometry, there are two nonequivalent time-dependent deflection angles and lens elements about the x and y axes. Evaluation of the optical elements is further complicated by the integration over the z axis path of the probe laser. The temperature impulse response may be simplified somewhat in the case where the focused beam waist diameters are much smaller than the dimensions of the flow cell. In this case, the term $(9V/16b^2)(1 - x^2/b^2)$ is effectively constant (Weimer and Dovichi 1985a–c). The impulse response for the two collinear deflection angles, evaluated at the probe displacements for maximum signal, are

$$\theta_{x,\text{impulse}}(x_0, t) = -\left(\frac{dn}{dT}\right)\frac{8x_0}{\pi w^4(t)\rho C_P}\int_{-b}^{b} e^{-2[K^2t^2(1-z^2/b^2)^2+x_0^2]/w^2(t)}\,dz$$

$$\theta_{y,\text{impulse}}(y_0, t) = -\left(\frac{dn}{dT}\right)\frac{8}{\pi w^4(t)\rho C_P}$$

$$\times\int_{-b}^{b}[y_0 - Kt(1 - z^2/b^2)]\,e^{-2[y_0-Kt(1-z^2/b^2)]^2/w^2(t)}\,dz \quad (12)$$

and the photothermal lens strengths are

$$\frac{1}{f_{x,\text{impulse}}(t)} = \left(\frac{dn}{dT}\right)\frac{8}{\pi w^4(t)\rho C_P}\int_{-b}^{b} e^{-2K^2t^2(1-z^2/b^2)^2/w^2(t)}\,dz$$

$$\frac{1}{f_{y,\text{impulse}}(t)} = \left(\frac{dn}{dT}\right)\frac{8}{\pi w^4(t)\rho C_P}\int_{-b}^{b}$$

$$\times\left[1 - \frac{4}{w^2(t)}K^2t^2(1 - z^2/b^2)^2\right]e^{-2K^2t^2(1-z^2/b^2)^2/w^2(t)}\,dz \quad (13)$$

where $K = (qV/16b^2)(1 - \chi_0^2/b^2)$. These integrals are not analytic and must be evaluated using numerical techniques.

Weimer and Dovichi (1985a) have measured and modeled continuous laser–excited collinear photothermal lens signals produced by flowing sample through a square-cross-section sample cell. The theoretical signal was calculated by numerically integrating both over the cell path, producing the impulse-response inverse focal lengths produced by the laminar flow, and over time, producing the time-dependent temperature change that occurs for continuous excitation. In addition, this double numerical integration was fit to measurement data using nonlinear regression and the model parameters were evaluated. Results from their experiments are illustrated in Figure 8.2. The two most apparent features of these data are that (1) the photothermal lens signal decreases with increasing flow rate, and (2) the effective time constant decreases. The diminished signal strength was attributed to a shortening of the effective pathlength with increasing flow rate. As the flow rate increases, the lens formed in the center of the sample cell is swept out faster. Only that lens formed near the windows contributes to the signal at the highest flows. The decrease in effective time constant is due simply to the increased heat transfer with increased mass transfer.

Weimer and Dovichi (1985a) suggest that the collinear configuration is not as useful as the crossed-beam experimental arrangement for determining

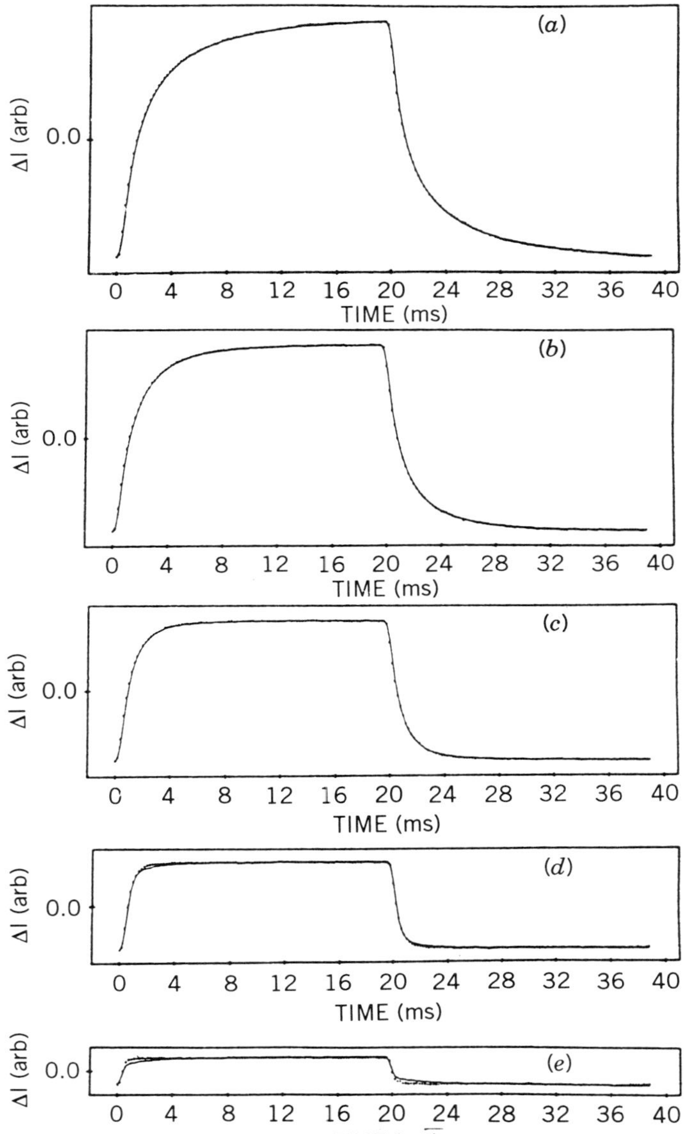

Figure 8.2. Time-resolved crossed-beam photothermal lens signal for flowing samples. Flow rates are 0 (a), 0.22 (b), 0.31 (c), 0.71 (d), and 2.9 (e) mL/min. The 200 data points shown as dots were from an average of 100 cycles and the smooth line is that of model regression. (Reprinted with permission from Weimer and Dovichi, 1985a. Copyright 1985, American Chemical Society.)

flow rates. With perpendicular flow, excitation, and probe laser propagation directions, the probe laser monitors only a small cross section of the flow profile. Subsequently, the flow appears to be isotropic to the probe. In this case the equations for bulk flow can be used (equations 8 and 9 for pulsed, equation 10 for continuous excitation). This technique has been used to measure fluid velocities and fluid velocity profiles in gases and liquids (Sell 1984, 1985, 1989; Weimer and Dovichi 1985b; Rose et al. 1986). Weimer and Dovichi (1985c) performed a study to find the optimum probe laser beam displacement for continuous-excitation crossed-beam photothermal lens apparatus. They found that the maximum signal occurs when the probe laser is displaced downstream from the excitation source. This effect was

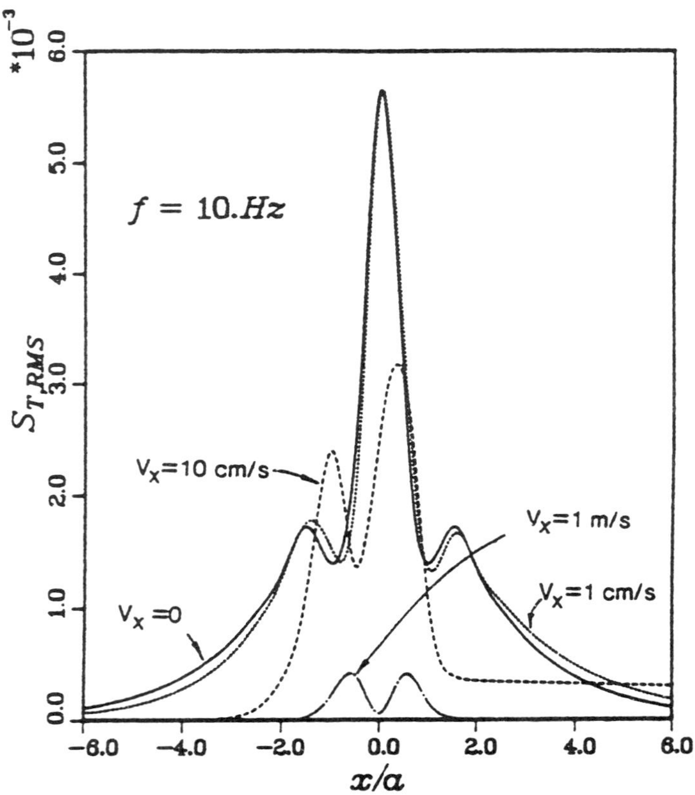

Figure 8.3. Calculated root-mean-square crossed-beam photothermal lens signals for chopped excitation at 10 Hz and several different bulk flow velocities. The downstream shift in the probe laser displacement for maximum signal is most apparent in the 10 cm s^{-1} flow velocity data. (Reprinted with permission from Vyas and Gupta 1988. Copyright 1988, Optical Society of America)

modeled by Vyas and Rose (1988). One of the several model predictions of the later is shown in Figure 8.3. In this figure the excitation laser is modulated at 10 Hz. The root-mean-square signal strength is plotted as a function of displacement for several fluid velocities. The maximum signal probe laser displacement is seen to shift from zero at the lower velocities downstream to finite displacements at higher fluid velocities. The a factor used in this figure is the excitation laser beam waist radius.

8.3.3. Applications

Many of the papers addressing instrumentation and signal processing for applications of photothermal spectroscopy to flowing sample analysis are summarized in Tables 8.3 to 8.6. Most instruments used in flowing sample analysis are based on two-laser, excitation and probe, designs with chopped continuous excitation and lock-in amplifier signal processing. I have attempted to put all detection limit reports in units of absorbance. However, sometimes the absorbance limit was not apparent. This is especially true for microcolumn HPLC and capillary electrophoreses work. In these cases, mass detection limits are given. Detection limits in gas chromatography studies (Table 8.4) are consistently estimated to be in the 10^{-7} AU range. The ease with which this small absorbance can be detected is presumably due to low matrix absorbance in this media. Besides the low detection limit in supercritical CO_2, flow injection analysis detection limits seen in Table 8.3 is not that impressive, only 10^{-4} AU or so. The reason for the low detection limits in flow injection is not certain. I suspect that the low detection limits observed for liquid chromatography studies could probably be obtained in the flow studies. Detection limits for liquid chromatography range from 10^{-4} to 10^{-7} AU. These are consistently better than that obtained using conventional HPLC absorbance detectors and are consistently below the intrinsic solvent absorbance. The reason for this is presumably that the time-dependent absorbance changes can be observed in the presence of the relatively high intrinsic solvent absorbance. Lock-in amplifiers are the most common means of signal processing. They are apparently capable of the 10^3 dynamic reserve required to measure the 10^{-7} absorbance in aqueous solution.

8.3.3.1. Gas Chromatography and Flowing Gas Analysis

There have been relatively few applications of photothermal spectroscopy to gas chromatography effluent detection. Nickolaisen and Bialkowski examined pulsed laser photothermal lens signals obtained in flowing gas (1985) and liquid (1986a) samples. They concluded that the sample flow did not

Table 8.3. Flow Injection Analysis Studies

Technique	Excitation Source	Sample Pathlength or Volume	Signal Processing	Analyte	Matrix	Detection Limits	Ref.
Photothermal lens	Ar^+ laser. 514.5 nm, 160 mW	1 cm. 8 and 70 μL	Digital signal processing	Co^{2+}	H_2O	4.2×10^{-5} AU	1
Photothermal lens	Kr^+ laser, 647.1 nm, 50 mW	1 cm, 8 μL	Digital signal processing	Azulene	Supercritical CO_2	2×10^{-7} AU	2
Cross-beam photothermal lens	Ne-Ne laser, 632.8 nm, 2 mW	1 mm, 40 μL	Lock-in amplifier	Bromophenol blue	Ethanol	7.2×10^{-5} AU	3
Photothermal lens	N_2 laser, 337.1 nm, 20 μJ	4 mm, 0.3 mL	Digital signal processing	2-mercaptopyridine	H_2O	4.7×10^{-3} AU	4
Photothermal lens	Ar^+ laser, 514.5 nm, 20 mW	8 mm, 12 μL	Lock-in amplifier	Tb^{3+}: 4-(2-pyridylazo)-resorcinol	$H_2O:CH_3OH$	4×10^{-4} AU	5

References:
1. Dovichi, N. J. and Harris, J. M. *Anal. Chem.* **53** 698 (1981).
2. Leach, R. A. and Harris, J. M. *Anal. Chem.* **56** 2801 (1984).
3. Yang, Y. and Hairtell, R. E. *Anal. Chem.* **56** 3002 (1984).
4. Nickolaisen, S. L. and Bialkowski, S. E. *Anal. Chem.* **58** 215 (1986).
5. Chartier, A.; Fox, C. G.; and Georges, J. *Analyst* **118** 157 (1993).

Table 8.4. Gas Chromatography Studies

Technique	Excitation Source	Sample Pathlength and Volume	Signal Processing	Analytes	Carrier	Detection Limits	Ref.
Photothermal interferometry	CO_2 laser, 9–11 μm, 160 mW	20 cm, 1.4 mL	Lock-in amplifier	SF_6, hydrocarbons	N_2	4.6×10^{-7} AU	1
Photothermal lens	CO_2 laser, 9–11 μm, 20 mJ	5.5 cm, 2.1 mL	Digital signal processing	Hydrochlorofluorocarbons, alcohols	Ar	2.5×10^{-7} AU	2
Photothermal interferometry	CO_2 laser, 9–11 μm, 160 mW	5 cm, 350 μL	Lock-in amplifier	SF_6, hydrocarbons	N_2	1.1×10^{-7} AU	3
Photothermal deflection	CO_2 laser, 9–11 μm, 20 mJ	2.5 cm, 314 μL	Digital signal processing	Hydrofluorocarbons carbons	Ar	1×10^{-7} AU	4

References:
1. Lin, H. B.; Gaffney, J. S.; and Campillo, A. J. *J. Chromatogr.* **206** 205 (1981).
2. Nickolaisen, S. L. and Bialkowski, S. E. *Anal. Chem.* **57** 758 (1985).
3. Fung, K. H. and Gaffney, J. S. *J. Chromatogr.* **363** 207 (1986).
4. Nickolaisen, S. L. and Bialkowski, S. E. *J. Chromatogr.* **366** 127 (1986).

Table 8.5. Liquid Chromatography Studies

Technique	Excitation Source	Sample Pathlength 2nd Volume	Signal Processing	Analytes	Column Solvent	Detection Limits	Ref.
Photothermal lens	Ar^+ laser, 458 nm, 190 mW	1 cm, 18 μL	Digital signal processing	Nitroanalines	5-μm ODS, 25 cm, $H_2O:CH_3OH$	1.5×10^{-5} AU	1
Photothermal lens	Ar^+ laser, 458 nm, 90 mW	1 cm, 8 μL	Lock-in amplifier	Nitroanalines	10-μm ODS, 25 cm, $H_2O:CH_3OH$	1.2×10^{-6} AU	2
Photothermal lens	Ar^+ laser, 458 nm, 75 mW	1 mm, 0.5 μL	?	Nitroanalines	10 μm ODS micro, 25 cm $H_2O:CH_3CN$	2×10^{-6} AU	3
Photothermal lens	Ar^+ laser, 458 nm 800 mW	100 and 200 μm	Gated integrator	NBD-amines, nitroanalines	ODS-OTLC, 20 μm, 7 m, $H_2O:CH_3OH$	1×10^{-4} AU	4
Crossed-beam photothermal lens	Ar^+ laser, 488 nm, 100 mW	200 μm	Lock-in amplifier	DABSYL-amino acids	ODS, micro, 10 cm, $H_2O:CH_3OH$	1×10^{-7} AU	5
Photothermal lens	Ar^+ laser, 458 nm, 100 mW	1 mm	Lock-in amplifier	Nitroanalines	?	1×10^{-6} AU	6
Photothermal lens	Ar^+ laser, 458 nm, 30–75 mW	1 mm i.d.	Lock-in amplifier	*o*-Nitroanaline	10-μm ODS, 25 cm, $H_2O:CH_3OH$	?	7
Differential photothermal lens	Ar^+ laser, 458 nm, 70 mW	?	Lock-in amplifier	*o*-Nitroanaline	10-μm ODS, 25 cm, $H_2O:CH_3OH$	×2 improvement	8
Photothermal	Ar^+ laser, 514.5 nm, 90 mW	1 cm	High-frequency lock-in amplifier	Benzopurpurin 4B	5-μm ODS, 25 cm, $H_2O:CH_3CN$	4×10^{-6} AU	9

Photothermal lens	He-Cd laser, 442 nm, 10 mW	200 μm	Lock-in amplifier	NBD-amines, nitroanalines	10-μm ODS, 25 cm, $H_2O:CH_3OH$	3×10^{-5} AU	10
Photothermal deflection	Ar^+ laser, 488, 458 nm, 0.2–1.25 W	1–10 mm, 0.5–8 μL	Lock-in amplifier	Acid sulfonate dyes	5-μm ODS, micro, 50 cm, $H_2O:CH_3OH$	8×10^{-8} AU	11
Cross-beam photothermal	He–Cd laser, 442 nm, 3 mW	80 μm	Lock-in amplifier	2,4-Dinitrophenyl hydrozone ketones	5-μm ODS, micro, 40 cm, $H_2O:CH_3CN$	120 fmole	12
Crossed-beam photothermal lens	Ar^+ laser, 458 nm, 150 mW	80 μm	Lock-in amplifier	DABSYL-amino acids	5-μm ODS, micro, 80 cm, $H_2O:CH_3CN$	0.75 fmole	13
Crossed-beam photothermal lens	XeCl pumped dye, laser, 308 nm, 30 μJ	500 μm	Lock-in amplifier	Nitropyrenes	5-μm CPS, micro, 7 m C_6H_{12}/C_3H_8OH	9×10^{-5} AU	14
Photothermal lens	Ar^+ laser, 514.5 nm 35 mW	1 cm, 8 μL	Lock-in amplifier	Benzopurpurin 4B, methyl orange	5-μm ODS, 25 cm, $H_2O:CH_3CN$	6×10^{-6} AU	15
Photothermal lens and deflection	Ar^+ laser, 458 nm, 100 mW	8 mm	Lock-in amplifier	*o*-Nitroaniline	10-μm ODS, 25 cm, $H_2O:CH_3OH$	Similar LOD deflection and lens	16
Differential photothermal lens	Ar^+ laser, 514.5 nm, 35 mW	2 at 6.2 mm, 5 μL	Lock-in amplifier/ digital signal processing	Benzopurpurin 4B, fatty acids	10-μm ODS, 25 cm, $H_2O:CH_3CN$	2×10^{-7} AU	17
Circular dichroism photothermal lens	Ar^+ laser, 514.5 nm, 6 mW	5 mm, 10 μL	Lock-in amplifier	$\pm Co(en)_3^{3+}$	5-μm ODS, 25 cm, H_2O	$\Delta A = 1.9\times10^{-6}$	18
Circular dichroism differential photothermal lens	Ar^+ laser, 488 nm, 36 mW	5 mm	Lock-in amplifier	Fatty acids, $\pm Co(en)_3^{3+}$	ODS	2×10^{-7} AU	19

Table 8.5. (Continued)

Technique	Excitation Source	Sample Pathlength 2nd Volume	Signal Processing	Analytes	Column Solvent	Detection Limits	Ref.
Photothermal lens	Ar^+ laser, 488 nm, 34 mW	1 cm, 7.5 μL	Special analog processing circuit	Chloro-4-dimethyl aminoazobenzenes	3-μm ODS, 10 cm, CH_3OH	2×10^{-7} AU	20
Photothermal lens	Ar^+ laser, 488 nm, 40 mW	1 cm, 8 μL	Lock-in amplifier	Carotenoids	5-μm ODS, micro, 25 cm, $CH_3OH:THF$	2.4×10^{-5} AU	21

References:

1. Leach, R. A. and Harris, J. M. *J. Chromatogr.* **218** 15 (1981).
2. Buffett, C. E. and Morris, M. D. *Anal. Chem.* **54** 1824 (1982).
3. Buffett, C. E. and Morris, M. D. *Anal. Chem.* **55** 376 (1983).
4. Sepaniak, M. J.; Vargo, J. D.; Kettler, C. N.; and Maskarinec, M. P. *Anal. Chem.* **56** 1252 (1984).
5. Nolan, T. G.; Hart, B. K.; and Dovichi, N. J. *Anal. Chem.* **57** 2702 (1985).
6. Morris, M. D. *Proc. SPIE* **426** 116 (1983).
7. Pang, T.-K. and Morris, M. D. *Appl. Spectrosc.* **39** 90 (1985).
8. Pang, T.-K. and Morris, M. D. *Anal. Chem.* **57** 2153 (1985).
9. Skogerboe, K. J. and Yeung, E. S. *Anal. Chem.* **58** 1014 (1986).
10. Yang, Y.; Hall, S. C.; and De La Cruz, M. S. *Anal. Chem.* **58** 758 (1986).
11. Collette, T. W.; Parekh, N. J.; Griffin, J. H.; Carreira, L. A.; and Rogers, L. B. *Appl. Spectrosc.* **40** 164 (1986).
12. Nolan, T. G.; Bornhop, D. J.; and Dovichi, N. J. *J. Chromatogr.* **384** 189 (1987).
13. Nolan, T. G. and Dovichi, N. J. *Anal. Chem.* **59** 2803 (1987).
14. Kettler, C. N. and Sepaniak, M. J. *Anal. Chem.* **59** 1733 (1987).
15. Erskine, S. R.; Foley, C. M.; and Bobbitt, D. R. *Appl. Spectrosc.* **41** 1189 (1987).
16. Yang, Y. and Ho, T. V. *Appl. Spectrosc.* **41** 583 (1987).
17. Erskine, S. R. and Bobbitt, D. R. *Appl. Spetrosc.* **43** 668 (1989).
18. Xu, M. and Tran, C. D. *Anal. Chem.* **62** 2467 (1990).
19. Rice, P. D.; Thorne, J. B.; and Bobbitt, D. R. *Proc. SPIE* **1435** 104 (1991).
20. Rosenzweig, Z. and Yeung, E. S. *Appl. Spectrosc.* **47** 1175 (1993).
21. Chartier, A. and Georges, J. *Anal. Chim. Acta* **284** 311 (1993).

Table 8.6. Capillary Electrophoresis Studies

Technique	Excitation Source	Sample Pathlength	Signal Processing	Analytes	Column	Mass Detection Limits	Ref.
Crossed-beam photothermal lens	Ar^+ laser, 458 nm, 130 mW	50 μm	Lock-in amplifier	DABSYL-amino acids	115 cm, 50 μm i.d.	37–450 amol	1
Crossed-beam photothermal	KrF laser, 248 nm, 10 μJ	50 μm	Lock-in amplifier	PTH-amino acids	34 cm, 50 μm i.d.	0.5 fmol	2

References:
1. Yu, M. and Dovichi, N. J. *Anal. Chem.* **61** 37 (1989).
2. Waldron, K. C. and Dovichi, N. J. *Anal. Chem.* **64** 1396 (1992).

affect signal magnitudes, although the shapes of the signals were altered. Pulsed laser photothermal lens detection was applied successfully for chlorofluorocarbon effluent detection in gas chromatography (Nickolaisen and Bialkowski 1986b). In laminar flow conditions the thermal perturbation travels and can be monitored with a separate probe laser. Sontag and Tam (1985, 1986) and Sell (1985, 1989) have used this effect to monitor flow velocity and temperature.

8.3.3.2. Liquid Phase

Perhaps the most successful area of analytical application has been the use of photothermal spectrometry for the detection of effluents in liquid chromatography and in gel and capillary electrophoresis. With a few caveats, photothermal spectrometry is almost ideally suited for chromatographic effluent detection. The low volumes found in capillary liquid and gas chromatography, and capillary zone electrophoresis, coupled with the necessarily low absorbances of minute quantities of analytes, present a difficult measurement problem using conventional techniques. Photothermal spectrometry is favored in small volumes and is definitely sensitive enough to detect the small amounts of analytes needed in the high-resolution separation techniques.

Leach and Harris (1981) were the first to use photothermal spectroscopy as a detector for liquid chromatography. They used single-laser photothermal lens design with a low-volume 1-cm cell to reach absorption coefficient detection limits of 10^{-3} m^{-1}. Buffett and Morris (1982) used a two-laser design and lock-in amplifier detection to obtain a detection limit of about 10^{-4} m^{-1}. Several studies using different apparatuses followed (Morris and Fotiou 1989). They all apparently have about the same absorption limits of detection. Sepaniak et al. (1984) used photothermal methods for sensitive open tubular liquid chromatography effluent detection. Kettler and Sepaniak (1987) used a pulsed laser for open tubular liquid chromatography effluent detection. Dovichi (1990) has found similar absorption detection limits using capillary electrophoreses for separation. The high sensitivity of photothermal spectroscopy does not alone ensure that low analyte concentrations can be measured. Analyte absorbance must still be measured in the presence of the mobile phase. Pang and Morris (1985) used a differential measurement capabilities of the two-cell photothermal lens spectrometer for improved effluent detection.

In a crossed-beam refraction study, a 10-μJ pulsed KrF excimer laser operating at 248 nm was used to excite phenylthiohydantion (PTH) amino acid derivatives (Waldron and Dovichi 1992). The electrophoresis medium was a 35 m$\underline{M}$ sodium dodecyl sulfate in aqueous buffer. Waldron and

Dovichi point out that this is not a particularly good solution for photothermal detection, and better detection limits would have been found using a binary solvent system. The mass detection limits were somewhat higher than that obtained for the DABSYL derivatives (e.g., 500 attomole for glycine). However, the absorbance detection limit was about 2×10^{-6} AU. In addition to the fact that photothermal spectroscopy was applied for detection in a separation technique, these studies illustrate two aspects of photothermal spectroscopy: (1) photothermal spectroscopy is ideally suited for measurement of low-volume samples, and (2) low detection limits can be obtained using solvents with unfavorable thermooptical and optical absorption properties using pulsed laser excitation sources.

In addition to the capillary electrophoresis amino acid detection described above (Yu and Dovichi 1988, Waldron and Dovichi 1992), one of the more successful applications of photothermal spectroscopy to chromatographic effluent detection techniques is that reported by Erskine and Bobbitt (1989a, b). Their apparatus used a continuous Ar^+ laser to excite the flowing HPLC effluent and a chopped probe laser to detect the resulting differential photothermal lens produced as the effluent flowed through two sample cells plumbed in series. The apparatus was based on the differential absorption technique of Dovichi and Harris (1980). However, several insightful features were incorporated into this HPLC detector. Chopping the probe laser at high frequencies allowed photothermal lens signal noise reduction through lock-in amplifier processing, while the continuous laser excitation allowed for maximum thermal lens elements to be produced in the sample cells (Rice et al. 1991). The sequential flow of the effluent through the sample cells produced a differential response that dramatically lowered noise due to residual absorbance and compensated for excitation laser power fluctuations. In addition, chromatographic peaks were reconstructed by integrating the differential response. This significantly increased the signal-to-noise ratio. Direct absorption coefficient detection limits obtained with this apparatus were 2×10^{-5} m^{-1} in 75:25 acetonitrile:water eluent solutions using 35-mW 514.5 excitation of benzopurpurin 4B analyte. Indirect analyte measurements were also demonstrated by adding pontacyl carmine 3B to 95% acetonitrile, 5% dilute phosphoric acid eluent. The eluent absorption coefficient was 1.3 m^{-1}, and it did not interact strongly with the stationary phase. A mixture of fatty acids, which do not have visible absorption near the excitation laser wavelength, were separated on a ODS column and detected in the differential photothermal lens apparatus. Detection occurs because the fatty acids displace the eluent when they elute from the column. The detection limit for fatty acid measurement was 15 ng of decanoic acid injected. This limit corresponds to 270 pg in the 90-nL optical detection volume (Rice et al. 1991).

8.4. EXCITATION AND RELAXATION KINETICS

Photothermal spectroscopy has been to study excited-state energy transfer kinetics. Since the photothermal spectroscopy signal is a measure of the energy that evolves subsequent to or during sample excitation, it is particularly well suited for use in solving energy transfer problems. Studies of the type utilizing the energy measurement capabilities of photothermal spectroscopy include the determination of vibrational relaxation rate constants, fluorescence and intersystem crossing quantum yields, optical saturation and bleaching photodynamics, and the kinetics of triplet, or other trap states, relaxation to the ground state.

There are three methods used in performing energy transfer studies. The first method is based on the competative nature of the rates of excited-state relaxation through radiative and nonradiative processes. Both continuous and pulsed laser excitation can be used in this method. The signal produced by a given photothermal apparatus is calibrated with samples of known absorbance and heat yield. The signal magnitude is subsequently measured for samples with known absorbance but unknown heat yields. The heat yield is found from the ratio of the standard to the unknown signal magnitude. Luminescence quantum yields are subsequently determined with knowledge of the Stokes shifted luminescence wavelengths. For example, the fluorescence quantum yield is

$$\phi_F = \left(1 - \frac{A_r S}{A S_r}\right) \frac{\nu_{\text{ex}}}{\langle \nu_{\text{em}} \rangle} \tag{14}$$

where A and A_r are the sample and reference compound absorbances at the excitation laser frequency, S and S_r are the photothermal signals obtained from the sample and reference compounds, ν_{ex} is the excitation laser frequency, and $\langle \nu_{\text{em}} \rangle$ is the average emission frequency obtained from the fluorescence spectrum. Brannon and Madge (1978), Madge et al. (1979, and Shen and Snook (1989a, b) have applied this method to determine fluorescence quantum yields using a continuous laser excitation in a single-laser photothermal lens apparatus. Chartier et al. (1990a, b) showed that this method does not necessarily work when pulsed laser excitation is used. This is because excited states can couple into metastable triplet states, thus decreasing the total heat availible on the short time scale of the transient signal measurement.

The second method relies on a time-resolved measurement of the transient photothermal signals produced upon pulsed sample excitation. In this method, excited-state relaxation kinetics may be obtained directly from the transient photothermal signal if the time scale for relaxation is between the characteristic thermal decay time constant and the time required for acoustic relaxation. If the relaxation rate is slower than thermal diffusion, the relaxation rate may still be obtained using deconvolution. Quantum yields can be obtained from the transient signal as well. This method has been applied to the measurement of rotational and vibrational relaxation rate constants of gas-phase species and to determine quantum yields and triplet-state relaxation rates of organic dye molecules in condensed phases.

The third method utilizes excitation irradiance-dependent photothermal signals to elucidate excited-state relaxation kinetics. The method is based on the fact that excitation of both ground and excited states can compete with relaxation. Data interpretation with this method is highly model dependent. However, it can be used to circumvent the intrinsic acoustic-limited density change limitation of time-resolved signal analysis. This method has been used to study rates of vibrational relaxation in gas-phase species and rates of excited-state relaxation in condensed-phase organic dye species.

8.4.1. Relaxation Kinetics and Quantum Yield Studies

Pulsed laser–excited photothermal lens spectroscopy owes its conception to studies of excited-state relaxation rate processes. With rapid impulse excitation, relaxation of the excited state produces a density change that is represented mathematically as the convolution of the rate of heat production due to excited-state relaxation with the time-dependent impulse-response function of the density change. Barker and Rothem (1982) and Barker and Toselli (1989) considered in detail the effects that limit the use of time-resolved photothermal spectroscopy for determining excited-state relaxation rate constants in the gas phase by solving fluid dynamics equations which account for the density change. Their treatment takes into account excited-state relaxation, mass and thermal diffusion, and the sound speed–limited density change. Conclusions regarding the sample parameters that would allow measurable relaxation rates were drawn from test calculations. Many of these were discussed in Chapter 3. The conclusions are straightforward. To be able to use time-dependent photothermal spectroscopy to measure relaxation rates, the time for excited-state relaxation must be greater than that required for acoustic relaxation time, and less or on the order of time scales for mass and thermal diffusion. Some flexibility is afforded by the adjustable experimental parameters. For example, time scales for different relaxation processes may be affected by gas pressure or the scale of the excitation beam. In the gas phase, mass and thermal diffusion coefficients decrease with increasing pressure, while the excited-state relaxation rate constant generally increases with pressure. In both solution and gas phases, the sound speed is, to a first-order approximation, independent of density. However, the times required for acoustic relaxation are a linear function of the spatial scale (e.g, beam waist radius) of excitation. Smaller excitation scales result in faster acoutic relaxation time scales.

As shown in Chapter 3, the time-dependent density change is comprised of acoustic or density wave and thermal diffusion components which, to a first-order approximation, are linear additive terms. The radial density change that occurs due to an excited-state relaxation rate process, $R'(t)$, is

$$\begin{aligned} \delta\rho(r,\, t) &= R'(t) * \delta\rho_{\text{impulse}}(r,\, t) \\ \delta\rho(r,\, t) &= R'(t) * [\delta\rho_{\text{diffusion}}(r,\, t) + \delta\rho_{\text{acoustic}}(r,\, t)] \end{aligned} \tag{15}$$

For measurement times greater than that required for acoustic wave relaxation (e.g., $t > w/c$, where w is the Gaussian beam waist radius and c is the sound velocity, or Λ/c, where Λ is the grating plane spacing in a thermal grating), the acoustic-wave density change term can be neglected relative to the slower thermal diffusion term. Moreover, the density change is directly related to the temperature change and

$$\begin{aligned} \delta T_{\text{impulse}}(r,\, t) &= -\frac{1}{\rho\beta}\, \delta\rho_{\text{diffusion}}(r,\, t) \\ \delta T(r,\, t) &= R'(t) * \delta T_{\text{impulse}}(r,\, t) \end{aligned} \tag{16}$$

If the rate of heat production does not depend on the coordinates (e.g., is a linear combination of first-order processes), operations required to calculate the photothermal signals do not affect the rate term, and the signals can be calculated by convolution of the heat production rate with the signal impulse response. For example, the photothermal lens and deflection signals are expressed by

$$\begin{aligned} \theta(t) &= R'(t) * \theta_{\text{impulse}}(t) \\ f^{-1}(t) &= R'(t) * f^{-1}_{\text{impulse}}(t) \end{aligned} \tag{17}$$

If, in addition, the excited-state relaxation time is shorter than the characteristic relaxation time of the signal, the convolution can be approximated by the product of the time-integrated heat production rate with the photothermal spectroscopy signal impulse response

$$\begin{aligned} \theta(t < t_c) &= R(t)\theta_{\text{impulse}}(t) \\ f^{-1}(t < t_c) &= R(t)\, f^{-1}_{\text{impulse}}(t) \end{aligned} \tag{18}$$

The latter equation is used primarily for interpreting the rates of excited relaxation from the rising portion of the pulsed laser–excited photothermal spectroscopy signals.

Braslavsky and Heibel (1993) point out that there can be other contributions to the density change that do not arise from thermal sources. These changes could result in errors in the kinetic and quantum yield parameters determined using photothermal spectroscopy. Nonthermal density changes occur when volume differences between the ground- and excited-state species arise upon excitation. They may have a significant contribution to the total density change in solution-phase studies but should not affect gas-phase results because molecular volume does not strongly affect pressure and density. Density changes due to excited-state volume change in strongly interacting solution-phase samples are expected to be proportional to the number density of excited-state species. For measurements on time scales longer than the acoustic relaxation time, the density change is

$$\delta\rho(r, t) = R'(t) * \delta\rho_{\text{diffusion}}(r, t) + \sum_{i=1}^{n} N_i(r, t)\, \delta\rho_i \tag{19}$$

where $N_i(r, t)$ is the time-dependent number density of species in the ith excited state and $\delta\rho_i$ is the effective density change that occurs due to a change in molecular volume. The number densities are related to the integrated rate of heat production, and in general, the relative contributions to the thermal diffusion and volume change terms cannot be separated based on kinetic effects alone. They can be separated by taking measurements at various temperatures, in particular, in aqueous solution.

It is interesting to note that photothermal lens signals were used to study rotation- and vibration-state relaxation times even before a theory describing the signal was developed. Excited-state relaxation rate studies require the excited-state lifetime to be longer than the acoustic relaxation time constant. In addition, data analysis is much easier if the relaxation time is less than the characteristic thermal decay time. Because of the wide variablity in collision rates possible by varying gas pressures, these conditions can usually be met when studying relaxation of low-lying excited vibrational states in the gas phase. Grabiner et al. (1972) first described the use of pulsed infrared laser–excited photothermal lens spectroscopy for measurement of vibrational and rotational relaxation times. In this pioneering study, vibrational relaxation rate constants for CH_3F, CH_3Cl, and C_2H_4 were obtained. Plots of the photothermal lens signal rise time versus partial pressure of the collision partner yielded lines with slopes that are the second-order vibrational relaxation rate constants. They found that kinetic cooling prior to heating due to relaxation back to the ground state can also occur. Kinetic cooling results from the rapid excited-state equilibration between two (or more) close-lying states. With pulsed excitation, molecules are first excited into the a single rotational–vibrational state. These excited-state molecules subsequently equilibrate among accessible excited rotational and vibrational states, sometimes at the cost of translational energy. Rotational and V–V energy transfer can occur more rapidly compared to V–R/T relaxation (back to the ground state) because the efficiency of the process and because the energy gap between the two excited-states is typically smaller than that required to return to the ground state.

Grabiner et al. found that the transient photothermal lens signal for low pressures of CH_3F in Ar exhibited behavior that could be attributed to an initial cooling followed by longer-time-scale heating. The lowest-lying CH_3F vibrational state, ν_3, was excited using a Q-switched CO_2 laser and a He–Ne laser with a pinhole and photomultiplier tube was used to monitor the resulting transient photothermal lens signal. It was postulated that translational cooling was due to a $\nu_3 \rightarrow \nu_6$ transition, a process with an energy debt of -147 cm^{-1}. The longer-time relaxation was then due to the combined ν_3 and ν_6 vibrational relaxation to the ground state, releasing 1048 and 1195 cm^{-1} each, respectively. CH_3Cl and C_2H_4 did not exhibit translational cooling. The fact that only heating was observed may be due to the fact that these species were excited to vibrational states that were not the lowest lying. V–V relaxation among the excited vibrational states would preferentially occur by populating the lower-lying states, subsequently heating the sample. In a later study, Siebert et al. (1974) used the same technique to measure relaxation rate constants in CD_4, SO_2, and OCS. CD_4 exhibited rapid kinetic cooling followed by heating that could be explained in much the same fashion as for CH_3F. Here, too, the lowest-lying vibrational state was excited. SO_2 and OCS exhibited contrary behaviors. Although in both cases, molecules are initially excited to higher-lying vibrational states, only SO_2 exhibited rapid kinetic cooling. Other gas kinetic relaxation studies are summarized in Table 8.7.

Solution-phase studies have been confined to measurements of excited electronic state relaxation and photolysis since vibrational relaxation rates are extremely rapid in condensed phases. Compared to the gas-phase studies, the experimentalist has less control over relaxation time scales. In the gas-phase studies, collision partner partial pressures can usually be adjusted such that excited-state relaxation occurs in times between the acoustic and thermal relaxation limits. On the contrary, in solution-phase studies, time scales for vibrational relaxation are usually much faster than the acoustic relaxation limit and excited-electronic-state relaxation rates are similarly fast, unless they are spin forbidden. Although the experimental and theoretical framework for performing pulsed laser–excited studies of condensed-phase relaxation processes was in place since the time of the vibrational relaxation work, the first condensed-phase relaxation rate study was not reported until 15 years after the initial reports by George Flynn and co-workers.

Terazima and Azumi (1987) used a method for interpreting time-resolved pulsed visible wavelength laser–excited photothermal signals produced from condensed-phase organic dye molecules proposed by Rossbroich et al. (1985). This method is applicable to molecules with a relatively long-lived metastable state such as a triplet state. Organic dye molecules with relatively long-lived triplet states produce photothermal signals that consist of a relatively fast component and a long-lived triplet-state relaxation component (Figure 8.4). The relative amount of heat produced by these two components is interpreted in terms of the quantum yield for triplet-state formation. In addition, since the rate of heat production due to the relaxing triplet state could be directly observed in the photothermal lens signal, the first-order rate of excited-state relaxation could be measured directly for the time-dependent signal. This method has subsequently been used to measure triplet quantum yields and relaxation times for a number of compounds. The method relies on there being an experimentally observable difference in relaxation rates between singlet and triplet states.

Upon excitation, organic dye molecules typically produce higher-lying vibrational levels of an excited singlet state. Vibrational and/or electronic state energy in excess of the lowest vibrational state of the first excited

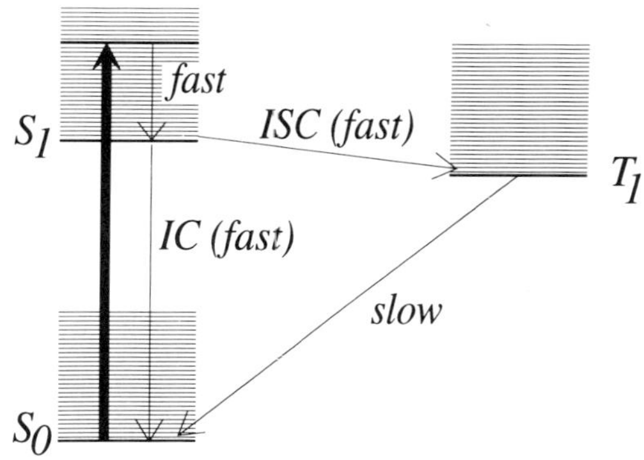

Figure 8.4. Typical energy-level diagram for an organic dye molecule. The dark arrow shows excitation into excited vibrational levels of the first excited singlet state, S_1. Vibronic relaxation of the excited state occurs on rapidly producing ground vibrational state S_1. Molecules in S_1 can undergo rapid internal conversion (IC), fluorescence (not shown), or intersystem crossing (ISC), entering the triplet system, T_1. The metastable T_1 state relaxes back to the ground state on a relatively long time scale.

Table 8.7. Gas-Phase Energy Transfer Studies

Technique	Experiment	Signal Processing	Substance	Effect	Ref.
Photothermal lens	Q-switched CO_2, 9.6 or 10.4 μm, He-Ne probe	Multichannel averaging	CH_3F, CH_3Cl, C_2H_4; Ar buffer	V–V; V–R/T	1
Photothermal lens	Q-switched CO_2, 9.6 or 10.4 μm, He-Ne probe	Multichannel averaging	CD_4, SO_2, OCS; Ar buffer	V–V; V–R/T	2
Photothermal lens	N_2-dye, He-Ne probe	Multichannel averaging	Azulene; N_2, He, Ar, Ne, Kr, Xe buffer	IC; V–V, R/T	3
Photothermal lens	TEA-CO_2, 9.6–10.6 μm, He-Ne probe	Boxcar averager	CCl_2F_2, $C_2Cl_3F_3$, CH_3OH He, Ne, Ar buffer	V–R–T	4
Photothermal lens	N_2-dye, He-Ne probe	Multichannel averaging	Azulene; Kr buffer	IC; V–V, R/T	5
Photothermal interferometry	TEA-CO_2, 9.4–10.6 μm, He-Ne probe	Multi-channel averaging	SF_6	V–R/T	6
Photothermal lens	TEA-CO_2, 9.4–10.6 μm, He-Ne probe	Multichannel averaging	CO_2; CO_2, CO, N_2 buffer	V–V; V–R/T	7
Photothermal lens	TEA-CO_2, 9.4–10.6 μm, He-Ne probe	Multichannel averaging	CO_2; CO_2, CO, N_2 buffer	V–V; V–R/T	8
Photothermal lens	TEA-CO_2, 9.3 μm, He-Ne probe	Oscilloscope trace	Cyclobutanone	Infrared photolysis	9

Photothermal interferometry	TEA-CO_2, 9.2–10.6 μm, He-Ne probe	Multichannel averaging	—	V–R/T	10
Photothermal interferometry	TEA-CO_2, 9.2–10.6 μm, He-Ne probe	Multichannel averaging	SF_6	V–R/T	11
Photothermal interferometry	TEA-CO_2, 9.2–10.6 μm, He-Ne probe	Multichannel averaging	—	V–R/T	12
Photothermal interferometry	TEA-CO_2, 9.2–10.6 μm, He-Ne probe	Multichannel averaging	Highly excited CF_2HCl	V–R/T	13
Photothermal interferometry	TEA-CO_2, 9.2–10.6 μm, He-Ne probe	Multichannel averaging	Highly excited CF_2HCl	V–R/T	14

References:
1. Grabiner, F. R.; Siebert, D. R.; and Flynn, G. W. *Chem. Phys. Lett.* **17** 189 (1972).
2. Siebert, D. R.; Grabiner, F. R.; and Flynn, G. W. *J. Chem. Phys.* **60** 1564 (1974).
3. Trevor, P. L.; Rothem, T.; and Barker, J. R. *Chem. Phys.* **68** 341 (1982).
4. Xing-Xao, M. and Zhu-De, X. *Chem. Phys. Lett.* **98** 563 (1983).
5. Barker, J. R. *J. Phys. Chem.* **88** 11 (1984).
6. D'Ambrosio, C.; Bruzzese, R.; Ferrigno, A.; and Salimeno, S. *Opt. Commun.* **55**, 159 (1985).
7. Bailey, R. T.; Cruickshank, F. R.; Pugh, D.; and Middleton, K. M. *J. Chem. Soc. Faraday Trans.* 2 **81** 255 (1985).
8. Bailey, R. T.; Cruickshank, F. R.; Middleton, K. M.; and Pugh, D. *J. Radioanal. Nucl. Chem.* **101** 383 (1986).
9. Guckert, J. R. and Carr, R. W. *J. Phys. Chem.* **90** 4286 (1986).
10. Pan, C. Z.; Bruzzese, R.; Salimeno, S.; and Velotta, R. *Opt. Commun.* **59** 183 (1986).
11. Pan, C. Z.; Bruzzese, R.; Salimeno, S.; and Velotta, R. *J. Opt. Soc. Amer.* **B4** 452 (1987).
12. Achasov, O. V.; Bruzzese, R.; de Lisio, C.; and Salimeno, S. *Opt. Commun.* **64** 234 (1987).
13. Salimeno, S.; Bruzzese, R.; de Lisio, C.; D'Ambrosio, C.; Velotta, R.; and Tosa, V. *Infrared Phys.* **29** 473 (1989).
14. Tosa, V.; Labuda, S.; Bruzzese, R.; de Lisio, C.; and Salimeno, S. *J. Chem. Phys.* **91** 4134 (1989).

singlet state, S_1, is rapidly lost, within time scales on the order of picoseconds (Guiliano and Hess 1967), producing a vibrationally "cool" excited S_1 state. The S_1 state relaxes to the ground, S_0, state either through radiative fluorescence or internal conversion, or may undergo intersystem crossing to produce a ground triplet state, T_1. Vibrational relaxation is rapid in both T_1 and S_0 states. Inverse intersystem crossing, (i.e., relaxation of T_1 to S_0) is generally slow compared to vibrational relaxation. Although the exact path of relaxation may differ slightly from molecule to molecule (e.g., vibronic relaxation may occur through excited triplet electronic states), the end result is the same; rapid vibronic relaxation produces a certain fraction of species in T_1 which subsequently relax to S_0 over relatively long time scales. The photothermal signals produced from triplet relaxation is often observed to occur on a longer time scale than that of the more rapid vibrational relaxation.

The relatively time-dependent photothermal lens signals produced by fast vibronic relaxation and slow triplet relaxation processes can be related to the quantum yield for intersystem crossing by accounting for the heat produced in both fast and slow processes, and then solving for the intersystem crossing efficiency. For species that do not undergo photolysis, the heats produced by the individual relaxation processes are

$$\begin{aligned} U_{\mathrm{IVR}} &= N_{\mathrm{ex}}(E_{\mathrm{ex}} - E_S) \qquad \text{(fast)} \\ U_{\mathrm{IC}} &= N_{\mathrm{ex}}(1 - \phi_{\mathrm{ISC}} - \phi_F)E_S \qquad \text{(fast)} \\ U_{\mathrm{ISC}} &= N_{\mathrm{ex}}\phi_{\mathrm{ISC}}(E_S - E_T) \qquad \text{(fast)} \\ U_F &= N_{\mathrm{ex}}\phi_F Y_F E_S \qquad \text{(fast)} \\ U_{T_1 \to S_0} &= N_{\mathrm{ex}}\phi_{\mathrm{ISC}} E_T \qquad \text{(slow)} \end{aligned} \tag{20}$$

where U_{IVR}, U_{IC}, U_{ISC}, U_F, and $U_{T \to S}$ are energy densities due to excited-state vibronic relaxation, internal conversion, intersystem crossing, fluorescence, and triplet relaxation respectively; N_{ex} is the number density of excited species; $E_{\mathrm{ex}} = h\nu$ is the excitation energy; E_S and E_T are the energies of the excited singlet, S_1, and triplet, T_1, states, respectively; ϕ_{ISC} and ϕ_F are the triplet state and fluorescence quantum yields; and Y_F is the fractional heat yield due to molecules undergoing fluorescent transitions to vibrationally excited S_0 states. Assuming that the fractional heat yield is unity, the triplet-state quantum yield is (Terazima and Azumi 1987)

$$\phi_T = \frac{U_{\mathrm{slow}}(E_{\mathrm{ex}} - \phi_f E_S)}{U_{\mathrm{tot}} E_T} \tag{21}$$

where U_{slow} is the energy density produced by slow $T_1 \to S_0$ relaxation and U_{tot} is the total energy density found from the sum of the individual components listed above. Signal magnitudes produced in time-resolved

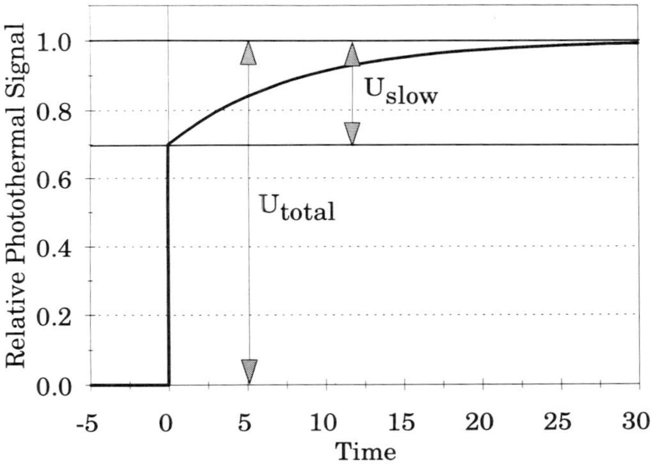

Figure 8.5. Time-resolved photothermal signal produced when there is a slow heat production step due to metastable state production. The U are energy densities, proportional to the photothermal signals. The total signal is comprised of fast and slow time-dependent components.

photothermal spectroscopy are proportional to the energy densities (Figure 8.5). So by taking the ratio of the slow to total signal magnitudes, these magnitudes can be related directly to the energy densities, and thus to the quantum yield. Fluorescence and triplet-state quantum yields based on this method of measurement would not be accurate in cases where there is a large change in molecular volume upon irradiation (Braslavsky and Heibel 1993).

The triplet-state lifetime can also be determined by fitting the slow-rising component of the transient signal to an exponential function. In this case the contribution to the density from molecular volume changes should not affect the result. However, the triplet lifetime is sometimes difficult to interpret, due to the strong influence that trace species and dissolved O_2 have on the lifetime of the triplet state. The intrinsic triplet lifetime and the quenching constant for trace species and dissolved O_2 can be obtained by measuring the exponential lifetime of the slow heat production portion of the transient photothermal signal and then plotting the inverse lifetime as a function of quencher concentration. The resulting Stern–Volmer plot should be linear, with a slope equal to the quenching constant and an intercept equal to the intrinsic (i.e., in the absence of the quencher) lifetime. This method has been applied extensively to determine the kinetics of triplet-state relaxation in benzophenone solutions (Isak et al. 1989; Poston and Harris 1990, 1991; Cambron and Harris 1993). Benzophenone is a model molecule

Table 8.8. Solution-Phase Quantum Yield and Kinetics Studies

Technique	Experiment	Signal Processing	Substance	Effect	Ref.
Single-laser photothermal lens	Ar^+, 496.5 nm, 100 mW	Transient waveform recorder	Fluorescein	ϕ_F	1
Single-laser photothermal lens	He-Ne, 632.8 nm	Oscilloscope	Cresol violet	ϕ_F	2
Time-resolved pulsed-laser photothermal lens	N_2-dye laser, 520 nm, He-Ne probe	Oscilloscope	Erythosin and TPP-sensitized $O_2(^1\Delta_g)$ production and quenching in several solvents/acceptors	ΔH and τ_{02} for $O_2\,^1\Delta$ reaction	3
Pulsed-laser photothermal lens	Nd:YAG-dye laser, 322 nm He-Ne probe	Multichannel averaged transient signal	Anthracene, ZnTPP, TPP-sensitized O_2 $(^1\Delta_g)$ production and quenching	ϕ_Δ for $O_2\,^1\Delta$, ϕ_{ISC}	4
Pulsed-laser Photothermal lens-laser photothermal lens	N_2-dye laser, He-Ne probe	Multichannel averaged transient signal	Pthalazine, anthracene, quinoxaline, benzophenone, isoquinoline; benzene and ethanol solvent	ϕ_{ISC}, τ_T	5
Time-resolved pulsed-laser photothermal lens	N_2-dye laser, He-Ne probe	Multichannel averaged transient signal	Pyridazine, 3-methyl pentane solvent	ϕ_{ISC}	6
Time-resolved two-photon pulsed-laser photothermal lens	N_2-dye laser, He-Ne probe	Multichannel averaged transient signal	Pyridine, quinoxaline, cyclohexane solvent	ϕ_{ISC}, τ_T	7

Time-resolved pulsed-laser photothermal lens	N_2-dye laser, He-Ne probe	Multichannel averaged transient signal	Keto,enol-7-hydroxy-quinoline, methanol solvent	$\Delta H_{keto\text{-}enol}$	8
Continuous two-laser photothermal lens	Ar^+ 488 nm (40 mW), He-Ne probe	Oscilloscope	Sodium fluorescein, H_2O solvent	ϕ_F	9
Single-laser photothermal lens	Ar^+ 488 nm (40 mW)	Oscilloscope	Sodium fluorescein, H_2O and ethanol solvents	ϕ_F	10
Time-resolved pulsed-laser photothermal lens	Nd : YAG, 355 nm, 100 μJ, He-Ne probe	Multichannel averaged transient signal	Benzophenone, CH_3CN + KI	τ_T, k_Q	11
Time-resolved pulsed-laser photothermal deflection	Nd : YAG-dye laser, 322 nm, He-Ne probe	Multichannel averaged transient signal	Benzophenone, CH_3CN + ethanol hydrogen, abstraction kinetics	τ_T, k_Q	12
Time-resolved pulsed-laser photothermal lens	N_2-dye laser, He-Ne probe	Boxcar averager	Erythrosin, H_2O + surfactants and ethylene glycol + O_2 acceptor	τ_T, τ_{02}	13
Pulsed-laser photothermal lens	N_2-dye laser, He-Ne probe	Boxcar averager	Fluorescein, eosin, erythrosin, proflavine, acradine orange, phenol red in H_2O and ethanol	ϕ_F, ϕ_{ISC}	14
Time-resolved pulsed-laser photothermal deflection	Nd : YAG-dye laser, 322 nm, He-Ne probe	Multichannel averaged transient signal	Benzophenone, CCl_4 + biphenyl acceptor, energy transfer kinetics and acceptor triplet energy	τ_T, k_Q, E_T	15

Table 8.8. (Continued)

Technique	Experiment	Signal Processing	Substance	Effect	Ref.
Pulsed-laser photothermal lens	N_2-dye laser, He-Ne probe	Boxcar averager	Fluorescein, phenol red in H_2O at several pH	pH dependent, ϕ_F	16
Pulsed-laser photothermal lens	N_2-dye laser, He-Ne probe	Boxcar averager	Fluorescein, eosin, erythrosin, H_2O solvent	ϕ_F, ϕ_{ISC}	17
Time-resolved pulsed-laser photothermal lens and diffraction	Excimer-dye laser, He-Ne probe	Multichannel averaged transient signal	C_{60}-sensitized $O_2\,{}^1\Delta_g$ formation in benzene	ϕ_Δ for $O_2\,{}^1\Delta$	18

References:

1. Brannon, J. H. and Magde, D. *J. Phys. Chem.* **82** 705 (1978).
2. Madge, D. and Brannon, J. H. *J. Phys. Chem.* **83** 696 (1979).
3. Fuke, K.; Ueda, M.; and Itoh, M. *J. Am. Chem. Soc.* **105** 1091 (1983).
4. Rossbroich, G.; Garcia, N. A.; and Braslavsky, S. E. *J. Photochem.* **31** 37 (1985).
5. Terazima, M. and Azumi, T. *Chem. Phys. Lett.* **141** 237 (1987).
6. Terazima, M. and Azumi, T. *Chem. Phys. Lett.* **145** 286 (1988).
7. Terazima, M. and Azumi, T. *Chem. Phys. Lett.* **153** 27 (1988).
8. Terazima, M. and Azumi, T. *J. Amer. Chem. Soc.* **111** 3824 (1989).
9. Shen, J. and Snook, R. D. *Chem. Phys. Lett.* **155** 583 (1989).
10. Shen, J. and Snook, R. D. *Anal. Proc.* **26** 27 (1989).
11. Isak, S. J.; Komorowski, S. J.; Merrow, C. N.; Poston, P. E.; and Eyring, E. M. *Appl. Spectrosc.* **43** 419 (1989).
12. Poston, P. E. and Harris, J. M. *J. Amer. Chem. Soc.* **112** 644 (1990).
13. Chartier, A.; Georges, J.; and Mermet, J. M. *Spectrochem. Acta* **46A** 1737 (1990).
14. Chartier, A.; Georges, J.; and Mermet, J. M. *Chem. Phys. Lett.* **171** 347 (1990).
15. Poston, P. E. and Harris, J. M. *J. Photochem. Photobiol. A* **60** 51 (1991).
16. Chartier, A.; Georges, J.; and Mermet, J. M. *Spectrochem. Acta* **47A** 1505 (1991).
17. Chartier, A. and Georges, J. *J. Phys. IV* **1** c7 (1991).
18. Teramiza, M.; Hirota, N.; Shinohara, H.; and Saito, Y. *J. Phys. Chem.* **95** 9080 (1991).

since the excitation and relaxation dynamics are well known, it has unit intersystem crossing quantum yield, and the triplet-state lifetime is nearly ideal for study by the evolution of the time-dependent photothermal signal. Cambron and Harris (1993) have considered the effect of triplet–triplet inhalation and other second-order relaxation terms in the benzophenone relaxation kinetics. These and other solution-phase studies are summarized in Table 8.8. This table is by no means comprehensive but serves to give an idea as to the utility of photothermal spectroscopy to solution-phase excited-state dynamics measurements.

8.4.2. Photodynamic Irradiance Dependent Signal Studies

Studies of irradiance-dependent photothermal signal magnitudes may be used to determine relaxation rate constants and excited-state optical absorption coefficients. Basically, the excited-state relaxation constant controls the rate at which molecules can return to the ground state and subsequently be reexcited. When the excitation irradiance is high enough, the rate of optical excitation exceeds the rate at which molecules return to the ground state. In this case the photothermal signal magnitude produced for a given excitation irradiance is not as high as that predicted by the linear absorption model. The decreased absorption at higher excitation irradiances is either optical saturation or bleaching, depending on whether or not stimulated emission occurs. By determining the irradiance at which optical saturation or bleaching occurs, and relating that irradiance to the excitation rate, the relaxation rate can be estimated. The irradiance at which the rate of excitation equals the rate of excited-state return can be determined from the irradiance versus photothermal signal data. The relationship between relaxation rate and signal magnitude is complicated by the fact that the spatial profile of excited-state species is distorted over that of the excitation source (Twarowski and Kliger 1977, Long and Bialkowski 1984, 1985; McGraw and Harris 1986).

The fact that excitation laser irradiance could affect the magnitude of the photothermal signal was noticed by Long and Bialkowski (1984, 1985) when using a pulsed TEA–CO_2 laser to measure gas-phase fluorocarbon concentration using photothermal lens and deflection spectroscopies. The nonlinear irradiance-dependent photothermal signal behavior was later used to discriminate between nonlinear absorbers (Bialkowski and Long 1987). Most recently, Bialkowski and He (1992, 1994) have used the nonlinear irradiance-dependent photothermal lens signals to determine excited-state relaxation rate constants for ethylene with a number of collision partners. In the gas phase, stimulated emission of the excited state can compete with relaxation out of the state initially excited from the ground state. Subsequently, the rate constant for relaxation out of the initially excited-state may be deduced. Using a four-level model for ethylene, Bialkowski and He were able to estimate both the rate at which ethylene was decoupled from the excitation radiation and the overall rate at which molecules relaxed to the ground state through a series of V–V/R/T energy transfer steps. It is of interest that the rate of excited-state de-population determined by competitive stimulated emission was greater than that which could be determined by monitoring the rise of the acoustic relaxation-limited transient photothermal signal.

8.4.3. Optical Bleaching in Organic Dye Molecules

A larger body of literature addresses optical bleaching that can occur when using pulsed lasers to excite certain organic dye molecules. Vibrational relaxation is extremely fast in condensed phases. The rapid vibrational relaxation "decouples" the molecules from the optical radiation. Thus optical saturation is unlikely, although bleaching can occur. Electronic excited singlet states can couple into the triplet manifold, producing metastable T_1. The metastable T_1 state constitutes a trap that prevents molecules from returning rapidly to the ground state. Molecules trapped in the triplet manifold may not absorb radiation, and the singlet transition is subsequently bleached. However, the T_1 state may also absorb optical radiation, somewhat masking the optical bleaching effect. Under moderate irradiance conditions, the energy absorbed by organic dye molecules can often be determined using a simple kinetic model that takes into account the major relaxation pathways (Guiliano and Hess 1967). The model shown below is based on the provisions of Guiliano and Hess but invokes a steady-state assumption regarding the first excited singlet.

Giuliano and Hess argue that most excited organic dye molecules relax through a series of fast and slow steps and that the mechanisms and the time scales for these steps are nearly the same for all species studied. The influence of the fast steps on the overall relaxation rates can be neglected relative to the slower one since the time scales are difference by several orders of magnitude. The model they propose can be used to describe optical bleaching in most organic dye molecules. A short summary of their arguments is given below. The energy-level diagram illustrating the major kinetic pathways is shown in Figure 8.5 to guide the reader.

Absorption of optical energy by the ground-state singlet, S_0, results in promotion to a vibrationally excited state of the excited electronic state, S_1. Vibronic overlap favors excitation into vibrationally excited S_1 states (i.e., the Franck–Condon principle). Vibrational energy in excess of the ground vibrational state is rapidly lost to the solvent through internal vibrational relaxation (IVR), producing a low-vibrational-state S_1 species. Molecules in S_1 may directly return to the ground S_0 state through internal conversion (IC) or fluorescence emission. The combined effect of these relaxation mechanisms is a first-order rate with a relaxation rate constant, k_{10}. Excitation of S_0 to higher electronic states (e.g., S_2, S_3, etc.) may also occur using more energetic excitation wavelengths. These higher excited states relax rapidly and couple to S_1 through IC and IVR processes. Times required for vibronic relaxation are generally on the order of picoseconds, whereas the lower vibrational states of S_1 species can take several nanoseconds to return to S_0. Vibronic relaxation is essentially instantaneous compared to the slower relaxation steps. In terms of the rate model, excitation to high vibrational levels of S_1, or higher-lying S_n states, will produce ground vibrational state S_1 species fast enough that the rate process does not have to be taken into account. Additionally, because the initially excited states relax so rapidly, stimulated emission resulting in optical saturation generally does not have to be considered. In terms of the heat produced, heat equal to the difference between the photon energy, hv, and the energy difference between the ground vibration states of S_0 and S_1 will be available as heat on a time scale much shorter than that typically required for thermal diffusion. The remainder of the energy may become available at longer times and/or may be partially lost through the emission of fluorescence photons.

S_1 species may also relax into the triplet spin manifold through intersystem crossing (ISC). The ISC is generally a function of quenching species present either in solution (intermolecular) or associated with the molecule (intramolecular). Although bimolecular relaxation is possible in high concentrated solutions (Cambron and Harris, 1993), we assume for simplicity that the intermolecular quencher concentrations do not change in time. Thus the relaxation rate is first order with a characteristic rate constant of k_{13}. Rapid triplet-state IVR produces molecules in low vibrational states of the first triplet, T_1. Lower-vibrational-state T_1 species generally do not couple back to S_1 because the S_1-T_1 energy gap is typically much larger than kT and IVR is fast in T_1. However, the T_1 state may absorb light producing vibrationally excited triplet states. As with S_n states, the excited triplet states relax to T_1 very rapidly through IC and IVR processes. Eventually, molecules in the trplit spin manifold will relax back to S_0 through either phosphorescence or inverse ISC. Triplet relaxation is modeled by a first-order rate with a rate constant, k_{30}, characterizing the combined effects of inverse ISC and phosphorescence.

In terms of the kinetic times scales needed for the model, molecules in higher triplet vibronic states will relax much more rapid than the slower S_1

and T_1 relaxations to S_0. For practical purposes, these relaxation steps are taken to be instantaneous compared to the slower processes of fluorescence and inverse ISC. Since molecules relax rapidly out of the excited triplet states, optical saturation cannot occur. In terms of the heat produced, all of the heat resulting from relaxation of optically excited triplet states will be transferred to the sample matrix on time scales less than typical thermal diffusion times. $T_1 \rightarrow S_0$ relaxation may result in rapid sample matrix heating in certain instances. But generally, triplet-state lifetimes are long enough that all of the heat is not available to the photothermal probe over the time scale for thermal diffusion.

The rate expressions for this model are given by

$$\begin{aligned} \frac{dN_0(t)}{dt} &= -\Omega N_0(t) + k_{10}N_1(t) + k_{30}N_3(t) \\ \frac{dN_1(t)}{dt} &= \Omega N_0(t) - k_{13}N_1(t) - k_{10}N_1(t) \\ \frac{dN_3(t)}{dt} &= k_{13}N_1(t) - k_{30}N_3(t) \end{aligned} \tag{22}$$

where $N_0(t)$, $N_1(t)$, and $N_3(t)$ (m^{-3}) are the time-dependent number densities of molecules in S_0, S_1, and the triplet states, respectively; $\Omega(s^{-1})$ is the rate constant for optical excitation; and the k (s^{-1}) are the relaxation rate constants. Ω is related to the singlet cross section, σ_1 (m^2), and the irradiance, E ($W\ m^{-2}$), through

$$\Omega = \frac{E\sigma_1}{h\nu} \tag{23}$$

The set of coupled differential equations is easily solved for steady-state conditions. The dynamic solution can be found when the irradiance is constant. Solution for an initial condition of no population in excited states, a conservation condition wherein the sum of the ground- and excited-electronic-state number densities is the total number density and for constant irradiance can be found in Guiliano and Hess (1967).

A simple solution may be obtained by imposing a steady-state approximation on the number density of species in S_1. This approximation is equivalent to stating that the rate at which molecules enter the S_1 state is equal to the combined rates of relaxation to S_0 and T_1. Solving for a constant optical excitation rate and a steady-state S_1 population results in the singlet and triplet number densities

$$N_0(t) = N_{\text{tot}} \frac{k_{30} + \phi_T \Omega e^{-(k_{30}+\phi_T\Omega)t}}{k_{30} + \phi_T \Omega}$$
$$N_3(t) = N_{\text{tot}} \frac{\phi_T \Omega[1 - e^{-(k_{30}+\phi_T\Omega)t}]}{k_{30} + \phi_T \Omega} \tag{24}$$

where $\phi_T = k_{13}/(k_{10} + k_{13})$ is the triplet-state quantum yield and N_{tot} is the total number density of molecules, independent of energy level. The population in ground and triplet-states evolve exponentially in time. The exponential rate constant is related to the triplet-state parameters, $k_{3C} + \phi_T \Omega$. For a given excitation rate, the limiting steady-state populations are $N_0(t) = N_{\text{tot}} k_{30}/(k_{30} + \phi_T \Omega)$ and $N_3(t) = N_{\text{tot}} \phi_T \Omega/(k_{30} + \phi_T \Omega)$. For low-irradiance low-singlet-excitation rate conditions, the triplet-state population is insignificant at all times. Sample heating will occur primarily as a consequence of optical absorption by S_0. However, when the rate of excitation exceeds the rate of T_1–S_0 relaxation, the population in S_0 becomes depleted in favor of T_1. When this occurs, the heat generated by optical absorption by molecules trapped in T_1 may become significant if the triplet has a finite absorption cross section at the excitation source wavelength.

The absorbed optical energy is found by integrating over the excitation time. A rectangular time-dependent pulse is assumed for simplicity. This approximation is valid independent of the actual temporal pulse profile, for a pulse duration, τ_p(s), is less than the triplet relaxation time, $\tau_p \ll k_{30}^{-1}$. The energy absorbed by the singlet, Q_S (J), and triplet, Q_T (J), will result in energy densities, U_S (J m^{-3}) and U_T (J m^{-3}), respectively, of

$$U_S = \sigma_1 E \int_0^{\tau_p} N_0(t)\, dt \qquad U_T = \sigma_3 E \int_0^{\tau_p} N_3(t)\, dt \tag{25}$$

The energy densities produced through singlet and triplet state absorption are

$$U_S = \sigma_1 E N_{\text{tot}} \frac{(k_{30} + \phi_T \Omega) k_{30} \tau_p + \phi_T \Omega[1 - e^{-(k_{30}+\phi_T\Omega)\tau_p}]}{(k_{30} + \phi_T \Omega)^2}$$
$$U_T = \sigma_3 E N_{\text{tot}} \frac{(k_{30} + \phi_T \Omega) \phi_T \Omega \tau_p - \phi_T \Omega[1 - e^{-(k_{30}+\phi_T\Omega)\tau_p}]}{(k_{30} + \phi_T \Omega)^2} \tag{26}$$

where σ_3 is the absorption cross section of T_1. The total absorbed energy density is the sum, $U_{tot} = U_S + U_T$. All of the optical energy absorbed by T_1 is converted to heat because of efficient radiationless relaxation of the excited electronic states. On the other hand, optical energy absorbed by S_0 may not be recovered completely as heat. Some energy results in the production of triplet-state species, and some may be lost through fluorescence. While heat lost by intersystem crossing will be recovered upon triplet-state relaxation, that lost through fluorescence does not result in sample heating. Corrections for incomplete heat conversion are obtained using the singlet-state quantum yield and a factor that accounts for the finite heat yields of the singlet, $Y_H^{(S)}$, and triplet, $Y_H^{(T)}$, states due to fluorescence and phosphorescence, where superscript S and T heat yield terms are those for S_1 relaxing through singlet and triplet states, respectively. Modification to include incomplete heat recovery yields

$$U_S = \sigma_1 E N_{tot}[(1-\phi_T)Y_H^{(S)} + \phi_T Y_H^{(T)}]\, \frac{(k_{30}+\phi_T\Omega)k_{30}\tau_p + \phi_T\Omega[1-e^{-(k_{30}+\phi_T\Omega)\tau_p}]}{(k_{30}+\phi_T\Omega)^2} \tag{27}$$

for the energy density produced by singlet-state absorption. Also, $Y_H^{(S)}$ may be larger than unity when S_1 absorption occurs prior to intersystem crossing or internal conversion.

The temperature change is related to the energy density and the heat capacity of the sample through $\delta T = U_{tot}/\rho C_P$. Optical element strengths produced by the photothermal effect are obtained by substituting the spatial excitation irradiance profile for irradiance, E, in the equations above, then evaluating the spatial derivatives at the appropriate radial offset from the beam center. The resulting expressions are complicated without approximations. In many cases the duration of the pulsed laser will be less than the T_1 relaxation time (Chartier and Bialkowski 1995). For these cases, taking the limit where k_{30} approaches zero, an approximation valid when the excitation laser pulse duration is less than τ_T, results in the deflection angle and inverse focal length expressions

$$\theta = -\left(\frac{dn}{dT}\right)\frac{2N_{tot}h\nu H_p l}{w\rho C_P} \times e^{-1/2}[\sigma_1[\phi_T Y_H^{(T)} + (1-\phi_T)Y_H^{(S)}]e^{-\sigma_1\phi_T H_p/\sqrt{e}} + \sigma_3(1-e^{-\sigma_1\phi_T H_p/\sqrt{e}})]$$

$$\frac{1}{f} = \left(\frac{dn}{dT}\right)\frac{4N_{tot}h\nu H_p l}{w^2\rho C_P} \times [\sigma_1[\phi_T Y_H^{(T)} + (1-\phi_T)Y_H^{(S)}]e^{-\sigma_1\phi_T H_p} + \sigma_3(1-e^{-\sigma_1\phi_T H_p})] \tag{28}$$

where l is the probe laser beam interaction length and H_p (photons m^{-2}) is the integrated photon irradiance given by

$$H_p = \frac{1}{h\nu}\frac{2\Phi_0\tau_p}{\pi w^2} = \frac{1}{h\nu}\frac{2Q}{\pi w^2} \tag{29}$$

for the pulsed Gaussian excitation beam with pulse energy Q (J).

For small signals, the photothermal signals are proportional to the optical element strengths. Both lens and deflection signals can be expressed in the form

$$S(H_p) = AH_p[e^{-H_p/H_T} + f_T(1 - e^{-H_p/H_T})] \tag{30}$$

where $S(H_p)$ is the integrated photon irradiance-dependent signal magnitude. A is identified with the signal magnitude that would be obtained in the absence of triplet-state trapping and absorption. H_T (photons m^{-2}) is the photon irradiance required to produce singlet-state bleaching via triplet-state trapping of molecules. f_T is the fractional amount of heat produced by T_1 absorption relative to that of S_0. For photothermal deflection, these terms are related to the photophysical, thermal, and optical parameters by

$$A = -\left(\frac{dn}{dT}\right)\frac{2N_{\text{tot}}h\nu H_p l}{w\rho C_p}\, e^{-1/2}\sigma_1[\phi_T Y_H^{(T)} + (1-\phi_T)Y_H^{(S)}]$$

$$H_T = \frac{\sqrt{e}}{\sigma_1\phi_T} \tag{31}$$

$$f_T = \frac{\sigma_3}{\sigma_1[\phi_T Y_H^{(T)} + (1-\phi_T)Y_H^{(S)}]}$$

Similar factors may be derived for the photothermal lens signal. The main difference being the lack of the $e^{1/2}$ term in H_T.

A predicted photothermal deflection signal produced as the integrated irradiance exceeds that required for optical bleaching is shown in Figure 8.6. Parameters for this plot are for erythrosin in pH 9 solution (Bowers and Porter 1967). Erythrosin has a unit triplet-state quantum and the triplet absorption cross section is about $0.25\sigma_1$ at 526 nm. Since $\phi_T = 1$, the A parameter is directly proportional to σ_1, and f_T is just the ratio of the absorption cross sections, $f_T = \sigma_3/\sigma_1$. H^T is simply $e^{1/2}/\sigma_1$. Based on the S_0 molar extinction coefficient of about 10^5 $\underline{M}^{-1}$ cm^{-1}, the absorption cross section is about 4×10^{-20} m^2. H_T is thus about 4×10^{19} photons m^{-2}, corresponding to an integrated irradiance, in energy units, of 16 kJ m^{-2}

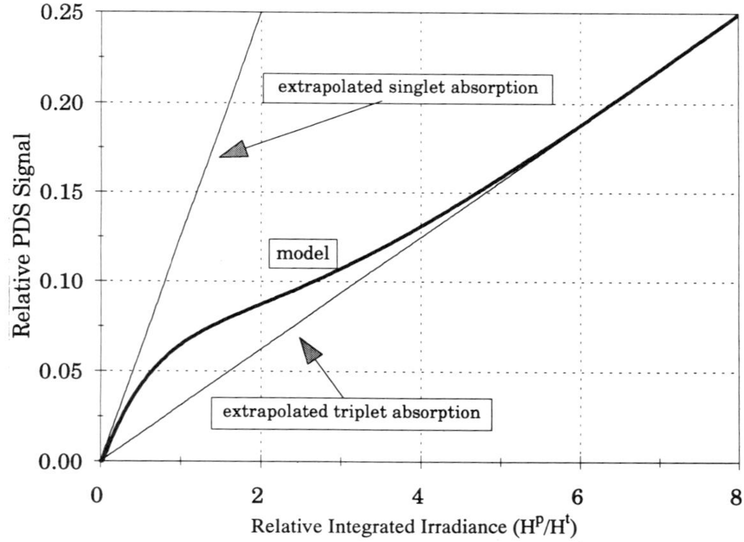

Figure 8.6. Effect of optical bleaching on the pulsed-laser-excited photothermal deflection signal calculated for erythrosin. The laser pulse duration is assumed to be less than the triplet-state lifetime. (Reprinted with permission from Chartier and Bialkowski 1995. Copyright 1995, American Chemical Society.)

with 526-nm excitation. For a laser beam focused to a modest beam radius of 8 μm, this corresponds to about 1.6 nJ of pulse energy. The focused beam radius and pulse energy are well within practical experimental parameters. One should expect to observe the effects of triplet-state trapping even with these moderate irradiation conditions. The figure shows the model calculation, the curve, and the relative signals extrapolated from the low- and high-irradiance portions of the model. The extrapolated low-irradiance line, when calibrated against a suitable standard absorber, yields a slope proportional to σ_1. The slope of the line extrapolated from the high irradiance portion of the model signal is proportional to σ_3. Significant deviation from the low-irradiance slope is apparent for relative integrated irradiances greater than about $0.5H_T$.

It is also possible to have singlet-state bleaching with continuous-wave laser excitation. Using a 40 mW average power continuous-wave Ar^+ laser, typical of the air-cooled lasers used in photothermal spectroscopy, one can expect about 10 mW on a single laser line. The irradiance produced in a 8 μm focused beam is $\sim 10^8$ W m^{-2}. For erythrosin in water, the triplet state lifetime is about 10 μs (Chartier, et al. 1990a, 1991), and the singlet state bleaching irradiance, $E_S = h\nu/\sigma_1\tau$, is about 1.6×10^6 W m^{-2}. It is apparent that bleaching may also affect continuous-laser excited photothermal signal.

This model is independent of the identity of the trap state. The trap state could equivalently be an isomer produced by excitation or a photodegradation product produced by reaction of the triplet. However, in cases where the trap state or species produces heat that is not transferred to the sample matrix in times that are short compared to thermal diffusion, the amplitude, A, and fractional absorption, f_T, terms are modified through the heat yield terms. A similar modification has to used when the triplet-state lifetime is long compared to characteristic thermal diffusion times. In this case, all of the optical energy absorbed by S_0 and coupled into the triplet system would not be recovered in the form of heat on time scales that produce a measurable signal. The singlet energy accounting term can be modified to account for this effective loss by adjusting the triplet relaxation route heat yield term, $Y_H^{(T)}$. In the limiting case where T_1 relaxation takes much longer than the characteristic thermal diffusion time the yield from intersystem crossing is proportional to the S_1-T_1 lower vibrational-state energy difference, ΔE_{T1-S1}. In this case the fractional heat yield from S_1 species relaxing via the triplet state is $Y_H^{(T)} = 1 - \Delta E_{T1-S1}/h\nu$. Thus an experimental determination of f_T may be used to deduce ΔE_{T1-S1} when the singlet and triplet absorption cross sections are known.

8.4.4. Optical Bleaching Effects in Pulsed Laser Photothermal Spectroscopy

McGraw et al. (1987) developed a similar model to describe photothermal diffraction signals produced when optical bleaching or photochemical change produces species that may subsequently absorb the excitation radiation. The model was applied to the determination of nonlinear absorption parameters for I_2 and Sudan Red dye. The determinations are based on measurements of the photothermal diffraction at a series of angles as a function of excitation irradiance. Diffraction occurs at different angles when the normally cosine photothermal grating is distorted by nonlinear absorption processes. McGraw and Harris used this technique to determine the three-photon absorption cross section for chlorobenzene (1986). Zhu and Harris applied this model for the determination of the absorption cross section for the photoisomer of DODCI (1988), the photoisomerization of *trans*-β-carotene (1989), and the optical bleaching and metastable state absorption in cyanine dyes (1990). Poston and Harris (1990) applied this model to determination of the triplet-state absorption cross section of benzophenone using photothermal deflection spectroscopy. Cambron and Harris used the model in their examinations of triplet–triplet annihilation kinetics in benzophenone using photothermal lens spectroscopy.

Several papers have reported errors associated with the use of pulsed excitation lasers, versus continuous ones, in quantitative analysis of dye molecules and for the determination of fluorescence quantum yields (Terazima et al. 1989; Chartier et al. 1990a, b, 1991; Chartier and Georges 1991; Ecker et al. 1992). Many of these studies found pulsed laser–excited photothermal signals that were different than expected for linear absorption. In particular, Terazima et al. found that pulsed N_2 laser–excited photothermal lens signals were consistently less than expected for many organic dye compounds, although some compounds produced signals that were greater than expected. In their attempts to determine the source of the error, they corrected for finite fluorescence and photoisomerization and performed extensive tests for transient absorption and multiphoton excitation. The corrections and test did not account for the absorbance error. They proposed that the calculated sample absorbance errors could be accounted for by a mechanism wherein a fraction of the absorbed optical energy is dispersed as vibrational energy in the solvent. In these experiments, the 5-μJ 337.1-nm laser pulse would produce an integrated irradiance of about 800 J m^{-2} ($H_p \sim 1.2 \times 10^{21}$) with the estimated 50-$\mu$m beam waist radius. Molecules possessing ground-state absorption cross sections greater than 8×10^{-22} m^2 ($\varepsilon \sim 2 \times 10^3$ $\underline{M}$ cm^{-1}) would exhibit a high degree of ground-state bleaching with these experimental conditions.

Chartier et al. also found that the calculated pulsed laser–excited photothermal lens signals for erythrosin, acridine orange, proflavin, and eosin dyes were much different than those measured. However, they utilized a pulsed dye laser with energies from 5 to 50 μJ, and an excitation beam waist radius of about 170 μm can be estimated from their photothermal lens decay. It is likely that significant optical bleaching was occurring during sample irradiation. Ecker et al. found that the 514-nm pulsed laser–excited photothermal lens signal for erythrosin was only about 0.24 that expected based on a Co^{2+} standard signal. They used the same optical apparatus as that of Chartier et al. The integrated irradiance was 10 to 100 times larger than that required for bleaching. Interestingly enough, the reported signal strength of ~ 0.24 is expected if bleaching was occurring and the triplet state was absorbing since $\sigma_1 \sim 0.25\sigma_3$ at this wavelength (Bowers and Porter 1967).

It is clear that pulsed lasers must be used with a degree of caution when performing analytical measurements with photothermal spectroscopy. Optical saturation of gas-phase species and optical bleaching of condensed-phase species can produce signals that are different than expected based on linear absorbing calibration standards. On the other hand, the nonlinear absorption behavior provides an opportunity to the analytical chemist. The nonlinear energy-dependent signal can be used to aid in speciation and may yield a wealth of information concerning excited-state absorption cross sections and relaxation rate constants.

REFERENCES

Barker, J. R. and Rothem, T. *Chem. Phys.* **68** 331 (1982).

Barker, J. R. and Toselli, B. M. In *Photothermal Investigations in Solids and Fluids*, Sell, J. A. ed., Academic Press, New York (1989).

Bialkowski, S. E. and He, Z.-F. *Anal. Chem.* **60** 2674 (1988).

Bialkowski, S. E. and He, Z.-F. *SPIE Proc.* **1637** 134 (1992).

Bialkowski, S. E. and He, Z.-F. *SPIE Proc.* **2138** 140 (1994).

Bialkowski, S. E. and Long, G. R. *Anal. Chem.* **59** 873 (1987).

Bialkowski, S. E.; Gu, X.; Poston, P. E.; and Powers, L. S. *Appl. Spectrosc.* **46** 1335 (1992).

Bicanic, D.; Kruger, S.; Torfs, P.; Bein, B.; and Harren, F. *Appl. Spectrosc.* **43** 148 (1989).

Bowers, P. G. and Porter, G. *Proc. Roy. Soc.* **299A** 348 (1967).

Brannon, J. H. and Madge, D. *J. Phys. Chem.* **82** 705 (1978).

Braslavsky, S. E. and Heibel, G. E. *Chem. Rev.* **92** 1381 (1993).

Buffet, C. E. and Morris, M. D. *Anal Chem.* **54** 1824 (1982).

Cambron, R. T. and Harris, J. M. *J. Phys. Chem.* **97** 13598 (1993).

Carter, C. A.; Brady J. M.; and Harris, J. M. *Appl. Spectrosc.* **36** 3 (1982).

Chartier, A. and Bialkowski, S. E. *Anal. Chem.* **67** 2672 (1995).

Chartier, A. and Georges, J. *J. Phys. IV, Coll. C7, Suppl. III* **1** 681 (1991).

Chartier, A.; Georges, J.; and Mermet, J. M. *Chem. Phys. Lett.* **171** 347 (1990a).

Chartier, A.; Georges, J.; and Mermet, J. M. *Spectrochim. Acta* **46A** 1737 (1990b).

Chartier, A.; Georges, J.; and Mermet, J. M. *Spectrochim. Acta* **47A** 1505 (1991).

Cussler, E. L. *Diffusion Mass Transfer in Fluid Systems*, Cambridge University Press, New York (1984).

Davis, C. C. *Appl. Phys. Lett.* **36** 515 (1980).

Davis, C. C. and Petuchowski, S. J. *Appl. Opt.* **20** 2539 (1981).

Dovichi, N. J. *CRC Crit. Rev. Anal. Chem.* **17** 357 (1987).

Dovichi, N. J. *Rev. Sci. Instrum.* **61** 3653 (1990).

Dovichi, N. J. and Harris, J. M. *Anal. Chem.* **51** 728 (1979).

Dovichi, N. J. and Harris, J. M. *Anal. Chem.* **52** 2338 (1980).

Dovichi, N. J. and Harris, J. M. *Anal. Chem.* **53** 689 (1981).

Ecker, A.; Georges, J.; and Mermet, J. M. *Appl. Spectrosc.* **46** 557 (1992).

Erskine, S. R. and Bobbitt, D. R. *Appl. Spectrosc.* **43** 668 (1989a).

Erskine, S. R. and Bobbitt, D. R. *Anal. Chem.* **61** 910 (1989b).

Fournier, D.; Boccara, A. C.; Amer, N. M.; and Gerlach, R. *Appl. Phys. Lett.* **37** 519 (1980).

Grabiner, F. R.; Siebert, D. R.; and Flynn, G. W. *Chem. Phys. Lett.* **17** 189 (1972).

Guiliano, C. R. and Hess, L. D. *IEEE J. Quant. Electron.* **3** 358 (1967).

Gupta, R. In *Photothermal Investigations in Solids and Fluids*, Sell, J. A., ed., Academic Press, New York (1989).

Higashi, T.; Imasaka, T.; and Ishibashi, N. *Anal. Chem.* **55** 1907 (1983).

Hineman, M. F.; Rodriguez, R. G.; and Nibler, J. W. *J. Chem. Phys.* **89** 2630 (1988).

Imasaka, T.; Higashi, T.; and Ishibashi, N. *Anal. Chim. Acta* **245** 191 (1991).

Isak, S. J.; Komorowski, S. J.; Merrow, C. N.; Poston, P. E.; and Eyring, E. M. *Appl. Spectrosc.* **43** 419 (1989).

Kettler, C. N. and Sepaniak, M. J. *Anal. Chem.* **59** 1733 (1987).

Leach, R. L. and Harris, J. M. *J. Chromatog.* **218** 15 (1981).

Long, G. R. and Bialkowski, S. E. *Anal. Chem.* **56** 2806 (1984).

Long, G. R. and Bialkowski, S. E. *Anal. Chem.* **57** 1079 (1985).

Madge, D.; Brannon, J. H.; Cremers, T. L.; and Olmsted, J. *Phys. Chem.* **83** 696 (1979).

McGraw, D. J. and Harris, J. M. *Phys. Rev. A* **34** 4829 (1986).

McGraw, D. J.; Michaelson, J.; and Harris, J. M. *J. Phys. Chem.* **86** 2536 (1987).

Miyaishi, K.; Imasaka, T.; and Ishibashi, N. *Anal. Chem.* **54** 2039 (1982).

Mori, K., Imasaka, T.; and Ishibashi, N. *Anal. Chem.* **55** 1075 (1983).

Morris, M. D. and Fotiou, F. K. In *Photothermal Investigations in Solids and Fluids*, Sell, J. A. ed., Academic Press, New York (1989).

Nickolaisen, S. L. and Bialkowski, S. E. *Anal. Chem.* **57** 758 (1985).

Nickolaisen, S. L. and Bialkowski, S. E. *Anal. Chem.* **58** 215 (1886a).

Nickolaisen, S. L. and Bialkowski, S. E. *J. Chromatogr.* **366** 127 (1986b).

Pang, T. J. and Morris, M. D. *Anal. Chem.* **57** 2153 (1985).

Poston, P. E. and Harris, J. M. *J. Amer. Chem. Soc.* **112** 644 (1990).

Poston, P. E. and Harris, J. M. *J. Photochem. Photobiol. A* **60** 51 (1991).

Rice, P. D.; Thorne, J. B.; and Bobbitt, D. R. *Proc. SPIE* **1435** 104 (1991).

Rose, A.; Vyas, R.; and Gupta, R. *Appl. Opt.* **25** 4626 (1986).

Rossbroich, G.; Garcia, N. A.; and Braslavsky, S. *J. Photochem.* **31** 37 (1985).

Sell, J. A. *Appl. Opt.* **23** 1586 (1984).

Sell, J. A. *Appl. Opt.* **24** 3725 (1985).

Sell, J. A. In *Photothermal Investigations in Solids and Fluids*, Sell, J. A., ed., Academic Press, New York (1989).

Sepaniak, M. J.; Vargo, J. D.; Detter, C. N.; and Maskarinec, M. P. *Anal. Chem.* **56** 1252 (1984).

Shen, J. and Snook, R. D. *Chem. Phys. Lett.* **155** 583 (1989a).

Shen, J. and Snook, R. D. *Analytical Procedings* **26** 27 (1989b).

Siebert, D. R.; Grabiner, F. R.; and Flynn, G. W. *J. Chem. Phys.* **60** 1564 (1974).
Sontag, H. and Tam, A. C. *Opt. Lett.* **10** 436 (1985).
Sontag, H. and Tam, A. C. *Can. J. Phys.* **64** 1121 (1986).
Terazima, M. and Azumi, T. *Chem. Phys. Lett.* **141** 237 (1987).
Terazima, M.; Horiguchi, M.; and Azumi, T. *Anal. Chem.* **61** 883 (1989).
Twarowski, A. J. and Kliger, D. S. *Chem. Phys.* **20** 253 (1977).
Vyas, R. and Gupta, R. *Appl. Opt.* **27** 4701 (1988).
Vyas, R.; Monson, B.; Nie, Y.-X.; and Gupta, R. *Appl. Opt.* **27** 3914 (1988).
Waldron, K. C. and Dovichi, N. J. *Anal. Chem.* **64** 1396 (1992).
Weimer, W. A. and Dovichi, N. J. *Anal. Chem.* **57** 2436 (1985a).
Weimer, W. A. and Dovichi, N. J. *Appl. Opt.* **24** 2981 (1985b).
Weimer, W. A. and Dovichi, N. J. *Appl. Spectrosc.* **39** 1009 (1985c).
Weimer, W. A. and Dovichi, N. J. *Anal. Chem.* **60** 662 (1988).
Yu, M. and Dovichi, N. J. *Anal. Chem.* **61** 37 (1989).
Zhu, X. R. and Harris, J. M. *Chem. Phys.* **124** 321 (1988).
Zhu, X. R. and Harris, J. M. *J. Phys. Chem.* **124** 321 (1989).
Zhu, X. R. and Harris, J. M. *Chem. Phys.* **142** 301 (1990).

CHAPTER

9

PHOTOTHERMAL SPECTROSCOPY OF HETEROGENEOUS SAMPLES

The operating principles for photothermal spectroscopy of solids and heterogeneous materials is considerably different from that of the homogeneous samples discussed up to this point. Heat will diffuse within the various layers or areas of the solid and may also be transferred to gas or liquid phases surrounding the sample. This complex thermal diffusion must be modeled to obtain a description of the resulting temperature change. Apparatus used to measure the photothermal signals is also considerably different from that used for homogeneous fluid and solid analysis. Photothermal spectroscopy can be used for both surface and bulk analysis. Transparent or semitransparent solid samples may be profiled in terms of the thermal conductivities and absorption coefficients within the respective layers. With a spatially selective excitation source (e.g., a focused laser source) a three-dimensional description of the solid material can be obtained.

The potential complexity of thermal diffusion, temperature change, and refractive index change that can occur in solids necessitates a different tactic in describing the photothermal spectroscopy signals. First, spatial anisotropy of the refractive index resulting in the photothermal signals does not necessarily come from that of the excitation source. It is more likely due to that of the sample. Second, one cannot solve thermal diffusion equations for a specific geometry. Geometry is dictated by the heterogeneous nature of the sample and the excitation source. It is better, then, to examine thermal diffusion and the resulting temperature and index changes in a general way. Instead of solving the thermal diffusion equation for a specific geometry as was done for homogeneous samples, one takes the experimental data, and knowledge gained from solutions obtained under specific geometric constraints, to infer the three dimensional structure of the material under study.

In this chapter some specific solutions to the thermal diffusion equation applicable to heterogeneous sample analysis are given. Ways to perform the measurements and interpret the data are subsequently explored for samples with well-defined geometries. The chapter ends up with a short survey of the analytical applications of the photothermal techniques used for heterogeneous sample analysis.

9.1. TYPES OF HETEROGENEITY

Several types of heterogeneity should be considered when analyzing solid samples. Consider any solid material in contact with a gas or liquid medium. The gas or liquid will couple energy out of the solid heated with optical radiation. The solid material absorbs optical radiation, heats up, and subsequently transfers heat within itself and to the fluid coupling medium. The solid may be internally heterogeneous in optical absorbance, thermal conductivity, or both. The surface may be flat or can be irregular, adding to the geometrical heterogeneity. The thickness of the sample may vary, thereby causing spatially heterogeneous thermal diffusion. Crystals or other features within the solid may vary in their thermal contact, adding to the heterogeneity of thermal diffusion. The material may scatter the excitation radiation differently and the scattering may also be spatially variant. Considering all the different ways that the absorption of light and subsequent diffusion of heat can vary within a sample, it would seem that a description of the photothermal signals expected for a heterogeneous solid sample would be impossible. However, the effects that these heterogeneity features have on the photothermal signal can be modeled, if only individually.

Perhaps the most compelling reason to overcome the various problems associated with interpretation of photothermal signals resulting from these heterogeneities is that the technique not only gives a quantitative estimate of the heterogeneously distributed optical absorption coefficient but may simultaneously yield quantitative estimates of thermal conductivity distributions. The latter can be used to evaluate the material and the material dimensions even if the sample is opaque. Images of the thermal properties of a heterogeneous sample may be used to locate faults in the material or faults in the manufacturing process. In practice, the optical absorption and/or thermal conductivity of solids measured with photothermal spectroscopy are not strictly quantitative. As with using photothermal spectroscopy to measure analyte concentration, the experimental apparatus must be calibrated in order to go from measurements to accurate data. This task is difficult enough for homogeneous-sample measurements. It is possible for heterogeneous samples only with control over the sample thermal properties and elaborate mathematical treatment of the data. In the first part of this chapter we address the theory for the temperature change and photothermal deflection spectroscopy signal production.

9.2. APPARATUS FOR PHOTOTHERMAL DEFLECTION

Although photothermal lens spectroscopy can be used for transparent solid samples, it is more common to use photothermal deflection and sometimes photothermal diffraction spectroscopy in the analysis of solid materials. Basic apparatus used for photothermal deflection spectroscopy is shown in Figure 9.1. Either a laser or a more conventional light source can be used to excite the sample. The spatial anisotropy required to produce a gradient in the refractive index is produced by the interface between the solid sample and the coupling fluid. The solid sample heated by the optical source increases in temperature. Heat is subsequently transferred into the coupling fluid. A temperature gradient, and subsequently a refractive index gradient, form in the coupling fluid. The probe laser beam passing near and tangent to the sample surface experiences the refractive index gradient and the beam is subsequently deflected.

Probe beam deflection due to optical sample heating is detected by any one of the several means discussed previously. The distance over which the probe laser beam interacts with the index gradient is limited to either the size of the sample or the area illuminated by the excitation source. The excitation source is normally chopped or pulsed to yield information regarding the thermal properties of the sample. Thermal imaging is performed by raster scanning the excitation and probe beam over the surface of the sample. An understanding of how the raw deflection signal is

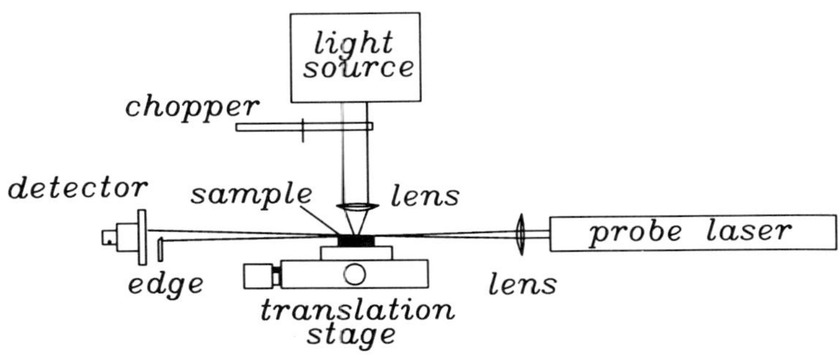

Figure 9.1. Photothermal deflection apparatus showing the solid sample, in coupling fluid, probe laser, and sample excitation source.

interpreted requires some knowledge of how heat is transferred in the heterogeneous samples. Several ideal cases are shown below.

9.3. SURFACE ABSORPTION

9.3.1. Thermal Diffusion at Surfaces

A common thermal diffusion problem encountered in the photothermal spectroscopy of heterogeneous samples is that which arises from optical excitation of a layer suspended between two dissimilar materials. A thermal discontinuity can arise between the two materials. The absorbing layer could be between two immiscible liquids or at the boundary between a liquid and a solid. As an example, consider a flat, transparent, solid sample support in contact with a gas or liquid thermal coupling fluid, with a thin absorbing layer on the surface. Optical absorption occurs only at the surface. The solid has a specific heat of $C_{P,s}$ and a density of ρ_s, while the coupling fluid has a specific heat of $C_{P,f}$ and a density of ρ_f. The optical absorption coefficient is given by κ, the unitless surface absorbance. The absorbing surface is assumed to be in good thermal contact with both the solid and the fluid, and to make a negligible contribution to the thermodynamic properties of the sample. The sample is assumed to be at thermal equilibrium prior to irradiation.

The impulse response of the temperature change for this sample can be found for both the solid and the fluid. Impulse excitation of the surface will produce heat equal to $q_H(0,0) = H\kappa\delta(z)$, where H (J m^{-2}) is the radiant exposure or integrated irradiance and $\delta(z)$ is a delta function describing the spatial characteristic of the absorbed energy. This heat is distributed into both the solid and the fluid:

$$H\kappa = \delta H_s + \delta H_f \tag{1}$$

where the two energies correspond to those in the solid and the fluid, respectively. These may exist in any relative proportion. Thermal diffusion will occur into both the solid and the coupling fluid. Equations describing thermal diffusion into the solid and the fluid are of a form given for the one-dimensional diffusion point source. The thermal diffusion solutions for energy partitioning are simply the two one-dimensional diffusion equation solutions integrated over the delta function time and space source function

$$\begin{aligned} \delta T_s(z, t) &= \delta H_s \, \delta T_{\text{point},s}(z, t) \\ \delta T_f(z, t) &= \delta H_f \, \delta T_{\text{point},f}(z, t) \end{aligned} \tag{2}$$

A distinction between the solid and fluid point sources exists because of the different thermal diffusion coefficients. The solid and the fluid are in contact at thermal equilibrium. The condition for thermal equilibrium is that the boundary temperatures are equal at all times,

$$\delta T_s(0, t) = \delta T_f(0, t) \tag{3}$$

at the surface ($z = 0$). Equating the two diffusion solutions results in the relationship

$$\frac{\delta H_s}{\rho_s C_{P,s} D_{T,s}^{1/2}} = \frac{\delta H_f}{\rho_f C_{P,f} D_{T,f}^{1/2}} \tag{4}$$

This equation shows that the energy is partitioned between the two phases according to the heat capacity and thermal diffusion coefficients. This makes sense considering the fact that the heat flux away from the interface is proportional to both the temperature gradient and the thermal conductivity. Heat partitioning can be found by combining the isothermal relationship with the energy conservation

$$\begin{aligned} \delta H_s &= \kappa H \frac{\rho_s C_{P,s} D_{T,s}^{1/2}}{\rho_s C_{P,s} D_{T,s}^{1/2} + \rho_f C_{P,f} D_{T,f}^{1/2}} \\ \delta H_f &= \kappa H \frac{\rho_f C_{P,f} D_{T,f}^{1/2}}{\rho_s C_{P,s} D_{T,s}^{1/2} + \rho_f C_{P,f} D_{T,f}^{1/2}} \end{aligned} \tag{5}$$

9.3.2. Temperature Change from Pulsed Excitation

Equations describing the time-dependent temperature in the solid and fluid layers are obtained using the impulse-response solution for one-dimensional thermal diffusion as discussed in Chapter 3. When combined with the instantaneous heat partitioning, the temperature changes for surface absorption are

$$\begin{aligned} \delta T_s(z, t) &= \frac{\kappa H}{\rho_s C_{P,s} D_{T,s}^{1/2} + \rho_f C_{P,f} D_{T,f}^{1/2}} \frac{1}{(\pi t)^{1/2}} e^{-z^2/4D_{T,s}t} \\ \delta T_f(z, t) &= \frac{\kappa H}{\rho_s C_{P,s} D_{T,s}^{1/2} + \rho_f C_{P,f} D_{T,f}^{1/2}} \frac{1}{(\pi t)^{1/2}} e^{-z^2/4D_{T,f}t} \end{aligned} \tag{6}$$

(Figure 9.2). These equations show that thermal diffusion at an interface is characterized by (1) a $t^{-1/2}$ time-dependent temperature at the surface, (2)

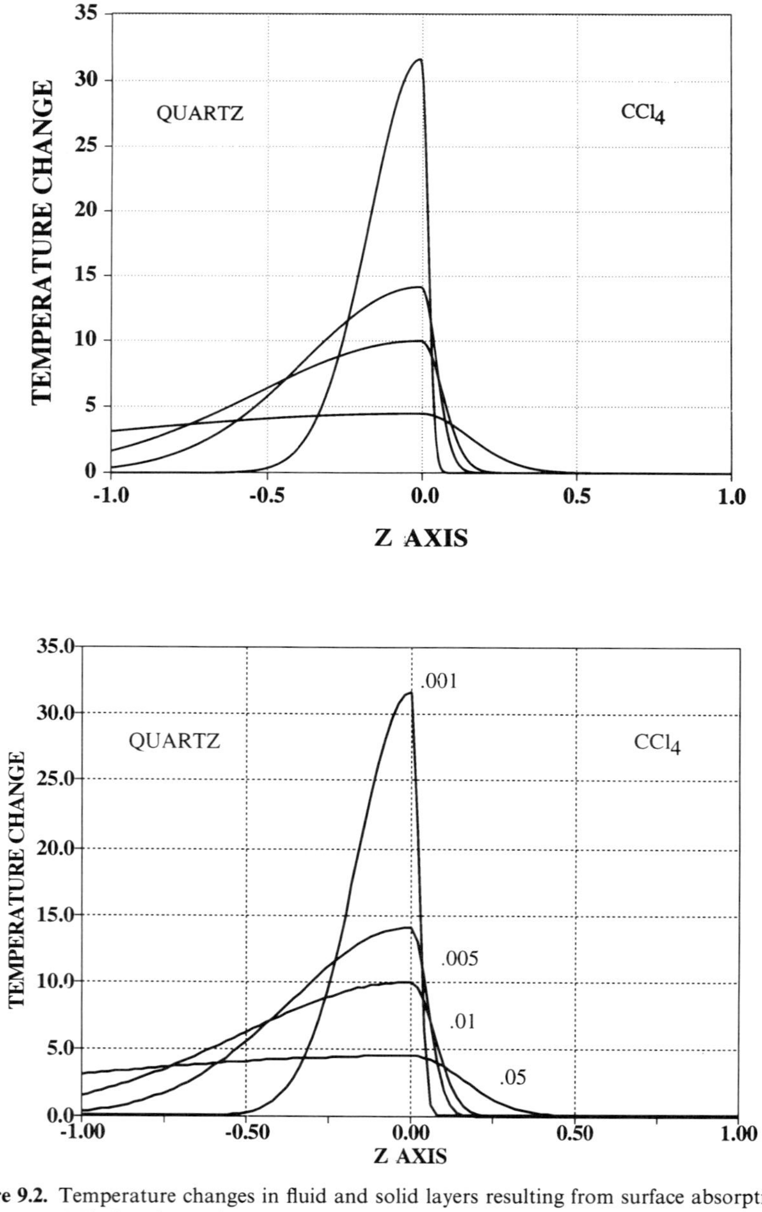

Figure 9.2. Temperature changes in fluid and solid layers resulting from surface absorption at the quartz–CCl_4 interface. Time and space are in units relative to D_{Tf}. Curves are shown for times of 0.01, 0.005, 0.001, and 0.05 $D_{T,f}/z^2$.

a surface temperature decrease and initial energy partitioning which are both proportional to the heat capacities and thermal diffusion coefficients of both the solid and the fluid, (3) a Gaussian-like spatially dependent diffusion term which is characteristic of one-dimensional diffusion, and (4) a diffusion length in a particular medium given by $l_D = (4D_T t)^{1/2}$, where D_T represents either the solid or the fluid. Photothermal spectroscopy usually monitors the temperature-dependent refractive index in the fluid phase. The impulse response of this sample is thus

$$\delta T_{\text{impulse},f}(z, t) = \frac{\kappa}{\rho_s C_{P,s} D_{T,s}^{1/2} + \rho_f C_{P,f} D_{T,f}^{1/2}} \frac{1}{(\pi t)^{1/2}} e^{-z^2/4D_{T,f}t} \tag{7}$$

9.3.3. Temperature Change from Continuous Excitation

For continuous sources, the temperature in the fluid is the convolution of the time-dependent irradiance with the impulse response

$$\delta T_f(z, t) = E(t) * \delta T_{\text{impulse},f}(z, t) \tag{8}$$

If irradiance is constant, the temperature change is

$$\delta T_{f,\text{cw}}(z, t) = \frac{\kappa E_0}{\rho_s C_{P,s} D_{T,s}^{1/2} + \rho_f C_{P,f} D_{T,f}^{1/2}} \left\{ \left(\frac{t}{\pi}\right)^{1/2} e^{-z^2/4D_{T,f}t} - \frac{z}{D_{T,f}^{1/2}} \operatorname{erfc}\left[\frac{z}{2(D_{T,f}t)^{1/2}}\right] \right\} \tag{9}$$

Here the subscript cw is used to indicate continuous-wave excitation and erfc is the usual error function integral,

$$\operatorname{erfc}(x) = \frac{2}{\pi^{1/2}} \int_x^{\infty} e^{-x'^2} dx' \tag{10}$$

The temperature increases continuously with time as in the case of the Gaussian source in a homogeneous sample. This can be seen by noting that at $z = 0$,

$$\delta T_{f,\text{cw}}(0, t) = \frac{\kappa E_0}{\rho_s C_{P,s} D_{T,s}^{1/2} + \rho_f C_{P,f} D_{T,f}^{1/2}} \left(\frac{t}{\pi}\right)^{1/2} \tag{11}$$

9.3.4. Temperature Change from Periodic Excitation

The solution for periodic excitation can be obtained from the temperature impulse-response function. For periodic excitation at frequency, ω, the response of the temperature perturbation is obtained from the convolution

$$\delta T_{\mathrm{osc},f}(x, t) = E_0 \int_0^\infty \delta T_{\mathrm{impulse},f}(x, t') e^{i\omega(t-t')}\, dt' \tag{12}$$

where E_0 ($\mathrm{W\,m^{-2}}$) is the peak irradiance of the oscillating source. This integral is similar to the Laplace transform,

$$\mathscr{L}\{\delta T_{\mathrm{impulse},f}(x, t)\} = \int_0^\infty \delta T_{\mathrm{impulse},f}(x, t) e^{-\zeta t}\, d\zeta \tag{13}$$

Subsequently, the relationship

$$\delta T_{\mathrm{osc},f}(x, t) = E_0 e^{i\omega t} \mathscr{L}\{\delta T_{\mathrm{impulse},f}(x, t)\}|_{\zeta = i\omega} \tag{14}$$

may be used to find the response for an oscillating excitation source. Laplace transform pairs are tabulated in many books. For the case at hand,

$$\delta T_{\mathrm{osc},f}(z, t) = \frac{\kappa E_0}{\rho_s C_{P,s} D_{T,s}^{1/2} + \rho_f C_{P,f} D_{T,f}^{1/2}} \left(\frac{2}{\omega}\right)^{1/2} \frac{1}{i+1}\, e^{-z(i+1)(\omega/2D_{T,f})^{1/2} + i\omega t} \tag{15}$$

This result is the steady-state periodic temperature change that is obtained since the integration limits on the convolution extend to infinite time. The exponential oscillation term does not evolve in time. The result shows that the magnitude of the temperature change will reach a steady-state value and that this steady-state magnitude decreases with increasing frequency.

It is now easy to calculate the temperature change that would be observed with real periodic excitation with irradiance

$$E(t) = E_0(1 + e^{i\omega t}) \tag{16}$$

The time-dependent temperature change is simply the sum of the continuous and the oscillating temperature changes:

$$\delta T_f(z, t) = \delta T_{f,\mathrm{cw}}(z, t) + \delta T_{f,\mathrm{osc}}(z, t) \tag{17}$$

The temperature change is thus the sum of a component that continues to increase the temperature of the sample and an oscillating component that reaches a steady-state value.

The periodic temperature change magnitude is

$$|\delta T_f(z,\omega)| = \frac{1}{2^{3/2}} \frac{\kappa E_0}{\rho_s C_{P,s} D_{T,s}^{1/2} + \rho_f C_{P,f} D_{T,f}^{1/2}} \frac{\mu_f}{D_{T,f}^{1/2}} e^{-z/\mu_f} \tag{18}$$

where $|\cdot|$ is used to indicate the magnitude, and the thermal diffusion length, μ_i (m), has been used. This parameter is defined as (Aamodt and Murphy 1981, 1983)

$$\mu_i = \sqrt{\frac{2D_{T,i}}{\omega}} \tag{19}$$

The diffusion length is the characteristic distance that a thermal wave at angular frequency ω propagates into the coupling fluid or, similarly, into the solid. It is apparent that the periodic temperature change is directly proportional to the absorption coefficient and that it decreases exponentially with distance from the surface. The exponential distance constant is the diffusion length. This in turn is a function of the excitation oscillation frequency. The diffusion distance increases with decreasing frequency. While the distance-dependent temperature-change magnitude is easily interpreted in theory, in practice it may be difficult to determine the absorption coefficient based on signal magnitude since it depends on the distance.

In addition to the signal magnitude, the phase angle may be analyzed. The tangent phase angle is found from

$$\tan\theta = \frac{\Re\{e^{i\omega t}\delta T_{\text{osc},f}(z,t)\}}{\Im\{e^{i\omega t}\delta T_{\text{osc},f}(z,t)\}} \tag{20}$$

which is the real part of the transforms product divided by the imaginary part. The phase angle for periodic excitation is

$$\theta = \frac{\pi}{4} - \frac{z}{\mu_f} \tag{21}$$

The temperature-change phase differs from the excitation by $\pi/4$ at the surface due to the energy-integrating nature of the absorption process. The phase angle is a function of distance from the surface sample and the diffusion length but does not depend on the absorption coefficient. The distance may be found from the phase-sensitive temperature measurements.

9.4. THERMAL DIFFUSION IN VOLUME-ABSORBING SAMPLES

Another common example of a one-dimensional thermal diffusion problem is volume absorption accompanied by heat transfer to an adjacent layer in good thermal contact with the sample. This situation arises in the heat transfer from weakly absorbing sample cell windows, in an optically thin solid sample in contact with a gas or liquid thermal coupling fluid, and in an absorbing liquid in contact with the air above. In all these cases the system is composed of a sample, with thermal and optical properties noted by the subscript s, and a liquid or gas coupling fluid, with properties noted by the subscript f. Excitation of the sample results in heating of the sample volume followed by heat transfer to the fluid. The coupling fluid is assumed to have a negligible absorbance. The temperature change of the fluid can be detected with photothermal methods.

9.4.1. Volume Temperature Change for Pulsed Excitation

The differential equations are first solved for an impulse excitation. Other time-dependent excitations can be obtained by convolution of the impulse response with the time dependence of the excitation source. Impulse excitation of an optically thin sample will produce heat equal to $q_H(0, z) = H\alpha_s$, where α_s is the sample absorption coefficient. The differential equations used to model this case are

$$\begin{aligned} \frac{\partial \delta T_f(z, t)}{\partial t} - D_{T,f} \frac{\partial^2 \delta T_f(z, t)}{\partial z^2} &= 0 \\ \frac{\partial \delta T_s(z, t)}{\partial t} - D_{T,s} \frac{\partial^2 \delta T_s(z, t)}{\partial z^2} &= \frac{\alpha H \delta(t)}{\rho_s C_{p,s}} \end{aligned} \tag{22}$$

The solution to this thermal diffusion problem is given by (Carslaw and Jaeger 1986)

$$\begin{aligned} \delta T_f(z, t) &= \alpha_s H \frac{D_{T,s}^{1/2}}{\rho_s C_{P,s} D_{T,s}^{1/2} + \rho_f C_{P,f} D_{T,f}^{1/2}} \operatorname{erfc}\left[\frac{z}{2(D_{T,f}t)^{1/2}}\right] \\ \delta T_s(z, t) &= \alpha_s H \frac{D_{T,s}^{1/2}}{\rho_s C_{P,s} D_{T,s}^{1/2} + \rho_f C_{P,f} D_{T,f}^{1/2}} \\ &\quad \times \left\{1 + \frac{\rho_f C_{P,f} D_{T,f}^{1/2}}{\rho_s C_{P,s} D_{T,s}^{1/2}} \operatorname{erf}\left[\frac{z}{2(D_{T,s}t)^{1/2}}\right]\right\} \end{aligned} \tag{23}$$

The impulse response of the sample is the temperature observed in the fluid for unit H excitation

$$\delta T_{\text{impulse},f}(z, t) = \frac{\alpha_s D_{T,s}^{1/2}}{\rho_s C_{P,s} D_{T,s}^{1/2} + \rho_f C_{P,f} D_{T,f}^{1/2}} \operatorname{erfc}\left[\frac{z}{2(D_{T,f}t)^{1/2}}\right] \tag{24}$$

This result is substantially different from that for surface absorption. By analyzing the impulse response of heat generation in the fluid, one can determine the location of the optical absorption. An exponential inverse time response corresponds to a surface absorption, while an error function response occurs for volume-absorbing samples.

9.4.2. Periodic Excitation of Volume Absorbers

The signal produced by an oscillating excitation source is

$$\begin{aligned}\delta T_f(z, t) &= E_0 \int_0^t [1 + e^{i\omega(t-t')}]\, \delta T_{\text{impulse},f}(z, t')\, dt' \\ &= \delta T_{\text{cw},f}(z, t) + \delta T_{\text{osc},f}(z, t)\end{aligned} \tag{25}$$

The continuous part of the temperature change must increase with time since the $\delta T_{\text{impulse},f}$ impulse increases. The oscillating part of the temperature change is found from Laplace transform tables. The oscillating temperature change is

$$\delta T_{\text{osc},f}(z, t) = \frac{\alpha_s E_0 D_{T,s}^{1/2} e^{i\omega t}}{\rho_s C_{P,s} D_{T,s}^{1/2} + \rho_f C_{P,f} D_{T,f}^{1/2}} \frac{-1}{i\omega} e^{-z(i+1)/\mu_f} \tag{26}$$

The magnitude of this periodic temperature change is

$$|\delta T_{\text{osc},f}(z, \omega)| = \frac{\alpha_s E_0 D_{T,s}^{1/2}}{\rho_s C_{P,s} D_{T,s}^{1/2} + \rho_f C_{P,f} D_{T,f}^{1/2}} \frac{1}{\omega} e^{-z/\mu_f} \tag{27}$$

and the phase angle is

$$\phi = \frac{\pi}{2} - \frac{z}{\mu_f} \tag{28}$$

As with the surface absorbing sample, the magnitude of the periodic temperature change is a linear function of the small absorption coefficient and varies exponentially with distance from the sample–coupling fluid interface. Differences in the frequency-dependent phase angle between this bulk and the surface absorption temperature changes may serve as a means to discriminate where the solid sample is absorbing (Nie and Bertrand 1989). There is a $\pi/2$ phase shift at $z = 0$, at the interface between the absorbing and fluid layers. This result is in contrast with that for the surface absorbing sample that exhibited a phase shift of $\pi/4$ at the surface. Both the impulse response and the phase angle change, depending on where the optical absorption takes place. However, the impulse response is probably more easily interpreted since the absolute phase angle requires knowledge of the absolute distance. Thus a combination of impulse and periodic excitation may be used to characterize the location of the absorption and to quantitate the absorption coefficient.

9.5. TEMPERATURE CHANGE IN LAYERED SAMPLES

Photothermal spectroscopy of surface and bulk absorbing samples can be used to quantitate the absorbance of the solid sample. Another interesting sample type is that of the layered sample. The layered sample will in general have a different photothermal signal magnitude because of the different heat capacities and thermal diffusion coefficients of the layers. There are many types of layered samples. For example, a painted surface, an oxide or chemically treated layer on substrate, or film on a liquid surface are all examples of layered samples.

A common model used to predict photothermal temperature changes in these types of samples is one where the absorbing sample is in thermal contact with a transparent backing material (Figure 9.3). This sample then

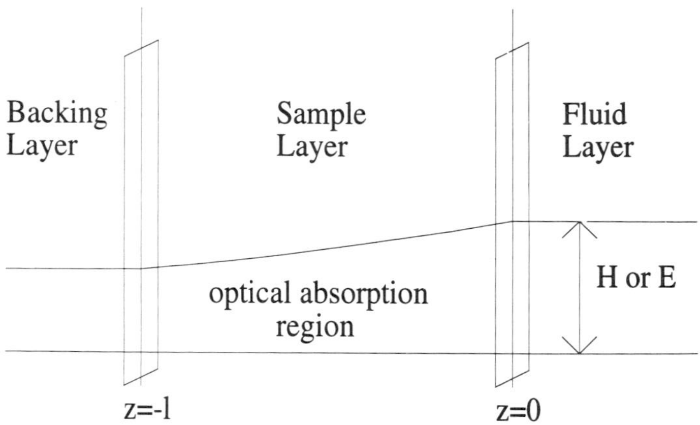

Figure 9.3. Three-layer sample.

has two solid or liquid layers, sample and backing. The coupling media must also be accounted for in the equations that describe thermal diffusion. The coupling media could be the air above the sample or a fluid in which a solid sample is submerged. The model is commonly used in material analysis. Solutions are discussed by Aamodt and Murphy (1981, 1983), Jackson et al. (1981), and Fournier and Boccara (1989). The procedures used for the surface and volume absorbing samples can be used here, too. The differential equations are solved for impulse excitation and the impulse solution is converted to frequency using a Laplace transform.

The three differential equations describing the layered sample are

$$\begin{aligned} &\frac{\partial \delta T_f(z,t)}{\partial t} - D_{T,f}\frac{\partial^2 \delta T_f(z,t)}{\partial z^2} = 0 && z \geqslant 0 \\ &\frac{\partial \delta T_s(z,t)}{\partial t} - D_{T,s}\frac{\partial^2 \delta T_s(z,t)}{\partial z^2} = \frac{\alpha H \delta(t) e^{\alpha z}}{\rho_s C_{p,s}} && -l \leqslant z \leqslant 0 \\ &\frac{\partial \delta T_b(z,t)}{\partial t} - D_{T,b}\frac{\partial^2 \delta T_b(z,t)}{\partial z^2} = 0 && z \leqslant -l \end{aligned} \tag{29}$$

where the subscripts f, s, and b correspond to the fluid, sample, and backing. The interface between the sample and the fluid is taken as the origin of the coordinate system. The fluid layer is of semi-infinite extent, $z \geqslant 0$, the sample extent is $-l \leqslant z \leqslant 0$, and the backing is also semi-infinite, with $z \leqslant -l$. The semi-infinite conditions will hold as long as the real fluid and backing dimensions are longer than the thermal diffusion length. The differential equations are solved with the boundary conditions (Carslaw and Jaeger 1986)

$$\begin{aligned} \delta T_f(0,t) &= \delta T_s(0,t) & \delta T_s(-l,t) &= \delta T_b(-l,t) \\ \delta T_f(\infty,t) &= 0 & \delta T_b(-\infty,t) &= 0 \\ \kappa_f \frac{\partial T_f(0,t)}{\partial z} &= \kappa_s \frac{\partial T_s(0,t)}{\partial z} & \kappa_s \frac{\partial T_s(-l,t)}{\partial z} &= \kappa_b \frac{\partial T_b(-l,t)}{\partial z} \end{aligned} \tag{30}$$

The κ_i (W m^{-1}K^{-1}) are thermal conductivities of medium i and are related to the thermal diffusion coefficient in the usual fashion, $\kappa_i = D_{T,i}\rho_i C_{P,i}$. The first equations are temperature balance conditions. These state that the temperatures on both sides of an interface are equal. The second set are the energy flux balance conditions. Energy flux must balance across an interface (i.e., energy is not lost at the interface). These boundary conditions are required at all interfaces, independent of how many interfaces there might happen to be.

The solution is found by first taking the Laplace transform of time,

$$
\begin{aligned}
&\zeta\delta T_f(z,\zeta) - D_{T,f}\frac{\partial^2 \delta T_f(z,\zeta)}{\partial z^2} = 0 && z \geqslant 0 \\
&\zeta\delta T_s(z,\zeta) - D_{T,s}\frac{\partial^2 \delta T_s(z,\zeta)}{\partial z^2} = \frac{\alpha H e^{\alpha z}}{\rho_s C_{P,s}} && -l \leqslant z \leqslant 0 \qquad (31)\\
&\zeta\delta T_b(z,\zeta) - D_{T,b}\frac{\partial^2 \delta T_b(z,\zeta)}{\partial z^2} = 0 && z \leqslant -l
\end{aligned}
$$

and then finding the coefficients of the general solution,

$$
\begin{aligned}
\delta T_f(z,\zeta) &= F_1(\zeta)e^{-z(\zeta/D_{T,f})^{1/2}} + F_2(\zeta)e^{z(\zeta/D_{T,f})^{1/2}} \\
\delta T_s(z,\zeta) &= S_1(\zeta)e^{-z(\zeta/D_{T,s})^{1/2}} + S_2(\zeta)e^{z(\zeta/D_{T,s})^{1/2}} + \frac{\alpha H e^{\alpha z}}{(\zeta/D_{T,s}-\alpha^2)\rho_s D_{T,s}} \qquad (32)\\
\delta T_b(z,\zeta) &= B_1(\zeta)e^{-z(\zeta/D_{T,b})^{1/2}} + B_2(\zeta)e^{z(\zeta/D_{T,b})^{1/2}}
\end{aligned}
$$

based on the boundary conditions. From the semi-infinite boundary conditions, the temperature change is zero at $\pm\infty$, and F_2 and B_1 must be zero. The remainder can be found through a tedious calculation which is aided by the use of a symbolic language processor. The solutions are

$$
\begin{aligned}
F_1(\zeta) &= \frac{-\alpha H}{\kappa_s(\zeta/D_{T,s}-\alpha^2)} \\
&\quad\times \frac{2(r_b-r_\alpha)e^{-\alpha l} + (r_\alpha-1)(r_b+1)e^{l(\zeta/D_{T,s})^{1/2}} - (r_\alpha+1)(r_b-1)e^{-l(\zeta/D_{T,s})^{1/2}}}{(r_b+1)(r_f+1)e^{l(\zeta/D_{T,s})^{1/2}} - (r_b-1)(r_f-1)e^{-l(\zeta/D_{T,s})^{1/2}}} \\
S_1(\zeta) &= \frac{-\alpha H}{\kappa_s(\zeta/D_{T,s}-\alpha^2)} \\
&\quad\times \frac{-(r_\alpha+r_f)(r_b-1)e^{-l(\zeta/D_{T,s})^{1/2}} + (r_b-r_\alpha)(r_f+1)e^{-\alpha l}}{(r_b+1)(r_f+1)e^{l(\zeta/D_{T,s})^{1/2}} - (r_b-1)(r_f-1)e^{-l(\zeta/D_{T,s})^{1/2}}} \qquad (33)\\
S_2(\zeta) &= \frac{-\alpha H}{\kappa_s(\zeta/D_{T,s}-\alpha^2)} \\
&\quad \frac{(r_\alpha+r_f)(r_b+1)e^{l(\zeta/D_{T,s})^{1/2}} - (r_b-r_\alpha)(r_f-1)e^{-\alpha l}}{(r_b+1)(r_f+1)e^{l(\zeta/D_{T,s})^{1/2}} - (r_b-1)(r_f-1)e^{-l(\zeta/D_{T,s})^{1/2}}} \\
B_2(\zeta) &= \frac{\alpha H e^{-\alpha l + l(\zeta/D_{T,b})^{1/2}}}{\kappa_s(\zeta/D_{T,s}-\alpha^2)} \\
&\quad\times \frac{(r_\alpha+1)(r_f+1)e^{l(\zeta/D_{T,s})^{1/2}} - (r_\alpha-1)(r_f-1)e^{-l(\zeta/D_{T,s})^{1/2}}}{(r_b+1)(r_f+1)e^{l(\zeta/D_{T,s})^{1/2}} - (r_b-1)(r_f-1)e^{-l(\zeta/D_{T,s})^{1/2}}}
\end{aligned}
$$

where the r_i are unitless ratios:

$$r_\alpha = \alpha\sqrt{\frac{D_{T,s}}{\zeta}} \qquad r_b = \frac{\kappa_b}{\kappa_s}\sqrt{\frac{D_{T,s}}{D_{T,b}}} \qquad r_f = \frac{\kappa_f}{\kappa_s}\sqrt{\frac{D_{T,s}}{D_{T,f}}} \tag{34}$$

The temperature change is normally measured in the coupling fluid. It is noteworthy that the temperature in the fluid layer is related to that at the surface since $F_1(\zeta) = \delta T_s(0, \zeta)$.

9.5.1. Periodic Excitation of Layered Samples

The complexity of the result result does not lend itself to determination of the impulse response. However, the time-dependent solution for oscillating excitation can be obtained from the inverse Laplace transform. In this case one need only replace the Laplace transform variable, ζ, with $i\omega$ and multiply by $\exp(i\omega t)$. The coefficient for the coupling fluid layer is

$$F_1(t) = \frac{\alpha E e^{i\omega t}}{\kappa_s(\sigma_s^2 - \alpha^2)} \frac{2(r_b - r_\alpha)e^{-\alpha l} + (r_\alpha - 1)(r_b + 1)e^{l\sigma_s} - (r_\alpha + 1)(r_b - 1)e^{-l\sigma_s}}{(r_b + 1)(r_f + 1)e^{l\sigma_s} - (r_b - 1)(r_f - 1)e^{-l\sigma_s}} \tag{35}$$

where

$$\sigma_s = \sqrt{\frac{i\omega}{D_{T,s}}} = \frac{i+1}{\mu_s} \tag{36}$$

and $\mu_s = (2D_{T,s}/\omega)^{1/2}$ is often called the thermal diffusion length. It is the distance through which a thermal wave at frequency ω will penetrate the solid absorbing phase of the sample. The time-dependent oscillating temperature in the fluid is obtained by combining the boundary condition factor with the exponential term

$$\delta T_{\text{osc},f}(z, t) = \frac{\alpha E e^{-(z\sigma_f - i\omega t)}}{\kappa_s(\sigma_s^2 - \alpha^2)} \times \frac{2(r_b - r_\alpha)e^{-\alpha l} + (r_\alpha - 1)(r_b + 1)e^{l\sigma_s} - (r_\alpha + 1)(r_b - 1)e^{-l\sigma_s}}{(r_b + 1)(r_f + 1)e^{l\sigma_s} - (r_b - 1)(r_f - 1)e^{-l\sigma_s}} \tag{37}$$

where

$$r_\alpha = \frac{\alpha}{\sigma_s} \qquad \sigma_f = \sqrt{\frac{i\omega}{D_{T,f}}} = \frac{i+1}{\mu_f} \tag{38}$$

Like μ_s, μ_f is the distance that a thermal wave of frequency ω can penetrate into the fluid layer of the sample. The larger the thermal diffusion coefficient, the greater will be the distance that wave can travel before being significantly attenuated. For a real, modulated light source, there will also be a temperature change due to the continuous part of the excitation.

Because of the complex nature of σ, each of the exponential terms with σ in the argument is a product of real, $\exp(\pm l/\mu)$, and complex $\exp(\pm il/\mu)$, components. The real component reflects the influence that the sample thickness, relative to the thermal diffusion length, has on the magnitude of the temperature change. The complex term is transcendental. The four transcendental term together with the r_α influence the relative phase of the thermal wave as a function of absorbing layer thickness. The temperature in the fluid layer is also influenced by the real and complex components relative to z through $\exp(-z/\mu_f)$ and $\exp(-iz/\mu_f)$ terms. The magnitude of the thermal wave decreases exponentially with distance away from the surface and with increasing excitation frequency. And since μ_f is proportional to $D_{T,f}^{1/2}$, the thermal wave extends further into the fluid layer for fluids with larger thermal diffusion coefficients.

The foregoing expression of the fluid layer temperature change can be used to determine the absorption coefficient of the solid layer as well as thicknesses of the solid and backing layers and relative thermal diffusion coefficients. This type of analysis is performed by varying the excitation frequency while monitoring the periodic temperature change, or as with photothermal deflection spectroscopy, the periodic temperature gradient, in the fluid layer. Data collection is facilitated by using phase-sensitive lock-in amplifier signal processing. Essentially, two sets of data are produced by monitoring both the phase and the magnitude of the resulting photothermal signal as a function of excitation frequency. Data analysis is generally performed by comparing these data to model predictions.

Although the result above is exact, it is difficult to interpret due to its complexity. Fournier and Boccara (1989) examine an interesting limit to this result. In cases where the backing can be neglected—either a sample without backing or one where the thermal diffusion length, μ_s, is much smaller than the sample thickness, l—the temperature change in the fluid reduces to

$$\delta T_{\text{osc},f}(z, t) = \alpha E \frac{D_{T,s}^{1/2}}{\rho_f C_{P,f} D_{T,f}^{1/2} + \rho_s C_{P,s} D_{T,s}^{1/2}} \frac{e^{-z(\sigma_f - i\omega t)}}{\omega(\alpha/\sigma_s + 1)} \tag{39}$$

For small α, or high frequencies and thus small μ_s, $\alpha/\sigma_s \ll 1$ and the result converges to that of the absorbing volume in contact with a fluid. For large

α or low frequencies, $\alpha/\sigma_s \gg 1$, and the result is similar to that found for the surface absorbing sample; the difference being that there is no α absorption coefficient dependence in the limiting behavior of the above. The temperature change observed in the coupling fluid is a function of the distance the optically induced temperature change extends into the sample. This distance can change with either the absorption coefficient or the thermal diffusion distance, which in turn is controlled by the periodic excitation frequency.

The oscillating temperature-change magnitude for the two-layer sample is

$$|\delta T_{\text{osc},f}(z, \omega)| = \alpha E \frac{D_{T,s}^{1/2}}{\rho_f C_{P,f} D_{T,f}^{1/2} + \rho_s C_{P,s} D_{T,s}^{1/2}} \frac{2^{1/2} e^{-z/\mu_f}}{\omega \sqrt{\alpha^2 \mu_s^2 + 2\alpha\mu_s + 2}} \tag{40}$$

As with the surface and volume absorbing samples, the magnitude of the temperature change decreases exponentially with distance from the sample surface and with oscillation excitation frequency. In this case there is an additional complicating factor due to the absorption coefficient and sample thermal diffusion distance. This complex temperature-change behavior may be elucidated with frequency-dependent temperature magnitude measurements if the distance, z, is known. The phase shift is given by

$$\theta = -\frac{z}{\mu_f} - \tan^{-1} \frac{\alpha\mu_s + 2}{\alpha\mu_s} - \omega \tag{41}$$

The temperature-change phase shift is a linear function of distance. At zero frequency, or for very large α, the phase angle is $z/\mu_f - \pi/4$, while for high frequency or very small α, the angle is $z/\mu_f - \pi/2$. The differential phase shift is the same as that found between the surface and volume absorbing samples. These two extreme phase shifts can be used to determine the exponential factor, z/μ_f, required to find the absorption coefficient from the temperature-change amplitude measurement. However, temperature-change magnitude measurement may not be required since the phase-angle change is also a function of the absorption coefficient.

Measurement of the excitation frequency-dependent temperature-change phase angle allows determination of the relative absorption coefficient of the sample. The relative depth and thermal diffusion coefficient of the absorbing layer may be deduced by examining the excitation frequency-dependent phase angle. If the fluid-layer characteristics are known prior to sample analysis, say by calibration with a surface absorbing sample, the parameters of the inverse tangent term, $\alpha\mu_s$, can be deduced. The layered sample result can be used both for the surface absorbing sample and for the volume absorbing two-layer sample. These solutions are obtained by taking appropriate limits of the three-layer sample results. For example, the limit as $\alpha \Rightarrow \infty$ results in equations that are nearly equivalent to the surface absorbing sample solution, while taking the limits $l \Rightarrow \infty$ and $\alpha \Rightarrow 0$ will result in a description of the volume absorbing sample. The model has been extended to four-layer samples. The four-layer model is useful for samples that have nonabsorbing coatings between the absorbing layer and the fluid.

9.5.2. Pulsed Excitation of Thick-Layered Samples

The impulse-response may be obtained in the limit when $l \Rightarrow \infty$. The inverse Laplace transform of the temperature change in this limit yields the impulse response

$$\delta T_{\text{impulse},f}(z, t) = \alpha H \frac{D_{T,s}^{1/2}}{\rho_f C_{P,f} D_{T,f}^{1/2} + \rho_s C_{P,s} D_{T,s}^{1/2}} \times \exp\left[\alpha z \left(\frac{D_{T,s}}{D_{T,f}}\right)^{1/2} + \alpha^2 D_{T,s} t\right] \times \text{erfc}[\alpha (D_{T,s} t)^{1/2} + z(4 D_{T,f} t)^{-1/2}] \tag{42}$$

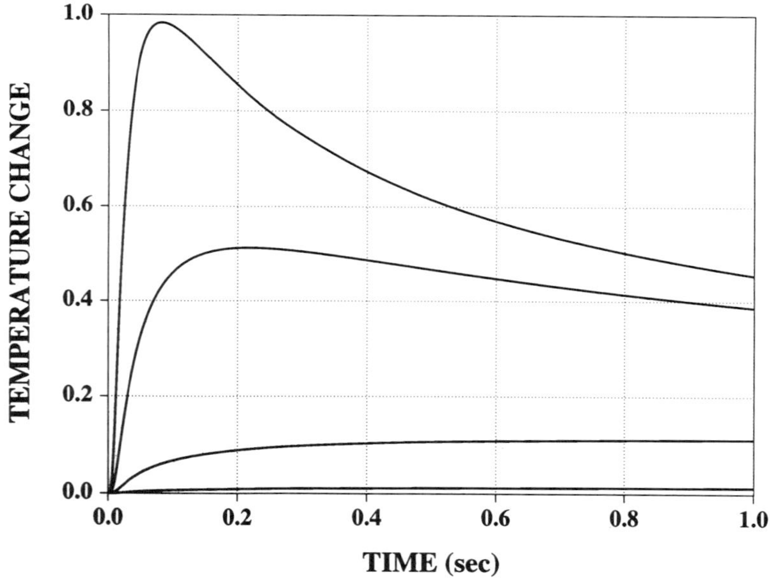

Figure 9.4. Impulse-response curves for the temperature change in the fluid layer. Thermal properties for quartz solid with a CCl_4 coupling layer were used with $z = 0.1$ mm. Absorption coefficients from top to bottom are $\alpha = 100$, 10, 1, and 0.1 cm^{-1}.

This limit corresponds to the two-layer sample since the thickness of the absorbing layer is infinitely thick. In fact, for very small α, this equation is equal to the impulse response for the volume absorbing sample. Figure 9.4 shows a series of impulse-response curves for the temperature change in the fluid layer. Each time-dependent curve corresponds to a different absorption coefficient. In addition to the signal magnitude change, the time dependence of the impulse response changes. This feature of layered sample analysis complicates the analysis of solids using pulsed excitation photothermal spectroscopy. The time evolution of the temperature change is seen more easily in Figure 9.5. Here all the data are scaled to the same maximum value (0.95) for clarity. The temperature changes arising from impulse excitation of samples with absorption coefficients less than about 10^{-2} have nearly equivalent impulse responses. The low-absorbance samples exhibit an impulse response that is a maximum at infinite time, and the magnitude of the signal is proportional to the absorption coefficient.

The reason for this complex behavior is discussed by Carslaw and Jaeger (1986). After impulse excitation, the temperature in the low-absorption-coefficient samples is nearly independent of distance into the sample. The heated sample thus represents a source of temperature change with an

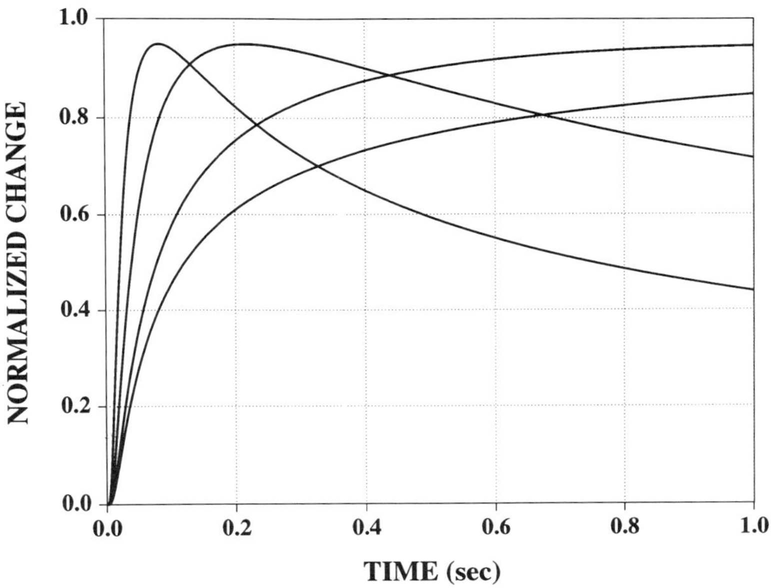

Figure 9.5. Scaled impulse response from Figure 9.4. Smaller sample absorption coefficients produce time-dependent signals that maximize at a later time.

infinite extent (in this approximation). Thermal diffusion will move heat toward the fluid, increasing the temperature in the fluid until the temperature of both phases are equal. On the other hand, impulse excitation of samples with a relatively high absorption coefficient will produce a temperature change in the sample that is highest at the interface, decreasing with distance into the sample. Thermal diffusion will transfer heat to both the coupling fluid and solid sample. This competitive heat flow produces a fluid temperature change that initially increases, then decreases with time as the sample comes to thermal equilibrium.

9.6. SURFACE POINT SOURCE

The thermal diffusion problems discussed up to this point have been one-dimensional in that the sample was assumed to be illuminated over a large area, of infinite extent in the x and y dimensions, and was homogeneous. Consider next the case where there is a heterogeneous absorption on the surface of the sample. The space-dependent surface absorption can be found by raster scanning the sample in the x and y dimensions while excitation and probe beams remain at a fixed position relative to the photothermal deflection detector. Photothermal deflection is detected along the z axis. The signals thereby produced will be a superposition of thermal point sources. These point sources, in turn, must be convoluted with the excitation source profile to yield the photothermal deflection response. The differential equations describing the time-dependent thermal diffusion are much more complicated in this case because the heat flows in both radial, or (x, y)-, and longitudinal, z-axis directions. Although this is the simplest three-dimensional problem conceivable for heat transfer at a sample interface, the results are complicated, represented only by integral equations.

The equations describing the thermal diffusion for an infinitesimal point-source absorption feature embedded in the surface of a solid, subscript s, sample of infinite extent, in contact with a coupling fluid, subscript f, also of infinite extent, are

$$\frac{\partial}{\partial t}\,\delta T_f(r,z,t) - D_{T,f}\left(\frac{\partial^2}{\partial r^2} + \frac{1}{r}\frac{\partial}{\partial r} + \frac{\partial^2}{\partial z^2}\right)\delta T_f(r,z,t) = 0 \qquad z \geqslant 0$$

$$\frac{\partial}{\partial t}\,\delta T_s(r,z,t) - D_{T,s}\left(\frac{\partial^2}{\partial r^2} + \frac{1}{r}\frac{\partial}{\partial r} + \frac{\partial^2}{\partial z^2}\right)\delta T_s(r,z,t) \tag{43}$$

$$= \frac{\kappa H}{\rho_s C_{P,s}}\,\delta(t)\,\delta(z)\,\delta(r) \qquad z \leqslant 0$$

where κ is the surface absorption coefficient. The surface absorption coefficient can be a function of space, but this will not affect the point-source thermal response. These equations are similar to those describing a one-dimensional surface-absorption case. The difference is the radially dependent gradient term. Moreover, the boundary conditions are the same as those of the one-dimensional case since radial terms are orthogonal to the interfaces. The equations may be solved by transforming the radial and time dimensions, then solving the resulting ordinary differential equations for the z dependence. The Fourier transform in two-dimensional Cartesian coordinates is

$$f(\mathbf{k}_x, \mathbf{k}_y) = \frac{1}{2\pi} \int_{-\infty}^{\infty} \int_{-\infty}^{\infty} e^{i\mathbf{k}_x x + i\mathbf{k}_y y} f(x, y)\, dx\, dy \tag{44}$$

Transformation into polar cylindrical coordinates yields

$$f(\mathbf{k}_r) = \int_0^{\infty} J_0(\mathbf{k}_r r) f(r)\, r\, dr \tag{45}$$

where $J_0(k, r)$ is the Bessel's function of order zero. The transform given by this integral is a Hankel transform (Luikov 1968). Because of the relationship to the Cartesian coordinate Fourier transform, it has derivative and symmetrical transform properties useful for solving differential equations. For example, the inverse transform is given by

$$f(r) = \int_0^{\infty} J_0(\mathbf{k}_r r) f(\mathbf{k}_r) \mathbf{k}_r\, d\mathbf{k}_r \tag{46}$$

and Laplacian derivatives reduce to

$$\nabla^2 f(r) \Leftrightarrow -\mathbf{k}_r^2 f(\mathbf{k}_r) \tag{47}$$

Taking the temporal Laplace and spatial Hankel transforms results in the algebraic equations

$$\begin{aligned} (\zeta + D_{T,f}\mathbf{k}_r^2)\delta T_f(\mathbf{k}_r, z, \zeta) - D_{T,f} \frac{\partial^2 \delta T_f(\mathbf{k}_r, z, \zeta)}{\partial z^2} &= 0 \qquad z \geqslant 0 \\ (\zeta + D_{T,s}\mathbf{k}_r^2)\delta T_s(\mathbf{k}_r, z, \zeta) - D_{T,s} \frac{\partial^2 \delta T_s(\mathbf{k}_r, z, \zeta)}{\partial z^2} &= \frac{\kappa H}{\rho_s C_{p,s}} \qquad z \leqslant 0 \end{aligned} \tag{48}$$

The solution to these equations will be of the form

$$\delta T_f(\mathbf{k}_r, z, \zeta) = F_1(\mathbf{k}_r, \zeta) e^{-z(\zeta/D_{T,f} + \mathbf{k}_r^2)^{1/2}} + F_2(\mathbf{k}_r, \zeta) e^{z(\zeta/D_{T,f} + \mathbf{k}_r^2)^{1/2}}$$
$$\delta T_s(\mathbf{k}_r, z, \zeta) = S_1(\mathbf{k}_r, \zeta) e^{-z(\zeta/D_{T,s} + \mathbf{k}_r^2)^{1/2}} + S_2(\mathbf{k}_r, \zeta) e^{z(\zeta/D_{T,s} + \mathbf{k}_r^2)^{1/2}} + \frac{\kappa H}{\rho_s C_{P,s} D_{T,s} (\zeta/D_{T,s} + \mathbf{k}_r^2)} \tag{49}$$

As with the layered sample, the boundary conditions

$$\delta T_f(\mathbf{k}_r, \infty, t) = 0 \qquad \delta T_s(\mathbf{k}_r, -\infty, t) = 0$$
$$\delta T_f(\mathbf{k}_r, 0, t) = \delta T_s(\mathbf{k}_r, 0, t) \qquad \kappa_f \frac{\partial T_f(\mathbf{k}_r, 0, t)}{\partial z} = \kappa_s \frac{\partial T_s(\mathbf{k}_r, 0, t)}{\partial z} \tag{50}$$

are used to find the $F_{1,2}$ and $S_{1,2}$ functions. F_2 and S_1 must be zero due to the infinite z-axis temperature boundary conditions. Using the temperature equivalence at the surface and heat flux constraints, F_1 is

$$F_1(\mathbf{k}_r, \zeta) = \frac{\kappa H}{\rho_s C_{P,s}} \frac{1}{\sqrt{\zeta + \mathbf{k}_r^2 D_{T,s}} \left(\sqrt{\zeta + \mathbf{k}_r^2 D_{T,s}} + r_f \sqrt{\zeta + \mathbf{k}_r^2 D_{T,f}}\right)} \tag{51}$$

The temperature change in the fluid where, presumably, the photothermal deflection signal will be generated is

$$\delta T_f(\mathbf{k}_r, z, \zeta) = F_1(\mathbf{k}_r, \zeta) e^{-z(\zeta/D_{T,f} + \mathbf{k}_r^2)^{1/2}} \tag{52}$$

The impulse response of the surface absorbing sample is obtained from inverse transforms:

$$\delta T_f(r, z, t) = \int_0^\infty \mathscr{L}^{-1}[F_1(\mathbf{k}_r, \zeta) e^{-z(\zeta/D_{T,f} + \mathbf{k}_r^2)^{1/2}}] J_0(r\,\mathbf{k}_r) \mathbf{k}_r \, d\mathbf{k}_r \tag{53}$$

The transform may be obtained by expressing the inverse-space temperature change as a product, transforming the individual parts, then convoluting the results over time and space. This is essentially the same technique as that used by Carslaw and Jaeger to derive the point response of the interface. Although the form of $F_1(\mathbf{k}_r, \zeta)$ is relatively simple, the inverse transform does not yield an analytical result. Instead, integral equations describing the temperatures in the sample and coupling fluid result.

Insight into how surface thermal diffusion affects the photothermal signal may be gained by examining the static temperature change occurring for continuous excitation. This is found in the limit of the inverse space temperature change

$$\lim_{\zeta \to 0} \delta T_f(\mathbf{k}_r, z, \zeta) = \delta T_{f,\mathrm{cw}}(\mathbf{k}_r, z, t = \infty)$$

$$\delta T_{f,\mathrm{cw}}(\mathbf{k}_r, z, \infty) = \frac{\kappa H}{\rho_s C_{P,s}} \frac{e^{-z|\mathbf{k}_r|}}{\mathbf{k}_r^2 \sqrt{D_{T,s}}(\sqrt{D_{T,s}} + r_f \sqrt{D_{T,f}})} = \frac{\kappa H e^{-z|\mathbf{k}_r|}}{\mathbf{k}_r^2} \frac{1}{\kappa_s + \kappa_f} \tag{54}$$

There are three other limiting cases where useful insight may be obtained. The trivial case is when the sample and coupling fluid have identical thermal properties. The other two cases are for sample and coupling fluid thermal properties that are very different. The case where thermal conduction and diffusion of the sample and fluid layers are equivalent is the same as the case where the sample has negligibly small thermal conductivity. The resulting impulse response is simply the three-dimensional impulse response in a homogeneous medium for a point source, as given in Chapter 3. The other limiting cases occur when thermal diffusion in either sample or fluid layers predominates. These extreme cases bracket the range of possibilities for the general case given above. As an example, for a thermally insulating solid like an organic polymer and a relatively high thermal conductivity liquid or gas coupling fluid, the temperature change in the fluid layer is

$$\lim_{D_{T,f}/D_{T,s} \to \infty} \delta T_{\mathrm{fluid}}(r, t) \approx \delta T_0 \frac{e^{-(z^2 + r^2)/4D_{T,f}t}}{(4\pi D_{T,f} t)^{3/2}} \tag{55}$$

where δT_0 is the instantaneous temperature change of the point source. δT_0 for this type of problem is given in Section 9.3. This solution also holds when the thermal properties of the sample are equal to that of the coupling fluid.

When thermal diffusion through the sample predominates, the temperature change at the surface may be obtained from the inverse transform of the limiting form of $F_1(\mathbf{k}_r, \zeta)$ as $D_{T,f}$ (and κ_f) approach zero. This situation might arise, for example, when the absorption is embedded in a high thermal conductivity solid (e.g., a metal) and the fluid is a low-thermal-conductivity fluid such as an organic solvent. The surface temperature for this case is

$$\lim_{D_{T,s}/D_{T,f} \to \infty} \delta T_s(r, 0, t) \approx \delta T_0 \frac{e^{-r^2/4D_{T,s}t}}{(4\pi D_{T,s} t)^{3/2}} \tag{56}$$

The fluid layer impulse response for a surface temperature change is the inverse transform of the exponential part of the solution general solution

$$\delta T_{\text{point}}(r, z, t) = \frac{e^{-(z^2+r^2)/D_{T,f}t}}{(4\pi D_{T,f}t)^{3/2}} \tag{57}$$

Subsequently, the temperature change in the fluid layer can be obtained from the time and space convolution of $\delta T_{\text{surface}}(r, t)$ with the fluid-layer point-source response $\delta T_{\text{point}}(r, t)$.

An exact formula for the time- and space-dependent temperature change may not be necessary for certain measurements. The photothermal deflection signal for a probe beam passing tangent to the surface is calculated from

$$\theta(t, z) = \left| \int_{-\infty}^{\infty} \frac{\partial}{\partial z} \, \delta T_f(x, y, z; t) \, dx \right|_{y=0} \tag{58}$$

The integral over x is over the propagation direction of the probe ray. Evaluating at $y = 0$ is used to place the point absorber directly below the probe ray. Substituting $\mathbf{k}_r^2 = \mathbf{k}_x^2 + \mathbf{k}_y^2$ in the exact solution, and with the relationships

$$\int_{-\infty}^{\infty} f(x, y) \, dx = \mathscr{F}[f(x, y)]_{\mathbf{k}_x=0} \qquad f(x, 0) = \int_{-\infty}^{\infty} \mathscr{F}[f(x, y)] \, d\mathbf{k}_y \tag{59}$$

the photothermal deflection signal is

$$\theta(t, z) = \left(\frac{dn}{dT}\right) \frac{\kappa H}{\rho_s C_{P,s}} \frac{\partial}{\partial z} \int_0^t \int_0^{t-\tau} \frac{f(\tau_1, \tau_2; t) z e^{-z^2/4\tau_2 D_{T,f}}}{2\pi D_{T,f}^{1/2} \tau_2^{3/2} \sqrt{\tau_1(t - \tau_1 - \tau_2)}} \, d\tau_2$$

$$f(\tau_1, \tau_2; t) = \frac{1}{\sqrt{tD_{T,s} - \tau_2(D_{T,s} - D_{T,f})} - r_f\sqrt{tD_{T,s} - (\tau_1 + \tau_2)(D_{T,s} - D_{T,f})}} \tag{60}$$

The limiting cases also yield simplified photothermal deflection-angle results. When the thermal conduction of the sample can be neglected relative to that of the coupling fluid, the resulting deflection angle is

$$\theta(t, z) = \left(\frac{dn}{dT}\right) \frac{z e^{-z^2/4D_{T,f}t}}{8\pi D_{T,f}^2 t^2} \tag{61}$$

while that produced when thermal conduction in the coupling fluid is small compared to the sample is

$$\theta(t, z) = \left(\frac{dn}{dT}\right) \frac{\delta T_0 z}{2D_{T,s}^{1/2}(4\pi D_{T,f})^{3/2}} \int_0^t \frac{e^{-z^2/4\tau D_{T,f}}}{(t-\tau)^{1/2}\tau^{3/2}[\tau(D_{T,f}-D_{T,s})+tD_{T,s}]^{1/2}} \tag{62}$$

In both cases the probe beam offset to maximum deflection angle is time dependent. Although more complicated, the photothermal deflection angle for the relatively high thermal conduction sample is similar to the other limiting instance. The main differences are that the deflection angle will develop more slowly because of the small coupling fluid thermal diffusion rate and that the deflection angle will generally be lower in magnitude due to rapid heat transfer into the sample.

9.7. GAUSSIAN BEAM EXCITATION OF SURFACES

The methods for finding the temperature change for the surface point source can be applied to a situation where a Gaussian profile laser beam is used to excite a sample with surface absorption. In this instance, the differential equations describing the heat flow are

$$\frac{\partial}{\partial t}\,\delta T_f(r, z, t) - D_{T,f}\left(\frac{\partial^2}{\partial r^2} + \frac{1}{r}\frac{\partial}{\partial r} + \frac{\partial^2}{\partial z^2}\right)\delta T_f(r, z, t) = 0 \qquad z \geqslant 0$$

$$\frac{\partial}{\partial t}\,\delta T_s(r, z, t) - D_{T,s}\left(\frac{\partial^2}{\partial r^2} + \frac{1}{r}\frac{\partial}{\partial r} + \frac{\partial^2}{\partial z^2}\right)\delta T_s(r, z, t) = \frac{\kappa H(r)}{\rho_s C_{P,s}}\,\delta(t)\,\delta(z) \qquad z \leqslant 0 \tag{63}$$

$H(r)$ is the Gaussian-form integrated irradiance,

$$H(r) = \frac{2Q}{\pi w^2}\, e^{-2r^2/w^2} \tag{64}$$

Using the method of Laplace and Hankel transforms, and with the boundary conditions given for a point source, the temperature change in the fluid layer is

$$\delta T_f(\mathbf{k}_r, z, \zeta) = F_1(\mathbf{k}_r, \zeta)^{e^{-z(\zeta/D_{T,f}+\mathbf{k}_r^2)^{1/2}}}$$

$$F_1(\mathbf{k}_r, \zeta) = \frac{\kappa H(\mathbf{k}_r)}{\rho_s C_{P,s}} \frac{1}{\sqrt{\zeta + \mathbf{k}_r^2 D_{T,s}}(\sqrt{\zeta + \mathbf{k}_r^2 D_{T,s}} + r_f\sqrt{\zeta + \mathbf{k}_r^2 D_{T,f}})} \tag{65}$$

where $H(\mathbf{k}_r)$ is the Hankel transform of the Gaussian integrated irradiance

$$H(\mathbf{k}_r) = \frac{Q}{2\pi} e^{-k_r^2 w^2/8} \tag{66}$$

The perpendicular photothermal deflection signal can be obtained as in the case of the point source. The photothermal deflection signal is

$$\theta(t, z) = \left(\frac{dn}{dT}\right) \frac{\kappa Q}{\pi^2 w^3 \rho_s C_{P,s}} \frac{\partial}{\partial z} \int_0^t \int_0^{t-\tau} \frac{f(\tau_1, \tau_2; t) z e^{-z^2/4\tau_2 D_{T,f}}}{D_{T,f}^{1/2} \tau_2^{3/2} \sqrt{\tau_1(t - \tau_1 - \tau_2)}} d\tau_2$$

$$f(\tau_1, \tau_2; t) = \frac{1}{\sqrt{1 + 2(t - \tau_2)/t_{c,s} + 2\tau_2/t_{c,f}} - r_f \sqrt{1 + 2(t - \tau_1 - \tau_2)/t_{c,s} + 2(\tau_1 + \tau_2)/t_{c,f}}} \tag{67}$$

where the t_c are defined in the usual fashion. As might be expected, the magnitude of the photothermal deflection angle is inversely propo tional to the excitation laser beam waist radius and proportional to the pulse energy.

For surface imaging studies, it is important to realize that the photothermal deflection angle is comprised of two main parts: (1) the surface absorption coefficient, $\kappa(x, y)$, and (2) the thermal response, $\theta_0(z; t, t_{c,s})$:

$$\theta(x, y, z; t) = \kappa(x, y)\theta_0(z; t, t_c) \tag{68}$$

The functional form of these terms is apparent from equation 67. Either or both of these terms may change with position. Raster-scanning absorption imaging studies utilize the position dependence of $\kappa(x, y)$. Of course, resolution of the image experiment will be on the order of the beam waist radius. If the thermal properties of the sample change with space, $t_{c,s} = w^2/4D_{T,s}$ will be affected. This case is not strictly valid for the foregoing treatment since the heat was assumed to diffuse through a sample with homogeneous thermal properties in the derivation. However, to within the approximation that the thermal characteristics do not change that significantly, raster scanning the sample under the excitation laser while monitoring the time-dependent deflection angle should yield an image of the thermal properties of the sample. This technique, called thermal imaging, has been performed by depositing a spatially homogeneous thin layer of highly absorbing material onto the sample to be analyzed. Raster scanning while monitoring the deflection angle will yield information about the thermal properties of the sample since the absorbance is constant. However, the thin layer will affect the heat transfer characteristics. A more accurate accounting of the deflection signal takes the thermal properties of this layer into account.

9.8. EXCITATION OF LAYERED SAMPLES WITH GAUSSIAN BEAMS

Gaussian laser beams are often used to induce photothermal signals in solid samples. This technique is used for nondestructive material analysis of material within the sample. As with surface absorption detection, the sample is raster scanned in the x and y directions and the photothermal signal created by exciting an area on the sample is recorded. Thermal diffusion is more complicated in this case because there is radial diffusion in addition to that along the z axis both within and outside the sample. The thermal diffusion problem for this particular geometry has been examined in detail by Murphy and Aamodt (1980), Jackson et al. (1981), and Aamodt and Murphy (1981, 1983). More recent discussions can be found in Mandelis and Power (1988), Fournier and Boccara (1989), and Schweitzer and Power (1994).

The differential equations for the three-layer sample excited with an impulse Gaussian beam are

$$\frac{1}{D_{T,f}}\frac{\partial \delta T_f(r,z,t)}{\partial t} - \left(\frac{\partial^2}{\partial r^2} + \frac{1}{r}\frac{\partial}{\partial r} + \frac{\partial^2}{\partial z^2}\right)\delta T_f(r,z,t) = 0 \qquad z \geqslant 0$$

$$\frac{1}{D_{T,s}}\frac{\partial \delta T_s(r,z,t)}{\partial t} - \left(\frac{\partial^2}{\partial r^2} + \frac{1}{r}\frac{\partial}{\partial r} + \frac{\partial^2}{\partial z^2}\right)\delta T_s(r,z,t) = \frac{\alpha H(r)\delta(t)e^{\alpha z}}{D_{T,s}\rho_s C_{p,s}} \qquad -l \leqslant z \leqslant 0$$

$$\frac{1}{D_{T,b}}\frac{\partial \delta T_b(r,z,t)}{\partial t} - \left(\frac{\partial^2}{\partial r^2} + \frac{1}{r}\frac{\partial}{\partial r} + \frac{\partial^2}{\partial z^2}\right)\delta T_b(r,z,t) = 0 \qquad z \leqslant -l$$

(69)

where, as before, subscripts f, s, and b correspond to the fluid, sample, and backing. The z axis is the axis of propagation of the excitation source. $H(r)$ is the integrated irradiance for the Gaussian source. The differential equations are reduced by taking the Hankel and Laplace transforms in radial space and time. These operations result in

$$
\begin{aligned}
&(\zeta + D_{T,f}\mathbf{k}_r^2)\,\delta T_f(\mathbf{k}_r, z, \zeta) - D_{T,f}\,\frac{\partial^2 \delta T_f(\mathbf{k}_r, z, \zeta)}{\partial z^2} = 0 && z \geqslant 0 \\
&(\zeta + D_{T,s}\mathbf{k}_r^2)\,\delta T_s(\mathbf{k}_r, z, \zeta) - D_{T,s}\,\frac{\partial^2 \delta T_s(\mathbf{k}_r, z, \zeta)}{\partial z^2} = \frac{\alpha H(\mathbf{k}_r) e^{\alpha z}}{\rho_s C_{p,s}} && -l \leqslant z \leqslant 0 \\
&(\zeta + D_{T,b}\mathbf{k}_r^2)\,\delta T_b(\mathbf{k}_r, z, \zeta) - D_{T,b}\,\frac{\partial^2 \delta T_b(\mathbf{k}_r, z, \zeta)}{\partial z^2} = 0 && z \leqslant -l
\end{aligned}
\tag{70}
$$

These equations are nearly equivalent to those obtained from the transform of the one-dimensional thermal diffusion case. The differences are that there is an additional $\mathbf{k}_r^2$ term and that the source has spatial or $\mathbf{k}_r^2$ dependence. As for the one-dimensional case, a general solution can be formulated:

$$
\begin{aligned}
\delta T_f(\mathbf{k}_r, z, \zeta) &= F_1(\mathbf{k}_r, \zeta) e^{-zC_f} \\
\delta T_s(\mathbf{k}_r, z, \zeta) &= S_1(\mathbf{k}_r, \zeta) e^{-zC_s} + S_2(\mathbf{k}_r, \zeta) e^{zC_s} + \frac{\alpha H(\mathbf{k}_r) e^{\alpha z}}{(C_s^2 - \alpha^2)\rho_s D_{T,s}} \\
\delta T_b(\mathbf{k}_r, z, \zeta) &= B_2(\mathbf{k}_r, \zeta) e^{zC_b}
\end{aligned}
\tag{71}
$$

where the C_i are

$$
C_i = \sqrt{\mathbf{k}_r^2 + \frac{\zeta}{D_{T,i}}} \tag{72}
$$

The coefficients may be obtained from the boundary conditions. Applying the same boundary condition as the one-dimensional case, the coefficients are found to be the same as those of the one-dimensional case, with the exception that ζ/D_T is everywhere replaced by $\mathbf{k}_r^2 + \zeta/D_T$. The coefficients are

$$
\begin{aligned}
F_1(\mathbf{k}_r, \zeta) &= \frac{-\alpha H(\mathbf{k}_r)}{\kappa_s(C_s^2 - \alpha^2)}\,\frac{2(r_b - r_\alpha)e^{-\alpha l} + (r_\alpha - 1)(r_b + 1)e^{lC_s} - (r_\alpha + 1)(r_b - 1)e^{-lC_s}}{(r_b + 1)(r_f + 1)e^{lC_s} - (r_b - 1)(r_f - 1)e^{-lC_s}} \\
S_1(\mathbf{k}_r, \zeta) &= \frac{-\alpha H(\mathbf{k}_r)}{\kappa_s(C_s^2 - \alpha^2)}\,\frac{-(r_\alpha + r_f)(r_b - 1)e^{-lC_s} + (r_b - r_\alpha)(r_f + 1)e^{-\alpha l}}{(r_b + 1)(r_f + 1)e^{lC_s} - (r_b - 1)(r_f - 1)e^{-lC_s}} \\
S_2(\mathbf{k}_r, \zeta) &= \frac{-\alpha H(\mathbf{k}_r)}{\kappa_s(C_s^2 - \alpha^2)}\,\frac{(r_\alpha + r_f)(r_b + 1)e^{lC_s} - (r_b - r_\alpha)(r_f - 1)e^{-\alpha l}}{(r_b + 1)(r_f + 1)e^{lC_s} - (r_b - 1)(r_f - 1)e^{-lC_s}} \\
B_2(\mathbf{k}_r, \zeta) &= \frac{\alpha H(\mathbf{k}_r) e^{-l(\alpha - C_b)}}{\kappa_s(C_s^2 - \alpha^2)}\,\frac{(r_\alpha + 1)(r_f + 1)e^{lC_s} - (r_\alpha - 1)(r_f - 1)e^{-lC_s}}{(r_b + 1)(r_f + 1)e^{lC_s} - (r_b - 1)(r_f - 1)e^{-lC_s}}
\end{aligned}
\tag{73}
$$

where the r_i are unitless ratios defined by

$$r_f = \frac{\kappa_f}{\kappa_s}\frac{C_f}{C_s} \qquad r_b = \frac{\kappa_b}{\kappa_s}\frac{C_b}{C_s} \qquad r_\alpha = \frac{\alpha}{C_s} \tag{74}$$

The spatially dependent temperature change is found from the inverse Hankel transforms of the impulse response. The temperature change in the coupling fluid is

$$\delta T_f(r, z, \zeta) = \int_0^\infty F_1(\mathbf{k}_r, \zeta) e^{-zC_f} J_0(r\,\mathbf{k}_r)\mathbf{k}_r\, d\mathbf{k}_r \tag{75}$$

Integration does not result in an analytical expression but may be obtained numerically for a given Laplace parameter. The periodic temperature change produced using a modulated excitation source oscillating at ω is obtained without the need for the inverse Laplace transform. In this case

$$\delta T_f(r, z, \omega) = e^{i\omega t} \int_0^\infty F_1(\mathbf{k}_r, i\omega) e^{-zC_f} J_0(r\,\mathbf{k}_r)\mathbf{k}_r\, d\,\mathbf{k}_r \tag{76}$$

Finally, Schweitzer and Power (1994) have recently extended the method of determining photothermal beam deflection angles in layered samples to cases where the optical absorbance is not constant. They used a superposition of impulse-response functions for thin sections of thickness l_i and absorption coefficient α_i located at a distance z_i from the sample surface. For an absorption-depth profile given by the energy equation

$$Q_i = Q_0 e^{-\sum_{j=1}^{i} \alpha_j l_j} \tag{77}$$

where Q is the excitation pulse energy, the energy deposited in layer j is $\alpha_j Q_{j-1}$. The superposition of impulse-response functions, equation 57 convoluted with the exponential z-dependent energy deposition and Gaussian r-dependent energy deposition, for each layer results in a temperature change that can be expressed by

$$\delta T(r, z, t) = \delta T(r, t)\, \delta T(z, t) \tag{78}$$

where $\delta T(r, t)$ is the impulse response temperature change for Gaussian excitation source encountered in earlier chapters:

$$\delta T(r, t) = \frac{2}{\pi w^2(t)\rho C_p} e^{-2r^2/w^2(t)} \tag{79}$$

where $w^2(t) = w^2(1 + 2t/t_c)$, and the axial temperature change is

$$\begin{aligned}\delta T(z, t) = \sum_{j=1}^{N} \alpha_j Q_j e^{-\alpha_j z + \alpha_j^2 D_{T,s} t} \\ \times \{\mathrm{erf}[\alpha_j(D_{T,s}t)^{1/2} + (d_j - z)(4D_{T,s}t)^{-1/2}] \\ - \mathrm{erf}[\alpha_j(D_{T,s}t)^{1/2} + (d_{j+1} - z)(4D_{T,s}t)^{-1/2}]\}\end{aligned} \tag{80}$$

where the depth to a particular layer, d_i, is defined by

$$d_i = \sum_{j=1}^{i} l_j \tag{81}$$

The surface temperature change defined by this result is not accurate since it does not take into account the change in thermal properties at the solid sample–coupling fluid boundary. However, the fluid-layer temperature may be approximated by convolution of the time- and space-dependent surface temperature change with the impulse response for the temperature change in the coupling fluid. Schweitzer and Power have shown that this approach to calculating the temperature change yields results that agree with experimental measurements of stratified materials with known composition.

For optically thin layers, the z-component temperature change at the surface of the stratified sample is

$$\delta T(0, t) = \sum_{j=1}^{N} \alpha_j Q_j [\mathrm{erf}(d_j(4D_{T,s}t)^{-1/2}) - \mathrm{erf}(d_{j+1}(4D_{T,s}t)^{-1/2})] \tag{82}$$

which is equivalent to a superposition of impulse response for a volume absorbing sample. This equation closely resembles that of a transform. In fact, in the limit of infinitesimally thin layers

$$\delta T(0, t) = \frac{2}{(4\pi D_{T,s}t)^{1/2}} \int_0^{\infty} Q(z) e^{-z^2/4D_{T,s}t}\, dz \tag{83}$$

where $Q(z)$ is the energy deposited at a particular distance into the sample:

$$Q(z) = \alpha(z) Q_0 e^{-\int_0^z \alpha(z)\,dz} \tag{84}$$

It should be possible to determine $Q(z)$, and subsequently $\alpha(z)$, from the time-dependent temperature-change response of a sample using one of the algorithms developed for inverse problems (e.g., Stanley et al. 1993).

9.9. DEFLECTION ANGLES WITH OSCILLATING GAUSSIAN EXCITATION

The probe beam breaks the cylindrical symmetry since the probe beam passes through the heated coupling fluid perpendicular to the excitation beam. Using $\mathbf{k}_r^2 = \mathbf{k}_x^2 + \mathbf{k}_y^2$, and for probe beam propagation along the x axis, the impulse response of the photothermal deflection angle in the z-direction is

$$\begin{aligned}\theta(z, t) &= -\left(\frac{dn}{dT}\right)\frac{\partial}{\partial z}\int_{-\infty}^{\infty} \mathscr{L}^{-1}[F_1(\mathbf{k}_y, \zeta)e^{-z(\zeta/D_{T,f}+\mathbf{k}_y^2)^{1/2}}]\,d\mathbf{k}_y \\ &= \left(\frac{dn}{dT}\right)\int_{-\infty}^{\infty} \mathscr{L}^{-1}[F_1(\mathbf{k}_y, \zeta)C_f e^{-zC_f}]\,d\mathbf{k}_y\end{aligned} \tag{85}$$

The inverse Laplace transform does not have to be taken to obtain the deflection angle for an oscillating source:

$$\theta(z, \omega) = \left(\frac{dn}{dT}\right) e^{i\omega t} \int_{-\infty}^{\infty} F_1(\mathbf{k}_y, i\omega) C_f e^{-zC_f}\,d\mathbf{k}_y \tag{86}$$

The spatial anisotropy introduced by the Gaussian excitation source produces a temperature gradient in both the z and y directions for a probe ray propagating along the x axis. Subsequently, there will be both perpendicular and transverse or parallel deflection, corresponding to directions normal and tangent to the surface. These orthogonal deflection angles are illustrated in Figure 9.6. The Gaussian distribution will also produce a thermal lens element, although it is difficult to measure since the beam also deflects. Aamodt and Murphy (1981, 1983) show that for any probe ray offset to the

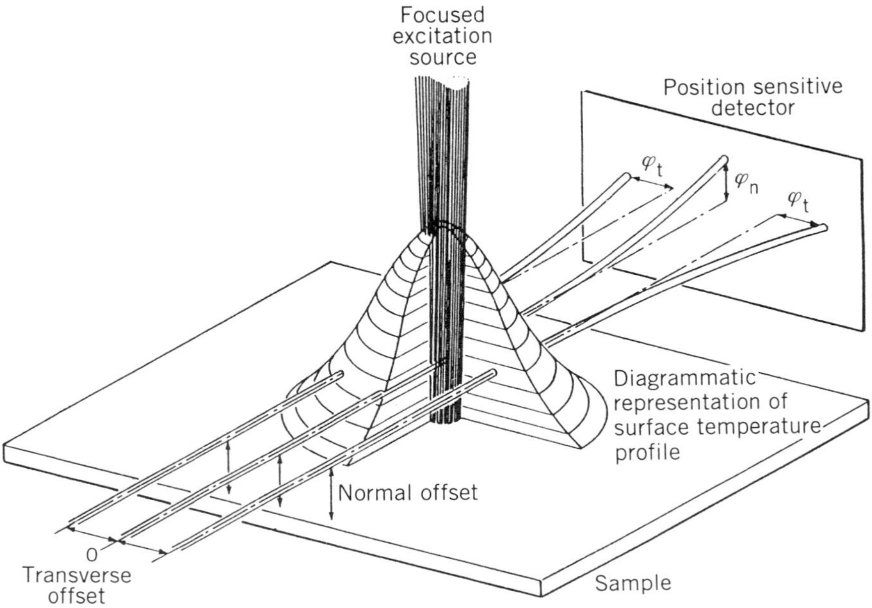

Figure 9.6. Illustration of the two photothermal deflection angle components. (Reprinted with permission from Aamodt and Murphy 1983. Copyright 1983, American Physical Society.)

excitation beam, the perpendicular and transverse deflection angles are

$$\theta_{\perp}(y, z, \omega) = \left(\frac{dn}{dT}\right)\frac{4}{w}\int_{0}^{\infty} F_1(\mathbf{k}_y, i\omega)C_f e^{-zC_f}\cos\frac{2y\mathbf{k}y}{w}\,d\mathbf{k}_y$$

$$\theta_{||}(y, z, \omega) = \left(\frac{dn}{dT}\right)\frac{8}{w^2}\int_{0}^{\infty} F_1(\mathbf{k}_y, i\omega)e^{-zC_f}\sin\frac{2y\mathbf{k}_y}{w}\,\mathbf{k}_y\,d\mathbf{k}_y \tag{87}$$

An experimental plot of the perpendicular and transverse deflection angles as a function of probe beam offset is shown in Figure 9.7.

$F_1(\mathbf{k}_y, i\omega)$ is the surface temperature change of the sample. Scanning the excitation to probe beam offset, the y coordinate, while monitoring deflection angles as a function of excitation frequency allows determination of the thermal characteristics of the sample, at least those local to the surface. If the spatial distribution of absorption or thermal properties of the sample varies slowly compared to the beam radius of the excitation source, raster scanning the sample while performing beam deflection measurements

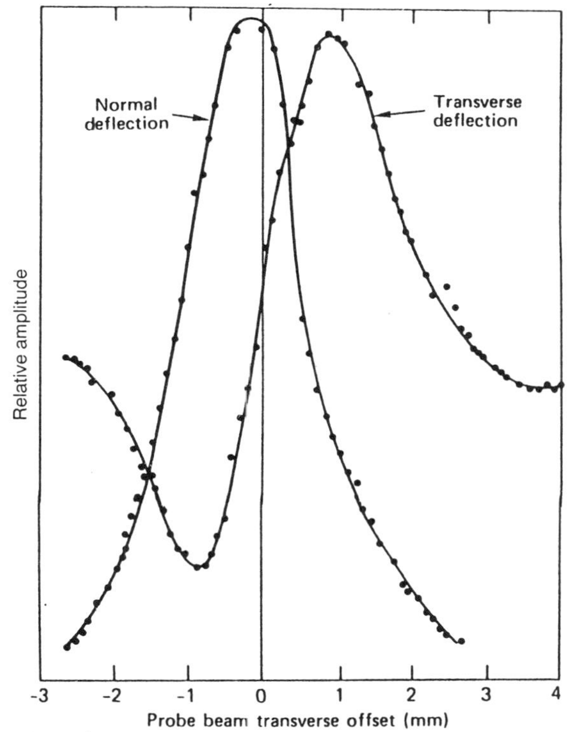

Figure 9.7. Experimental normal and transverse optical beam deflection signal magnitudes as a function of probe beam offset for a thin silicon wafer. (Reprinted with permission from Aamodt and Murphy 1983. Copyright 1983, American Physical Society.)

should, in principle, allow both absorption and thermal imaging of the sample. The additional information supplied by scanning the probe across the excitation region should allow the absorption and thermal characteristics of the sample to be obtained through inversion of $F_1(\mathbf{k}_y, i\omega)$. Aamodt and Murphy performed extensive numerical evaluations to determine under what conditions, and to what degree, optical and thermal parameters can be obtained separately. They conclude that not only can optical absorption and thermal conductivity be obtained, but the heat capacity of the sample can also be obtained under certain conditions. In addition, they point out that the scanning probe beam technique is sensitive to heterogeneous heat transfer within sample. Anisotropic heat transfer will distort the otherwise symmetric spatially dependent photothermal deflection angle. This effect is advantageous for the study of thermal defects in ceramic and metals.

9.10. PHOTOTHERMAL REFLECTION

Another optical element can be formed at or near the surface of a heated solid sample. This optical element can be monitored by the angle at which a probe laser is reflected from the sample. As such, it may be called photothermal reflection spectroscopy (Mandelis and Power 1988). For a reflecting sample, a probe laser directed next to the region excited by the excitation source will be reflected at an angle proportional to the temperature gradient. The reflection angle change is a consequence of both deformation due to sample deformation (expansion), and a change in the optical (e.g., reflectance or phase), properties of the sample surface with temperature. Signal interpretation is difficult, owing to the fact that the reflection angle is a function of both the deformation geometry and change in optical constants on the surface of the sample. Both effects are proportional to the temperature at the surface of the sample. However, by measuring reflectivity and reflection angle changes independently, corresponding to the sum and difference signals of a bicell detector, surface deformation and optical constant changes can be obtained. Rothenberg (1988) has used a picosecond excitation and probe laser with the latter technique to measure surface deformations on silicon that form in the picosecond time scale and relax on the order of several hundred picoseconds.

9.11. EXPERIMENT DESIGN FOR PHOTOTHERMAL DEFLECTION

Several types of experimental apparatus have been employed for solid sample analysis. They tend toward two main categories: those used to determine bulk properties and the analysis of stratified materials, and those used for lateral optical and/or thermal imaging. Apparatus for measuring bulk properties is based on probe laser beam deflection. The probe laser is directed across the surface of the sample, and one of several methods for spatial filtering and developing the electronic signal is used to estimate the deflection angle. The two most common methods are the use of an spatial edge filter followed by a photodetector and the commercial position-sensing detectors which produce signals that are compensated for probe laser beam power variations. Although the frequency response of position-sensing detectors is limited compared to that of the spatial edge filter photodetector combination, they are apparently adequate when used with oscillating sources for frequency-dependent measurement. Both laser and conventional light sources have been used in bulk property measurements. Measurements are obtained either as a time series, an impulse response, or a function of

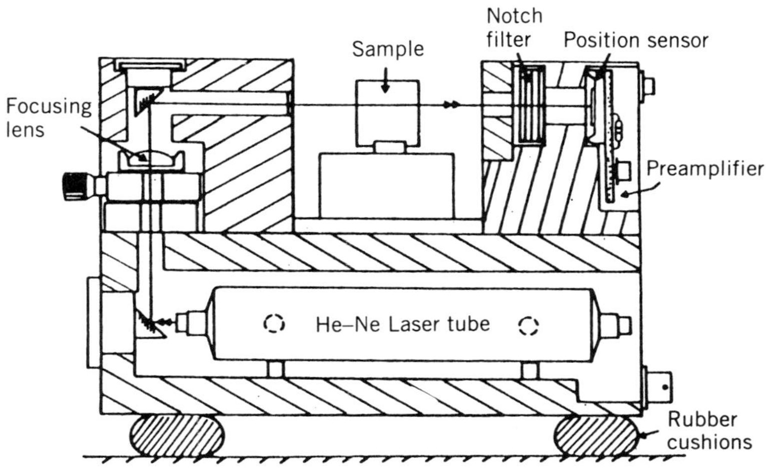

Figure 9.8. Compact apparatus for photothermal deflection spectroscopy. Probe laser, sample and position-sensing detector and rigidly mounted to reduce environmental interference. (Reprinted with permission from Charbonnier and Fournier 1986. Copyright 1986, American Physical Society.)

excitation frequency. Excitation frequency-dependent response utilizes a phase-sensitive lock-in amplifier so that both magnitude and phase information can be obtained.

Charbonnier and Fournier (1986) describe the construction and utilization of compact apparatus for photothermal beam deflection. Their apparatus, illustrated in Figure 9.8, is a marked improvement over the usual benchtop design, due to its relative rigidity and compactness. The probe, either a helium–neon or diode laser, sample compartment, and position-sensing detector are all housed in a compact aluminum frame. The number of optical positioning components is minimized to reduce the influence of vibrations on the measurements. The shorter, enclosed beam path of the probe laser results in reduced influence from air currents and mechanical vibrations. The position-sensing detector element can detect both transverse and perpendicular probe beam deflection. Absorption and thermal imaging can be performed using a raster scan sample holder. Overall, they found that this apparatus produced data with a SNR value over an order of magnitude better than that obtained with the benchtop apparatus. The SNR was not limited by environmental noise, but rather, by shot noise. This apparatus has been used with FTIR and conventional sources.

Although it is possible to use air or other gas above the sample as a coupling fluid for FTIR photothermal beam deflection studies, isolation of

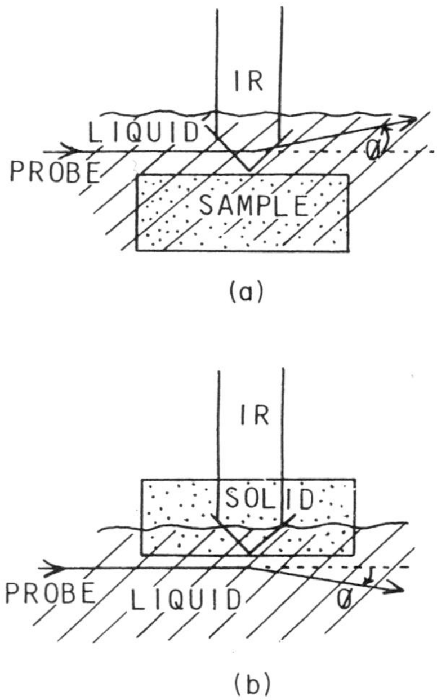

Figure 9.9. Two sample excitation/coupling fluid schemes. Scheme (*a*) requires infrared excitation light to pass through the coupling fluid, which will absorb certain wavelengths. Scheme (*b*) is a reverse-mirage technique that uses thermal coupling to transfer energy absorbed by the solid sample to the coupling fluid. (Reprinted with permission from Palmer and Smith, 1986. Copyright 1986, National Research Council of Canada.)

the sample from environmental noise presents a major problem. The use of liquid coupling fluid largely overcomes this problem. The problem with liquids is that they have only moderate transmission in the mid-infrared. Palmer and Smith (1986) describe a reverse-mirage detection probe beam geometry for bulk solid analysis utilizing a rapid-scan FTIR source. A diagram of the probe laser beam geometry is shown in Figure 9.9. In the normal beam deflection scheme, the probe beam passes above the sample, on the same side as that excited, as shown in part (*a*) of the figure. For sufficiently thin samples with relatively good thermal conductivity, heat will diffuse through and cause a temperature increase on the back side of the sample. In this geometry, the infrared excitation source does not have to pass through the coupling fluid, and subsequently is not attenuated prior to the solid sample. Coupling fluids that produce large photothermal enhancements but do not have good infrared transmission may be utilized as long as they have finite transmission at the probe laser wavelength (Figure 9.10).

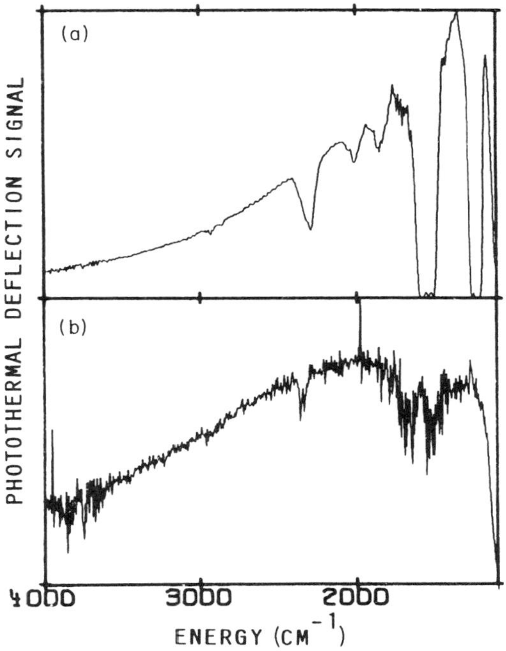

Figure 9.10. Single-beam FTIR photothermal deflection spectra of carbon black in a polymer reference for 2048 scans. Curve (*a*) is the spectrum obtained in CCl_4; curve (*b*) used air as the coupling fluid. The liquid phase coupling fluid produces superior SNR but absorbs the excitation source energy. (Reprinted with permission from Palmer and Smith, 1986, Copyright 1986, National Research Council of Canada.)

Apparatus for absorption and thermal imaging analysis uses the same basic features as those used for bulk measurements. However, since the spatial resolution of the analysis is proportional to the dimension of the excitation source, focused lasers are more commonly employed for excitation. Photothermal deflection signals can be collected perpendicular or parallel to the surface normal. A set of data required for complete optical and thermal sample analysis according to the model proposed by Aamodt and Murphy requires scanning over the sample at different relative probe-to-excitation laser beam offsets and at several periodic excitation source frequencies. This process can be very time consuming. Charbonnier et al. (1987) developed a CCD-based photothermal deflection angle detector that may speed up the process of sample analysis by simultaneously measuring deflections across one dimension of the sample (also see Fournier and Boccara 1989). The device is capable of determining the magnitude and phase of the oscillating photothermal deflection angle produced with a periodic source imaged to a line. Only the perpendicular beam deflection angle can be measured with this apparatus. A two-dimensional image of the absorption and thermal properties of the sample can be obtained by raster scanning the sample in one dimension. Mandelis (1987) and, most recently, Schweitzer and Power (1994) reinforce the connection between impulse response and frequency-phase techniques, pointing out that they yield identical information.

9.12. APPLICATIONS TO CHEMICAL ANALYSIS

9.12.1. Bulk Properties

The most promising application of photothermal deflection is for infrared analysis of solid materials. Fournier et al. (1982) reported on the use of FTIR to modulate the wavelengths of a 250-W quartz-halogen lamp source to excite amorphous silicon, Nd^{3+}:$CaWO_4$ and $NiSO_4$ samples. Although a FTIR was used, the spectral range was from 26,000 to 4000 cm^{-1}, thereby convering the visible and near-IR spectrum. A linear polarizer followed by quarter- or half-wave retardation devices are used to generate circular and linear polarized light for circular and linear dichroism spectra.

M. J. D. Low and co-workers have used mid-infrared FTIR modulated photothermal beam deflection to study an enormous number of samples, ranging from black inks on paper to mouse skulls (Morterra and Low 1982, 1985a–d; Varlashkin and Low 1986a–c; Low 1986; Low and Morterra 1983a,b, 1987). In addition to demonstrating the utility of this technique for difficult analysis, these papers discuss many practical aspects of using photothermal spectroscopy for infrared analysis. With one exception (Varlashkin and Low 1986c) these studies used air as the coupling fluid. Varlashkin et al. (1986) compare data obtained with photothermal beam deflection to that obtained using a photoacoustic detector and found that the photothermal data exhibit lower SNR. This was presumably due to the large amount of environmental noise associated with the FTIR spectrometer. Low's (1986) account of the role of vibration noise in the practical aspects of FTIR photothermal deflection spectroscopy should be on the reading list of anyone who plans to delve into real-world sample analysis. This humorous account reflects the frustrations of a scientist working with an ultrasensitive method in a noisy environment. Many of us working with photothermal spectroscopy have had these feelings.

Another problem with using FTIR as an excitation source for photothermal spectroscopy is that the modulation frequency is different for each wavelength. Since the photothermal deflection angle magnitude is a function of the modulation frequency, the spectrum obtained by Fourier transform of the data will be a function of the interferometer mirror velocity and the wavelength. This problem is solved by calibrating the instrumental response using a suitable "optically black" absorber. Chopped or frequency-modulated step-scan interferometers can also remedy this problem. With these instruments, all wavelengths will be modulated at the same frequency. The step-scan FTIR frequency-modulation technique has been used to depth-profile stratified polymers using photoacoustic spectroscopy detection (Palmer et al. 1993).

Several studies address bulk visible absorbance or thermal properties obtained with visible excitation sources. Many of these have been reported in books dedicated to photothermal spectroscopy of solids (Mandelis 1987; Hess 1989a,b). The excellent chapters by Fournier and Boccara (1989) and Kuo et al. (1989) also review literature related to solid measurement. Some studies have reported on the determination of bulk thermal diffusivity (e.g., Suber et al. 1988, Figari 1991). Applications to surface kinetics are apparently not successful, owing to the slow response time of liquid coupling fluids (Matsujima et al. 1984a,b).

9.12.2. Applications to Gel and Thin-Layer Chromatography

The detection of absorption in thin-layer chromatography is an area that has received much attention by M. D. Morris and co-workers. A history of these sudies is given by Morris and Fotiou (1989). Initial studies (Peck and Morris 1986a,b, 1988) used photothermal lens spectroscopy to detect optical absorption in polyacrylamide gel plates. Proteins were separated by electrophoreses, then stained using coomassie brilliant blue dyes or a silver colloid treatment. Since the gels are transparent, the photothermal lens signal was obtained in transmission mode. The plate was scanned through the excitation region while measuring the photothermal lens response. This resulted in the plate-position-dependent optical absorbance measurement required for protein analysis. One problem found for the coomassie brilliant blue dye was that the photothermal signal was not linear with protein concentration below a certain limit. This resulted in a detection limit which was only 1/10 that obtained by eye. Silver colloid treatment was not successful, due to the large residual absorbance left in the gel by the treatment.

Measurement of optical absorption on thin-layer chromatography (TLC) plates was accomplished using photothermal deflection spectroscopy (Chen and Morris 1984a,b; Matsujima et al. 1984a,b; Peck et al. 1985; Fotiou and

Morris 1986a). The opaque properties of TLC plates circumvent the use of a transmission experiment such as photothermal lens spectoscopy. The deflection apparatus is made by placing the TLC plate on a translation device and focusing a pulsed or continuous wave laser onto the sample. The probe laser, also stationary, probes the photothermal response as the plate is translated. The technique shows good detection limits, in the area of picograms to nanograms, depending on the absorbance of the analyte. A technique using photothermal deflection for detection of proteins separated on polyacrylamide gels plates was reported by Jager et al. (1989). Proteins were dyed with coomassie brilliant blue R350 and the gel plates were air dried prior to measurement. The dried gels were emersed into a carbon tetrachloride coupling fluid prior to deflection measurement. The detection limit was a factor of 4 better than that obtained using a conventional laser-based absorption scanner.

Fotiou and Morris (1986b, 1987a,b) and Treado and Morris (1990) developed a spatially multiplexed photothermal deflection apparatus for reading TLC plates. A Hadamard mask is used to change the irradiation pattern of the excitation radiation on the TLC plate. The photothermal deflection angle response produced by passing the probe laser beam just above the plate is measured as a function of Hadamard mask position. The space-dependent optical absorbance of plate is obtained from the Hadamard transform of the mask translation distance versus photothermal response data. The one-dimensional Hadamard mask is a pseudo-random collection of slit apertures that transmits roughly half of the radiation. The slits allow optimum exposure of the relatively long lines produced in TLC. Both pulsed and continuous laser illumination has been used. Critical evaluation revealed that the technique has limited spatial resolutions, due to limited imaging resolution of a mask with small apertures (Treado and Morris 1988). Of course, the ultimate resolution is obtained using diffraction-limited focused-laser-beam TLS plate reader apparatus. A compressed linear absorbance range results from the infamous "multiplex disadvantage." Relatively small absorbance areas of the plate would not be able to be detected in the transform.

The main advantage of this technique is the lower irradiance on the plate, which reportedly reduced optical damage over that of the scanning laser spot TLS plate reader. On the other hand, exposure times are increased to obtain the spatial information required to obtain the space-dependent optical absorbance. The integrated irradiance is probably the same as that of earlier apparatus where the plate is scanned under the focused excitation laser, and there should be little, if any, difference in effects produced by linear photolysis effects. Since the exposure is spatially multiplexed over the time of photodamage, accurate estimation of the space-dependent absorbances would be difficult for signals produced from multiplexed irradiance when exposure-dependent photodamage occurs.

9.12.3. Other Applications to Chemical and Biological Sample Analysis

Heterogeneous-phase photothermal spectroscopy has been used primarily for nondestructive material analysis and the analysis of thermal properties of solids. Although these are very interesting applications, a review of this literature is beyond the scope of this book. There are some notable applications that can be characterized as heterogeneous-sample chemical analysis. Photothermal spectroscopy has been widely applied to study semiconductors (Amato et al. 1992; Mandelis 1987). Photothermal deflection spectroscopy has found widespread use in electrochemistry (Mandelis 1987, 1989). It has been used for in situ characterization of electrodes, to monitor electrochemical reactions at the electrode interface (corrosion and film growth), to aid in elucidation of surface energy transfer physics, and to measure diffusion of electroactive species.

Some new applications are emerging. X.-Z. Wu et al. (1993) report an interesting application to the determination of enthalpies of reaction. J. Wu et al. (1990, 1991a,b) have measured optical absorbance in cells and of analytes adsorbed onto a microparticle. Using a conventional light source, they were able to measure absorption spectrum differences in normal and leukemia blood cells with photothermal deflection. The lamp was focused to

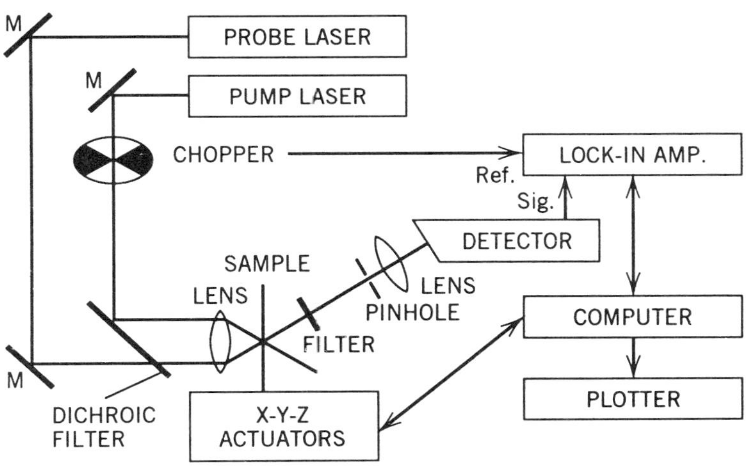

Figure 9.11. Apparatus used for crossed-beamed photothermal lens imaging microscopy in transparent solids. (Reprinted with permission from Burgi and Dovichi, 1987. Copyright 1987, Optical Society of America.)

a relatively large area at the sample and a focused probe laser was passed directly over the top of the particle or cell. The significant point regarding this work is that the large thermal gradient required to produce a reasonable deflection signal was produced by the small dimensions of the particle or cell. The optimum particle size was found to be on the order of 10 μm for a 5-μm-diameter probe laser beam. Larger particle or probe beam sizes produced signals of diminished magnitude. The demonstrated applicability of photothermal spectroscopy by Wu et al. to microparticle and cell analysis is an open door to applications of this technique to biological and medical diagnostics.

Finally, Burgi et al. (1984) and Burgi and Dovichi (1987) developed a scanning photothermal microscope suitable for tissue analysis. A diagram of the apparatus used in the later study is illustrated in Figure 9.11. The excitation laser was a 4-mW 422-nm helium–cadmium laser, and the probe

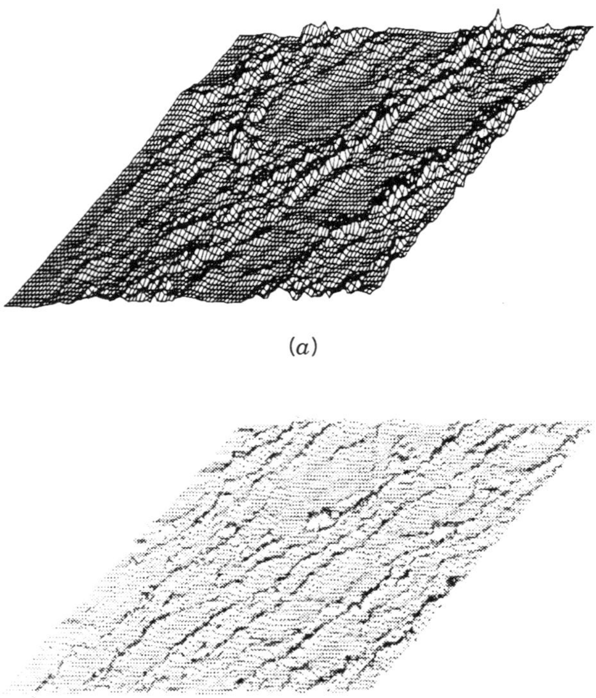

(*a*)

(*b*)

Figure 9.12. Crossed-beam photothermal lens microscope images of stained plant cell walls. Part (*a*) is the signal magnitude and part (*b*) is the phase angle. (Reprinted with permission from Burgi and Dovichi, 1987. Copyright 1987, Optical Society of America.)

was a 2-mW 632.8-nm helium–neon laser. The excitation and probe laser beams enter the microscope objective and are focused at the same position. Because they enter displaced from one another, they will cross within a set volume in the sample. Since generation of the photothermal signal requires the proximity of both excitation and probe lasers, the microscope should be able to image in three dimensions. They found that because of the small sample and focused beam sizes, the lasers power sometimes had to be reduced to below 1 mW to avoid sample damage.

Results obtained using a 40× microscope objective of plant stem material stained with safranin and fast green dyes are illustrated in Figure 9.12. The two-dimensional images are produced for a 100 μm × 67 μm area of the sample. The step size of the digital-controlled raster scanning actuators is 1 μm. The amplitude image corresponds to sample regions where the stain was located. The dye preferentially stained regions of high lignin content (e.g., the cell walls). The phase image gives an indication of the thermal properties of the sample. Thermal properties were not determined quantitatively, although differences between the phase and amplitude images suggested that they could be. Perhaps the most interesting aspect of this microscope was the three-dimensional imaging possibility. This aspect was not demonstrated, but the excitation-probe intersection angle of about 150° using the 100× microscope objective suggested that three-dimensional images of the optical absorption and thermal properties of the sample could be obtained at a resolution of about 1 μm^3.

REFERENCES

Aamodt, L. C. and Murphy, J. C. *J. Appl. Phys.* **52** 4903 (1981).

Aamodt, L. C. and Murphy, J. C. *J. Appl. Phys.* **54** 581 (1983).

Amato, G.; Benedetto, G.; and Boarino, L. *Proc. IEEE* **139** 161 (1992).

Burgi, D. S. and Dovichi, N. J. *Appl. Opt.* **26** 4665 (1987).

Burgi, D. S.; Nolan, T. G.; Risfelt, J. A.; and Dovichi, N. J. *Opt. Eng.* **23** 756 (1984).

Carslaw, H. S. and Jaeger, J. C. *Conduction of Heat in Solids*, 2nd ed., Clarendon, Oxford (1986).

Charbonnier, F. and Fournier, D. *Rev. Sci. Instrum.* **57** 1126 (1986).

Charbonnier, F.; Fournier, D.; and Boccara, A. C. In *Photoacoustic and Photothermal Phenomena*, Hess, P. and Pezl, J.; eds., Springer-Verlag, New York (1987).

Chen, T. I. and Morris, M. D. *Anal. Chem.* **56** 19 (1984a).

Chen, T. I. and Morris, M. D. *Anal. Chem.* **56** 1674 (1984b).

Figari, A. *Measurement Sci. Technol.* **2** 653 (1991).

Fotiou, F. K. and Morris, M. D. *Appl. Spectrosc.* **40** 700 (1986a).

Fotiou, F. K. and Morris, M. D. *Appl. Spectrosc.* **40** 704 (1986b).

Fotiou, F. K. and Morris, M. D. *Anal. Chem.* **59** 185 (1987a).

Fotiou, F. K. and Morris M. D. *Anal. Chem.* **59** 1446 (1987b).

Fournier, D. and Boccara, A. C. In *Photothermal Investigations in Solids and Fluids*, Sell, J. A., ed., Academic Press, New York (1989).

Fournier, D.; Boccara, A. C.; and Badoz, J. *Appl Opt.* **21** 74 (1982).

Hess, P., ed. *Photoacoustic, Phothermal and Photochemical Processes in Gases*, Springer-Verlag, New York (1989a).

Hess, P., ed. *Photoacoustic, Photothermal and Photochemical Processes at Surfaces and Thin Films*, Springer-Verlag, New York (1989b).

Jackson, W. B.; Amer, N. M.; Boccara, A. C.; and Fournier, D. *Appl. Opt.* **20** 1333 (1981).

Jager, B.; Carr, R. J. G.; and Goward, C. R. *J. Chromatogr.* **472** 331 (1989).

Kuo, P. K., Favro, L. D.; and Thomas, R. L. In *Photothermal Investigations in Solids and Fluids*, Sell, J. A., ed., Academic Press, New York (1989).

Low, M. J. D. *Appl. Spectrosc.* **40** 1011 (1986).

Low, M. J. D. and Morterra, C. *Carbon* **23** 311 (1983a).

Low, M. J. D. and Morterra, C. *Carbon* **21** 275 (1983b).

Low, M. J. D. and Morterra, C. *Appl. Spectros.* **41** 280 (1987).

Luikov, A. V. *Analytical Heat Diffusion Theory*, Academic Press, New York (1968).

Mandelis, A., ed. *Photoacoustic and Thermal Wave Phenomena in Semiconductors*, North-Holland, New York (1987).

Mandelis, A. In *Photothermal Investigations in Solids and Fluids*, Sell, J. A. ed., Academic Press, New York (1989).

Madelis, A. and Power, J. F. *Appl. Opt.* **27** 3397 and 3408 (1988).

Matsujima, T.; Lloyd, L. B.; and Eyring, E. M. *Appl. Spectrosc.* **38** 804 (1984a).

Matsujima, T.; Sharda, A. N.; Lloyd, L. B.; Harris, J. M.; and Erying, E. M. *Anal. Chem.* **58** 2877 (1984b).

Morris, M. D. and Fotiou, F. K. In *Photothermal Investigations in Solids and Fluids*, Sell, J. A., ed., Academic Press, New York (1989).

Morterra, C. and Low, M. J. D. *Spectrosc. Lett.* **21** 257 (1982).

Morterra, C. and Low, M. J. D. *Carbon* **21** 283 (1983); **23** 301 (1985b); **23** 335 (1985c); **23** 525 (1985d).

Morterra, C. and Low, M. J. D. *Appl. Spectrosc.* **38** 807 (1984).

Morterra, C. and Low, M. J. D. *Langmuir* **1** 320 (1985a).

Murphy, J. C. and Aamodt, L. C. *J. Appl. Phys.* **51** 4580 (1980).

Palmer, R. A. and Smith, M. J. *Can. J. Phys.* **64** 1086 (1986).

Palmer, R. A.; Chao, J. L.; Dittmar, R. M.; Gregoriou, V. G.; and Plunkett, S. E. *Appl. Spectrosc.* **47** 1297 (1993).

Peck, K. and Morris, M. D. *Anal. Chem.* **58** 506 (1986a).

Peck, K. and Morris, M. D. *Anal. Chem.* **58** 2876 (1986b).

Peck, K. and Morris, M. D. *Appl. Spectrosc.* **42** 513 (1988).

Peck, K.; Fotiou, F. K.; and Morris, M. D. *Anal. Chem.* **57** 1359 (1985).

Rothenberg, J. E. *Opt. Lett.* **13** 713 (1988).

Schweitzer, M. A. and Power, J. F. *Appl. Spectrosc* **48** 1054 (1994).

Stanley, B. T.; Bialkowski, S. E.; and Marshall, D. B. *Anal. Chem.* **65** 259 (1993).

Suber, G.; Bertolotti, M.; Sibilia, C.; and Ferrari, A. *Appl. Opt* **27** 1807 (1988).

Treado, P. J. and Morris, M. D. *Appl. Spectrosc.* **42** 1487 (1988).

Treado, P. J. and Morris, M. D. *J. Chromatogr.* **511** 341 (1990).

Varlashkin, P. G. and Low, M. J. D. *Appl. Spectrosc.* **40** 393 (1986a), **40** 507 (1986b), **40** 1170 (1986c).

Varlashkin, P. G.; Low, M. J. D.; Parodi, G. A.; and Morterra, C. *Appl. Spectrosc.* **40** 636 (1986).

Wu, J.; Kitamori, T.; and Sawada, T. *Appl. Phys. Lett.* **57** 22 (1990).

Wu, J.; Kitamori, T.; and Sawada, T. *J. Appl. Phys.* **69** 7015 (1991a).

Wu, J.; Kitamori, T.; and Sawada, T. *Anal. Chem.* **63** 217 (1991b).

Wu, X.-Z.; Shindoh, H.; and Yamada, M. *Anal. Chem.* **65** 834 (1993).

INDEX